Recent Developments in Mobile Technology

Recent Developments in Mobile Technology

Edited by **Catherine Hall**

Willford Press
New York

Published by Willford Press,
118-35 Queens Blvd., Suite 400,
Forest Hills, NY 11375, USA
www.willfordpress.com

Recent Developments in Mobile Technology
Edited by Catherine Hall

International Standard Book Number: 978-1-68285-304-7 (Hardback)

Printed in China.

Contents

Preface

The world is advancing at a fast pace like never before. Therefore, the need is to keep up with the latest developments. This book was an idea that came to fruition when the specialists in the area realized the need to coordinate together and document essential themes in the subject. That's when I was requested to be the editor. Editing this book has been an honour as it brings together diverse authors researching on different streams of the field. The book collates essential materials contributed by veterans in the area which can be utilized by students and researchers alike.

Mobile technology is a prominent field of telecommunications engineering. It has revolutionized the modern communication and information systems. This book is a compilation of researches and case studies on important topics like m-learning applications, mobile computing, wireless and ad-hoc networks, pilot projects and applications, etc. It incorporates theoretical as well as practical approaches for discerning the latest developments and emerging trends in the field of mobile technology. The contributions of internationally renowned experts compiled in this book, will provide innovative insights to the students and researches engaged in this field.

Each chapter is a sole-standing publication that reflects each author´s interpretation. Thus, the book displays a multi-facetted picture of our current understanding of applications and diverse aspects of the field. I would like to thank the contributors of this book and my family for their endless support.

Editor

1

Building Mathematics Cellular Phone Learning Communities

W. Daher

Al-Qasemi Academic College of Education, Baka, Israel & An-Najah National University, Nablus, Palestine

Abstract—**Researchers emphasize the importance of maintaining learning communities and environments. This article describes the building and nourishment of a learning community, one comprised of middle school students who learned mathematics out-of-class using the cellular phone. The building of the learning community was led by three third year pre-service teachers majoring in mathematics and computers. The pre-service teachers selected thirty 8th grade students to learn mathematics with the cellular phone and be part of a learning community experimenting with this learning. To analyze the building and development stages of the cellular phone learning community, two models of community building stages were used; first the team development model developed by Tuckman (1965), second the life cycle model of a virtual learning community developed by Garber (2004). The research findings indicate that a learning community which is centered on a new technology has five 'life' phases of development: Pre-birth, birth, formation, performing, and maturity. Further, the research finding indicate that the norms that were encouraged by the preservice teachers who initiated the cellular phone learning community resulted in a community which developed, nourished and matured to be similar to a community of experienced applied mathematicians who use mathematical formulae to study everyday phenomena.**

Index Terms—**Cellular phone, mobile learning, mathematics, learning community, middle school students, preservice teachers**

I. Introduction

A. Mathematics mobile education

Mobile and wireless devices, especially mobile phones, are increasingly common amongst the younger generation. This provides new possibilities, opportunities and challenges for the educational environment [3], especially for math. These possibilities, opportunities and challenges are examined in recent studies, for example [4] and [5] examined the use of mobile phones in learning mathematics amongst university pre-service teachers. Reference [5] studied learning processes and experiences within a mobile phone learning environment. In doing so they aimed to examine how socio-cultural and situated learning aspects were reflected in these processes and experiences. They found that the contribution of the mobile phone environment "lies not only in making dynamic mathematical application more available, but also in supporting the execution of tasks that are closer to the students' experiences and more relevant to them, which has the potential to enhance experiential learning. Further, the authors reported that the participants' learning experiences motivated their further learning.

Reference [6] examined the perceptions of middle school students regarding the use of cellular phone midlets and web applets in learning mathematics and how they differentiated between the two technological tools. Reference [6] reported that the students were aware of the different aspects of the tools, and paid attention to their following characteristics: their availability, their portability, the collaborative aspect of using them, their communicative aspect, the size of their interface, and their usability. The students used these aspects to describe their experience with the tools in learning mathematics, to differentiate between them, and to decide which tool they would use in their future learning and how they would use each tool. For example the students said they would use the web applets in their future learning because there are more applets than midlets that involve mathematical topics. A higher percent of students preferred the cellular phone as a learning tool because of its portability and communicability.

Reference [7] examined the perceptions of middle school students regarding using the cellular phone to learn mathematics by carrying out authentic activities out-of-the-classroom. They found that what encouraged the students to participate in the new learning experience were the novelty of the experience and the use of the cellular phone. On the other side, the students perceived the following characteristics of carrying out authentic activities out-of-the-classroom to learn mathematics using the cellular phone: (1) learning mathematics independently, (2) learning mathematics collaboratively and with equal relations with their teachers (3) learning mathematics in a humanistic environment, (4) learning the mathematics of real world phenomena, (5) learning visual mathematics, (6) performing different mathematical tasks, (7) learning mathematics easily and effectively.

Reference [8] examined the conditions that influence the middle school learning of mathematics using the cellular phone, and, at the same time, the consequences of such learning. They found that what affected the students' learning in the cellular phone environment were: (1) the characteristics and technologies of the cellular phone – like its portability and built-in camera, (2) the mathematical activities requirements and topics, (3) the learning setting (inside or outside the classroom), (4) the intention of the researchers who participated in the teaching processes, (5) the involvement of the school principal and the coordinating teacher. The consequences of the mathematics learning in the cellular phone environment were: (1) the students connected mathematics with real life, and (2) the students developed a new approach to mathematics where they looked at it as applied science. Further, the authors reported that during the carrying out of the activities the students (1) were in control of their learning and worked as applied mathematicians (examining a phenomenon and relating mathematical formulae to it).

B. *What is a learning community?*

Reference [9] describes learning communities as "collections of individuals who are bound together by natural will and a set of shared ideas and ideals". The authors add that communities depend on autonomous and independent individuals who influence each other within the learning process. Reference [10], as reported by [11], says that the word community implies both a meaning and a feeling. Bauman mentions understanding, common values, belonging, engagement, considerations and the collaboration between the members as some of the characteristics of a community.

Reference [12] says that most communities, whether online or traditional, share a number of qualities and characteristics: Communities are held together by distinct operating norms, members are distinguished by their formal and informal roles, trust must be built to ensure quality interactions, and a shared sense of purpose serves as the glue that bonds the community together. Regarding learning communities, Kaplan stresses that they must possess the previously mentioned characteristics and further engage people in learning processes over time.

Reference [13] distinguishes between a learning environment and a learning community, saying that a learning environment is not necessarily a learning community. Reference [13] stresses that "for a community to emerge, a learning environment must allow learners to engage each other intentionally and collectively in the transaction or transformation of knowledge", so individuals, in a learning community should take advantage and initiate processes for engaging ideas, negotiating meaning and learning collectively.

Reference [14], depending on the literature, points at essential steps to ensure creating and maintaining a community. These steps include: creating value, facilitation, establishing operating norms, creating an environment of trust, supporting individuality, and empowerment. These steps are described in Table 1 below as in [14].

TABLE I.
STEPS TO ENSURE THE CREATION AND MAINTENANCE OF A COMMUNITY

Steps	Description
Creating Value	Establishing good relationships amongst the student group. Using group techniques to foster healthy and meaningful discussion. Ensuring that appropriate spaces are established for work.
Facilitation	To enable facilitation, the coordinator should be able to connect with community members on a human level.
Establishing operating norms	Establishing operating norms helps the participants know how to behave and what to expect.
Creating an environment of trust	Techniques for developing trust include involving all participants in the development of shared expectation and the development of operating norms. They also include ensuring that people have the opportunity to get know each other and feel comfortable communicating with each other.
Supporting individuality, and empowerment	Supporting individuality and empowering community members maintain a community.

C. *The life cycle of communities: A model of group development*

Reference [1] describes a model of group development that consists of four stages: forming, storming, norming and performing. Reference [15] added a fifth stage: adjourning. These stages are described below depending on [1] and [15].

Forming: Groups initially start through orientation that is accomplished primarily through a testing that serves to identify the boundaries of both interpersonal and task behaviors. At the same time, dependency relationships with leaders, other group members, or pre-existing standards are established. So, the group process of *forming* consists of orientation, testing and dependence.

Storming: The second stage is characterized by conflict and polarization around interpersonal issues, with concomitant emotional responding in the task sphere. These behaviors serve to resist group influence and task requirements.

Norming: In the third stage, resistance is overcome by in-group feeling and cohesiveness, new standards evolve, and new roles are adopted. In the task realm, intimate, personal opinions are expressed.

Performing: In the fourth stage, roles become flexible and functional, and group energy is channeled into the task. Structural issues have been resolved, so structure can now become supportive of task performance.

Adjourning: The fifth stage involves dissolution of the community and entails bringing closure to the project on which the community is based. The process can be stressful, particularly where the dissolution is unplanned

D. *The life cycle of a virtual learning community:*

Reference [2] describes the stages of the life cycle of a virtual learning community:

Pre-birth: This stage occurs when the development, software, and policies of the community are established. Garber said that, in academic contexts, the elements of this stage are generally established before the community is formed.

Formative: During the formative stage, new members are brought into the community and the community's identity develops. Reference [16] described this stage as the one at which new acquaintances are made, similarities between members are identified, and communication is recurrent. Reference [17] said that at this stage the community needs nurturing to develop the community's identity.

Maturity: This stage occurs when the community begins to function independently of direct guidance. Reference [13] identified this stage as the one at which the purpose, shape, and operations of the community have been settled, and there is a less central role for the facilitator. Reference [16] pointed that at this stage camaraderie may be established through long-term and/ or intense associations.

Metamorphosis: This stage occurs when the community becomes something that it originally was not. Some members may resist this change and attempt to prevent it. Reference [18] says that for some communities, a natural decline, leading to death, may be a best option for the community at this stage. This could be so to enable active members of the community to join or initiate new communities which are still nourishing.

Death: This is the final stage of a virtual learning community, when members leave, discussion slows to the point that there is not enough participation to motivate them to return, and/ or the community has served its purpose. The death of an online community may be natural, as for example at the end of an online course, or unnatural, as when a high authority forces a community to break up.

E. The research rationale

Reference [19] pointed at the advantages of building learning community for students, saying that research provides evidence that strong feelings of community may not only increase participation, but also the flow of information among all learners, availability of support, commitment to group goals, cooperation among members, and satisfaction with group efforts. Reference [20] found that building a virtual 'artificial intelligence' enabled an Algebra learning community to sustain active participation of students and teachers, and to intensify their communication and interaction. Reference [21] pointed at the importance of building learning communities at schools, saying that learning communities do not merely represent congeniality, but rather dig deeply into learning. Reference [21] added that learning communities engage school students in disciplined inquiry and continuous improvement in order to “raise the bar” and “close the gap” of student learning and achievement.

Reference [22] indicates that it is necessary to create learning communities when given the task of teaching students new technologies. This is the case in the current research, where the preservice teachers built a students’ learning community to make the students use the cellular phone as a new technology for learning mathematics.

Little research has been done about building learning communities that learn mathematics with the cellular phone. This research attempts to examine how such a community developed and the phases it went through. This will give teachers an idea how to build a successful mathematics leaning communities; using technological tools, and what to expect in doing so.

F. Research question

What are the stages of building a mathematics learning community that evolves in a cellular phone environment and which is characterized by solving authentic problems?

II. Research Methodology:

A. Research setting and sample

The experiment took place in an Arab middle school in Israel. It was led by three third year pre-service teachers majoring in mathematics and computers. The pre-service teachers were part in the year before of a project led by their mentor and the writer of this article, and which integrated the cellular phone in teaching and learning the function topic which is part of the mathematics curriculum of grade nine. These pre-service teachers participated in preparing activities appropriate for learning mathematics with the cellular phone and in directing the students when carrying the activities. The three pre-service teachers carried out the experiment as their final project in the mathematics didactics course.

They selected thirty 8th grade students to participate in the project. The selection was done on the basis of volunteering from the side of the students and the ownership of appropriate cellular phones. All the learning was done by carrying out outdoor activities that involved exploring the mathematics of real life phenomena. The students utilized the various characteristics and technologies of the cellular phone to do such exploration (its portability, its camera, its stopper, its midlets, etc.).

At the beginning, the students carried out activities suggested by the preservice teachers. Later in the experiment, when they had carried out the real world activities suggested by the students, the students started to develop activities themselves. They did that by suggesting real world activities that they judged to be executable with cellular phones. The students usually started from a specific suggestion and developed it further till they considered the activity to be worth carrying out. All the activities are described below.

B. The Mathematical Software

The middle school students worked with cellular phone software (midlets) which support the learning of algebra and geometry. In order to perform the activities assigned to them, the students were required to use the algebraic midlets and various tools and technologies embedded in their cellular phones. The participants used the midlet "Fit2Go" which enables the user to draw specified points that fit a mathematical problem and then to fit a linear or a quadratic function for them.

The students downloaded the midlets from ‘Math4Mobile’ site [23] with the help of each other and the help of the preservice teachers. They used one of three methods to do that: (1) they went to the Math4Mobile site directly from their phones and downloaded the midlets (2) they downloaded the midlets to their computers and then used the phone’s connectivity software and a cable or an infra-red link to copy the downloaded midlets to their phones (3) they received the downloaded midlet courtesy of another student via Bluetooth.

Multipliers can be especially confusing. Write “Magnetization (kA/m)” or "Magnetization (10^3 A/m).” Figure labels should be legible, about 10-point type.

C. Activities

At the beginning, the preservice teachers required the students to carry out specific real world activities. The activities suggested by the preservice teachers were:

1. Finding the relation between the weight and the height of the members of a group.
2. Finding the relation between the number of sons and the number of daughters in families.
3. Finding the relation between the circumference of the trunk of a tree and the circumference of one of its branches.
4. Finding the relation between the radius of car's tyres and its height.
5. Finding the relation between the time of filling a container and the height of the water in it. The stu-

dents worked with different shapes of containers, one at a time.

6. Finding the relation between the height of a person and his/her running speed.
7. Finding the relation between the time that passes starting from the throwing of a ball and the distance that this ball passes until it halts.
8. Finding the relation between the weight of a person and the number of fingers in his/her hand. This activity was given to introduce the constant function to the students.

After the students carried out the activities suggested by the preservice teachers, they were requested to suggest by themselves authentic activities that they could carry out with cellular phones. The activities suggested by the students were:

1. Finding the relation between the temperature of the water in a container and the time required for a cube of ice to melt in that water.
2. Finding the relation between the circumference of a rock and its height.
3. Finding the relation between the length of a tree leaf and its width.
4. Finding the relation between the height of a person and the width of his/her step.
5. Finding the relation between the time that elapses from planting a plant and the plant's height.
6. Finding the relation between the time elapsing from the moment of lighting a candle and its height.

D. Data collecting tools

The preservice teachers used various means to collect data regarding the participants' learning of mathematics with cellular phones: observations, videoing, blogwriting and interviews.

The current research used all the data that the preservice teachers collected to analyze the students' building of mathematical knowledge when solving authentic problems using the cellular phone. In addition, the current research used additional data: diaries that the preservice teachers were requested to keep, so that they can reflect on the flowing of their experiment.

E. Data analysis

Coding the learning community building stages:

To determine the learning community building stages, I used the previously mentioned two models of community building stages; first the team development model developed by [1], and second the life cycle model of a virtual learning community developed by [2].

F. The relevance of the data analysis to the setting and theme of the research

The goal of this research was finding the stages of building a learning community of middle school students who solve authentic problems in a cellular phone environment. Regarding the stages of building a learning community, [24] used the model in [1] to examine the emergence of virtual teams in the business world. Reference [25] used the same model to examine the stages of students' group development in university setting. Regarding the use of the model in [2] to examine the development of the cellular phone mathematics learning community, it hasn't been used empirically in many researches, if any, but it suits the intention to examine the building development of the cellular phone learning community for it came to address similar phenomena.

III. Findings

The cellular phone learning community described here went through the following stages when carrying out outdoor mathematics activities.

A. The building phases of cellular phone mathematics learning community:

1) First phase: Pre-birth

This phase consisted of three stages:

Stage 1: Design of technology

The first stage was the design and development carried out by university designers of a cellular phone environment for the learning of mathematics. In our case, this environment was Math4Mobile.

Stage 2: Spreading the news about the technology

This stage consisted of the following: the university developers spread news about their new learning environment to their colleagues and fellows. One of these colleagues and fellows was a pedagogic mentor in a teacher training college who decided to try the new technology in the training schools where his preservice teachers practice teaching mathematics.

Stage 3: Participating in an experiment which utilizes new technology for teaching and learning

The pre-service teachers participated as students in an experiment in which their mentors used the cellular phone technology in teaching mathematics for grade nine in the training schools.

Examples from the preservice teachers' writing on this stage, as written in their diaries and which describe the preservice teachers' thoughts and wondering at this stage (translated from Arabic):

Preservice teacher: I wondered at the beginning whether the cellular phone would be an efficient tool in teaching mathematics.

Preservice teacher: The students had fun learning mathematics with the cellular phone.

Preservice teacher: the mathematics teacher started to take interest in the cellular phone after few lessons, and she expressed her idea that using the cellular phone would motivate students to learn mathematics.

2) Second phase: birth

This phase consists of two stages:

Stage 1: Deciding to use the technology in a project and research

The first stage of birth happened when the pre-service teachers decided to carry out a project concerned with learning mathematics in a cellular phone environment, so that they could research some aspects of its use in the mathematics classroom.

Examples from the preservice teachers' writing on this stage:

Preservice teacher: Watching our pedagogic mentor teach mathematics to middle school students with the cellular phone encouraged me to try teaching the same way myself.

Preservice teacher: When I saw that Suma (a pseudonym) decided to research how teaching the functions topic with the cellular phone would benefit grade eight students cognitively I decided to join her and examine how such teaching would benefit the affective aspect of learning.

Stage 2: management issues

The second stage of birth consisted of the following: contacting a school principal to make an agreement regarding the use of cellular phones in a mathematics classroom, and arranging with the mathematics teachers in the school regarding the students who would participate in the project.

Examples from the preservice teachers' writing on this stage:

Preservice teacher: One of the mathematics teachers didn't cooperate with us, saying that cellular phones are not for learning, but the mathematics coordinator in the school encouraged eight grade students to join our project.

3) Third phase: Formation

This phase consisted of two stages:

Stage 1: Wondering, answers and expectations

The first stage of the formation phase, which occurred in the experiment described here, consisted of the following happenings: some students volunteered to join the project, but part of them were not satisfied with this decision, the preservice teachers met student volunteer and presented the project to them, the students wondered about the nature of the project and asked questions which the preservice teachers answered. The students developed some expectations from the project.

Examples from the students' writing on this sub-phase:

Student: I hesitated if to join the project because I didn't want to learn more mathematics.

Student: I wondered how the cellular phone could be used for learning mathematics, for I knew it's only good for music, fun and communication.

Student: According to the teachers' description, this learning is going to be fun.

Examples from the preservice teachers' writing on this sub-phase:

Preservice teacher: the students had a lot of questions regarding the new project. We assured them everything was going to be all right. They didn't seem to believe us one hundred percent, but it was obvious that our answers encouraged them to stay in the project.

Stage 2: Setting the stage technologically for learning and beginning the community dynamics

The second stage of the formation phase, which occurred in our case, consisted of the following happenings: (1) The students had some difficulty downloading the midlets to their cellular phone, but with the help of the preservice teachers and the help of each other, they succeeded to download the midlets to their cellular phones. (2) The students got themselves familiar with the new environment.

Examples from the students' writing on this sub-phase:

Student: I found difficulty downloading the midlet to my cellular phone, but the teacher showed me how.

Student: I sent the midlet to other students using my Bluetooth.

Examples from the preservice teachers' writing on this sub-phase:

Preservice teacher: It was wonderful to watch the students help each other to download the midlets. This student collaboration made our interference unneeded.

4) Fourth phase: Performing and norming

The performing phase, which occurred in our case, consisted of the following: (1) the students performed the outdoor activities on their own, deciding how to perform the activity and which task every member of the group will perform; (2) During this performing of outdoor activities, the students got to know each other and to know the preservice teachers; the students turned to the preservice teachers for solutions to all of their problems, and the students got to know that the preservice teachers are but advisors to them but not the source of knowledge; and (3) They found learning mathematics outdoors with the cellular phone to be fun

Examples from the students' writing on this phase:

Student: at the beginning, we discussed how to carry out the activity, and then everyone chose which task to do. These steps are so important for they prepare for the following tasks.

Student: What astonished me is the collaboration between us and our teachers. This collaboration was on equal terms; unlike the situation in our school.

Student: the teachers directed us how to carry out the experiment by ourselves.

Examples from the preservice teachers' writing on this phase:

Preservice teacher: the students seemed dependent on us at the beginning, but we didn't give answers for them, only directions or leading questions. Gradually, the students learned to depend on themselves in carrying out the activities.

5) Fifth phase: Maturity

The maturity phase, which occurred in our case, consisted of the following happenings: (1) the students got more confident of themselves and gradually got less dependant on the preservice teachers (2) the students started on their own to suggest mathematical activities that could be carried outdoors with the help of cellular phones.

Examples from the students' writing on this phase:

Student: We enjoyed suggesting outdoor activities to carry out with cellular phone. We felt that we are teachers and students at the same time. We were happy when the teachers liked our activities.

Examples from the preservice teachers' writing on this phase:

Preservice teacher: It astonished us that the students started to suggest interesting activities to study using the cellular phone. In a way, this made us feel that our presence was unnecessary.

IV. Discussion:

Development phases of the cellular phone mathematics learning community:

Reference [26] says that the growth of the community depends on the emergence of new issues to be discussed or on setting new challenges for the community. This is true generally, but specifically the beginning of the cellular learning community described here was the emergence of new mathematical software for the cellular phone. The cellular phone would not have become a learning tool for mathematics, especially linear and quadratic functions, if it had not included a mathematical software component, in our case midlets.

The cellular mathematics learning community described here had five 'life' phases of development: Pre-birth, birth, formation, performing, and maturity. Reference [15] identified four 'life' phases of a learning community: forming, storming, norming and performing, while [2] identified three 'life' phases of learning community: pre-birth, formation and maturity. The third phase of the cellular phone learning community consisted itself of two stages: first stage - wondering, answers and expectations, and the second stage - setting the stage for the learning technology and the beginning of the community dynamics. So, the development of this cellular learning community has common factors with other learning communities and specific factors which distinguish it from other communities, like the wondering sub-stage of the second stage of the formation phase. What is also specific for this community life development is that its maturity phase is characterized by an interesting educational phenomenon: the students turning into mathematicians who look for real life phenomena to investigate mathematically.

Let's now discuss every phase of the life cycle of the cellular phone learning community:

A. *The pre-birth phase:*

The pre-birth phase constituted of three stages that involve issues related to design, communication and previous use of the new technology which is the tool of learning in the community. Only one sub-phase – the design of technology – is necessary, while the other two are possible but not necessary. The preservice teachers could have come upon the technology by accident and decided to experience it in their training classrooms. The previous scenario is possible but far from being likely because probably preservice teachers are open to what they learn and witness, but not to what they come upon by chance. On the other hand, inservice teachers are more likely to take a decision to try a new technology in their classrooms because they are more independent and powerful regarding their teaching decisions than preservice teachers. though the sub-phases 'spreading the news by the academic faculty about the technology' and 'witnessing of the technology as tool for teaching and learning' by the learning community initiators' are not necessary, they speed the birth of the learning community and at the same time speed its subsequent life phases. Spreading the news about the new technology speeds the birth of the learning community because otherwise it may take the learning community initiators a long time to know about the technology, and this may delay their consideration to initiate it. On the other hand, witnessing the technology as tool for teaching and learning gives the learning community initiators a detailed picture of what could be done with the technology in teaching and learning, which helps the initiators in subsequent phases of the learning community's life, for example it helps them manage the project (an issue related to the birth phase), set the stage technologically for learning and beginning the community dynamics (an issue related to the formation phase), or decide which learning roles of students to encourage (an issue related to the performing and norming phase).

B. *The birth phase:*

The birth phase constituted of two sub-phases: (1) Decision to use the technology in a project and as a subject for research (2) deciding how to manage the project. The two sub-phases are critical for the actual birth of the learning community because without any one of them no learning community would have come into life. One of the questions that could be asked regarding the first sub-phase is: what makes some preservice teachers decide to experiment teaching with a technology? Probably two issues are involved here: (1) witnessing a previous use of technology (2) the characteristics of the preservice teachers. The characteristics of the preservice teachers involve three aspects: (1) their tendency to use technology in their daily lives (2) their tendency to use technology in their teaching (3) their tendency to try new educational innovations in their teaching. This possible influence of the students' characteristics on their learning decisions is supported by the literature, for example [27] found that the characteristics of students influenced the efficacy of their face-to-face and computer supported collaborative learning.

The development of the second-phase of the birth phase depends not only on the preservice teachers (the initiators of the learning community in our case), but also on school principals and mathematics teachers. A positive responsiveness from the side of a school principal and mathematics teachers would speed the formation of the learning community, while a negative responsiveness would slow that formation and probably frustrate the new technology adopters, in our case the preservice teachers.

C. *The formation phase:*

The formation phase constituted of two stages. The first stage was the first encounter between the students and the preservice teachers, but it was more critical for some students than for others because they needed to decide if to join the new experiment and the learning community. It was important for the preservice teachers at this stage to convince the hesitating students to join the new experiment and they succeeded to do that, probably because they themselves believed in the importance and advantages of using the cellular phones in teaching and learning mathematics. The second stage of the formation phase is the first encounter between the students and the new technology, in our case cellular phone midlets. The students overcame the difficulties they confronted in this first encounter by collaborating with the preservice teachers and with each other. This success of the students in their first encounter with the midlets' technology by collaborating agrees with the finding of [28] who reported that students found colleagues a valuable source of information when acquiring new technology skills. Reference [29] also reported that friends were a major source of information about new technologies. Talking about collaboration among the students, the second sub-phase was the beginning of collaboration in the learning community and the base on which further collaboration was done in subsequent phases.

D. Performing and norming phase:

In the performing and norming phase, the students performed the learning activities in collaboration and on their own. Performing the learning activities, they needed to know their different roles in the learning community and the nature of the mutual relations between them and the preservice teachers and among themselves. In this phase, the students also got to know that the preservice teachers are not the source of knowledge but directors for them in their collaborative building of the mathematical knowledge. Thus the students' collaboration with the directing of the preservice teachers made the students' learning successful. This success in using technology in carrying out collaboratively learning activities is also reported by other researchers, for example [30] found that students' learning with technology was enhanced as a result of the collaboration and communication between the students who worked in pairs to study math and science.

E. The maturity phase:

In this phase the students matured as learners of mathematics and investigators of the mathematical aspects of real life phenomena. Thus they started to behave as applied mathematicians in the learning community; i.e. using mathematical theories and techniques to solve real-world problems. What could have influenced the students to become real mathematicians who want to research new real life phenomena is their appreciation of authentic learning experiences [29].

V. Conclusions:

Reference [22] suggests learning communities as a tool which helps perform the task of teaching students new technologies. In this research, the preservice teaches' intention was not teaching new technologies but using new technologies to teach different mathematics outdoors, where the students take the responsibility on their learning. Doing so, they helped to build a learning community which flourished step by step going through life phases known in the life cycle of other social and leaning communities. The cellular learning community described here had five 'life' phases of development: Pre-birth, birth, formation, performing, and maturity. These are the phases of the life cycle of a learning community which lives a new experiment, in our case integrating the cellular phone in teaching mathematics outdoor. The described development of the cellular learning community has common factors with other learning communities and specific factors which distinguish it from other communities, like the wondering sub-stage of the second stage of the formation phase. What is also specific for the development of the community life is its maturity phase which is characterized by an interesting educational phenomenon: the students turning into mathematicians who look for real life phenomena to investigate mathematically.

This study supports the suggestion made by [31] that teachers should be flexible in the ways in which they present new technologies. The flexibility of the preservice teachers who initiated the cellular phone learning community resulted in a community which developed and matured to be more similar to a community of experienced mathematicians than to a community of middle school students who study mathematics depending on their teacher. It is important to note that this community worked with untraditional mathematics consisting of authentic activities outside the classroom. Thus the new technology represented by the cellular phone enabled new opportunities and learning contexts for the mathematics student. We can conclude that the mentioned new technology holds hope for mathematics learning and teaching.

References

[1] Tuckman, B. W. (1965). Developmental sequence in small groups. *Psychological Bulletin*, 63, 384-399. The article was reprinted in Group Facilitation: A Research and Applications Journal, 3, 2001 and is available as a Word document.

[2] Garber, D. (2004). Growing virtual communities. *International Review of Research in Open and Distance Learning*. 5(2).

[3] Cobcroft, R., Towers, S., Smith, J. and Bruns, A. (2006). Mobile learning in review: Opportunities and challenges for learners, teachers, and institutions. In *Proceedings of Online Learning and Teaching (OLT) Conference 2006*, 21-30, Queensland University of Technology, Brisbane.

[4] Botzer, G., & Yerushalmy, M. (2007). Mobile applications for mobile learning. *Proceedings for "Cognition & Exploratory Learning in Digital Age*" (CELDA), Algrave, Portugal.

[5] Genossar, S., Botzer, G., & Yerushalmy, M. (2008). Learning with mobile technology: A Case Study with Students in Mathematics Education. *Proceedings of the CHAIS conference*, Open University.

[6] Daher, W. (2009). Students' Perceptions of learning mathematics with cellular phones and applets. *International Journal of Emerging Technologies in learning*, (1), 23-28.

[7] Baya'a, N., & Daher, W. (2009). Learning mathematics in an authentic mobile environment: the perceptions of students. *International Journal of Interactive Mobile Technologies*, 3 (Special Issue, IMCL 2009), 6-14.

[8] Baya'a, N., & Daher, W. (2010). Middle school students' learning of mathematics using mobile phones: Conditions and consequences. *Journal of Interactive Learning Research*, 21(2), 165-185. Chesapeake, VA: AACE.

[9] Kowch, E., & Schwier, R. (1997). Characteristics of technology-based virtual learning communities.

[10] Bauman, Z. (2001). Community - seeking safety in an insecure world. Cambridge: Polity.

[11] Carlén, U. (2005). Typology of online learning communities. *International Journal of Web Based Communities.*

[12] Kaplan, S (2002). Building communities - strategies for collaborative learning, *Learning Circuits*, American Society for Training & Development.

[13] Schwier, R. (2002). Shaping the metaphor of community in online learning environments.

[14] Australian National Training Authority (2003). *What are the conditions for and characteristics of effective online learning communities?* (Version 1.00), Australian Flexible Learning Framework Quick Guides series.

[15] Tuckman, B. W., & Jensen, M. C. (1977). Stages of small group development revisited. *Group and Organizational Studies*, 2, 419-427.

[16] Brown, R. (2001). The process of community-building in distance learning classes. *Journal of Asynchronous Learning Networks* 5(2).18-34.

[17] Preece, J. (2000). *Online communities: Designing usability, supporting* sociability. Chichester, UK.: Wiley.

[18] Schwier, R.A. (2001). Catalysts, emphases and elements of virtual learning communities: Implications for research and practice. *The Quarterly Review of Distance Education*, 2(1), 5-18.

[19] Rovai, A. (2002). Building a sense of community at a distance. *International Review of Research in Open and Distance Learning 3(1).*

[20] Lee, F. L. (2001). Towards a virtual AI-enabled Algebra learning community. Paper presented in *International Conference on Computers in Education*, 2001, Seoul, Korea.

[21] Fullan, M. (2005). Learning communities writ large. In R. DuFour, R. Eaker, R. DuFour (Eds.), *On Common Ground,* 209–223. Bloomington, Indiana: National Education Service.

[22] Riel, M., & Fulton, K. (2001). The role of technology in supporting learning communities. *Phi Delta Kappan*, 82(7), 518-523.

[23] Yerushalmy, M., Weizman, A. (2007). *Math4Mobile mobile environment.* The University of Haifa.

[24] Matthews, J. and Gladstone, B. (2000). Extending the group: a strategy for virtual team formation." *Industrial and Commercial Training*, (32), 24-29.

[25] Runkel, P., Lawrence, M., Oldfield, S., Rider, M., & Clark, C. (1971). Stages of group development: An empirical test of Tuckman's hypothesis. *The Journal of Applied Behavioral Science*, 7, (2), 180-193.

[26] Aceto, S., Dondi, C., & Marzotto, P. (2010). Pedagogical innovation in new learning communities: An in-depth study of twelve online learning communities.

[27] Solimeno, A.; Mebane, M.E.; Tomai, M. & Francescato, D. (2008) The influence of students and teachers characteristics on the efficacy of face-to-face and computer supported collaborative learning. *Computers & Education,* 51(1), 109-128.

[28] Jones, M. C., & Berry, R. L. (2000). Knowledge about information technology - a cross-cultural comparison. *Journal of Education for Business*, 173-176.

[29] Dooling, J. O. (2000). What students want to learn about computers? *Educational Leadership*, 58 (2), 20-24.

[30] Venville, G.; Wallace, J.; Rennie, L. & Malone J. (2000). Bridging the boundaries of compartmentalized knowledge: student learning in an integrated environment. *Research in Science & Technological Education*, 18 (1), 23-35.

[31] Kuperstein, J., & Gentile, C. (2001, October). Calling students back: using technology to support engaged learning.

AUTHORS

W. Daher is with Al-Qasemi Academic College of Education, Baka, Israel & An-Najah University, Nablus, Palestine.

2

Using Mobile Devices and Gaming as a Means of Building Vocabulary

Dr. J. B. Redd, Dr. D. Schmidt
Iowa State University, Ames, IA, USA

***Abstract*—Mobile devices are gaining momentum and popularity in the daily lives of today's youth. These devices embrace and encourage ubiquitous learning through their ease of portability and access to various activities that can allow for learning to occur. Learning theories and cognitive techniques are explored as a means of providing a mobile educational experience that is individualized, motivational, and beneficial for transfer to later word learning and reading comprehension.**

***Index Terms*—mobile devices, game theory, gaming in education, vocabulary**

I. Introduction

"The digital world has placed the great library resources of the world at anyone's fingertips" --([46] p. v)

This article provides a review of the literary foundation for the basis of studying vocabulary acquisition using mobile devices. The main areas discussed include mobile devices, theoretical frameworks, game theory and design, games in education, and vocabulary learning techniques as they relate to vocabulary acquisition by high school students. Each of these areas plays an instrumental role in providing the background required to study the impact mobile technologies can have on vocabulary development.

A. *Mobile Devices*

Technology is constantly evolving and this change is moving into the educational realm by creating opportunities for enhanced learning that is characterized as mobile learning, seamless learning, and ubiquitous learning [78]. This learning can be accomplished through the use of mobile devices. Mobile devices are small portable computing devices that can be used in a variety of environments [61]. This includes handheld devices, such as, personal digital assistants (PDA), portable media players (iPod Touch), and handheld game consoles (Nintendo DS). They possess the features of an operating system, a processor unit, a microbrowser, input/output devices, memory, and batteries [40]. These devices are making their way from outside to inside classrooms. They provide a way for students to construct knowledge and meaning from experiences by accomplishing learning tasks by using a portable tool [78]. Many of these mobile devices are characterized as miniature classroom computers that allow students to partake in the "anytime, anywhere" learning movement [82]. Mobile devices may be useful in creating a way for secondary students to transfer learning from inside to outside the classroom seamlessly.

The use of mobile devices, such as the iPod Touch, is becoming more common among secondary students. For example, [1] found that 69% of iPod Touch users are between 13-24 years of age with 46% of the users between 13 and 17. Similarly, [52] found that 74% of teens (ages 12-17) own an iPod or Mp3 player. As mobile devices get more advanced, the possible educational applications improve, which can lead to a re-organization of the learning environment [39]. Reference [27] notes there are many ways to incorporate the use of mobile devices and library activities, such as searching for and reading information (i.e. Kindle App) or interacting with the environment (i.e. Google Goggles). Reference [21] describes a variety of ways of using mobile devices, for example, composing podcasts and movies. Mobile devices allow secondary students to engage in educational activities using the devices they are accustomed to.

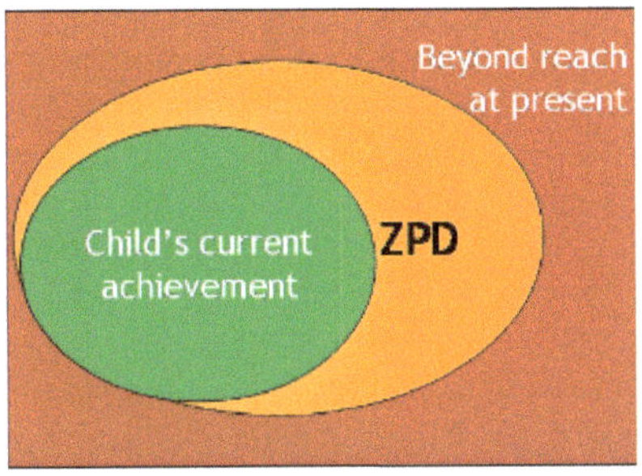

Figure 1. Three stages of the Zone of Proximal Development [5]

II. Theoretical Frameworks Related to Learning with Mobile Devices

Behind the scenes of these mobile devices are theories at work that describe the way with which learning can occur. Three learning theories, Zone of Proximal Development, Social Learning Theory, and The Law of Effect, offer an explanation as to how learning can occur using this type of portable device.

A. *Zone of Proximal Development*

One-way knowledge is constructed via a mobile device is based upon Vygotsky's Theory of Zone of the Proximal Development. This theory highlights the space between the actual level of problem solving and the level that can be achieved through guidance and support [98] (see Fig. 1).

This theory can be broken down into three stages that look at what the learner can do alone, the level he wants to achieve, and the scaffolding and support necessary to achieve that level. The first stage is *independent*. This is the learner's present knowledge and skill level and where he feels comfortable and will have success achieving and solving problems on his own. The next stage is where the zone of proximal development rests. The learner through guidance and collaboration is able to work on something that is just above the level where he can complete it independently, but not so difficult it causes confusion or boredom [64]; [98]. The ultimate goal of providing these supports is to build up the learner's knowledgebase so he can later perform the tasks independently. The third stage is *frustration*. This is what happens when a learner attempts to move beyond his current knowledge level without the proper support. Mobile devices have the ability to engage players individually while in the zone of proximal development.

While using mobile devices, players move beyond their current skill level with the mobile device acting as a scaffold [8]; [28]. This theory suggests learners work within the state of instruction and development [15]. By using the mobile device with a vocabulary game, players are introduced via *instruction* to words that are currently beyond their knowledge base but within their grasp. Then through interaction with these words, provided by strategically placed *scaffolds*, knowledge of the vocabulary words is attained. Reference [8] describes, "Knowledge within a discipline is important, but solving problems that encourage students to go beyond their current skill and knowledge level is critical to effective instruction" (para. 20). This can apply to learning vocabulary words with a mobile device since the game is designed to provide support beginning with meeting the player at his present knowledge level. Then through these scaffolded experiences, knowledge levels advance beyond what they were before. Slowly, the scaffolds are phased out, leaving the learner with the ability to apply the knowledge gained with support into scenarios without any support [15]; [64]. Ultimately, it is the learner's goal to possess knowledge on the particular subject area *independently,* thus allowing for the cycle to once again begin.

Typically, games designed for mobile devices push players beyond their current knowledgebase slowly and with support. For example, the Playstation 3 game, *Brain Challenge,* starts at the level where the player is currently. Then through a self-adjusting difficulty system, the questions on topics, such as logic, math, and concentration adjust depending on the responses and choices made by the player [73]. The Zone of Proximal Development seeks to engage learners at the optimal instructional level in a safe and supported manner. Another theoretical model that explores the way learning occurs with a mobile device is based on social learning theory.

B. *Social Learning Theory*

The use of mobile devices also coincides with Bandura's Social Learning Theory. This theory notes that the actions of a person are strongly influenced by their consequences [6]. After discovering the effects of performing a behavior, the learner mentally begins to anticipate the possible consequences of the behavior [6]. Then matching behaviors are performed with the concept of reinforcement in mind. Learning in this manner takes into consideration four essential components. These include attention, retention, reproduction, and motivation [16]; [26]; [43]; [62].

These four essential components may be applied to learning with a mobile device. First, the amount of *attention* paid to the task and the steps needed to complete it are essential for success while playing games. For example, having knowledge that a certain behavior will reap certain rewards will cause extra attention to be paid to models and important components deemed necessary in order to achieve the reward [6]. Second, the images and text encountered on the device aid in *remembering* the content experienced. Memory is used to organize and mentally code the items, which were paid attention to and experienced [51]. Third, based on the feedback from the game the user's behavior is modified and *reproduces* the correct responses for success. This can also be considered imitative learning because by observing and focusing on the feedback provided, decisions are made that will attempt to maximize the rewards earned [6]. Finally, there are *motivational* incentives when it comes to games. This can be in the form of intrinsic factors, such as satisfactory feelings of learning something new by means of curiosity, exploration, and manipulation, or extrinsic factors, such as incentives, rewards and punishments, co-operation, and completion [16]. Another theoretical framework that can explain how learning transpires through trial and error while using a mobile device is based on Thorndike's theory of law of effect.

C. *Law of Effect*

The theory of law of effect is based on choosing the path of least resistance toward a goal that results in the greatest satisfaction [95]. The theory explains that behaviors are modified based upon experiences and their consequences [96]. This theory is best exemplified by trial and error learning [86]. The luxury of using a mobile gaming device is the repeated opportunities to engage in an activity. Then if the strategy used was not successful or did not, for example, earn the most points, the player has the ability to try it again and modify the original behaviors based upon the feedback received. Increasingly, the level of satisfyingness felt while completing the activity is the goal [96]. Reference [26] describes, "The learner is rewarded or punished depending upon what he does; the 'effect' (satisfaction or annoyance) contingent upon how the learner responds to the environment" (p. 15). The interaction allows for the learner to try out different actions and experience rewards and punishments based upon those decisions. These can cause an increase in the motivation and attention the learner may exhibit while playing a game [26]. Once the learner has discovered the behavior principle behind the learning, which can occur through a combination of positive and negative reinforcements, the learner can continue to learn in new contexts [12]; [26]; [87]. Playing a game on a mobile device can create new educational possibilities of learning.

The theoretical foundations for learning on a mobile device can be constructed through the combination of these three theories. The Zone of Proximal Development explains how these devices can bridge the gap of learning through well-constructed scaffolds. Social Learning Theory illustrates the impact that attention, motivation, and feedback can have on the player. Finally, theory of law of effect highlights the importance of trial and error learning opportunities and the effects of feedback in the games. Together, these three theories lay the foundation for learning on a mobile device. Also of importance is the programming that goes into the development of gaming applications. The construction of a game is based on the game theory, which includes the foundational elements, such as choices, rules, and goals, that are instrumental to the proper unfolding of a game [47].

III. Game Theory

Game theory defines the core structure of a game by developing and providing numerous sequences of decisions and choices pre-programmed into the gaming structure that will allow the player to progress based upon their own actions. This theory develops gaming sequences based upon utility, or the probability that a selection will be made by the player [81]. Reference [13] describes the following:

Within game theory, the structural elements of a game exist as rules, turns, collaboration and competition, where winning, or fun, is modeled as numerical payoff. Game theory explains *how playing (a game) works*, and defines games as an interactive process striving toward payoff. (p. 7)

This theory is based on decisions. As decisions are made, they should be well-informed and based on prior knowledge learned through experiences in the game [47]. Inside of a game there are decision trees that create various paths based on the choices made by the player [74]; [81]. Each decision causes a chain reaction leading to a different outcome.

As the player progresses, his interactions within the game help to evolve into strategies that maximize the chance of success at reaching predetermined goals [10]; [11]; [20]. For example, the Leapster Explorer Learning Game, *Mr. Pencil Saves Doodleburg,* introduces the player on how to write numbers and letters and to create art and recognize shapes [50]. Reproducing the correct symbols allows Mr. Pencil to progress though the game and save the town. There are rules built into the game that funnel the player's actions to have specific consequences. This means there are multiple paths to success solely dependent on the actions and decisions of the player. For example, if the letters are not correctly traced, the player has to start again, and when correctly traced, the player proceeds to more difficult activities.

Game theory explains how the achievement principle applies to learning with games. Reference [31] notes learners while gaming are impacted by the intrinsic rewards. The rewards are customized based on the player's movements. As the levels shift, the effort and skill mastery required by the player also does [38]. The rewards signify the growing achievements earned by the player. Every player's experience is different with the same game [20]. A common algorithm implemented into the design of a game is based on the SuperMemo Method. This method offers a custom, adaptive spaced repetition program adapted to suit the learner's abilities [92]. Information is strategically placed in increments that complement the curve most commonly associated with forgetting (see Fig. 2).

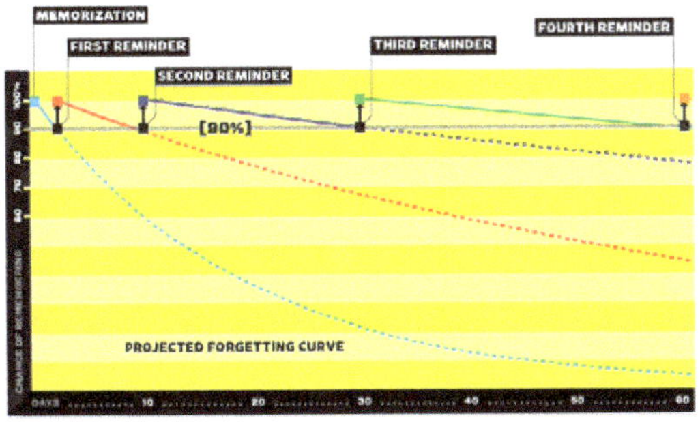

Figure 2. Game algorithm that replicates the forgetting curve [103]

The game begins at the player's knowledge level and then moves through playful, free movements within a well-defined, rigid structure at a level and pace that best suits the player [13]. Reminders are spaced after the first introduction of material in order to reduce forgetting. This is accomplished through practice and easily refined through active integration, meaningful use, and repetition [88]. Reference [31] concludes that a virtual world provides the player with numerous opportunities for practice that is not boring since it allows for the player to create their own experiences. Through these interactions with the games by means of images, text, and actions, players' skills and knowledge are then constructed. The design and development involved in the creation of the game creates an environment conducive for learning.

IV. Game Design

The design of a game can take on two formats [56]. The most common is a linear format. This is where the player, after making a certain response, receives a reinforcement or reward, and then progresses onto the next frame. The second format involves branching. This is when the next page to be encountered varies upon the action performed by the player. A branching game is well designed when no two students follow the exact same pathway through the learning experience [46]. Based on this format, the player has multiple ways to make progress in the game by making choices based on individual strengths and learning styles [31].

There is an overarching, basic design, though, that games must follow. Reference [54] notes that games should possess three essential characteristics. The first, challenge, means that games will shift in complexity based upon the user's actions. There is a goal to be reached and when play begins, the user is uncertain of the outcome. The second, fantasy, creates mental images. These images allow the learner to relate prior knowledge toward the understanding of new items [54]. Fantasy involves evoking the senses through the use of images, plot structure, and sound effects. The third essential characteristic is curiosity. By having a game that possesses the "optimal level of informational complexity," the users are engaged and not overwhelmed [54], p. 42.

Games also need to adjust to the player's skill level, offer elements of surprise, and appeal to emotions [33]; [57]; [89]. The design of a game begins with a clear goal structure that offers scoring and feedback along the way [14]; [57]. The feedback mechanism offers valuable information to the player, which will shape his later decisions and actions within the game [30]. The feedback helps a player progress by offering corrective information and emotional support [93]. Reference [35] notes games allow learning to change by providing opportunities to fail safely and create memory knowledge that can improve later recall and performance.

The knowledge built into memory is a key component of learning and the design of a game needs to take this into account. Memory involves the activities of acquiring, retaining, and recalling information [16]. The act of learning involves taking new information presented and intertwining it with previous knowledge in a meaningful way [25]; [41]. Then, the act of analyzing, synthesizing, and developing critical thinking skills evolves as the player moves through the game and takes the information previously encountered and remembered and builds upon it

[24]. Reference [26] found the following, "Learning is a continuous process of combining the familiar with the new." (p. 114). The amount and extended opportunities of practice are instrumental components of the design of a game to aid in memory and the learning process. The use of gaming in education can allow for these types of opportunities to occur.

V. Gaming in Education

Historically, video games were only seen as a form of entertainment, but as the technologies progress, these tools are becoming increasingly more popular as new mediums for education [85]. Reference [45] defines a game as "purposeful, goal-oriented, rule-based activity that the players perceive as fun" (p. 14). Mattel released the first handheld game, *Football*, in 1977 [45]. This first game was primitive in the design and artificial intelligence as the range of actions allowed in the game was quite limited. Over the years, this type of game has advanced in complexity and design so that it is now used on cell phones and other portable devices. For example, some games today have motion sensors and touch screens that react based on the player's movements (i.e., Wii, iPods). Mobile devices have increased in processor power and screen resolution and allow for gamers to play for a few minutes here and there since they are always connected [45].

As the evolution of mobile devices dramatically increases every year, so do the possible applications within the educational realm. Technology in and of itself can create activities that allow students to master and experience things that could not be discovered in another manner. Reference [89] discusses the possibilities of educators using video games as a model to improve learning environments by providing clear goals, challenging students, allowing for collaboration, using criterion-based assessments, giving students more control over the learning process, and incorporating novelty into the environment. In regard to video games as a means of learning, [59] concludes digital games allow for the unfolding of actions and interactions to happen naturally, slowly revealing the path to success as opposed to traditional games, which state the rules up front. Reference [59] goes on to explain that digital gaming environments need to support the varying pace of the learners. This is necessary to build up the player's gaming confidence by encouraging and motivating them with the drive to continue to play and succeed. Also, using games allows learners to experience things that would be difficult to replicate in another manner. For example, the game, *Go, Diego, Go: Safari Rescue* for the Wii, allows the player to go on an African safari rescue with the game's characters. Together, they save the animals from the mischief of a magician while at the same time collecting stamps and photos of the different locations visited [68]. Another game, Sid Meir's *Civilization Revolution* for the Nintendo DS, allows players to construct an entire empire from the very beginning. This includes activities such as waging war, conducting diplomacy, and discovering new technologies [69]. Games give players the opportunity to experience things that otherwise would not be possible.

Gaming provides a method of instruction in education that may allow for greater learning outcomes. Reference [70] implemented a computer memory concepts activity with 88 high school students. Approximately, half of the students experienced the activities by means of a game and the others used non-gaming techniques. Based upon a computer memory knowledge pre-test and post-test, the gaming students found the approach more effective and motivational in the memorization of concepts. Reference [97] studied the effect that a 3D geography game they developed had on 24 upper elementary students. They reported significant learning gains made by students while participating in the game (based on pre- and post-achievement tests) and a dramatic increase in motivation to engage in the game-based learning environment.

Understanding the possibilities that gaming can provide in education is vital to the success of this method of teaching. Reference [77] concludes when preservice teachers are instructed on possible uses of digital games in education, the view of how these games might be effective shifts dramatically in favor of the game's potential positive impact on learning. Games that connect educational needs and games elements (visualization and problem solving) can lead to the most engagement by learners [4]. They also need to take into account the content, which encourages and promotes the learning of new material.

A video game's content is of great importance. [3] note that learning through video games is dependent upon the situation that it simulates as opposed to the characteristics of the computer itself. This means that it is not the hardware but rather the software. It is through the design and development of the features of the game that can promote learning opportunities [3]. The mobile device may be the technological tool, but it is material that comprises the game that harvests the potential learning possibilities. Reference [44] states the combination of digital games and quality learning content engages the learner during the process of learning. Then it is the interactions in this gaming medium that allows for the mastery of material in a personalized manner.

Using a mobile gaming device allows the player to master the skills of the game while learning about the content of the game itself. Reference [91] states that game playing requires the ability to read and interpret the rapidly changing images encountered. While in the gaming environment, the player needs to comprehend and analyze the various amounts of information he encounters. This helps to refine reading skills. Also, developing an understanding of the rules of the game is an important component of playing the game [9]. Reference [71] found that by making sure students understand what is expected of them, students often do better. It is not just understanding the game's rules, but it also requires active participation in comprehending and internalizing what is required of them. The active engagement with the game creates the opportunity for learning to occur.

A. *Gaming and Active Learning*

Using games in education provides teachers with ways that allow students to become active participants in their educational experience. Reference [29] describes games as "action-and-goal-directed preparations for, and simulations of, embodied experience" (p. 201). Games provide the medium for learning, but in order to gain the full experience submersed in a gaming environment, the player needs to become an active participant–a producer–playing toward the completion of goals/win state [29]. Games can enhance learning by providing an active method of building knowledge that was not previously possible [101]. The use of a game on a mobile device creates new and possibly even more powerful methods of educating a generation of students that have grown up accustomed to using the technology accessible to them.

Games give the player a chance to participate in a series of activities from a particular perspective toward the accomplishment of a goal in the manner they see fit [29]; [93]. While using games, players are actively working toward the completion of a series of tasks toward a goal. Reference [60] defines, "Active learning refers to techniques where students do more than simply listen to a lecture. Students are DOING something including discovering, processing, and applying information" (para. 1). This means students are in charge of their learning; they are involved and engaged [100]. Reference [49] found that only 30% of students in college level physics courses of a traditional nature had a grasp on fundamental acceleration concepts whereas with active learning activities using computer software and other tools, this percentage increased to 75%. The use of games as a form of active learning can also be exemplified by the game, *River City*. This game encourages small group collaboration to determine the cause of illness in the 19th century. Based on realistic historical, sociological, and geographical data, together the group members actively look for solutions to the outbreak by formulating hypotheses, conducting experiments, and recommending possible solutions [94]. Gaming allows users to participate in the construction of their own knowledge by connecting concepts and applications. Reference [48] remarks games encourage active learning, collaboration, and interactivity. Games allow for collaboration, practice, discussion and negotiation within a virtual environment. Through hands-on manipulation of this virtual environment, students are actively shaping their knowledge and learning at the same time.

The idea of active learning through game usage creates an environment where the learner is in charge of his actions: learning by doing [23]; [91]. The game provides the format for the learning and by its basic design "all aspects of the learning environment are set up to encourage active and critical, not passive learning" [31]. p. 49. Reference [17] found that computer-based video game playing improved fact differentiation/recall processes as well as problem-solving skills by allowing for the exploration of multiple problem solutions. Part of the learning that evolves when using a gaming technology is the aspect of problem solving in addition to learning the basic facts.

Active learning using a mobile device allows each individual student to have a different experience with the game. Reference [58] notes video games allow the players to navigate different scenarios facing the consequences of varying actions. By making decisions the learner is in control of what happens next. The mobile device provides the means for an individualized educational experience that allows the learner to take on a more active role where learning does not occur by just listening but rather by doing [80]. Through active engagement, students experience the curricular concepts "metal representations of coherent classes of entities", being taught by completing activities that require doing and then reflecting upon them [63], p. 11. These experiences have the capabilities to be transferred to new learning situations.

B. *Gaming and Transfer*

The use of gaming in education can create possibilities of transfer from one situation to another. Reference [48] concludes, "Creating opportunities for students to practice applying the material, such as in a game or simulation, can bridge the distance between learning concepts presented in a classroom and using the information to solve a problem met outside of the school" (p. 54). It is this transfer of knowledge that can have such a profound impact. When participating in these games, students gain a skill set that will help them work with others and apply the conceptual knowledge to new situations [48]. As the player progresses in a game, the concepts behind the game play are mastered as well as the content. By understanding the principles, procedures, and general rules behind the game, students can recognize the meanings of specific and factual events, which can lead to transferring these ideas into new settings [26]; [34]. A helpful component to transferring ideas and skills is the motivation that the game provides.

C. *Gaming and Motivation*

Games are a way to extend motivational factors that typically apply to their recreational use to that which can apply in the classroom. Using games in educational environments allow students to be taught in innovative ways that can increase their motivation while teaching content. Reference [58] concluded that games could address the *systemic deficiencies* that are present in today's traditional teaching formats. She notes that 86% of students who leave the engineering major cite poor teaching, predominantly lecture format, by faculty of undergraduate science and engineering classes as an influential factor. By transforming the way we teach, learning itself is transformed. Reference [58] reported that lectures had the lowest rated learning outcomes in today's classrooms, and using games increases the motivation of the students. For example, in the collegiate classroom, the games Geography Explorer and Virtual Cell increased students' learning outcomes by 15%-40% and 30%-63% respectively when compared to the lecture format. She concluded that video games provide one means of motivating students to learn. Reference [44] notes education needs to move away from the traditional lecture format toward one that emphasizes teamwork, independence, and responsibility. They found the use of mobile devices and a digital economy game as a learning mode led to an increase in positive student emotions and knowledge transfer as opposed to the traditional lecture method and the use of economy case studies. Gaming is a technology the students of this generation are accustomed to and instead of ignoring or avoiding it, needs to become implemented into the curriculum and pedagogy of educators [99]. Using teaching techniques that appeal to and motivate learners in class is key in today's classrooms.

Creating something that will appeal to and motivate the learners in today's classrooms is one of the possibilities available by using video games. Reference [83] discovered while adding interactive multimedia into his teaching repertoire to help promote active learning, students were more motivated to engage in class activities in an authentic and spontaneous manner. Reference [57] conducted a survey of computer game preferences and found that not one single game can appeal to everyone, but some factors that contribute to intrinsically motivating instruction include personally meaningful goals, performance feedback, uncertain outcomes, scorekeeping, and randomness. In regards to increasing motivation, [57] concludes that *personalization* must be present. This involves a gaming environment that responds to the learners' actions as well as provide ample opportunity for reflecting upon their actions. A game can become personally motivating by allowing the player to create an account and each time he enters, he receives a personal message. He can also pull up his individual achievements previously attained on the game, for example, scores. Also, every gaming experience is different since the outcomes are based on the individual's actions. Games have the capabilities to move at the individual's pace and at their skill level, which can be an instrumental learning tool when mastering certain curricular concepts, such as vocabulary.

VI. Vocabulary Development Related to Gaming

Mastery of new vocabulary is a complex process, involving acquiring the words of a language [71]. The development of vocabulary is a major curricular concept addressed throughout the K-12 school curriculum. Academic vocabulary consists of content-specific, school-task, and literary words [36]. A strong vocabulary has been linked as a catalyst for improving reading fluency [34]; [84]; [88]. In order to strengthen vocabulary, certain techniques might be implemented. This involves immersion in rich environments that provide rich and varied language experiences [32]; [76]; [102]. Reference [7] found that students immersed in a vocabulary-rich environment, who were provided with word instruction and learning strategies, ended up developing larger, more expansive vocabularies. The use of a computer game can mimic an immersive environment that permits the player to have multiple interactions with vocabulary words. An immersive environment allows for active inquiry and exploration in a defined space [18]; [79].

Immersive experiences in the gaming realm might be helpful for building a foundation for learning vocabulary. A word-rich environment is a catalyst for an increase in vocabulary usage and reading fluency [88]. Reference [53] notes, "Children learn language from their language experiences – there is no other way" (p. 168). The players, by taking on an active role by playing the game, master the game's components as well as the vocabulary based content of the game itself. Through interactions with different word activities, situations are created where learners can move at their own pace to master new vocabulary.

Reference [12] notes that games and word play provide a context where students can enjoy the process of learning. By playing games, various types of interactivity and engagement occur, such as player positioning, narrative, and choice [24]. Reference [67] reported the use of games contributed to the learning of vocabulary as long as the game allowed the student the opportunity to interact with the word through practice and review in an atmosphere optimal for learning (pleasant). The choice of the game should reflect the needs of the students, such as, proficiency level, cultural contest, and timing [68]. Engagement is important in learning scenarios in order to permit the learners to extract meaning and make sense of the information [22]; [48]; [75]. Reference [104] found that when engineering undergraduate students were given one of two conditions to learn English vocabulary–experimental (web sites with games) or control (face-to-face activity-based lessons)–those in the group using games statistically outperformed the control group. Also, many students found that the gaming format provided a context that made remembering of the vocabulary words easier.

A. *Word Activities*

Another key element that increases vocabulary learning is teaching individual words by providing concise and student friendly definitions [32]; [72]; [76]. Reference [55] found that when high school students were presented with concise vocabulary information through Short Message Service (SMS) messages as opposed to detailed print materials, more vocabulary was recognized on the post-test. The use of easily understandable definitions also has an impact. In a high school science classroom, [105] concluded that the use of definitions linked to related ideas and concepts ultimately lead to a better understanding of the words, which in turn impacted the overall understanding of the concepts they helped define. It is a combination of concise and understandable definitions that may contribute to the retention and learning of new vocabulary words.

Reference [32] notes it is important to engage students in a variety of word activities. This involves designing activities that take into account the multiple facets of a word and its meaning and how the word relates to other words. Reference [65] found that high school students involved in varied exposure to vocabulary words in meaningful contexts outperformed their peers on a vocabulary post-test. It is important that within vocabulary building activities the word is not learned as a stand-alone item, but rather to learn its connections with other words [90]; [105]. Within the gaming environment, this can be done by implementing various activities, such as, finding a word's synonyms and antonyms. Reference [37] found a significant difference in post-test scores between a group of elementary students who received vocabulary instruction with only electronic book activities (animations about six different vocabulary words) and those who received the electronic book along with feedback, prompting, and supplemental information. They concluded that it was important to use supplemental tools (e.g., synonyms and asking questions), not just books for engaging students in a variety of word learning formats.

Another important component to building a larger vocabulary involves repeated encounters with words. Reference [32] explains students should work with a word more than once. Building up a strong vocabulary is not something that can happen in just one day through one encounter. It needs to occur over a long period of time, incrementally, within more than one context [32]; [71]; [76]. This repeated exposure will help build word precision, which means developing an understanding of when it would be appropriate to use the word [72]. By interacting with a word multiple times via multiple activities, learning of words and the relationships between them occurs. Reference [106] concluded that L2 (English as a Second Language) students who received periods of reading and interactive vocabulary instruction showed an increase in vocabulary knowledge. This finding may provide an understanding regarding the way words and vocabulary are mastered. By understanding the skills needed to learn vocabulary, word development has the possibility to transfer into other scenarios. Reference [71] defines transfer as a movement beyond the texts to which the original learning is tied. This can be done by teaching word learning strategies within a contextual environment, such as a game [32]. Reference [66] explains the instruction of vocabulary should emphasize skill development and strategy building to allow for the growth of independent word learning. It is essential to not just teach words in isolation, but instead to teach the concepts and skills behind learning and mastering new words with the ultimate goal of transitioning the new words into one's vocabulary. By introducing functional and skill-based strategies for word learning in the English classroom, transfer applications emerge for application in other content areas within secondary classrooms [34].

B. *Vocabulary Acquisition*

New words are learned in a variety of ways in the secondary school context, for example, through reading literature [7]; [84]; [102]. A main staple of most high school literacy curriculum involves vocabulary acquisition [19]; [42]. Understanding and maintaining a strong vocabulary is essential for comprehending texts [34]. Unfortunately, with estimates at 88,500 word families known by ninth grade, there is no way they could all be covered by one teacher, which is where supplementary out-of-class activities come in [66]. Reference [66] claims that due to the large number of words encountered in reading, the approach to teaching vocabulary must include methods that increase the capabilities for children in secondary schools to learn vocabulary on their own. Vocabulary instruction should use a variety of approaches, methods, and techniques [2].

VII. Conclusion

Through the implementation of a mobile learning device and a gaming tool, learning meets the needs of today's students of the learn anywhere and anytime movement. Vygotsky's Zone of Proximal Development, Bandura's Social Learning Theory, and Thorndike's Theory of Law of Effect, provide the theoretical base explaining how learning expands outside the restriction of the classroom building and one teacher's instructions by means of a mobile device. Students become actively in control of their own learning potential. The goal and intention of using this mobile device and game is to build up a student's word consciousness, which is "an awareness of and interest in words and their meanings" [7], p. 116. The use of word activities based on the secondary school curriculum and the way memory unfolds creates a supportive environment conducive for learning.

The design of the game holds great importance. It provides the structural element that houses the content material of vocabulary building. The design needs to take into account active learning essentials, game theory, future transfer implications as well as motivational principles. All of these elements together can encourage and contribute to the success of a player in the game.

References

[1] Admob. (2009). New research on the demographics and behavioral characteristics of iPhone and iPod touch users.

[2] Allen, J. (1999). *Words, words, words: teaching vocabulary in grades 4-12.* York, Maine: Stenhouse Publishers.

[3] Antonietti, A, & Mellone, R. (2003). The difference between playing games with and without the computer: A preliminary view. *The Journal of Psychology*, 137(2), 133-144.

[4] Amory, A., Naicker, K., Vincent, J., and Adams, C. (1999). The use of computer games as an educational tool: Identification of appropriate game types and game elements. *British Journal of Education*, 30(4), 311-321.

[5] Atherton, J. S. (2010). *Learning and Teaching; Constructivism in learning.* Retrieved on September 26, 2010, from http://www.learningandteaching.info/learning/constructivism.htm

[6] Bandura, A. (2006). Analysis of modeling processes. In A. Bandura (Ed.), *Psychological modeling: Conflicting theories,* (pp.1-62). New Brunswick, New Jersey: Transaction Publishers.

[7] Baumann, J. F., Ware, D., & Edwards, E. C. (2007). "Bumping into spicy, tasty words that catch your tongue": A formative experiment on vocabulary instruction. *The Reading Teacher*, 61(2), 108-122. doi: 10.1598/RT.61.2.1

[8] Benson, A. (1995). Review and analysis of Vygotsky's *thought and language.*

[9] Berger, A. A. (2002). *Video games: A popular culture phenomenon.* New Brunswick: Transaction Publishers.

[10] Birdwell, K. (2006). The cabal: Valve's design process for creating half-life. In K. Salen & E. Zimmerman (Eds.), *The game design reader,* (pp. 212-225). Cambridge, Massachusetts: The MIT Press.

[11] Bjork, S., & Holopainen, J. (2006). Game and design patterns. In K. Salen & E. Zimmerman (Eds.), *The game design reader,* (pp. 410-437). Cambridge, Massachusetts: The MIT Press.

[12] Blachowicz, C. L. Z., & Fisher, P. (2008). Attentional vocabulary instruction: read-alouds, word play, and other motivating strategies for fostering informal word learning. In A. E. Farstrup & S. J. Samuels (Eds.), *What research has to say about vocabulary instruction* (pp.32-55). Newark, DE: International Reading Association.

[13] Botturi, L., & Loh C. S. (2008). Once upon a game: Rediscovering the roots of games in education. In C. T. Miller (Ed.), *Games: Purpose and potential in education*, (pp.1-22). New York: Springer.

[14] Bowman, R. F. (1982). A "Pac-Man" theory of motivation: Tactile implications for classroom instruction. *Educational Technology, 22*(9), 14–17.

[15] Chaiklin, S. (2003). The zone of proximal development in Vygotsky's analysis of learning and instruction. In A. Kozulin, B. Gindis, V. S. Ageyev, & S. M. Miller (Eds.), *Vygotsky's educational theory in cultural context.* (pp. 39-64). New York: Cambridge University Press.

[16] Child, D. (1993). *Psychology and the teacher.* 5th ed. New York: Cassell.

[17] Chuang, T.-Y. & Chen, W.-F. (2009). Effect of computer-based video games on children: An experimental study. *Educational Technology*, 12(2), 1-10.

[18] Colella, V. (2000). Participatory simulations: Building collaborative understanding through immersive dynamic modeling. The Journal of the Learning Sciences, 9(4), 471–500.

[19] Common Core State Standards. (2010).English language arts standards: Language , grade 9-10.

[20] Crawford, C. (1984). *The art of computer game design.* Berkley, CA: Osborne/McGraw-Hill.

[21] Croy, K. (2007). An iPod Touch in Every Classroom: Even More Innovative Techniques for Using iPods to Reach Students.

[22] Dede, C. (2000). Emerging influences of informational technology on school curriculum. *Journal of Curriculum Studies*, 32(2), 281-303.

[23] Devine, D. J., Habig, J. K., Martin, K. E., Bott, J. P., & Grayson, A. L. (2004). Tinsel Town: A top management simulation involving distributed expertise. *Simulation and Gaming*, 35(1), 94-134.

[24] Dickey, M. D. (2005). Engaging by design: How engagement strategies in popular computer and video games can inform instructional design. *ETR&D*, 53(2), 67-83.

[25] Donovan, M. S., Bransford, J. D., & Pellegrino, J. W. (1999). *How people learn: Bridging research and practice.* Washington, D. C.: National Academies Press.

[26] Erickson, S. C. (1974). *Motivation for learning: A guide for the teacher of the young adult.* Ann Arbor: The University of Michigan Press.

[27] Fox, M. (2010). Mobile trends: Finding information on the web.

[28] Ganske, K. (2000). *Word journeys: Assessment-guided phonics, spelling, and vocabulary instruction.* New York: The Guilford Press.

[29] Gee, J. (2008). Learning theory, video games, and popular culture. In K. Drotner & S. Livingstone (Eds.), *The international handbook of children, media, and culture*, (pp.196-211). Sage: Los Angeles.

[30] Gee, J. P. (2008). Video games, learning, and "content". In C. T. Miller (Ed.), *Games: Purpose and potential in education*, (pp.43-53). New York: Springer.

[31] Gee, J. P. (2003). *What video games have to teach us about learning and literacy.* New York: Palgrave Macmillan.

[32] Graves, M. F. (2008). Instruction on individual words: One size does not fit all. In A. E. Farstrup & S. J. Samuels (Eds.), *What research has to say about vocabulary instruction* (pp.56-79). Newark: DE: International Reading Association.

[33] Grodal, T. (2003). Stories for eye, ear, and muscles: Video games, media, and embodied experiences. In M. J. P. Wolf & B. Perron (Eds.), *Video game theory reader* (pp. 129-155). New York: Routledge.

[34] Harmon, J. M., Wood, K. D., & Hedrick, W. B. (2008). In A. E. Farstrup & S. J. Samuels (Eds.), *What research has to say about vocabulary instruction* (pp.150-181). Newark: DE: International Reading Association.

[35] Harris, P. D. (2009, January). Immersive learning seeks a foothold. *T+D*, 40-45.

[36] Heibert, E. H. & Lubliner, S. (2008). The nature, learning, and instruction of general academic vocabulary. In A. E. Farstrup & S. J. Samuels (Eds.),*What research has to say about vocabulary instruction* (pp.106-129). Newark: DE: International Reading Association.

[37] Higgins, N., & Hess, L. (1999). Electronic books to promote vocabulary development. *Journal of Research on Computing in Education*, 31(4), 425. Retrieved from Academic Search Elite database.

[38] Holland, W., Jenkins, H., & Squire, K. (2003). Theory by design. In M. J. P. Wolf & B. Perron (Eds.), *The video game theory reader*, (pp. 25-46). New York: Routledge.

[39] Holzinger, A., Nischelwitzer, A., & Meisenberger, M. (2005, November). Lifelong-learning support by m-learning: example scenarios. *eLearn*, 11.

[40] Hu, W., Yeh, J., Chu, H., & Lee, C., (2005, September). Internet-enabled mobile handheld devices for mobile commerce, *Contemporary Management Research*, 1(1), 13-34.

[41] Hunter, M. (1995). *Retention theory for teachers.* Thousand Oaks, CA: Corwin Press.

[42] Iowa Department of Education. (n.d.). Reading – grades 9-12.

[43] Kirsh, S. J. (2010). *Media and youth: a developmental perspective.* Malden, MA: Wiley-Blackwell.

[44] Kittl, C., Edegger, F., & Petrovic, O. (2009). Learning by pervasive gaming: An empirical study. In Designing learning activities with mobile technologies. Innovative mobile learning H. Ryu & D. Parsons (Eds.). Information Science Reference: Hershey.(pp. 60-82).

[45] Klopfer, E. (2008). *Augmented learning: Research and design of mobile educational games.* The MIT Press: Cambridge, Massachusetts.

[46] Kolb, L. (2008). *Toys to tools: Connecting student cell phones to education.* Eugene, Oregon: International Society for Technology in Education.

[47] Kremers, R. (2009). *Level design: Concept, theory, & practice.* Natick, MA: A K Peters, Ltd.

[48] Kumar, R. & Lightner, R. (2007). Games as an interactive classroom technique: perceptions of corporate trainers, college instructors and students. *International Journal of Teaching and Learning in Higher Education*, 19(1), 53-63.

[49] Laws, P., Sokoloff, D., & Thornton, R. (1999). Promoting active learning using the results of physics education research.

[50] Leapfrog Enterprises. (2010). Leapster explorer learning game: Mr. Pencil saves doodleburg.

[51] Learning Theories Knowledgebase. (2010, September). Social Learning Theory (Bandura) at Learning-Theories.com.

[52] Lenhart, A. (2009). Teens, mobile, & games: An overview of Pew Internet data.

[53] Lieven, E. & Tomasello, M. (2008). Children's first language acquisition from a usage-based perspective. In P. Robinson & N. C. Ellis (Eds.), *Handbook of cognitive linguistics and second language acquisition* (pp.168-196). New York: Routledge.

[54] Loftus, G. R., & Loftus, E. F. (1983). *Mind at play.* New York: Basic Books, Inc.

[55] Lu, M. (2008). Effectiveness of vocabulary learning via mobile phone. Journal of Computer Assisted Learning, 24, 515–525.

[56] Lumsden, D. B. (1975). Adult learning and the applications of modern educational technology. In D. B. Lumsden & R. H. Sherron (Eds.), *Experimental studies in adult learning and memory* (pp.97-114). New York: John Wiley & Sons.

[57] Malone, T. W. (1984). Toward a theory of intrinsically motivating instruction. In D. F. Walker & R. D. Hess (Eds.), *Instructional software: Principles and perspectives for design and use* (pp.68-94). Belmont, CA: Wadsworth. (Original work published 1981).

[58] Mayo, M. J. (2007, July). Games for science and engineering education. *Communications of the ACM,* 50(7), 31-35.

[59] McFarlane, A. (2007). Learning and lessons from the world of games and play. In R. Andrews & C. Haythornthwaite (Eds.), *The sage handbook of e-learning research* (pp.119-137). London: Sage Publications.

[60] McKinney, K. (2010). Active learning.

[61] Microsoft Corporation. (2010). 1.1 Glossary.

[62] Moore, A. (1999). Albert Bandura.

[63] Murphy, G. L. (1991). Meaning and concepts. In P.J. Schwanenflugel (Ed.), *The psychology of word meanings* (pp.11-35). Hillsdale, New Jersey: Lawrence Erlbaum Associates.

[64] Murray, T. &, Arroyo, I. (2002). Toward measuring and maintaining the Zone of Proximal Development in adaptive instructional systems.

[65] Myers, J., & Shu-fen, C. (2009). A multiple-strategy-based approach to word and collocation acquisition. IRAL: International Review of Applied Linguistics in Language Teaching, 47(2), 179-207.

[66] Nagy, W. E., & Anderson, R. C. (1984). How many words are there in printed school English? *Reading Research Quarterly*, 19(3), 304-330.

[67] Nguyen, T., & Khuat, T. (2003). Learning vocabulary through games: The effectiveness of learning vocabulary through games. *Asian EFL Journal, 5*(4).

[68] Nintendo. (2010). Go, Diego, go: Safari rescue.

[69] Nintendo. (2010). Sid Meir's Civilization Revolution for the Nintendo DS. , from

[70] Papastergiou, M. (2009, January). Digital game-based learning in high school computer science education: Impact on educational effectiveness and student motivation. *Computers & Education*, 52(1), 1-12.

[71] Pearson, P. D., Hiebert, E. H., & Kamil, M. L. (2007, April/June). Theory and research into practice: Vocabulary assessment what we know and what we need to learn. *International Reading Association*, 42(2), 282-296.

[72] Phythian-Sence, C. & Wagner, R. K. (2007). Vocabulary Acquisition: A primer. In R. K. Wagner, A. E. Muse, & K. R. Tannenbaum (Eds.), *Vocabulary acquisition: Implications for reading comprehension*, (pp.1-14). New York: Guilford Press.

[73] Play.com (2010). Playstation 3: Brain Challenge.

[74] Poundstone, W. (2006). Game theory. In K. Salen & E. Zimmerman (Eds.), *The game design reader,* (pp. 382-408). Cambridge, MA: The MIT Press.

[75] Prensky, M. (2005, September/October). Enrage me or engage me: What today's learners demand, *EDUCAUSE*, 60-64.

[76] Pressley, M., Disney, L., & Anderson, K. (2007). Landmark vocabulary instructional research and the vocabulary instructional research that makes sense now. In R. K. Wagner, A. E. Muse, & K. R. Tannenbaum (Eds.), *Vocabulary acquisition: Implications for reading comprehension*, (pp.205-232). New York: Guilford Press.

[77] Ray, B., & Coulter, G. A. (2010). Perceptions of the Value of Digital Mini-Games: Implications for Middle School Classrooms, *Journal of Digital Learning in teacher Education*, 26(3), 92-100.

[78] Rogers, Y. (2009). How mobile technologies are changing the way children learn. In A. Druin (Ed.). *Mobile Technology for Children: Designing for interaction and learning*, (pp.3-22). Amsterdam: Morgan Kaufmann Publishers.

[79] Roussos, M., Johnson, A., Moher, T., Leigh, J., Vasilakis, C., Barnes, C., (1999). Learning and building together in an immersive virtual world.

[80] Ryu, H., & Parsons, D. (2009). Designing learning activities with mobile technologies. In H. Ryu & D. Parsons (Eds.), *Innovative mobile learning*, (pp.1-20). Information Science Reference: Hershey.

[81] Salen, K., & Zimmerman, E. (2004). *Rules of play: game design fundamentals*. Cambridge, MA: The MIT Press.

[82] Schachter, R. (2009, Nov/Dec). Mobile devices in the classroom. *District Administration,* 31-36.

[83] Schrand, T. (2008, Spring). Tapping into active learning and multiple intelligences with interactive multimedia. *College Teaching*, 78-84.

[84] Segalowitz, N., Watson, V & Segalowitz, S. (1995). Vocabulary skill: Since-case assessment of automaticity of word recognition in a timed lexical decision task. *Second Language Research*, 11(2), 121-136.

[85] Shreve, J. (2005). Let the games begin: Entertainment meets education.

[86] Simpson, E., & Stansberry, S. (2008). Video games and teacher development: Bridging the gap in the classroom. In C. T. Miller (Ed.), *Games: Purpose and potential in education*, (pp.163-184). New York: Springer.

[87] Skinner, B. F. (2006). The science of learning and the art of teaching. In B. A. Marlowe & A. S. Canestrari (Eds.), *Educational psychology in context: Readings for future teachers* (pp.45-55). Thousand Oaks, CA: Sage.

[88] Smith, T. B. (2008). Teaching vocabulary expeditiously: Three keys to improving vocabulary instruction. *English Journal*, 97(4), 20-25.

[89] Squire, K. (2003). Video games in education. International Journal of Intelligent Simulations and Gaming, 2(1).

[90] St. Clair Otten, A. (2003, July). Defning moment: teaching vocabulary to unmotivated students. *The English Journal*, 92(6), 75-78.

[91] Suoranta, J., & Lehtimaki, H. (2004). *Children in the information society*. New York: Peter Lang Publishing.

[92] SuperMemo World. (2010). SuperMemo method.

[93] Sweetser, P., & Wyeth, P. (2005, July). GameFlow: A model for evaluating player enjoyment in games. *ACM Computers in Entertainment*, 3(3), 1-24.

[94] The River City Research Team. (2007). Introducing River City.

[95] Thorndike, E. L. (1905). *The elements of psychology*. New York: A. G. Seiler.

[96] Thorndike, E. L. (1913). *Educational Psychology: The original nature of man.* New York: Teachers College, Columbia University.

[97] Tuzan, H., Yilmaz-Soylu, M., Inal, Y., and Kizilkaya, G. (2009, January). The effects of computer games on primary school students' achievement and motivation in geography learning. *Computers & Education*, 52(1), 68-77.

[98] Vygotsky, L. (1978). *Mind in society: development of higher psychological processes.* Cambridge: Harvard University Press.

[99] Weibe, G. (2007). Having fun is a good thing: Video games in education.

[100] Wellert, P. (2008). Breaking the technology barrier: Using technology in education.

[101] Wideman, H. H., Owston, R. D., Brown, C., Kushnirul, A., Ho, F., & Pitts, K. C. (2007, March). Unpacking the potential of educational gaming: A new tool for gaming research. *Simulation & Gaming*, 38(10), 10-30.

[102] Withrow, F. B. (2004). *Literacy in the digital age: Reading, writing, viewing, and computing.* Lanham, Maryland: ScarecrowEducation.

[103] Wolf, G. (2008). Want to remember everything you'll ever learn? Surrender to this algorithm read more.

[104] Yip, F. W. M., & Kwan, A. C. M. (2006). Online vocabulary games as a tool for teaching and learning English vocabulary. *Educational Media International*, 43(3), 233-249.

[105] Young, E. (2005). The language of science, the language of students: Bridging the Gap with Engaged Learning Vocabulary Strategies. Science Activities, 42(2), 12-17. Retrieved from Academic Search Elite database.

[106] Zimmerman, C. B. (1997, Spring). Do reading and interactive vocabulary instruction make a difference? An empirical study. *TESOL Quarterly*, 31(1), 121-140.

AUTHORS

Dr. **J. B. Redd** is with Iowa State University, Ames, Iowa, 50011 (e-mail: jredd@iastate.edu)

Dr. **D. Schmidt** is with Iowa State University, Ames, Iowa, 50011 (e-mail: dschmidt@iastate.edu)

An Adaptive and Cost-Effective Data Transmission Framework Using MMS Technology for Client-Server Application

A. Setyono[1], M. J. Alam[2] and C. Eswaran[3]
[1, 2, 3] Multimedia University, Cyberjaya, Malaysia
[1] Dian Nuswantoro University, Semarang, Indonesia

***Abstract*—Limitation of message size of the MMS technology is one of the important factors which make this technology less attractive. This limitation restricts the system developers in developing new mobile applications using MMS technology. This research attempts to improve the capacity of the maximum message size by implementing compression, splitting and cropping techniques for MMS application on the mobile phone devices. This paper presents the architecture to enhance the multimedia capabilities of the MMS technology and also the implementation details for developing client-server system applications. The modified MMS technology can be used to develop practical client-server applications.**

***Index Terms*—Multimedia message, Message size, MMS, client-server application**

I. Introduction

Multimedia messaging service (MMS) provides the user a complete knowledge about multimedia and has the potential to significantly increase the mobile communication usage [1]. MMS is considered as the best and proven messaging technology and it has surpassed email and SMS [2]. MMS experienced dramatic growth as for example in the USA, more than 10.3 billion MMS messages were sent in the first half of 2009 and experienced a growth of 4.7 billion from the same period of the previous year [3]. The initial standard for MMS size was up to a maximum of 30 KB. The standard was raised to 100 KB and the 3G standard is 300 KB [3], [4]. Currently, MMS supports an average message size of 600 KB with maximum message size limited only by the handset's and MMS Center (MMSC) capabilities. The standard is not used widely yet, because it must be supported by the availability of sophisticated mobile phone and MMSC capability [5].

Due to the authorization problem, we cannot change the message size that is received and forwarded to the recipient. We use the MMS technology standard that has been provided by the operator. We develop a data transmission method with improved MMS technology for client-server application. We attempt to make MMS technology more adaptive and cost-effective to overcome the limitations in the maximum size for developing client-server applications.

The main goals of this research are: (i) to explore data transmission for large multimedia file using MMS technology (ii) to develop data transmission by implementing the proposed techniques (iii) to design an adaptive and cost-effective client-server application and (iv) to implement the techniques for developing client-server application. The developed system will be used to implement the client-server application, with MMS application on the client side and Internet applications on the server side. Internet applications on the server side can be accessed by user with MMS application in mobile environment. This research is expected to provide a valuable contribution to the readers and researchers for developing mobile system applications using MMS technology. The following are the important contributions of this paper: (i) it describes about the mobile messaging system (ii) it develops transmission method for large multimedia file size (iii) it presents the design of an adaptive and cost-effective for client-server applications and (iv) it also implements and discusses about an adaptive and cost-effective transmission method for mobile system application.

The rest of the paper is organized as follows: Section 2 discusses data transmission method. The proposed design of client-server application and discussion is presented in Section 3. Implementation and simulation are presented in Section 4. Section 5 discusses the evaluation and analysis and finally conclusion and future work are given in Section 6.

II. DATA TRANSMISSION TECHNIQUES

MMS technology is the developed version of the SMS technology [2]. SMS has experienced tremendous success that cannot be exceeded by MMS [4]. In the early stages, many people expected that the MMS could replace SMS, as it has the advantage of sending multimedia data [6]. However, the growth of MMS technology was restricted because of reasons such as high-cost of sending multimedia message, high-prices of mobile devices and message size limitations [7]. In multimedia data transmission, MMS technology uses 2.5G or 3G network with speeds up to 384 Kbps [8]. In this work, we make some improvements on the MMS application by implementing a few techniques before data transmission to the recipient user.

The MMS server is more known as MMS center (MMSC) that is the center of the MMS system. MMSC performs a lot of functions, such as store MMS messages, forward MMS messages to external networks, send MMS messages as email using the SMTP protocol over the Internet, send MMS messages via the SMS for enabling users of non-MMS phones to receive MMS messages and

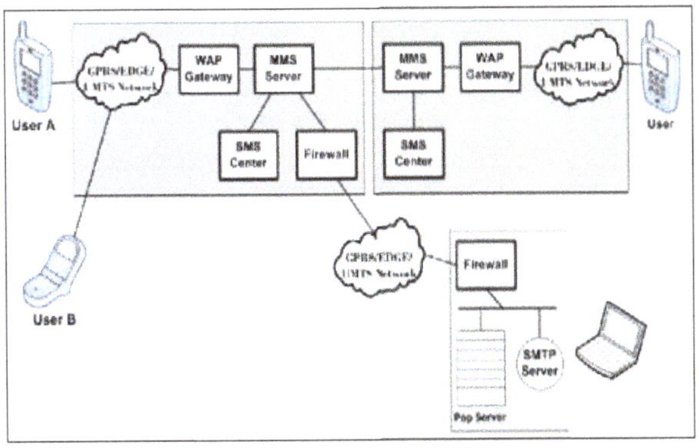

Figure 1. MMS architecture for data transmission

view the text as SMS messages and the whole message via Web or WAP and perform content adaptation [3]. The client-server applications have a client program which is running on one end of the system that requests and receives a service from server program running on another end system. This will be different if the client-server programs running on the computer, it is called a distributed application [9]. Figure 1 shows the MMS architecture for data transmission.

On the client side, we develop MMS application by implementing some techniques, such as compression, splitting, masking and cropping to get an adaptive and cost-effective in message delivery of the large multimedia file. Then, on the server side, we develop Internet application to use the data which is sent by the user to the mobile phone recipient which serves as the gateway between client and server. We implement merging and decompression techniques when downloading data from gateway to store into the database server.

Currently, MMS technology still uses 2.5G or 3G mobile network to transmit multimedia message, where MMS does not use higher technology, such as 3.5G or HSDPA. The development of MMS technology should be supported by the ability of MMSC that is provided by the operator and mobile phone features such as camera, voice recording and video. In future, we believe that MMS technology will be able to send larger data than now, and then implementation techniques on the MMS application will give a few contributions to advancement of MMS technology. The use of MMS technology is easier (like sending SMS) compared to making Internet connection on mobile phone to access Internet applications. Besides that, we can make the system more secured because the user sends the data without opening the system on the Internet to upload multimedia data directly. The developed system will be an integrated system. Table 1 shows the mobile network capability for MMS technology.

Currently, MMS technology performs resizing process to send larger multimedia message but the maximum size is determined by MMSC and MMS features on the mobile phone. As this process reduces the quality of image and file size, the user is not able to send larger multimedia data without degrading the quality. The quality of handsets is improving rapidly with high quality of camera and screen resolution which allow users to send large contents of multimedia message more easily. Good quality of data is essential for consumers, but it increases the possibility that the message will become large. Operator increases the message size and ensures that its MMSC has high capacity and can handle the growth of messages efficiently.

TABLE I.
MOBILE NETWORK CAPABILITY FOR MMS TECHNOLOGY

Technology	Data Rate	Frequency
Mobile Technology		
GSM	9.6 kbps	900/1800/1900MHz
GPRS	171.2 kbps	900/1800/1900MHz
EDGE	384 kbps	900/1800/1900MHz
3G / UMTS	2 Mbps	1885 – 2200 MHz
HSPDA	14.4Mbps	850/900 /2100MHz
HSUPA	5.76Mbps	850/900 /2100MHz
LTE	100 Mbps	900 /1700/1800//2100Mhz
WiMAX	1 Gbps	2 GHz to 11 GHz

In this research, we use data transmission method of the existing MMS protocol. We use mobile communications network that is provided by the service provider. We perform the development of MMS technology within the creation of MMS message by implementing some techniques in order to send large file size where the existing MMS technology cannot do it. This method is expected to be used in large multimedia message delivery

The developed system is a client server application. As we know that a mobile phone has many limitations compared to a personal computer (PC) or a notebook. Building the applications on mobile phone is rather complex considering its limitations. In developing the techniques to improve the ability of the MMS technology in sending the message, there are several factors that should be considered, such as resizing, converting, compressing, splitting, masking and cropping. These techniques can be used to overcome the problem of size limitation for data transmission. These techniques have advantages as well as limitations depending on the type of application. These techniques are briefly described below:

A. *Resizing*

This technique is widely used in the current MMS applications. Message that has a size in excess of the permitted capacity will be resized and as a result the quality will be reduced [10].

B. *Converting*

This technique is normally used in the MMS center to perform content adaptation process. For example, an audio file format allowed by the MMS center is AMR format. If we want to send an MP3 file, the MMS center must convert MP3 files into AMR before forwarding it to the recipient [11].

C. *Compressing*

Compression techniques can be used to overcome the problem of MM delivery of larger size files. By using the compression technique, we can reduce the file size, thus simplify the delivery process. This technique can also restore the quality as the original file is recovered using decompression techniques [12].

D. *Splitting*

This technique is used to split files, and then send these files one by one. The files are sent to the mobile phone recipient and then merged into a single file. This is similar to the technique used for sending an SMS [14].

E. *Masking*

Masking technique is used for masking a file. This technique is used to add a header for each split-file which can be received by MMSC. The masking technique can also be used to send the non-standardized file on MMS protocol by adding a header for each split-file message using the allowed image format, such as JPEG, PNG or GIF [14].

F. *Cropping*

Cropping is the process of removing portions of an image to focus or strengthen the composition. It means cutting part of object on image area to separate the objects from each other. This technique will reduce the image and file size [15].

In this research, we implement compression, splitting, masking and cropping techniques to develop MMS application. The compression technique is used to reduce the file size which is followed by a splitting process. The system splits the data if the compressed data file is still larger than the permitted MMS capacity. Masking technique is used to send audio or video files. The cropping technique is used to send a selected area of file (image) in a single message. The implementation of these techniques will lead to an adaptive and cost-effective client server application. We study and explore these techniques to improve the MMS technology.

III. DESIGN OF CLIENT-SERVER APPLICATION

A. *Design of Client Application*

We develop client-server application using Java programming language with its advantages and disadvantages. The system design on the client side is described through architecture design, mapping techniques and algorithms. MMS application is implemented by considering safety, efficiency, privacy, integrity and authentication. In terms of security, the message is encoded and split form, making it difficult for leaking of the information content in the message. Efficiency means that the user only pay the standard MMS tariff determined operator. The problem of privacy is also guaranteed because the user can not access the server directly, only the administrator can process data at the server. Integrity emphasizes that the information should not be altered without permission from the owner of the information, where the client-server application is already integrated and automated. If message delivery is not up to the server, then the server will send notification to the sender automatically thus making the system robust. Figure 2 shows the overview of the architecture on the client side.

There are two main parts of the development in the MMS application, namely, delivery of large image files, and large audio and video files. Figure 3 shows the various techniques that are used to develop the MMS application on the client side.

Then, we also discuss the design of an algorithm for sending large multimedia messages. We expect that the outcome of this design can provide an adaptive and cost-effective application. The MMS application is developed to send multimedia files of sizes which are larger than the allowed file sizes especially for image, audio and video data. They commonly have big file sizes so that we need to develop some techniques to transmit them from client to server side. We implement compression, splitting, cropping and masking techniques to enhance the MMS technology for transmitting multimedia files. We can use lossy or lossless compression, splitting based on pixel and bytes for image files and masking using image file for non-standard files that is not allowed to be sent using MMS technology. Figure 4 shows the step-by-step process of sending data from the sender to the recipient.

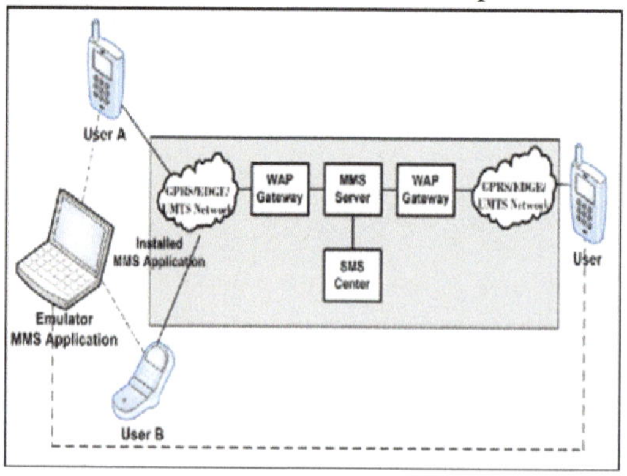

Figure 2. Overview of architecture on the client side

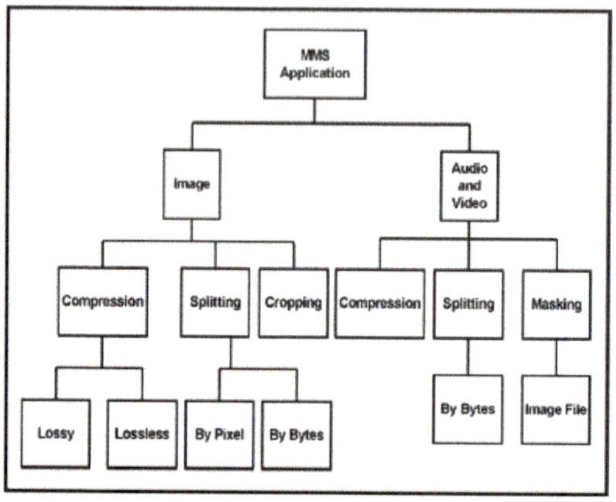

Figure 3. Techniques used to develop MMS application

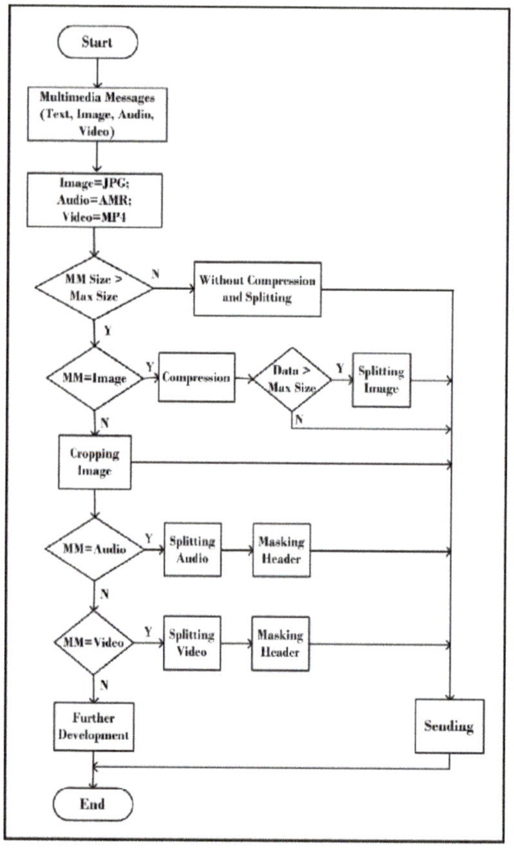

Figure 4. Flowchart for sending large multimedia file size

MMS technology can be developed by implementing the compressing and splitting techniques. The implemented results of these techniques will lead to a MMS framework that is more efficient (cost-effective) and also adaptive for various sizes and types of files. Figure 5 shows the development of MMS technology to send large image file size.

The other technique that is used to improve the MMS technology is cropping technique. Figure 6 describes the algorithm in developing MMS technology using cropping technique.

The cropping will reduce the file size thus making the transmission and storage problems easier. The data quality is preserved because the collected data is uncompressed and only resized. To obtain better results, the cropping technique will be combined with compression. The unselected area of the image is also transmitted as this infor mation may also be needed in some cases. The unselected area can be compressed with high compression ratios to reduce the file size, so that it can be transmitted together with the cropped file. Figure 7 describes the implementation of the MMS technology using compression and cropping techniques.

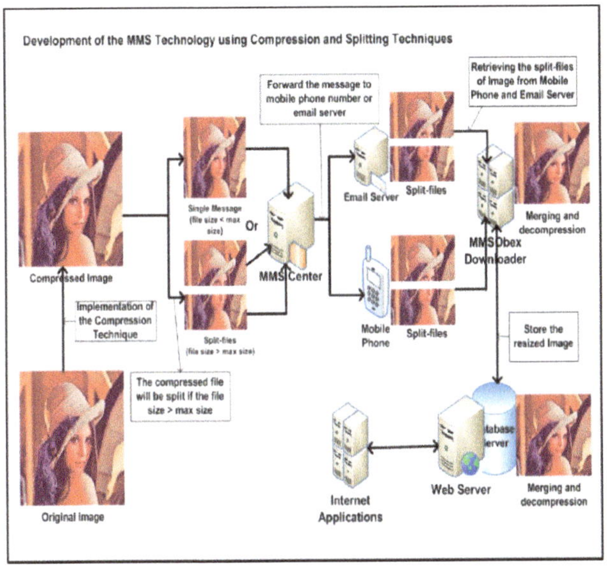

Figure 5. Development of MMS technology to send large image size

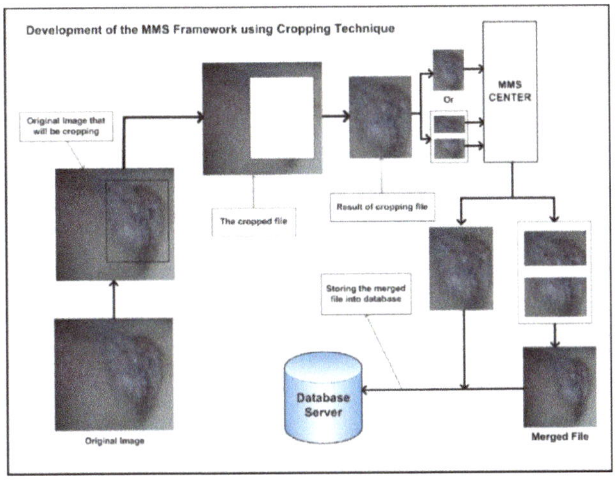

Figure 6. Development of the MMS using cropping technique

B. *Design of Server Application*

On the server side, we design an application for downloading, merging and decompressing data before it is stored into the database server. This framework can be used for a variety of Internet applications as client-server applications. Figure 8 shows the overview of the network architecture on the server side.

MMS data which is sent by the sender will be downloaded to move it to the server computer. Connection between mobile phone and the server computer must occur first. We use a data cable and Object Exchange (OBEX) downloader application. Figure 9 describes the algorithm for downloading data.

At the time of downloading the data, merging and decompression processes are performed. The development of server side application is not very complex as the client side. This is because the server computer can be easily customized to the application requirements which are not constrained by issues of memory and processing power.

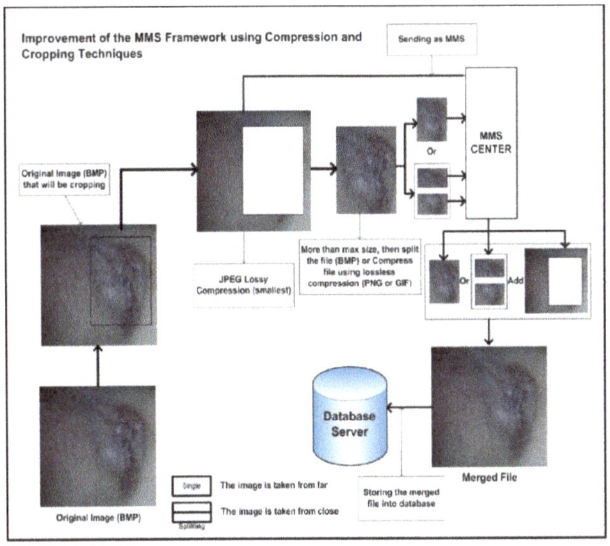

Figure 7. Improvement of the MMS using cropping technique

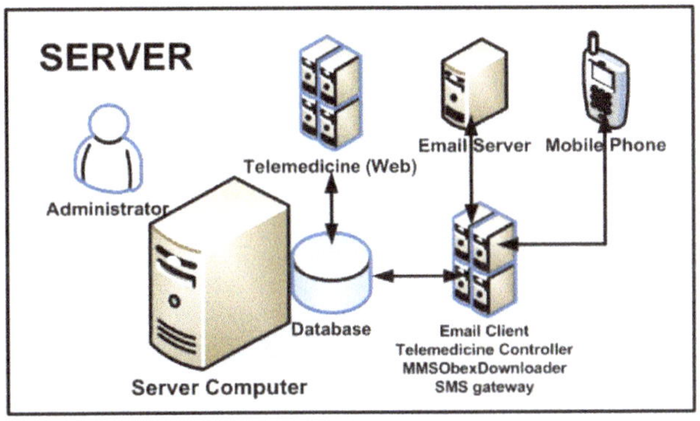

Figure 8. Overview of architecture on the server side

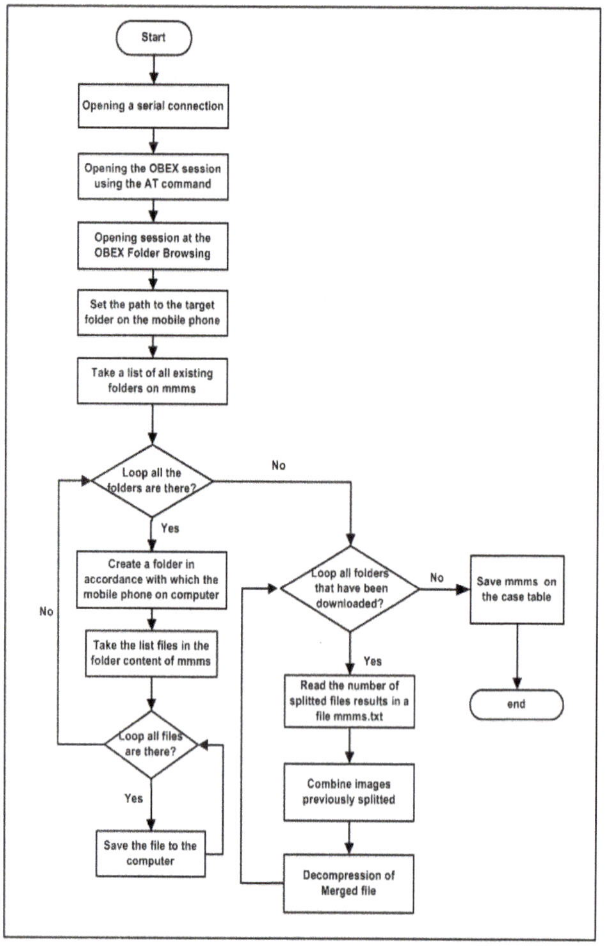

Figure 9. Improvement algorithm for downloading data

IV. Implementation and Simulation

In our previous paper [16], we have discussed about the implementation of compression and splitting techniques on the MMS application. In [17], we have discussed the implementation of MMS technology for streaming audio and video files. In this paper, we add cropping techniques for delivery of image files. This development aims to achieve an adaptive and cost-effective MMS framework for Internet applications. The cropping technique also ensures the good quality of the image which is critical for applications such as medical diagnosis. In general, the cropped part of the image may not be compressed as its size is normally small.

A. *Client Applications*

We use Java 2 micro edition (J2ME) programming language and Sony Ericsson mobile phone K630, J105i and J10i2 types to develop MMS application on the client side. This application can transmit multimedia files, such as images (JPEG, PNG, GIF and BMP), audio (AMR, WAV and AAC) and video (MP4, 3GP). Figure 10 shows MMS application for creating multimedia message.

We deliberately separate the MMS creation and sending applications in order to make the retransmission process of split files easier. The created messages are stored temporarily in the mobile phone till confirmation is received from the server. If confirmation is not received for any split-file, it will be retransmitted. The retransmitted split-file will be merged with other split-files belonging to the

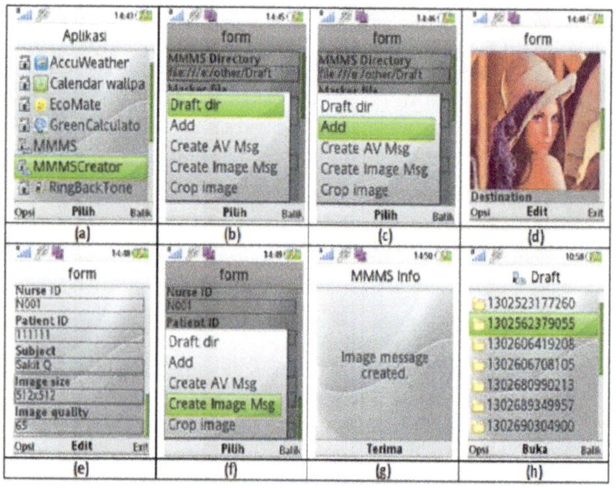

Figure 10. MMS application for creating multimedia message

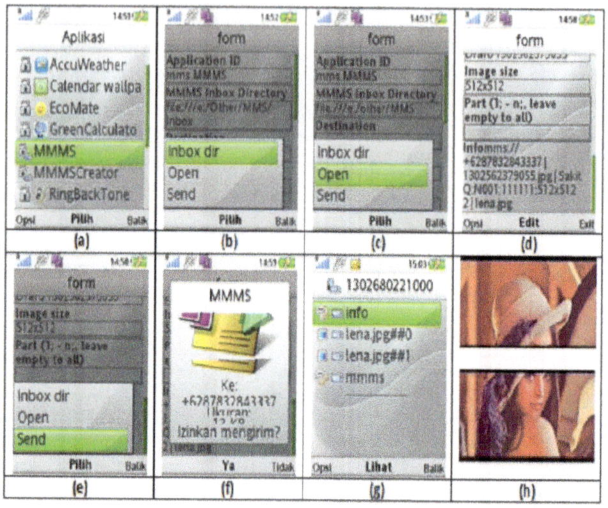

Figure 11. MMS application for sending multimedia message

same message identification. Sometimes, the sending process of split-files fails. This can happen due to disruption in network, disruption in MMS center or peak-traffic network of the operator used. This problem can be solved as follows: After server successfully downloads the whole message on the mobile phone gateway, it examines the successfully sent split files one by one using the information on the info.txt. If there any missing split files, the server will give a notification message using SMS technology to the sender and request for the missing files. Figure 11 shows the MMS application for sending multimedia messages.

We also develop cropping technique to answer the delivery address problems of the large image size. The cropping technique is used when only a small portion of the image area is needed. If we use resizing technique, it will certainly result in a significant loss of the quality. With the cropping technique, the image quality is preserved. The cropping technique results in a small file. This is almost the same as the one obtained with resize process but in this case the quality of image is preserved. This makes the developed system more adaptive and cost-effective. Figure 12 shows the implementation of cropping technique on the MMS application.

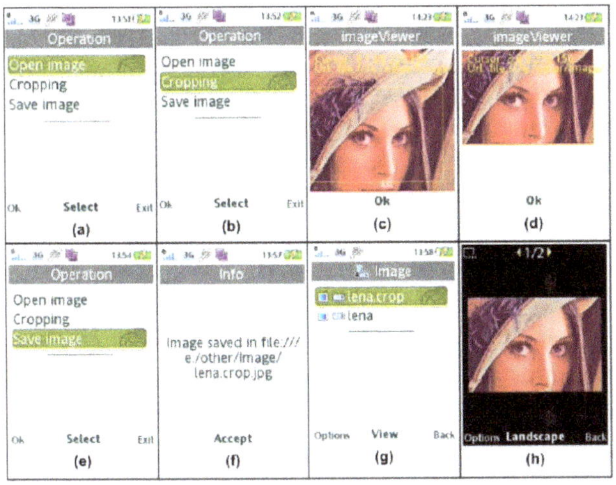

Figure 12. Implementation of cropping technique on the MMS

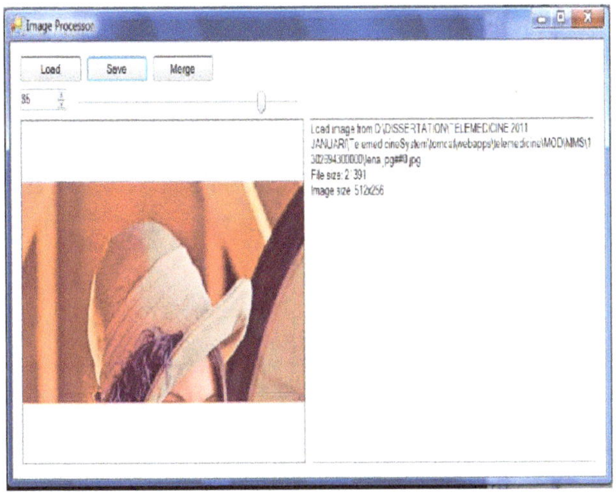

Figure 13. User interface for merging and decompression image

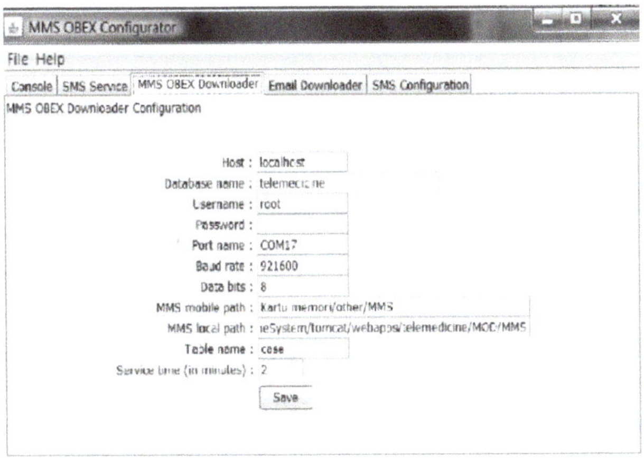

Figure 14. OBEX's User interface

B. *Server Applications*

On the client side, we use MMS application to send multimedia file to the server. For server application, we can develop any kind of Internet applications, such as telemedicine [18], e-commerce [19], e-learning [20], e-banking [21], etc. In this paper, we show how the merging and decompression processes occur on the server side. Figure 13 shows the user interface for performing merging and decompression of JPEG image file to bitmap (BMP) image file.

We make a small program to perform merging and decompression processes. Our intention is to simulate and find out the results of each step of the merging and decompressing process before it is stored into the database server and to get more accurate data. This application is a small independent desktop application which performs merging and decompressing data processes successfully. In developing client-server application, the process of downloading, merging and decompressing is performed automatically by using OBEX that is installed on the server computer. Figure 14 shows the user interface for downloading, merging and decompressing data file on the server computer.

After the multimedia data is successfully stored into the database server, the data can be used to develop any kind of Internet application.

V. Evaluation and Analysis

MMS technology will grow continuously in usage. Operators are facing challenges of taking appropriate measures to ensure that they keep pace with the rapid changes in order to optimize the users' experience and maximize the revenue benefit from multimedia messaging services. In the implementation of the development of a client-server application, we use a network that has been provided by the operator. In developing MMS application is adjusted to the system need. It improves the ability of the MMS technology by applying several techniques such as compression, splitting, masking and cropping.

These techniques can be used not only for MMS technology but also for web-based services and technology such as the Hypertext Transfer Protocol (HTTP), File Transfer Protocol (FTP), Electronic Mail (E-Mail) or streaming. The bandwidth limitation on the internet or mobile network requires performing pre-processing on the mobile phone before the data is transmitted. We develop data transmission framework using MMS technology. It is because MMS technology can use general Packet Radio Service (GPRS) network which is used widely in rural and remote areas, while the 3G technology and latest mobile technology have not covered these areas yet. The GPRS coverage area is same like Global System for Mobile communications (GSM) mobile network, where it can cover almost all the areas in a country.

The developed data transmission framework is not only to transfer larger file size through the network but also maintain the data quality. This research deals with implementation of techniques with an aim to achieve an adaptive and cost-effective method for transmitting data using MMS technology. The existing and the developed methods are then compared. The developed system is also adaptive to a variety of mobile internet applications.

We can use mobile phone device to access Internet applications easily, especially with the presence of smart phones that are flooding the market. Smart phones in general have wide display or screen therefore information can be seen more clearly than ordinary mobile phone. Web-based services use smart phones and provide facilities for mobile transactions. This method however requires large bandwidth and it needs the latest mobile technology such as Universal Mobile Telecommunications

System (UMTS) or High-Speed Downlink Packet Access (HSDPA). If the mobile network available is only GPRS, Internet connection will be very slow. With GPRS, web-based services cannot be used in rural and remote areas that have limited bandwidth. The proposed framework is suitable for areas which have limited bandwidth and it also provides advantages with respect to issues such as security, integrity, privacy, portability and adaptability. In the developed framework, the use of smart phones and web-based services method are accommodated if sufficient bandwidth is available. The proposed method can be used on ordinary mobile phone if the availability of the bandwidth is limited.

We measure the quality of the data before and after transmission in order to determine whether the data quality is preserved or not. The data quality is influenced by the compression ratio that is used. The best data quality is obtained without compression. We use peak-signal to noise ratio (PSNR) and universal image quality index (UIQI) to measure the quality of the image. Tables II and III show the values of the quality measures after merging and after decompression respectively

The simulation results show that the proposed method can be successfully implemented to send large amount of data using MMS application without degrading the quality. It is proved by the PSNR values are more than +30 dB and for splitting and cropping files before and after sending are identical.

The developed system is still a prototype system which can still be improved to adapt to the user needs. The main contribution of this research is to increase the capacity of files several times to transmit data from client to server. We have managed to simulate the developed application and installed in mobile phone device for using in multimedia message delivery. In this simulation, we use the maximum size 30 MB using J2ME. This is a constraint of the use of Java technology that only has maximum size 30 KB. This means that the ability of sending MMS can be increased several-fold if more memory is available on the mobile phone. Table IV shows the comparison of the existing and developed frameworks.

We choose to develop client-server application using MMS technology because MMS has some advantages, such as (i) MMS needs lower bandwidth and lower sending cost compared to smart phones for accessing server applications, (ii) it needs ordinary mobile phones which are cheaper (iii) MMS data is more secured and also protected from unauthorized user compared to sending data through smart phone to the Internet application directly. We have explained the techniques in detail for transmitting large multimedia file using MMS technology. The modified techniques of the MMS technology can be used to develop practical client-server applications.

VI. Conclusion and Future Work

In this study, we have proposed a solution to solve the problem of size limitation in multimedia data transmission using MMS technology. The limitation of the maximum size restricts the use of the MMS technology severely. We cannot transmit good quality of data by using the existing technology.

In this paper, we have proposed a new framework for transmitting data using MMS technology for client-server application by implementing some techniques, such as compression, splitting, masking and cropping. The implementation of these techniques leads to an adaptive and cost-effective data transmission system. This method can be used to develop any kind of client-server application where the client is in mobile environment. For future work, we will develop and improve cropping technique that can be combined with compression and active contour model to produce automatic cropping for selected image object.

TABLE II.
PSNR and UIQI Between Compression and Merging Data

Quality Factor (%)	Lena.JPEG (512x512) size= 69214 Bytes			
	PSNR		UIQI	
	On the Client	On the Server	On the Client	On the Server
	After Compression	After Merging	After Compression	After Merging
85	37.68	35.93	0.83	0.78
80	37.07	35.41	0.8	0.75
75	36.31	34.88	0.77	0.72
70	35.88	34.31	0.75	0.7
65	35.37	33.82	0.73	0.68
60	35.02	33.71	0.71	0.66
55	34.63	33.4	0.7	0.65
50	34.37	33.09	0.69	0.64

TABLE III.
PSNR and UIQI Between Compression and Decompression Data

Quality Factor (%)	Lena.JPEG (512x512) size= 69214 Bytes			
	PSNR		UIQI	
	On the Client	On the Server	On the Client	On the Server
	After Compression	After Merging & Decompression	After Compression	After Merging & Decompression
85	37.68	43.60	0.83	0.93
80	37.07	42.79	0.8	0.91
75	36.31	42.17	0.77	0.90
70	35.88	42.47	0.75	0.90
65	35.37	42.27	0.73	0.89
60	35.02	41.25	0.71	0.88
55	34.63	41.49	0.70	0.88
50	34.37	41.37	0.69	0.87

TABLE IV.
Comparison of the Existing and developed Framework

Item	The Existing	The Developed
Data Size	30 KB	> 30 KB
Language	Java 2 ME	Java 2 ME
Quality	Reduced	Without reduced
Technique	Resizing	Compression and Split
Method	Single message every sending	Multiple message one sending
Strengths	Automatically resizing before sending	- It can send data larger than the existing - Quality is not reduced
Weakness	- It cannot be modified - It cannot send data larger than 30 KB - Quality and size are reduced.	- Risk of failure is still high - The big file is split to be several file.

REFERENCES

[1] Article Snacth.com, http://www.articlesbase.com/technology-articles/mms-the-best-messaging-technology-410230.html

[2] G. L. Bodic, *Mobile Messaging Technologies and Services SMS, EMS, MMS,* Second Edition, Jhon Wiley & Sons, Ltd., 2005.

[3] Comverse making your newtwork smarter, MMS is growing in More Ways than One: The Dimensions and Implicetions of Surging MMS Growth, pp. 1-9, 2009.

[4] G. L. Bodic, *Multimedia Messaging Service: An Engineering Approach to MMS*, Jhon Wiley & Sons, Chichester, England 2003.

[5] Open Mobile Alliance (OMA), *Multimedia Messaging Service Requirement Candidate Version*, OMA-MMS-ENC-V1_3-20080128-C, Approved Version 1.3, 2008.

[6] D. R. P. Graham, *MMS Technology, Usage and Business Models*, England, Jhon Wiley & Sons Ltd., 2004.

[7] S. K. Samanta, J. Woods and M. Ghanbari, "Special Delivery: An Increase in MMS adoption", *IEEE Potentials*, pp. 12-16, 2009.

[8] A. Setyono, M. J. Alam and R. A. Al-Saqour, Evaluation and Analysis of Multimedia Messaging Service (MMS) Framework in Message Delivery, CITA'09, pp. 241-245, 2009.

[9] K. James, and W. Keith, *Computer Networking, a Top-Down Approach Featuring the Internet,* Addison Wesley, 2010.

[10] C. Salazar and T. D. Tran, "A Complexity Scalable Universal DCT Domain Image Resizing Algorithm", *IEEE Transactions On Circuits And Systems For Video Technology*, Vol. 17, No. 4, 2007.

[11] C. Saravanan, Color Image to Grayscale Image Conversion, Second International Conference on Computer Engineering and Applications, pp. 196-199, 2010.

[12] F. Ebrahimi, M. Chamik and S. Winker: *JPEG vs. JPEG2000: An Objective Comparison of Image Encoding Quality*, Genista Corporation, 1-9, 2005.

[13] A. Setyono, M. J. Alam and R. A. Al-Saqour, Split and Merge for Multimedia Messaging Service, The 2nd International Conference on Engineering and ICT, 2010.

[14] I. F. Jafar and K. A. Darabkh, "A Modified Unsharp-Masking Technique for Image Contrast Enhancement", *IEEE*, 2011.

[15] S. Bongwon, L. Haibin, B. Benjamin, and W. David, "Automatic Thumbnail Cropping and its Effectiveness", *ACM Letters* Chi Vol. 5 Issue 2, 2003.

[16] A. Setyono, M. J. Alam and C. Eswaran, Development and Implementation of Compression and Split Techniques for Multimedia Messaging Service Applications, International Conference on Intelligent Network and Computing, 2010.

[17] A. Setyono, M. J. Alam and C. Eswaran, Development of Streaming Application for Mobile Telemedicine System Using Multimedia Messaging Service Technology, International Conference on Computer and Communication Devices, 2011.

[18] A. A. Tahat, Mobile Personal Electrocardiogram Monitoring System and Transmission Using MMS, Proceeding International Caribbean Conference on Device, 2008.

[19] C. Lei-da, and S. Gordon, *Mobile Commerce Development Application.* Cybertech Publishing. USA, 2005.

[20] M. H. Hasan, N. S. Haron, N. Syafiqoh, and S. M. Yazid: Development of MMS-Based Examination Results System, Proceeding of the 9th WSEAS International Conference on Applications of Computer Engineering, pp. 157-163, 2010.

[21] M. H. Hasan and A. Khalid., "Development of MMS-Based Receipt System for Mobile Banking". *IEEE*, 2010.

AUTHORS

A. Setyono is a Ph.D student by research from Faculty of Information technology at Multimedia University, Malaysia. He is also a lecturer of Faculty of Computer Science at Dian Nuswantoro University, Indonesia. (e-mail: setyonoandik@ gmail.com).

M. J. Alam is a lecturer from Faculty of Information Technology at Multimedia University, Malaysia. His research interests are human computer interaction, image processing, medical image technology, 3D reconstruction, face recognition and gait recognition. (e-mail: md.jahangir.alam@mmu.edu.my).

C. Eswaran is a professor of Faculty of Information Technology and chairman of Center for Multimedia Computing at Multimedia University, Malaysia. He served as professor in the department of electrical engineering at Indian Institute of Technology, Madras, India. (e-mail: eswaran@mmu.edu.my).

Advances in Integrated Vehicle Health Monitoring Systems

Debopam Acharya[1] and Hyo-Joo Han[2]
[1]Computer Science, Georgia Southern University, GA U.S.A.
[2]Information Systems, Georgia Southern University, GA U.S.A.

***Abstract*—One of the fastest growing fields of study in defense market currently is Integrated Vehicle Health Monitoring (IVHM). These systems perform collection and analysis of data concerning operating parameters and damage information of vehicles in real-time and periodically send them to the base station for appropriate action. Recent advances in materials and sensor networks and technologies promise development of such systems for land, water, and aerial vehicles. These IVHM systems are of immense use in defense services which require their vehicles and systems to operate normally even under hostile and harsh environments. This work will discuss various issues related to IVHMs and prominent sensor technologies available to build such systems. It will also include the structure of a general purpose IVHM of a vehicle which can be used to monitor its inner operating parameters and damage information of its components.**

***Index Terms*—Integrated Vehicle Health Monitoring (IVHM), Sensor, Data Aggregation, Flat Routing, Hierarchical Routing**

I. Introduction

Integrated vehicle health monitoring (IVHM) system is fast developing into a very active field of research and development. Although vehicle health monitoring is not a new concept, recent advances in the field of microelectronics, nanotechnology, sensors and sensing technologies have brought about a renewed interest in the field of IVHM. An IVHM facilitates collection and analysis of data concerning operating parameters and damage information of vehicles in real-time. This leads to an increase in vehicle safety conditions, decrease in maintenance cost and significant increase in operational life of the vehicle. Continuous operational monitoring helps manage the frequency and extent of maintenance procedures for various components within a vehicle. The development of real time localized inspection methods adds the ability to pinpoint damage in vehicles thus reducing or even eliminating the need for lengthy periodic inspections covering large areas of a vehicle. A long term vision would be to develop systems that would monitor and collect operational data, and after analyzing, self repairs the defective part leading to a reliable ageless system. IVHMs are a crucial part of defense vehicles as they operate in stringent atmospheric conditions. IVHMs in defense vehicles require greater initial investment during the development phase with the returns being greater capability and high utilization for a reduced total ownership cost over the service life of a fleet of vehicles. Hence, the purpose of IVHM is to provide a balance between maximizing the operational benefit and minimizing the development cost. Vehicle health monitoring systems are also getting popular in automobiles and are already present in some form in high end car market segment.

The successful operation of IVHM depends on a network of wired or wireless sensors, data aggregation sensor units, a central processing unit, and logistics and maintenance personnel (Figure 1). An important component of IVHM systems is the array of sensors. Sensors must be small, lightweight and networked to an onboard data aggregation sensor unit. They should also be capable of withstanding vehicle's operating conditions and environment, insusceptible to electromagnetic interference, both reliable and accurate and require low power input. One of the core functions of an IVHM is continuous data collection from a large number of sensors. This is done by data aggregation sensor units. Efficiently managing this high volume of data is central to the performance of an IVHM. Since the volume of data collected over a period of time can be very large, it may not be possible for the data aggregation sensor units to communicate all data to the central processing unit. Hence, it is the job of the data aggregation sensor units to generate relevant information from this high volume data before sending them to the central processing unit. An important constituent of data aggregation is the routing protocol running across the sensor network inside the vehicle. We will discuss few routing protocols that may be helpful in IVHMs.

II. Countribution of this Work

IVHMs have attracted attention from researchers all over the world. Yet, to the best of our knowledge there is no paper that summarizes functionalities, threats monitored in a vehicle and prominent sensors and technologies of a vehicle health monitoring system all in one place. In this paper, we have described all of these in detail. There are only few papers related to IVHMs and most of them are vehicle specific (fighter aircraft, spaceships, ground defense vehicles). To the best of our knowledge, there is no paper that describes the components and architecture of a general purpose IVHM. This is essential as any person intending to develop even a vehicle specific IVHM will need it in the initial phase of development. Hence, we developed and described a general purpose IVHM architecture that describes the basic components an IVHM must have.

III. Functionality of an IVHM

The key requirements of an IVHM are that it should be able to monitor and identify damaging events, its extent, seriousness, characteristics, and respond in an intelligent manner to mitigate the effects of the damage or propose its repair. Although a pure monitoring system is expected only to report damage, it is expected that the ultimate objective of an IVHM should also be able to formulate some kind of response to the damage.

The functionality of an IVHM [2, 3] can be subdivided as follows:

i. **Detection of events:** This requires complete information of the environment in which the vehicle will be operating, the threats it will face, use of appropriate sensors and a strategy for using them to detect events and respond within a given time frame.

ii. **Characterization of the damage:** This may be a sub-process of event detection. It may use same or different sensors (active or passive) depending upon the IVHM system at hand.

iii. **Prioritization of the seriousness of the damage:** An important component of an intelligent response is to identify the damages comprehensively and identify those damages that require immediate attention. In addition to that, the IVHM system should be able to determine the urgency (both relative and absolute) with which response to those damages should be formulated.

iv. **Identification of the cause of the damage:** An IVHM system should be able to collect data from a vast array of sensors seamlessly. Absence of any data (from a sensor) may be detrimental to formulate an accurate response.

v. **Developing an effective response:** Creating an effective and accurate response will depend on a number of factors such as the range of possible solution mechanisms, the nature and extent of the damage, availability of repair services, available response time, etc. A response normally consists of a sequence of actions. Major damage may demand an immediate repair or action such as complete isolation of a section of the vehicle, followed by a quick damage evaluation and repair strategy.

IV. Threats Monitored by IVHMs

An IVHM may need to monitor several types of threats. Threats to which a vehicle will be subject may be loosely divided between those produced directly by the environment (external threats), and those that arise from within the vehicle and its structure (internal threats). It is useful to distinctively identify various types of threats. The usefulness of making such a distinction is that external threats will depend strongly on the type of vehicle and its mission (e.g. aircraft, space shuttle, space vehicle, battle tanks, etc.), whereas the internal threats will be broadly similar for different vehicle types. The specific nature, severity and frequency of occurrence will depend on the materials, structure, and operating conditions of the vehicle.

The external threats to aerial vehicles [1] may include radiation from various sources (ionizing, non-ionizing, low and high energy, solar and cosmic, etc.), meteoroids and space debris, gravitation and upper atmospheric affects (atomic oxygen, plasma, etc.). External threats to land vehicles include ground materials, stringent atmospheric conditions, war zones, and so on. Internal threats are those generated within the vehicle, or indirectly due to the external environment. They include the following:

i. **Materials from which the vehicle is fabricated:** This includes mechanical failure of materials (fracture, bond breaks, cracking, etc.), and degradation of material at surfaces and interfaces (wear, erosion, corrosion, adhesion, etc.).

ii. **Fuel and propulsion systems:** This may include engine failure (which may be due to material failure) or such things as fuel leaks or contamination.

iii. **Energy and power systems:** This may include generators (such as solar panels, nuclear, thermoelectric), batteries and storage devices, and the energy distribution system (wiring, insulation failures).

iv. **Electronic systems:** Guidance, Navigation and Control (GNC) systems, communications systems (internal and external), sensing and supervisory systems are some of the common electronic systems available in land and aerial vehicles.

v. **Life support systems (where applicable):** These may involve atmosphere (pressure, temperature, composition/contamination, explosive mixtures, etc.), water (quantity, purity, and contamination), food health systems (sensing, monitoring, and treatment).

Consequently it is clear that a comprehensive IVHM system will need to monitor broad range of measurement conditions and locations. Depending upon the vehicle, it may also have to measure a large variety of mechanical, electrical and chemical quantities, at a wide range of temporal and spatial scales.

V. Sensors and Technologies

A. Fiber Optic Sensors

Sensors are indispensible in today's IVHMs. Wireless sensor networks and devices [8] are used to monitor environments to detect changes and relay back information in real-time to a server. Current IVHM systems are aimed at replacing traditional network cables with 802.15.4 wireless links between groups of sensors and aggregation systems [16]. IVHM systems using such intra-vehicle wireless transmission are suitable for long duration missions. Considering the large aggregation of vehicle structural elements, it is natural that extremely large numbers of sensors will be required for on-board structural integrity assessment. Fiber optic sensors have been identified as the leading candidate technology for meeting this requirement with a minimal weight penalty. Fiber optic sensors require a light source such as a laser, light emitting diode, or incandescent light. One of the main advantages of fiber optic sensors is that numerous sensor sites can be multiplexed along a single optical fiber, mitigating the complexity and weight inherent with the wiring required for a large number of single ended sensors. Fiber optic sensors can be used to measure strain, temperature, and pressure and also can be used for bond-line integrity, acoustic emission, and corrosion monitoring. Fiber optic sensors have several advantages: they are highly sensitive, may be surface mounted or embedded, and are insusceptible to electromagnetic interference. Multiple Bragg grating sensing systems, known as multiplexing, can be incorporated into a single fiber [7]. They are lightweight and can be used in a number of component geometries. Fibers must be handled carefully due to their brittleness. Commercial fiber optic sensors are currently available with research continuing into higher sensitivity sensors.

Fiber optic sensors designs can be of two types: cavity based designs and grating based designs [10, 17]. Cavity based designs utilize an interferometric cavity in the fiber to create the sensor. Examples include the extrinsic Fabry-Perot interferometer (EFPI) and the intrinsic or fiber Fabry-Perot interferometer (IFPI or FFPI). Although these designs have been incorporated in different applications such as in high temperature and electro mechanical impedance environments, they do not have multiplexing capability in a single fiber. Hence, they are not suited for applications requiring large number of sensors.

Grating based designs use a photo-induced or heat-induced periodicity in a fiber's core refractive index to create a sensor whose reflected or transmitted wavelength is a function of this periodicity. Grating based sensors (e.g., Bragg gratings) can be easily multiplexed by using gratings of different wavelength as in the case of wavelength division multiplexing (WDM). Factors limiting the number of sensors in a single fiber include limited bandwidth of the source as well as that supported by the fiber, and the range over which the physical parameter of interest is being measured.

B. Piezoelectric Materials

Piezoelectric materials [7] can be used for IVHM in several ways. Piezoelectric materials have the ability to convert a mechanical stress into an electrical signal thus acting like a sensor or convert an electric pulse into a mechanical strain (like an actuator). Piezoelectric sensors can detect energies emanating from impact events and defect generation, including crack formation and delamination. In this manner, arrays of sensors are used to monitor or detect such events. Each event has a unique energy signal associated with it, and through extensive data collecting, testing and field monitoring, real-time events can be identified and monitored. There is always a certain amount of background noise, hence significant experience and expertise is required to accurately diagnose the source.

Arrays of sensors also allow a pinpoint determination of an energy source's location. Sensors have been able to detect cracks 0.005 inches in length from a distance of six inches on flat plate samples. Sensors can be placed farther apart for more realistic crack sizes. The problem with this sensing method is that although crack and delaminations are readily detected, the exact size of the defect cannot be well-determined in the field. Crack and delamination lengths are critical parameters in determining maintenance of components. Piezoelectric sensors are passive devices requiring no energy input, which is also the case with fiber optics. They may be used on metals or composites, and may be surface mounted (patches) or embedded. A second potential use for piezoelectric material is active defect detection. This method is currently under research and development, particularly for use with composite materials. Here, an electric pulse is sent to a piezoelectric actuator which produces Lamb waves within a structure. An array of piezoelectric sensors will pick up the resultant Lamb waves for processing. If a defect such as delamination or corrosion exists within the array of sensors, a change in the signal results, which is distinct from the original non-defect component. These systems are calibrated by recording the signals in a structure before it is placed in service. The sensors will be able to locate the position of the defect within the structure, and in this case, the defect size can be determined from the degree of signal change.

C. Active and Passive Ultrasonic Sensing

Ultrasonic sensing, applied in both active and passive modes, is another sensor technology area receiving considerable attention. Evaluation of transmitted active ultrasonic signals is a methodology that has been used to identify and assess damage [17]. However, such approaches use sensors that are scanned over the structure to provide a point-by-point representation of material properties and/or damage locations. Such scanning probe approaches are not currently feasible for continuous onboard monitoring. Therefore, the use of arrays of permanently attached or embedded ultrasonic transducers, which act both as transmitters and receivers, is being researched. Ultrasonic signals generated by one transducer are detected by neighboring transducers within an array. Damage between the transducers can be identified, and with advanced analysis methods, material along propagation paths that include reflections from structural boundaries can also be tested. The development of the Stanford Multi-Actuator Receiver Transduction (SMART) layer is an excellent example of recent efforts in this area [11]. Ongoing areas of research in active ultrasonic sensing technology for structural health monitoring include 1) the further improvement and characterization of miniaturized, rugged, embeddable sensors, 2) analysis methodologies for optimized sensor placement to enable characterization of damage throughout the entire structure rather than just along direct propagation paths, and 3) ultrasonic wave propagation modeling that occurs when these sensors are attached or embedded on thin-walled aerospace structures.

Passive ultrasonic monitoring, also known as acoustic emission (AE), utilizes an array of ultrasonic sensors. The sensor array is used to passively monitor acoustic signals generated by damage mechanisms such as crack growth. AE is widely used as a conventional method for off-line structural assessment, and can also be implemented to monitor a structure while in service. This capability makes it well suited for onboard structural health monitoring of aerospace vehicles. However, considerably more research and development is being undertaken to make AE a more viable technology for IVHM. Successful implementation of AE requires sensors having lighter weight, increased sensitivity, and increased ruggedness over those currently available. Additionally, reductions in size, weight, and power requirements of the associated AE monitoring instrumentation are also needed. Modern AE analysis methodologies are not only expected to locate and identify damage accurately, but also discriminate extraneous noise from signals indicating actual damage. Current efforts in this research area include development of AE multiplexing instrumentation systems that can miniaturize AE flight systems, the development of fiber optic AE sensors [12] and the development of Modal AE based analysis methods [13]. Another significant development is that of modeling approaches to better understand and predict AE propagation phenomena [14]. Such models are advantageous for a number of reasons and include characterization of AE transducers, optimization of sensor placement on a structure, and development of new and automated AE data analysis methods.

D. Microelectromechanical devices

Microelectromechanical devices (MEMs) incorporate an existing sensing technology, or multiple technologies into one device. Sensing technologies that are being transitioned to MEMs devices include strain gages, accelerometers, temperature sensors, humidity sensors, and corrosion sensors. In addition, wireless MEMs devices using radio frequency communications have been implemented. Such devices typically use a radio frequency identification chip (RFID) which is activated through a local processing unit. The RFID chip responds to the processing unit with a signal identifying the sensor as well as information from the sensing mechanisms. One downfall of MEMs devices is their susceptibility to electromagnetic interference.

E. Wireless Remote Sensor Systems

Several conventional sensors such as strain gages, thermocouples, and accelerometers are used for structural health monitoring. One major issue for such sensors is the need to route large numbers of wires to provide power and data communication. This is a difficult problem when retrofitting these sensors into existing structures, such as the aging aircraft fleet. To address this concern, a prototype adaptable vehicle health monitoring architecture has been developed [15] and flight tested. The architecture is self-contained and requires limited integration intrusion into existing systems. There are three operational levels to the architecture: one or more remote data acquisition units (RDAU) located throughout the vehicle; a command and control unit (CCU) located within the vehicle; and, a terminal collection unit (TCU) to collect analysis results from all vehicles.

The RDAUs are multi-sensor interfaces with an onboard miniature computer, programmable digital interface, nonvolatile solid-state memory and a wireless transceiver for communication with the command and control unit. Communication is achieved by using wireless radio frequency transceivers operating at 433MHz. The RDAUs were designed to withstand impact during aircraft landing while mounted on the main landing gear, and have been vibration tested up to acceleration amplitude of 20g at 2000 Hz. It was also designed to operate in non-environmentally controlled locations of the plane. The RDAU was thermally tested for temperatures ranging from -50°C to 55°C and pressure tested to simulate 50,000 ft altitude. Vibration tests verified that the remote data acquisition unit could operate at vibration levels representative of those experienced by commercial aircraft. During vibration testing, the final acceleration amplitude was 20g at 2000 Hz. The RDAU has a multiple channel programmable digital interface that allows user discretion in choosing type and number of sensors, sampling rate and sampling duration for each sensor chosen. Programmable data acquisition circuitry and expert systems trained to performance baselines in each RDAU allow the architecture to be adaptable for many types of vehicles and structures. Once a suite of sensors has been chosen for each RDAU and installed on the vehicle, a baseline of acceptable vehicle performance is established from measurements acquired when the vehicle is performing correctly. Each RDAU uses an embedded expert system trained to its respective baseline. The CCU is a computer-based subsystem that provides the communications, analysis repository, and user interface functions for the RDAUs. The CCU can also serve as a power management tool by regulating when individual or combinations of RDAUs are powered. A radio frequency (RF) wireless network of RDAUs can be controlled from a single CCU. The TCU provides the means to autonomously retrieve vehicle analysis results from all vehicle CCUs. The TCU performs analysis on results collected from all vehicles to identify any fleet-wide anomalies (e.g., all aircraft have the same faulty bearing at a similar location). The TCU develops the final summary of the vehicle health monitoring results that gets routed to the appropriate users (e.g., maintenance workers, airlines operations, etc.).

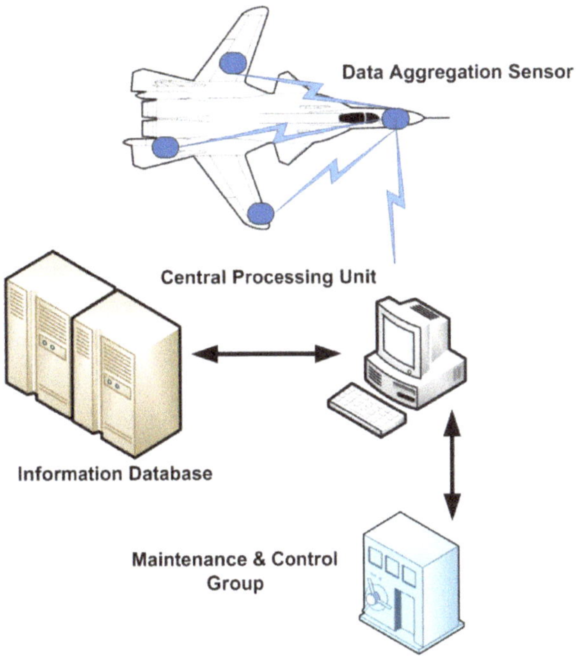

Figure 1. General Purpose Architecture of IVHM

VI. ARCHITECTURE OF A VEHICLE HEALTH MONITORING SYSTEM

A general purpose architecture of vehicle health monitoring system is given in Figure 1. Minor changes should be made to portions the architecture to suit vehicles in different terrains and conditions (land, air, water, civilian, military, etc.). Each individual component of the architecture is described below:

A. Array of Sensors

The architecture consists of an array of sensors in the vehicle. Sensor nodes [9] are hardware devices that produce measurable response to a change in physical conditions like temperature, pressure, humidity, acceleration, light intensity, etc. Functions of a sensor include sensing or measuring physical data of the area to be monitored. Signals sent from sensors either reach the destination directly or have to be aggregated by another intermediate sensor node. These data aggregation sensor nodes then transfer data to the server for further processing. Aggregation sensor node may also need to filter out unnecessary or redundant data and send only useful information to the server. Sensor nodes should be of small size, should consume extremely low energy, operate autonomous and unattended, and always stay adaptive to the environment. As wireless sensor nodes are micro-electronic sensor device, they can only be equipped with a limited power source. In an IVHM, sensor nodes will monitor operational (both structural and technical) conditions in the vehicle. The sensors send this maintenance log regularly to a data aggregation sensor. Various types of sensors may be fitted to a vehicle depending upon its function and operation. Sensor technologies used in IVHMs have been extensively discussed in section 4.

B. Sensor Data Aggregation

Data aggregation in sensor network is a widely discussed topic [4, 6]. When an event occurs, sensors gather data and communicate it to the end sensor node. As the number of nodes in the sensor network increases the probability of congestion also increases. This congestion leads to sub-optimal routing performance. Additionally, lots of packets get dropped and they need to be retransmitted again. Hence, the over-all response time increases. Data aggregation is a technique which tries to alleviate this congestion problem. It attempts to collect useful information from the sensors surrounding the event. It then transmits only the useful information to the end point thereby reducing congestion and its associated problems. In our architecture, data aggregation sensors have an important job of selectively aggregating useful information and transfer it to the vehicle monitoring center using wireless communication. It is also responsible for discarding any unnecessary data as it may lead to redundant data transfer to the monitoring station. Hence, only useful information rather than unnecessary huge volume of data will be transferred to the central processing unit. Routing protocols are the backbones of data aggregation schemes. We discuss two broad types of routing protocols in this chapter. Depending upon the conditions (type of vehicle, rate of data transfer, operating conditions, etc.) one or more of the schemes may be used.

1) Flat Routing

The first category of routing protocols is the multi-hop flat routing protocols. In flat routing based networks, each node typically plays the same role and sensor nodes collaborate to perform the sensing task. Presence of large number of identical nodes makes it infeasible to assign a global identifier to each node. This consideration has led to data-centric routing, where the data aggregation node sends queries to certain regions and waits for data from the sensors located in the selected regions. Data is normally requested using queries, and hence attribute based naming is necessary to specify the properties of data.

One of the popular types of flat routing protocols has been proposed by Heinzelman et al [4, 5]. They proposed a family of adaptive routing protocols called Sensor Protocols for Information via Negotiation (SPIN) that disseminate all the information at each node to every node in the network assuming all nodes as potential aggregation nodes. This allows a user to query any node and get the required information immediately. These protocols make use of the property that nodes in close proximity have similar data, and hence there is a need to only distribute data other nodes do not posses. The SPIN family of protocols uses data negotiation and resource-adaptive algorithms. Nodes running SPIN assign a high-level name to completely describe their collected data (called meta-data) and perform metadata negotiations before any data is transmitted. This ensures that there is no redundant data sent throughout the network. In addition, SPIN has access to current energy levels of the nodes and adapts the protocol it is running based on how much energy remains at these nodes. These protocols work seamlessly and broadcast information all over the network, even when a request to access data has not been made.

2) Hierarchical Routing

Hierarchical based routing methods are popular techniques that have advantages like scalability and efficient communication. Hierarchical routing is also utilized to perform energy-efficient routing in sensor networks as they are normally power constrained. In a hierarchical routing architecture, higher energy nodes can be used to aggregate and send information to the server, while low energy nodes can be used just to perform the sensing the target. This concept of creation of clusters where special tasks are assigned to cluster heads can greatly contribute to overall system scalability, lifetime, and energy efficiency. Hierarchical routing is an efficient method to lower energy consumption within a cluster as it leads to a net decrease in the total number of transmitted messages to the data aggregation node. Heinzelman et al [4, 5] introduced a hierarchical clustering protocol for sensor networks, called Low Energy Adaptive Clustering Hierarchy (LEACH). LEACH is a cluster-based protocol, which includes distributed cluster formation. LEACH randomly selects few sensor nodes as cluster heads (CHs) and rotates this role to evenly distribute the energy load among the sensors in the network. In LEACH, the CH nodes compress data arriving from nodes that belong to the respective cluster, and send an aggregated packet to the data aggregation node (in order to reduce the amount of information that must be transmitted). LEACH uses a TDMA/code-division multiple access (CDMA) MAC to reduce inter-cluster and intra-cluster collisions. Therefore, this protocol is most appropriate when there is a need for constant monitoring by the sensor network. Since this is the basic purpose of an IVHM, data aggregation node in such systems should use hierarchical protocols.

C. Central Processing Unit

The data aggregation sensor collects useful information and passes it to the central processing unit. The central processor has the important job of processing and analyzing incoming data. Suitable test cases in the central processing unit will use incoming information (from the data aggregation sensor) to generate results. These results will be compared with the simulation test results generated earlier during lab testing of equipments. The report generated will depict the current operation status of each component. All the reports should be organized and send to an information database. Maintenance and control personnel should be able to access the database anytime, anywhere.

D. Maintenance and Control Group

In the base station, the Maintenance and Control Group (MCG) support post-processing of data. Their activity includes usage trending and health prognostic functions that require vehicle and fleet historical data. The MCG collates historical condition, usage, maintenance, and diagnostic/prognostic data with current report (generated by the central processing unit) to produce accurate and optimum diagnostic and/or prognostic data for the maintenance personnel. This critical process of health assessment may incorporate Bayesian network models to describe failure probabilities of various subsystems. The report generated helps the maintenance personnel in deciding what to repair and when to perform the repair as a function of operational requirements and resource availability. The report also helps the personnel in generating an estimate of remaining useful life of a component or subsystem for given usage profiles.

VII. Conclusion

Recent technology advances in wireless sensor networks, developments in technologies such as MEMS, microelectronics, nanotechnology and wireless communication networks have lead to the development of integrated vehicle health monitoring systems. These systems are now an important component in most military vehicles on land, sea and air. The chapter discusses functionalities expected from an IVHM and identifies threats which an IVHM can monitor. In the longer term, these systems will provide a basis for the development of self-repairing, and perhaps even ageless, structures. In the shorter term, IVHM could allow more efficient design leading to the decrease in overall cost of redundant simulation and testing, and reduce maintenance cost and inspection requirements. All these will ultimately make the operating environment more secure for the passengers inside the vehicle.

References

[1] Abbott, D., Doyle, B., Dunlop, J. B., Farmer, A. J., Hedley, M., Herrmann, J., James, G. C., Johnson, M. E., Joshi, B., Poulton, G. T., Price, D. C., Prokopenko, M., Reda, T., Rees, D. E., Scott, D. A., Valencia, P., Ward, D., Winter, J. G.: Concepts for an Integrated Vehicle Health Monitoring System. In: Review of Progress in Quantitative Nondestructive Evaluation, Vol. 22, pp. 1606-14 (eds. D.O. Thompson and D.E. Chimenti), American Institute of Physics Conference Proceedings ,Vol. 657, pp. 1606-1614, 2003..

[2] Price, D., Scott, A., Edwards, G., Batten, A., Farmer, A., , Johnson, M., Lewis, C., Poulton, G., Prokopenko, M., Valencia, P., Wang, P.: An Integrated Health Monitoring System for an Ageless Aerospace Vehicle. In: 4th International Workshop on Structural Health Monitoring, pp. 310-318 , September 2003, Stanford, CA, USA

[3] Price, D., Batten, A., Edwards, G., A., Farmer, Gerasimov, V., Hedley, M., Hoschke, N., Johnson, M., Lewis, C, Murdoch, A., Prokopenko, M., Scott, A., Valencia, P., Wang. P.: Detection, Evaluation and Diagnosis of Impact Damage in a Complex Multi-Agent Structural Health Management System. In: 2nd Australasian Workshop on Structural Health Monitoring, pp 16-27, December 2004, Melbourne, Australia.

[4] Heinzelman, W., Chandrakasan, A., Balakrishnan, H.: Energy-Efficient Communication Protocol for Wireless Microsensor Networks. In: 33rd Hawaii International Conference on Systems Sciences, pp 1-10, January 2000, Hawaii, USA.

[5] Heinzelman, W., Kulik, J., Balakrishnan, H.: Adaptive Protocols for Information Dissemination in Wireless Sensor Networks. In: 5th ACM Annual International Conference on Mobile Computing and Networking, pp. 174–85, August 1999, Seattle, WA, USA.

[6] Kulik, J., Heinzelman, W., Balakrishnan, H.: Negotiation-Based Protocols for Disseminating Information in Wireless Sensor Networks. Wireless Networks. 8, 169--85 (2002).

[7] Advanced Materials, Manufacturing and Testing Information Analysis Center.

[8] Crossbow Technology: Low-Power Wireless Mote Solutions Overview.

[9] Crossbow Technology: Wireless.

[10] Steenkiste, R. J. and Springer, G. S., Strain and Tempertaure Measurement with Fiber Optic Sensors, Technomic Publishing (Lancaster, PA), pp. 3-4 (1997).

[11] M. Lin, 2nd International Workshop on Structural Health Monitoring, Stanford University Technomic, pp. 603-611 (1999).

[12] Duke, J. C., Cassino, C. D., Childers, B. A., and Prosser, W. H.: Characterization of an EFPI AE Sensor, submitted to Research in Nondestructive Evaluation (2002).

[13] Gorman, M.R, Progress in QNDE, Vol. 17, eds. D. O. Thompson and D.E. Chimenti, Plenum Press, New York, pp. 557-564 (1998).

[14] Prosser, W. H., Hamstad, M. A, Gary, J, and Gallagher, A.O.: Reflections of AE Waves in Finite Plates; Finite Element Modeling and Experimental Measurements, Journal of Acoustic Emission, Vol. 17(1-2), pp. 37-47 (1999).

[15] Woodard, S. E., Coffey, N. c., Taylor, B. D., Brett, R. R. and Woodman, K. L.: Autonomous Health Monitoring Architecture and Hardware, United States Patent Provisional Application and NASA Langley Research Center Invention Disclosure LAR 16407-1, Feb 2002.

[16] Prosser, W. H., Brown, T. L., Woodard, S.E., Fleming, G. A. and Cooper, E.G.: Sensor Technology for Integrated Vehicle Health Management of Aerospace Vehicles, in 29th Annual Review of Progress in Quantitative Nondestructive Evaluation (2003).

[17] Prosser, W. H., Allison, S. G., Woodard, S. E., Wincheski, R. A., Cooper, E. G., Price, D. C., Hedley, M., Prokopenko, M., Scott, D. A., Tessler A. and Spangler, J. L.: Structural Health Management for Future Aerospace Vehicles, in 2nd Australasian Workshop on Structural Health Monitoring, Melbourne, Australia, 2004.

Authors

Debopam Acharya (dacharya@georgiasouthern.edu) is an Assistant Professor of Computer Sciences at Georgia Southern University at Statesboro, GA. He holds a PhD from University of Missouri. He is a member of IEEE. His research interests include Context Aware Computing, Mobile Healthcare and Social Network Analysis.

Hyo-Joo Han (hhan@georgiasouthern.edu) is an Associate Professor of Information Systems at Georgia Southern University. She holds a PhD from New Jersey Institute of Technology. Her research interests include Pervasive Computing, Computer-Mediated Communication, Computer-Supported Cooperative Work, Education in I.T., and E-Commerce.

A Mobile Shared Workspace Supporting Healthcare Task Cooperative Planning

F.Lezzar, B.Belattar and A.Zidani
LaSTIC laboratory, University of Batna, Algeria

Abstract—**In a health care environment, in which task planning process is fundamentally a collaborative task that requires the participation of many health care professionals with different skills working simultaneously on same cases. However, most recent healthcare research has focused on methods mainly based on communication, rather than collaboration supports. We followed then a CSCW-based planning approach which constitutes an evolution of planning environments toward new shared workspaces supporting collaboration. Our work led us first, to analyse the related tasks in an Algerian maternity ward in order to highlight the vital collaborative medical tasks that need to be modelled. The following sections outline basic design concepts of our mobile collaborative planning system, which is designed to provide a flexible group interaction support for care coordination and continuity.**

Index Terms—**Collaborative planning, Healthcare task modelling, Mobile application, Synchronous/Asynchronous interaction**

I. Introduction

In healthcare planning, the need of collaboration is well recognized [1] and it seems obvious now that such task remains hard to be individually completed. We started therefore by analysing the planning process and attempted to understand how health care professionals may elaborate plans, and what happens when this work is done with a team of collaborators. It is also important to well identify which users are implied in such process. Such understanding will undoubtedly allow us to provide the adequate design by addressing the following interrogations:

- How medical staff members' collaborate?
- What means are required to improve the care process?
- Which artefacts are used to coordinate work? And how?
- From a collaboration point of view, what are the specific characteristics of collaborative medical activities?
- Which computer tools may provide the required assistance for the medical staff members and get them to work collaboratively?

It is practically impossible to develop a computer tool addressing all users' needs. Nevertheless, group work experiences provide us pertinent information to clarify some useful development ideas about suitable support tools. The experimentation of these tools thereafter, will unveil obstacles to overcome as well as perspectives to follow. Our approach is drawn in a direction which aims to favour collective thought and expertise. Therefore, as we will show it in the following sections, the planning task analysis will bring us an understanding to concretely increase the commitment of participants that may have a great impact on the whole related process.

The main objective of this paper is to present the mobile collaborative system CPMS supporting healthcare planning activities. In the following sections we will first introduce the notion of collaborative planning and review the main works achieved in this area. In section 3 we will present Android technology which is used to develop our collaborative system. Sections 4 and 5 expose our conceptual methodology and discuss the choices made as well as the software architecture designed for CPMS. We consider then details of the different architecture levels as well as the main functionalities supported. We will explicitly attempt to show that CPMS design is mainly focused on concepts of data sharing and exchange to favour collaboration between participants. Section 6 explains how our system is implemented. Finally, perspectives of the accomplished work are presented in the conclusion of the paper.

II. Healcare Tasks Planning

Hospital emergency wards such as Gynaecology and obstetrics maternities are extremely complex to manage and pose serious health risks to patients. Related tasks which are mainly focused around patient management are basically achieved through a cooperative way that involves several health care professionals.

Our observation study [2] has explicitly showed that most of the medical activities we supervised were group-based. Likewise the main deficiencies in patients monitoring precisely arose from the lack of coordination between the various members of the involved medical team, which thus constitutes a key factor as it has been so well confirmed several numerous studies carried out on this issue [3] [4].

Based on this observation, we believe that the task management should require from us a special attention. We must therefore address the issue of the targeted maternity ward under a new perspective, that of the medical staff needs, taking into account the economic and performance constraints as well as socio-health hospitals mission objectives of providing optimal care and well-being of patients. Providing a technological answer through cooperation, coordination and communication facilities seems to be the most appropriate initiative. However the past experiences reported in this area that work in-situ [1][5] should be first carefully analysed from a social point of view [6], and through a structurally opened cooperation vision that enables users to build their cooperation workspace structure in order to interact within it [7]. Consequently, we focused our interest on collaborative practices of patient care teams [8] as well as their organization [9] to better understand the usual manner with which tasks are actually performed.

Several collaborative medical care needs have been identified by a wide body of researches in informatics and medical science fields. There are common processes that are more difficult and complex in collaborative situations, because they need to integrate many parties. Such as decision making process that needs involvement of several persons to arrive to a decision, which can take long time [4]. In [10] authors showed that the collaborative nature of the executed process, determines the type of information management necessary for this process. Though, a poor structure of information can lead to coordination and communication breakdowns.

Maternity services are highly risky and still very hard to manage. They require coordination among several teams whose tasks achievement most of the time confront them to conflict situations [1]. The exploitation of information and communication technologies proves to be an effective approach if it is appropriately used [11]. It will enable us reduce the effects generated by the coordination problems that directly disrupt the patient's care chain and degrade their quality as noted Scupelli [12] through his study. Consequently, coordination breakdowns among the medical staff members inevitably that have an impact on the quality of care provided to patients and put them in a potentially vulnerable and dangerous situations should be significantly reduced with the availability of a medium of communication, cooperation, and coordination. Such an approach will provide collaborative tools that may effectively address medical staff vital needs and improve the quality of patients' care. Our research work falls then within the CSCW area (Computer Supported Cooperative Work). Thus, with a CSCW-based management strategy [13] we wish to provide an effective support of these activities enabling by the way finer planning features of the related tasks as well as providing real time mutual awareness around the occurring events within the maternity unit which constitutes a priority of our work.

CPMS (for Collaborative Planning Mobile System) design is intended with many pertinent objectives in mind such as to take in account healthcare professionals availability and mobility. CPMS supports two main interaction modes through which users may individually or collectively achieve their tasks.

According to these observations, we claim that a planning task should be meticulously analysed from a groupware perspective. Such work will probably provide us with a better understanding of the collaborative healthcare task planning process. Indeed, the complexity generated by scheduling process modelling comes from the fact that there are unexpected events. This kind of events plus a lack of coordination can generate undesirable results.

Furthermore, the complexity is increased with a team of healthcare professionals working together. Thus collaborative maintenance projects still difficult, even when sophisticated communication tools are used. The main reason is the lack of social contact due to the geographical separation of the experts [3] [15]. Besides, it is necessary to provide the adequate supports to maintain data consistency, concurrent accesses, group interaction, and coordination. Groupware concepts may help us to effectively reduce these problems, such as group awareness allowing experts to be aware of the others work and be warned about events that may arise within the hospital and within the shared workspace on the common handled artefacts.

III. ANDROID PLATEFORM

Android is based on Linux (Figure 1), what hides traditional OS abstractions because of to the middleware presented to application developers. Android focuses on applications. Much of the platform functionality is implemented as applications in the same way used by third-party developers. Android relies on Linux version 2.6 for core system services: memory management, network stack and security. The kernel acts also as an intermediate between the hardware and the rest of the software stack.

Android contains some basic Java libraries along with Android specific libraries and the Dalvik Virtual Machine (DVM). DVM allows the device to run multiple virtual machines (VMs), which allows every Android application to run in its own process. The Dalvik VM relies on the Linux kernel for functionalities such as low-level memory management and threading. For IPC (Inter Process Communication) Android provides a sophisticated message passing system, in which Intents are used to link applications. Intents are used in applications use for both inter-application and intra-application communication. : Intentions allow applications to have notifications on events. These can be sent by the terminal (incoming call, SMS, internet connection lost) or by another application wants to send information. The framework is located above the Android runtime and libraries.

It provides APIsthat enable developers to create rich applications. Android applications include an email client, SMS (Short Message Service) programs, a calendar, maps, a web browser, contacts and other programs written in the Java programming Language. There are four types of components used to construct applications; each type has a specific purpose:

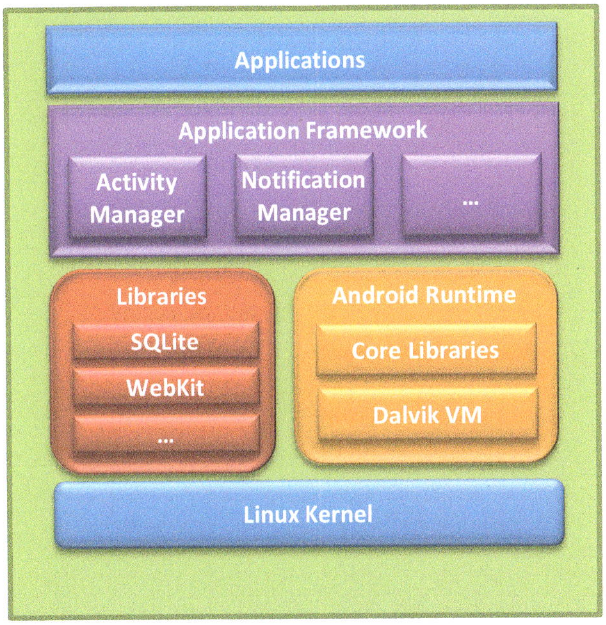

Figure 1. Architecture of Android platform

- **Activity**: is an application component that provides a screen that allows users to interact with the device applications in order to do something.
- **Content Provider**: manages data and controls the accessibility to the data for a certain application. The various applications on a device do not need to be independent. Indeed, they may need information stored in other applications (recovery of SMS, phone settings, etc...)
- **Services**: activities can be stopped at any time when the user decides to leave the application. The goal of services is to continue to run in background when the application is not "launched", (not visible on the terminal). This can allow to contact every minute a server to check for new items (emails, RSS, etc...). Background services provide RPC and call-backs interfaces that applications use to trigger actions or access data.
- **Broadcast receiver**: provide a generalized mechanism for asynchronous event notifications. So it acts as mailboxes for messages from other applications.

IV. Software Architecture

In this paper we developed a synchronous Android mobile groupware that enables real-time collaboration among collocated or geographically separated group members.

The first part of our software architecture presented in figure 2 contains the system database which is mainly characterized by its capacity to provide concurrency control management, reliable data for long time, security capabilities and data storage. The developed system database contains all the necessary data for management and for the collaborative planning process as the list of available and occupied resources, list of tasks future tasks to accomplish... We used a relational centralised data base for many reasons: a reduced complexity during the development process and also because the data volume is not important.

The second part contains all the software components. In the case of a real-time groupware, sharing data and events constitutes the most important aspect. Thus enabling data sharing requires that any event or data generated by one user has to be immediately delivered to all the other collaborators (in real- time). For better workspace awareness, fault tolerance, responsiveness, and replication of shared data objects are often used together with other operations on them like creation, updating, deletion and reading. There are many techniques to ensure awareness in a mobile collaborative system:

- **Poll:** The idea here is to periodically poll the server for new messages from the background. The more often the application poll, the most the application data will be in time. This kind of application is very easy to implement, but obviously we will never be actually in real-time. If interval polling is 10 minutes, we can get a message that is late of 9 minutes and 59 seconds. Polling more often than every 15 min will kill the battery pretty quickly.
- **SMS:** Android allows interception of SMS messages. The server sends a special SMS to our phone, whenever there is something new. The application intercepts all messages, looks for the ones from the server, and then displays the notification. This idea is easy to implement, and the system reminds fully real-time updated. The problem with this solution is that it costs a lot, especially if the notification number is important.

Figure 2. Software architecture

- **Persistent TCP/IP:** The phone initiates a long-lived mostly idle TCP/IP connection with the server and maintains it by occasionally sending keep-alive messages. Whenever there is something new on the server, it sends a message to the phone over the TCP connection. This technique provides a fully real-time updates. The Android OS is known to be able to kill services when it's running low on memory, so the notifications service can easily disappear. Another problem of battery life can appear that is related to maintaining an active connection.
- **Android Cloud to Device Messaging (C2DM):** C2DM[1] is a Google service that helps developers sending data from servers to their applications on Android devices. In the C2DM architecture, three parts are involved: the android application on an Android device. This latter must has at least an Android OS 2.2 version (in 2012 40.5% of Android devices have an OS version above than 2.2 [17]), has Android Market installed, and it must have at least one logged in Google account. An application server that developers set up as part of implementing C2DM in their applications. This part sends data to an Android application on the device via the C2DM server. And finally the Google C2DM servers that are involved in taking messages from the application server and sending them to the device. The message size limit is 1024 bytes, and Google limits the number of messages a sender sends in aggregate, and the number of messages a sender sends to a specific device.

The technique we need to develop a synchronous and real time collaborative system is called Push. The server contacts the mobile application once new data is available (notification), without pre-sending of a request from the application for an update which optimizes to minimum bandwidth and ensures real time update in the collaborative workspace. Our system CPMS combines the SMS solution and persistent TCP/IP to produce a relatively stable way of implementing push.

In our approach an SMS is sent only to inform a participant about a new event (in general when the schedule is changed...; the system notifies all the concerned persons to keep them constantly automatically informed). The server extracts the required information from the database and uses an SMS gateway to send messages to the staff members.

When a participant logs in the workspace, an android application that uses TCP/IP protocol takes over the responsibility of notifications and ensuring real time awareness. When a participant add, delete, edit a task in the schedule, this modification appears on the devices of all participants connected to CPMS in real time, persons who are not connected and involved in a new task for example, are notified by an SMS as we mentioned it before.

The second part defines two servers: Mobile Server and Displaying-Server [18]. The mobile server contains the server that notifies and sends (push) data to mobile devices. The displaying server is intended to display the schedule on eWhiteBoards (screens disposed on the appropriate locations in the hospital). After every modification of the schedule, all the eWhiteBoards are automatically updated. Also to significantly reduce users' cognitive overload such as nurses, surgeons..., the eWhiteBoard(s) can be configured to restrict the display only for pertinent information needed by each group [1] and decrease the amount of data on screens. Finally, the third part consists of the client devices.

The developed system allows an authorized physician to access at any location to the electronic patient record data, using a hand held device or even a desktop. Another version of the system for desktops has been developed for users who prefer working on big screens, or who are already in front of a computer. Using the mobile version, allow direct access of participants to consult, or modify the schedule.

V. SOFTWARE COMPONENTS

The components shown on the Figure 3 are launched in the mobile device. Our architecture is composed of several modules which are important for the collaborative scheduling task:

- **Interface module**: this module plays the role of a medium between the user and the system. The Interface module intercepts the local users' actions and sends them to the server via internet.
- **Scheduling module**: This module is the most important component; it provides the necessary tools to manage tasks and resources. List of the tasks is displayed in a table to the users with their related information such as priority, starting time...(Figure 4). Once a task is created, it will appear in real time on the other users (participants) mobile screens. So the authorized participants can modify it while the others get the latest information on their patients' state, and the cooperation process may naturally take place among the medical staff. When a user joins a work session a set of information is displayed on his screen, the shown information concerns only that user. After a creation or an update of a task, an SMS is sent to the persons assigned to that task.

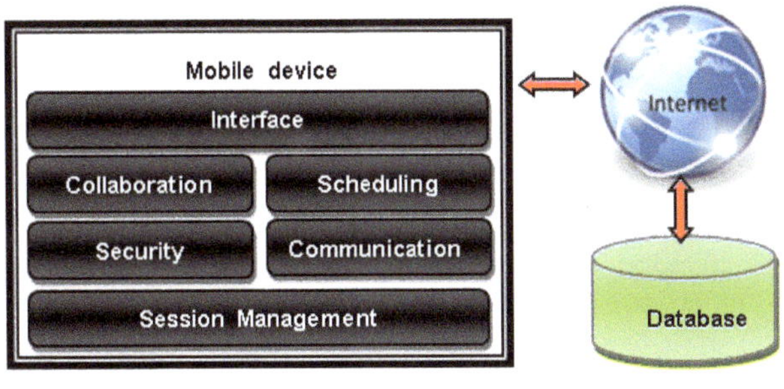

Figure 3. Components of the developed mobile architecture

Figure 4. The shared table tasks

[1] https://developers.google.com/android/c2dm/

- **Collaboration module**: to allow users to simultaneously work together, we designed several appropriate tools, such as the tasks shared table for example. Users share the same display which instantly shows any event that may occur in consequence of a user action. Such way provides users with real time awareness capabilities and enhances coordination among them. Furthermore, we take into account the ergonomic considerations while implicitly informing users about: "Who does what" through coloration use. Thus when a user starts editing a task, this latter is coloured by his specific colour (assigned once a user is registered). So users can easily identify the task author but also locate him.
- **Communication module**: In order to coordinate their work, users must exchange their ideas. CPMS supports both synchronous and asynchronous communications. Through asynchronous communication, users can post comments, new messages, express their opinions, give valuable suggestions to their colleagues. We wish through asynchronous communication improve the efficiency of the shared schedule. However, when two users or more are working at the same time, they use the instant messaging tool to communicate synchronously.
- **Collaborative diagnostic module**: Is a component intended for diagnostics elaboration of a given case under a collaborative way. Participants who are generally doctors can propose different diagnostics for a specific case and share their points of view with peers. Working under a collaborative way increases chances for faster diagnostic elaboration. The shared diagnostic is displayed as a tree expressing the whole doctors' points of view around the examined case.
- **Session Management module**: This component is intended to manage users' work sessions. During the registration of a user, the administrator can specifies his access rights on the schedule list (read/write). The Session Manager also manages users join/leave within the shared workspace, and any related event will be immediately notified the others such as latecomers...This module uses also a comonent clalled the Participant list (Figure 5). It shows the list of users with their prerogatives and roles, to increase awareness among the shared workspace.
- **Security module**: this module assures security of electronic medical records during collaboration process, by encrypting and decrypting data. We used elliptic curve cryptography technique [19][14]. This technique is a very efficient asymmetric algorithm and it is more adequate to mobiles, in which energy and calculus capacities are limited.

VI. TESTING

The next step is to test out system in a real medical context which is the maternity clinic where our study is done, to have users' opinions and to improve the collaborative process in CPMS . But before giving our application to users for tests, we made first another kind of test, to let users focus only on functionalities limitations and not on application bugs. For this purpose we used a set of tools available in the Android environment to make the following tests:

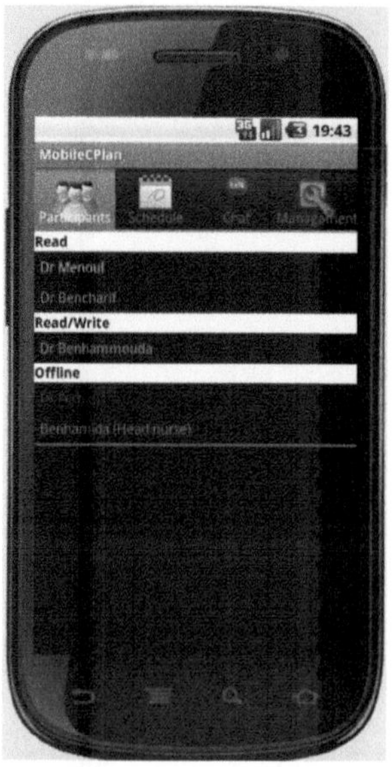

Figure 5. Participant list

- **Activities test**: to test activities. JUnit is a tool integrated with the Android development environment. JUnit generate several class tests based on the developed activities. We have first to identify activities of CPMS, and generate a testing class for each activity.
- **Event test**: Monkey is a tool used for verifying GUI application. Monkey tool generates random or deterministic event sequences such as touches, clicks, or gestures, as well as another number of events. Monkey can be used also for stress-test applications.
- **Type tests**: type bug are exception that appear in a log file.

VII. CONCLUSION AND PERSPECTIVES

This paper presents a study led in an Algeria maternity, to identify main coordination problems between medical staff members. In this study we found that coordination breakdowns can be solved by using the adequate communication technologies to improve coordination capabilities. We developed a mobile application which supports collaboration through Internet between several users.

To objectively measure the efficiency of CPMS we have developed and evaluate a first prototype. Tests the current version on a local network give rich ideas on collaboration and coordination opportunities provided.

We have discussed basic design concepts of our groupware mobile application CPMS. We have showed that it enables collaboration between several participants within a shared workspace and allows users' individual and collective actions on a common patient case as the care planning elaboration.

Care scheduling sharing has been widely discussed; its efficiency is increased over transparency while our approach allows us to concretely inform participants on their mutual actions. At the visual level, the interface shows explicitly the shared plan phases and significantly reduces participants' cognitive view and enables them to intuitively understand what is currently going on and get knowledge about their patients' states evolution as well as the next actions that should be achieved.

During a work session, the medical staff members may act on the shared care plan under a specific role, dynamically exchange messages and interact through a natural cooperation way. Such flexibility is motivated by the necessity to enable CPMS to support the dynamics implied by the care process.

Being conscious of the great interest of CPMS experimentation in effective context situations, we plan in the next step of our research work to collect more information about our targeted context with the medical staff members. It is of an extreme importance for us and represents a double objective. First, we can validate or forsake some technical choices among those we made during the implementation. Second, we will be able to determine with more precisions the appropriated adaptations we should apply to the supports provided in CPMS. To this end such as any software project we designed modular and extendable software architecture, in the sense that it allows design and integration of new components through an incremental way.

REFERENCES

[1] Ren, Y., S. Kiesler, S. Fussell, P. Scupelli. Multiple Group Coordination in Complex and Dynamic Task Environments: Interruptions, Coping Mechanisms, and Technology Recommendations. Journal of Management Information Systems / Summer 2008, Vol. 25, No. 1, pp. 105–130.

[2] F.Lezzar, A.Zidani, A.Chor_ (2012); A web application for supporting health care tasks with a groupware planning approach; International Conference on Information Technology and e-Services (ICITeS)

[3] Bardram JE, Hansen TR. Context-based workplace awareness concepts and technologies for supporting distributed awareness in a hospital environment. Computer Supported Cooperative Work. 2010;19:105–138

[4] Kuziemsky, C.E., Varpio, L. 2011. A Model of Awareness to Enhance Our Understanding of Interprofessional Collaborative Care Delivery and Health Information System Design to Support it, International Journal of Medical Informatics, forthcoming 2011

[5] Seagull, F.J., Plasters, C., Xiao, Y. and Mackenzie, C.F. (2003) Collaborative management of complex coordination systems: Operating room schedule coordination. Proc. of HFES'03.

[6] J. Cummings, S. Kiesler, "Coordination and success in multidisciplinary scientific collaborations," International Conferenceon Information Systems (ICIS), Seattle, WA: Association for Information Systems, 2003.

[7] M. Zacklad, "Communities of Action: a Cognitive and Social Approach to the Design of CSCW Systems," GROUP'03, November 9–12, 2003, Sanibel Island,Florida, USA, pp.190-197.

[8] Seffah A, Forbrig P, Javahery H (2004) Multi-devices "multiple" user interfaces: development models and research opportunities. J SystSoftw 73 (n 2): 287–3001

[9] S. R. Barley, W. H. Dutton, S. Kiesler, P. Resnick, R. E. Kraut, and J. A. Yates, "Does CSCW Need Organization Theory ?," Proceedings of the 2004 ACM conference on Computer supported cooperative work (CSCW'04), ACM Press, November 6–10, 2004, Chicago, Illinois, USA, pp.122-124.

[10] Reddy, M.S., Spence, P.R. Collaborative information seeking: A field study of a multidisciplinary patient care team. Information Processing and Management 2008; 44: 242–255

[11] Z. Niazkhani, et al., Evaluating the medication process in the context of CPOE use: The significance of working around the system, Int. J. Med. Inform. (2011).

[12] Scupelli, P., Xiao, Y., Fussell, S. R., Kiesler, S., & Gross, M. (2010). Supporting coordination in surgical suites: Physical aspects of common information spaces. Proceedings of the Conference on Human Factors in Computing Systems CHI10, NY: ACM Press

[13] K. Schmidt, and C. Simone, (1996)."Coordination mechanisms: Towards a conceptual foundation of CSCW systems design," Journal of Computer Supported Cooperative Work: The Journal of Collaborative Computing, 5 (2-3), 155-200

[14] Sahli Nabil, Benmohammed Mohamed: Security Ontology for Semantic SCADA. ICWIT 2012: 179-192

[15] Dourish, P. (2004). What we talk about when we talk about context. Personal and Ubiquitous Computing, 8(1), 19–30.

[16] Google's Android platform, http://developer.android.com.

[17] http://developer.android.com/about/dashboards/index.html

[18] M. Hertzum.: Electronic emergency-department whiteboards: A study of clinicians' expectations and experiences. I. J. Medical Informatics, Vol. 80, No. 9, pp. 618-630 (2011)

[19] N. Koblitz. A Family of Jacobians Suitable for Discrete Log Cryptosystems. In Shafi Goldwasser, editor, Advances in Cryptology - Crypto '88, volume 403 of Lecture Notes in Computer Science, pages 94 – 99, Berlin, 1988.

AUTHORS

F. Lezzar is now with LaSTIC laboratory of the department of computer sciences, university of Batna, Algeria (e-mail: lezzar.fouzi@gmail).

B.Belattar. is now with LaSTIC laboratory of the department of computer sciences, university of Batna, Algeria (e-mail: belattarb@yahoo.com).

A.Zidani is now with LaSTIC laboratory of the department of computer sciences, university of Batna, Algeria (e-mail: abdelmadjid.zidani@univ-batna.dz).

6

Agent Based Framework Architecture for Supporting Content Adaptation for Mobile Government

Hasan Omar Al-Sakran, Qasem Kharma, Irina Serguievskaia
King Saud University, Saudi Arabia

Abstract—**Rapid spread of smart mobile technology that supports internet access is transforming the way governments provide services to their citizens. Mobile devices have different capabilities based on the manufacturers and models. This paper proposes a new framework for adapting the content of M-government services using mobile agent technology. The framework is based on a mediation architecture that uses multiple mobile agents and XML as semi-structure mediation language. The flexibility of the mediation and XML provide an adaptive environment to stream data based on the capabilities of the device sending the query to the system.**

Index Terms—**Agent, Mobile-government, Content Adaptation, E-government, mediation.**

I. Introduction

Due to the latest advances in wireless mobile communication infrastructure and web technologies citizens, businesses, and public agencies are using wireless technologies in order to do their daily transactions.

One of the important transactions categories is governmental transactions such as submitting and following up applications, querying records, etc. These data are distributed over many ministries' systems that may use different software applications and apply different restrictions on accessing data. In other words, the data sources are heterogeneous in the structure and naming conventions. On the other hand, mobile technologies or what is called mobility nowadays offers many services and applications that enhance our lifestyle. The technology is fast growing and the number of users is increasing dramatically as well. In the past few years we have observed a rapid evolution of wireless technologies and a widespread of internet-enabled mobile devices. In the mid of 2005, the number of cellular phones per person was over three times higher than PCs, and most of sophisticated phones now have the processing power of a mid-1990s PC, and it is increasing [1]. Moreover, mobile phones are a relatively low-cost technology so practically everyone can afford it, and their usage is relatively simple comparing to Internet technology [4]. In 2006 about one billion people, worldwide, purchased a new handset [2]. The number of mobile users is increasing rapidly and in 2010, according to an international statistics, there were 5.3 billion mobile subscriptions, which means 77% of world population [3]. There are rapid advances in mobile technologies and services.

Mobile devices can be categorized into main three categories as following:

- Personal Digital Assistant (PDA): PDAs are small handheld devices which have some of the personal computers capabilities as well as telephone capabilities. PDAs offer many interesting functionalities such as: organizing personal schedules, multimedia support, recognizing text and voice input. They offer the user the ability to connect the Internet to check an e-mail or to search the web [5].
- Cellular Phones: In the last few years, cellular phones spread all over the world. Cellular phones range from devices with limited functionalities that are used for voice and short text message communications to advanced devices, like third generation phones (3G), which allow the user to connect to the Internet, to send or check e-mails and to open small web pages. For that, the flexibility of these devices is less than the flexibility of PDAs [5].
- Smartphone: Smartphone is hybrid devices that take some abilities from PDAs and other abilities from cellular phones. Smartphone can be used for text and voice communication, e-mail, web access and as media or video player.

Mobile devices have different operating systems and different capabilities such as different connection speed, different resolutions, etc. The major challenges of wireless network are low bandwidth, low quality connections, and the limitations of portable devices such as screen size, power consumption, processing, and memory.

The continuous improvement and development of mobile and wireless technologies provides new opportunities and challenges for mobile e-government. Using mobile technologies benefit the users in different ways. One of the benefits is the accessibility of the government services from anywhere and anytime. Use of wireless and mobile devices allows governments to provide citizens with efficient public services and to share information within and between agencies [6], [7]. To stay relevant, governments should take action and use the opportunities to enhance the usability and accessibility of their services through mobile devices. Without a doubt, mobile government is one of the most recent technology trends that use mobile phones to access government websites and services. The advancements in wireless technologies play the major role in growing of the number of worldwide mobile Internet subscribers who have a wireless access to various online recourses through mobile devices. Such facts reveal the importance of moving from E-government to M-government and they can be considered as the driving forces toward the adoption of M-government [8].

In this paper we propose a framework solving the aforementioned problems: accessing governmental heterogeneous data sources and manipulating different presentations for mobile devices. The solution is based on a mediation framework that is based on collaboration of mobile and stationary agents. These agents are used to get data from different ministries and capable of solving the problem of having different data structures and naming conventions. When a new device connects to the mediator, the device will promote its capabilities that will be taken into consideration by specific agent which will prepare the returning data with corresponding style sheet supporting the mobile device. Therefore, the framework is capable of handling heterogeneous data source and delivering data in different presentation.

The rest of the paper is organized as follows. Section 2 briefly reviews related work for mediation for content adaptation for mobile E-government. Section 3 presents an overview of M-government. Section 4 presents the details of the proposed architecture including agent technology. Section 5 concludes the paper and suggests direction for future work.

II. Related Work

Significant amount of research have been conducted to deal with content adaptation issues on mobile devices. Different approaches of content adaptation have been proposed to resolve these issues. Fudzee and Abawajy [9] classified these approaches into static adaptation which generates and stores multiple variants of the content and dynamic adaptation where the adaptation of the content is done on fly during the user's request based on his mobile device specifications. Adaptation can be performed at the client side, at the intermediary proxy server, or at the server side where the content is stored [10]. Shahidi et al. [11] proposed a method where a large number of adaptation mechanisms conducted at different locations in the network.

The advantage of the client-side adaptation is that users with more powerful devices can define preferences and determine the type and scale of the adaptation process. But not all devices are capable to deal with high-complexity applications, and the adaptation software should be generated for different classes of devices. With the server side approach the content at every web page can be adapted for specific needs of the requesting mobile device. The transformed content is fully controlled by the content provider. Unfortunately the content provider should maintain and update database of all client mobile device types and has to implement all possible content adaptations. The intermediary proxy server approach can be implemented regardless of the content providers and can be used for different contents types by applying set of general transformation rules. It is an efficient solution for mobile government and its implementation is relatively easy.

Different mediation frameworks have been proposed for designing the content of M-government services to resolve heterogeneous data sources integration and content adaptation. A number of authors suggested to customize Web content according to the mobile client contextual environments, including personal preferences, access condition, network connection, and device capabilities [12, 13, 14, 15, 16], while others proposed frameworks that capable of recognizing mobile devices' technologies, understanding their implications for M-government applications, and adapting of the content on mobile devices [17,18]. J. He et al. applied a rule based approach to guide content adaptations [19].

Dickson et al. [20] proposed a framework of three tiers of views: user interface view, data view, and process view. The process view serves as a mechanism for integrating user interface view and data view. Sheng [5] discussed some technical and policy considerations related to M-government. One of the important issues that were described in this paper is the content and presentation management. The authors suggested some useful guidelines to handle this issue exploiting techniques such as: utilizing Content Management Systems (CMS) to add a formal structure to the content and to adopt enterprise-wide web and content design standards, using the Extensible Markup Language (XML), Extensible Style sheet Language (XSL), and employing Simple Object Access Protocol (SOAP) technology.

Beside M-government issues and challenges, Sharma and Gupta [21] presented a web services architecture, while Abramowicz et al. discussed the design issues of user interface and the requirements of mobile user representations [22]. Amailef and Lu [23] present a framework of mobile government in emergency cases. The framework supports interactions between government, citizens, and other non-government agencies in emergency situations. Similar methods for content adaptation to support mobile devices have been suggested in [24, 25] where intermediary services adapt the content provided by a server to the context of the user.

Authors' interests lie in different research direction: implementing an agent technology to solve the content adaptation problem. Minjie LI et al. [26] use agents for content adaptation and solving the problem of temporary disconnection in wireless mobile network applications. Hongtao Zhu and Fangli Su [27] suggested an E-government integration model based on the semantic web, web services and intelligent agent technologies.

III. Mobile Government:

Governments have created new channels of government services delivery known as Mobile Government (M-government) aimed at engaging citizens and improving system efficiency [25].Mobile government (M-government) extends E-government benefits from its minimal requirement which is a personal computer (PC) to mobile devices by implementing wireless technologies. M-government is considered a supplementary approach of delivery of governmental services from distributed heterogeneous data sources using different transmission channels and technologies [28]. M-government is defined by Kushchu & Kuscu as "a strategy and its implementation involving the utilization of all wireless and mobile technologies, services, applications and devices for improving benefits to the parties involved in e-government including citizens, businesses and all government units" [29]. In this work we define M-government as the flexible provider of public services through mobile and wireless technologies to support users anytime and anywhere.

The M-government practices can be classified into informational, transactional and operational functions [30]. Informational functions are one way transmission of information from government to a user. These functions provide governmental information to citizens via publishing and broadcasting. It can also send alerts and notifications to a user through SMS or e-mails. Transactional functions are two way transmissions of information from government to a user and vice versa. This class of functions allows a user to interact with the M-government system, such as online procurement and payments. Operational functions aim to handle the internal governmental operations. It enables government employees to access some important information from remote locations through their mobile devices. In this work we are focusing on the content of the informational and operational functions and how this content must be personalized and adapted to meet the user preferences and technologies.

Saldhana [31] and Kumar et al. [32] stated that there are four different types of interaction levels where M-government operates represented as follow:

- MGovernment-to-government (mG2G) that is interactions between government agencies;
- MGovernment-to-employee (mG2E) that is interactions between government and its employees;
- MGovernment-to-citizen (mG2C) that is interactions between government and citizens; and
- MGovernment-to-business (mG2B) that is interactions between government and businesses.

The government services provided through mobile devices are called mobile services (m-service). This domain of M-government encompasses very critical services that would benefit citizens. The criteria of evaluating the mobile government services are: websites usability, SMS notifications and extra m-services. Main categories of usability issues are screen design, content, accessibility and navigation. The category of SMS notifications includes sending messages to and receiving notifications from ministries. Any services designed for mobile phone users not included in two first categories represent the category of extra m-services for example downloadable maps, m-books, m-brochures, etc.

L. Knutsen, et al. categorized mobile services as follows. First category is represented by the mobile content and information services (e.g. map, location-based services, news, and entertainment content downloads). Another category is messaging services such as short message service (SMS), multimedia messaging service (MMS) and email service. Transaction-based services such as mobile banking and airline reservations are yet another category of m-services [33].

Kuscu [34] and Su and Pei [35] identified many useful M-government applications that have been used. These applications include SMS for people with hearing disabilities, flood and fire notifications, search for missing children, automobile parking, tax declaration, mobile elderly care workers, garbage collection, reporting crime, etc.

In spite of all benefits and advantages discussed so far, implementing M-government is complex. Interoperability (variety of platforms, etc.) and usability (mobile devices limitations) are some of the challenges that the implementers of M-government are facing. Success of mobile government will depend largely on the number of its users. That's why implementation of user-friendly and easy access to information in alternative forms is a must.

IV. M-Government Agent-Based Framework

As we have mentioned previously, M-government informational functions can provide end users with published online information and can send alerts and notifications to users while operational functions enable government employees to access any needed information from remote locations. In the following sections, we introduce our adaptive M-government framework based on mediation architecture and highlight some useful guidelines that must be applied in order to design information content, which is provided by the M-government platform, displayed efficiently and effectively.

A. *Agent technology:*

Agent is intelligent computational reusable executable software on a network, which can carry out autonomous tasks or delegate the execution of these tasks to another agent or set of agents. It has a degree of control over its actions and does not always have to wait for commands. It adapts its behavior in response to the changing environment. A mobile agent has the property that its state and code can be moved to another server within the same network, and then continues executing from where it stopped on the previous server. The operation of agents is supported and managed by distributed agent platforms systems, an agent can migrate from server to server in heterogeneous environments. On each server an agent interacts with stationary services and other resources to accomplish its mission. It can communicate to anticipate, adopt and plan tasks. Its behavior consists of beliefs, desires, and interaction depending on the location of an entity within an agent-based system. Agent technology represents excellent solutions for solving problems in large-scale distributed systems, in our case for the e-government applications. E-government system by nature is distributed, and the integration of data from distributed ministries' servers takes a long time. Mobile agents can migrate to ministries' servers and process data locally, thus significantly reducing the response time. Such a system supports intelligent and distributed storage, allows an optimal personalized m-government environment, supports an intelligent decision-making, and ensures interoperability between different systems that are to be integrated into an operational heterogeneous e-government system. Besides that, it solves the problem of heterogeneity and low-bandwidth, reduces network traffic, process data locally instead of transmitting the data over a network and supports disconnected operation. Using agent components could accelerate development; enhance modularity, reusability, flexibility and reliability. When a mobile agent is dispatched there is no need to wait for it to return. The original connection to the network does not need to remain maintained while the mobile agents are out. The mobile agents can wait until the original connection is reestablished before attempting to return.

B. *System overview*

The architecture of proposed m-government agent-based system has a 3-tier structure involving the client mobile devices, the web server holding the mediator, and the ministry application servers as shown in Figure1.

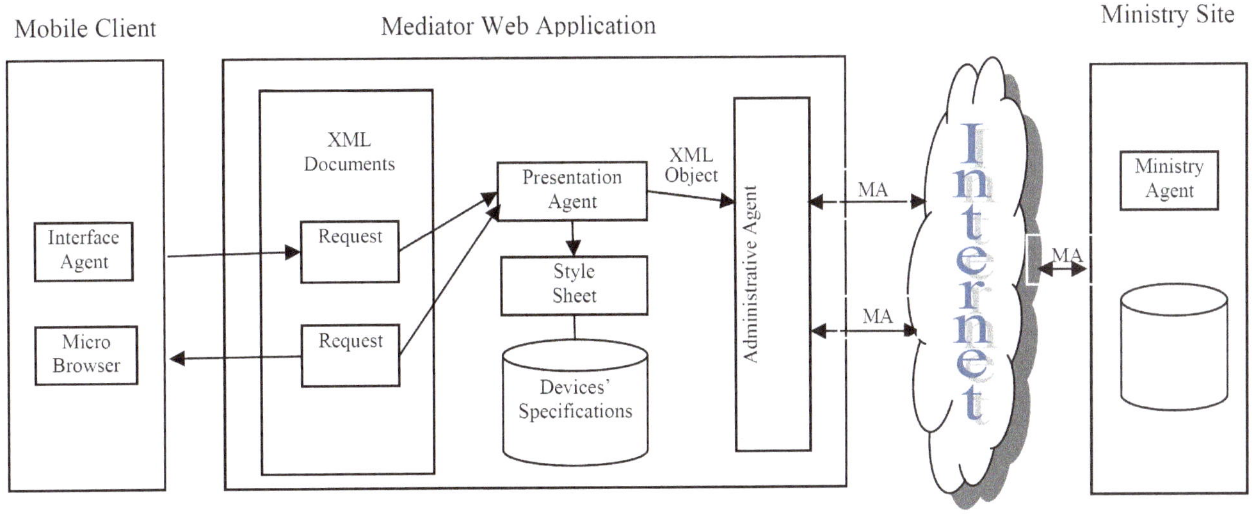

Figure 1. System Architecture

The main components of the architecture and their functionalities can be summarized as follows:

1. Client who accesses the Internet using a wireless Internet device via a micro browser. Mobile browsing is based on the Wireless Application Protocol (WAP). WAP is a protocol that was designed to allow the users to browse the Internet from their mobile device. This protocol can view the information that is written by Wireless Markup Language (WML). Pages need to be reformatted to narrow and tall view to fit small displays. This reformatting technique may affect the order of presenting the content. Client fills the request form and the browser submits it. The platform receives the requests of the client sends the request to the web server, and save the request in a temporary mobile agent.
2. Intermediate adaptation web server: This tier responsible for content adaptation. It is a key part in the process of designing M-government applications with regard to the adaptation of the content to meet the user preferences and the different capabilities and limitations of mobile devices and wireless technologies that are used by different users. This server is an intermediary server which placed between a mobile client and an application server, and it controls the adaptation process. It is in charge of content delivery. The server holds stationary agents (administrating and presentation agents), user profile, and device specification database. These stationary agents execute only on the web server where it begins execution, and their code cannot be changed during execution. The user profile is created when the user requests service(s) for the first time. It contains the client's device specification that provides the services requested by the user and the list of services based on user's preferences. The agent specifies which presentation type is most appropriate to the user according to predefined set of rules. As a result, each user can receive an adaptive content that meet his preferences and is compatible with his mobile device and wireless technology.
3. Ministry Application server: Ministries' sites provide the services to satisfy the requests from the incoming mobile agents through their ministry agents. By nature data sources are distributed and heterogeneous, and their structures are different. In order to manipulate those sources efficiently, data must be homogenized to resolve these differences. All the data will be converted into an XML object before being passed to the mobile agents.

C. Dedicated agents:

Agents need a running agent platform (agent execution environment) on every server they might move to (on mobile device, mediator, and ministries application servers) to send or receive mobile agents. Several mobile agent platforms have been proposed to include java agents such as IBM'Aglets, ObjectSpace's Voyager, Mitsubishi's Concordia, and General Magic's Odyssey. The combination of agent platform environments and the micro edition of Java (J2ME) technology provide a good solution for building very sophisticated mobile applications on most of mobile devices.

1. Interface Agent: it provides user-friendly interactive interface for mobile user. It is a static agent within the client mobile device.
2. Mobile agents: Mobile agent can move from one system to another. Mobile agents can be generated dynamically during the execution. It can reconfigure itself dynamically based on changes of the services.
3. Administrating agent: The administrating agent will generate and send one or multi mobile agents to the appropriate ministries sites. This agent is responsible for the connectivity to ministries' Databases, thus eliminating a connection per request. When mobile agents are back, the administrating agent will analyze the collected information based on client preference.
4. Presentation agent: With the emergence of heterogeneous devices, adaptation became unavoidable. It's main goal to enable the presentation of e-government content of on different mobile devices. The device context is information that is used to characterize the user mobile device. It is crucial to specify the user device's capabilities in M-government application because they can have a big impact on what content is appropriate. Some of the parameters that characterize mobile devices are the main factors that will be used to characterize this context, such as device type and device screen resolution. Nowadays mobile de-

vices can be connected to Internet through different wireless technologies. Each of them has different data transfer rate. As a result, we have to specify the type of wireless technology that will be used by the user to connect his device to the Internet; this is called the connectivity context. The agent takes responsibility for detecting the mobile device capabilities (memory, screen size and resolution), mobile browser, and information about users, such as location, time, and network performance using user profile and device specification databases. Based on these features the agent will create adaptive content for specific characteristics of mobile device. The presentation agent will convert the result to the XSL or other style according to the wireless device. Style adaptation is achieved with device-dependent style sheets (in the form of CSS, XSL or XSLT). CSS-style used to remove background shading and increase font sizes [36].

5. Ministry agent: receives the request from mobile agent and searches the local ministry's database. Initially the result is presented in XML document, and then converted into an XML object. The XML object is passed to the mobile agent who can go back to the web server and pass the result to administrating agent of the mediator. Then presentation agent converts the XML object into XML document again.

In case of reliable connection, the mobile client request is sent to the mediator and saved there. The mediator creates an agent or multiple agents to carry the request to the e-government application servers where it is processed and the result is send back by the agent(s). The presentation agent represents the result in the XSL or other style according to the requesting device's specifications and then delivers it back to the mobile device.

Agent technology is an effective approach to handle weak connectivity and disconnection problems in wireless networks (whether a mobile client sends to or receives data from the e-government servers). In such cases, the mediator saves the result in a temporary mobile agent. When the connection is re-established that agent migrates to the mobile device with the result. Then the browser of the mobile device displays the result. If the client device is offline at the time, the mediator will send email or SMS to the client. In case of no wireless connection, the interface agent saves the request of the client and as soon as the connection is re-established the interface agent via a mobile agent passes the request to the mediator where it will be processed in the same way as mentioned above.

V. CONCLUSIONS

The work presents a new framework for m-government system based on agent technology that improves the provision of M-government services by delivering personalized and adapted service content to the appropriate user efficiently and effectively. The proposed framework provides the ability to access e-government services over mobile devices which allow citizens to expand the capabilities of their devices. The agent approach described in this framework offers enough flexibility through a collection of dedicated agents. The proposed framework used a mediation architecture that can handle heterogeneous data sources, and adapt different presentation to the same data based on the characteristics of the connected client's device. The future work will be the implementation of this framework using Aglets platform with the micro edition of Java (J2ME).

REFERENCES

[1] Georgiev, T., Georgieva, E., & Trajovski, G.: "Transitioning from e-Learning to m-Learning: Present issues and future challenges", Proc. of the Seventh ACIS International Conference on Software Engineering, Artificial Intelligence, Networking, and Parallel/ Distributed Computing (SNPD '06), (2006).

[2] Misuraca, G.: " Exploring the potential of Mobile Internet Services in emerging countries: a look towards the future", Proc. of 12th EADI General Conference- Global Governance for Sustainable Development: The Need for Policy Coherence and New Partnerships, Geneve, (2008).

[3] Telecompaper, mobile subscribers in 2010 in the world, www.telecompaper.com

[4] Mengistu, D. & Zo, H. & Rho, J. J. (2009), M-Government: Opportunities and Challenges to Deliver Mobile Government Services in Developing Countries, 2009 Fourth International Conference on Computer Sciences and Convergence Information Technology, p.1445-1450

[5] Sheng, H.; Trimi, S.: "M-government: technologies, applications and challenges", Electronic Government, an International Journal, Vol. 5, No. 1, pp.1–18, (2008).

[6] Moon, M. J. (2004). From E-Government to M-Government? Emerging Practices in the Use of Mobile Technology by State Governments. Washington: IBM Center for The Business of Government.

[7] Khaled Amailef and Jie Lu, 2008,"m-Government: A Framework of Mobile-based Emergency Response Systems", Proceedings of 2008 3rd International Conference on Intelligent System and Knowledge Engineering

[8] Du Huiying, Lu Tingjie, Liu Jiale, Wu Miao,2010, "The consider on the construction of Mobile City combined with mobile computingtechnology: a case of mobile government", 2010 International Conference on Net working and Digital Society

[9] Fudzee, M.F.M.,Abawajy, J.:Aclassification for content adaptation system. In: Proceedings of iiWAS2008,pp. 426–429, Linz, Austria (2008)

[10] Velibor Adzic & Hari Kalva & Borko Furht "A survey of multimedia content adaptation for mobile devices", journal of Multimed Tools Appl (2011) 51:379–396

[11] Merat Shahidi, Nika Naghavi, A. Hamid Aghvami, Content Adaptor Selection Models in daptation Management Framework, 2011, 18th International conference on Telecommunications, IEEE, Cyprus May 2011.

[12] Stephen J.H. Yang , Jia Zhang, Angus F.M. Huang, Jeffrey J.P. Tsai, Philip S. Yu, "A Context-driven Content Adaptation Planner for Improving Mobile Internet Accessibility" , 2008 IEEE International Conference on Web Services.

[13] A. Pashtan, S. Kollipara, and M. Pearce. "Adapting Content for Wireless Web Service." IEEE Internet Computing, 7(5), 79-85, 2003.

[14] S. Toivonen, J. Kolari, and T. Laakko. "Facilitating Mobile Users with Contextualized Content." Artificial Intelligence in Mobile System Workshop, Seattle, WA, USA, October, 2003.

[15] B. Kurz, I. Popescu, and S. Gallacher. "FACADE – A Framework for Context-Aware Content Adaptation and Delivery." Second Annual Conference on Communication Networks and Services Research, 46-55, 2004.

[16] A. Krause, A. Smailagic, and D. P. Siewiorek. "Context-Aware Mobile Computing: Learning Context- Dependent Personal Preferences from A Wearable Sensor Array." IEEE Transactions on Mobile Computing, 5(2), Feb., 113-127, 2006.

[17] Kumar, M.; Sinha, O.: "M-government – Mobile Technology for E-government ", Proc. of 5th international conference on e-governance (ICEG-2007), Hyderabad, India, (2007).

[18] Guirguis, S.K. Hassan, M.A. 2010 "A smart framework for web content and resources adaptation in mobile devices", Advanced Communication Technology (ICACT), 2010 The 12th International Conference, 487 - 492 , Phoenix Park.

[19] Jiang He, Tong Gao, Wei Hao, I-Ling Yen, and Farokh Bastani, 2007, A Flexible Content Adaptation System Using a Rule-Based Approach, IEEE TRANSACTIONS ON KNOWLEDGE AND DATA ENGINEERING, VOL. 19, NO. 1, JANUARY 2007.

[20] Dickson K.W. Chiu, Dan Hong, S.C. Cheung, and Eleanna Kafeza, "Towards Ubiquitous Government Services through Adaptations with Context and Views in a Three-Tier Architecture", Proceedings of the 40th Hawaii International Conference on System Sciences – 2007

[21] Sushil K. Sharma and Jatinder N.D. Gupta. Web services architecture for m-government: issues and challenges. Electronic Government 1(4):462-474, 2004.

[22] Witold Abramowicz, Andrzej Bassara, Agata Filipowska, Marek Wiśniewski, Paweł Żebrowski. Mobile Implications for m-Government Platform Design. Cybernetics & Systems 27(2-3):119-135, 2006.

[23] Khaled Amailef and Jie Lu "m-Government: A Framework of Mobile-based Emergency Response Systems" Proceedings of 2008 3rd International Conference on Intelligent System and Knowledge Engineering.

[24] D. Malandrino, F. Mazzoni, D. Riboni, C. Bettini, M. Colajanni, and V. Scarano, "MIMOSA: Context-aware adaptation for ubiquitous web access ". Personal & Ubiquitous Copmut. 14(4), pp. 301–320, May 2010.

[25] Christos K. Georgiadis and Emmanouil Stiakakis,2010,"Extending an e-Government Service Measurement Framework to m-Governement Services, 2010 Ninth International Conference on Mobile Business / 2010 Ninth Global Mobility Roundtable.

[26] Minjie LI, Liqiang WANG and Ying HAO, "A Mobile Agents Based Framework in Wireless Network",The 2nd IEEE International conference on Information Management and engineering(ICIME),90-93, 2010.

[27] Hongtao Zhu and Fangli Su, "An Integrated Model in E-Government Based on Semantic Web, Web Service and Intelligent Agent",2010.

[28] Gasem

[29] Kushchu, I., & Kuscu, H. (2003). From E-Government to M-government: Facing the Inevitable, The 3rd European conference on E-Government (ECEG03), Trinity College, Dublin, p. 253-260

[30] Norris, D.F..; Moon, M.J.: "Advancing E-government at the grassroots: tortoise or hare?", Public Administration Review, Vol. 65, No. 1, pp.64–75, (2005).

[31] A. Saldhana. Secure E-Government Portals - Building a web of trust and convenience for global citizens. In W3C Workshop on e-Government and the Web. 2007. Washington DC, USA.

[32] Kumar,M., Hanumanthappa, M. & Reddy, B.L. "Security issues in m-government", Proc. of the International Conference on Global e-Security (ICGeS 2008), London, UK, CCIS 12, Springer-Verlag, 2008, pp. 265-273.

[33] L. Knutsen, et al., "Acceptance and perceptions of advanced mobile services: alterations during a field study," presented at the International Conference on Mobile Business, Sydney, Australia, 2005.

[34] Kuscu, M.; Kushchu, Ibrahim; Yu, Betty (2009), Mobile Computing,Introducing Mobile Government

[35] Chen Su and Zhou Pei 2010 "Application Model of Mobile E-Government in Wuhan Urban Circle", 2010 International Conference on Multimedia Information Networking and Security

[36] U. Erra, G. Iaccarino, D. Malandrino, and V. Scarano, "Personalizable edge services for web accessibility ". Universal Access Inf. Soc. 6(3), pp. 285–306, 2007.

AUTHORS

Hasan Omar Al-Sakran, Qasem Kharma, and **Irina Serguievskaia** are with King Saud University, Saudi Arabia.

Enterprise Mobile Tracking and Reminder System: MAE

Yoong Cheah Huei, Thet Swe, Lee Zheng Xian, Neo Shou Yee,
Ngee Ann Polytechnic, Singapore

***Abstract*—Mobile phones have made significant improvements from providing voice communications to advanced features such as camera, GPS, Wi-Fi, SMS, voice recognition, Internet surfing, and touch screen. This paper presents an enterprise mobile tracking and reminder system (MAE) that enables the elderly to have a better elder-care experience. The high-level architecture and major software algorithms especially the tracking in Android phones and SMS functions in server are described. The analysis of data captured and performance study of the server are discussed. In order to show the effectiveness of MAE, a pilot test was carried out with a retirement village in Singapore and the feedback from the elderly was evaluated. Generally, most comments received from the elderly were positive.**

***Index Terms*—mobile tracking, architecture, software algorithms, server performance study, pilot test, elderly feedback**

I. Introduction

Today mobile phones have advanced from providing voice calls to many built-in functions like touch screen, Wi-Fi capability, camera, speech to text and vice-versa, Gobal Positioning System (GPS), Accelerometer and other features. Mobile phones with the built-in functions are called smart phones. The prices of smart phones such as iPhone 4S, Android Galaxy, Windows Mobile 8, Blackberry 9900 and other older models are getting cheaper and more affordable. Hence, there is no surprise that the smart phones are replacing the older generation of voice communication mobile phones.

In the area of tracking, the study of power consumption in smart phones was studied [1-2], GPS information was extracted in Nokia N95 [3], social networking applications were studied in [4-7], features of Android for location-based function was reported in [8], [9] investigated the human behavior for location-based services, using location-based services in city maps was published [10], and middleware development for location-aware was published in [11].

Moreover, many research work has been done in other areas of smart phone applications as well, for example electronic payment [12], consumer services [13-14], medical [15-19], education [20-21], logistics [22], fault reporting [23], military [24], security [25], motion detection [26], cloud computing [27], protocols [28-29], and quality-of-service [30].

In this paper, we propose the Mobile Assistance for Elderly (MAE) system with the aim to utilize the rapid growth of smart phones and technologies so that a better elderly experienced is achieved through mobile-based platform. This system leverages on some of the available telecommunication technologies like GPS and Short Message Service (SMS) to help the elderly and their caregivers stay in constant engagement while allowing the elderly to lead a more independent and mobile life as they stay active socially. While the target caregiver is the nursing home, the system can be streamlined for eventual use for home caregivers to their elderly parents.

Section II presents the overall system architecture of MAE. Section III describes the client. Section IV describes the server, in particular the SMS module. Section V highlights the mobile-to-mobile tracking and server-to-mobiles tracking. Section VI discusses some results of our system performance study. Section VII evaluates the feedback from the elderly about the pilot test. Section VIII concludes this paper.

II. Overall System Architecture

Figure 1 shows the high-level overall architecture of MAE. The server consists of three major components: Database, ASP.Net web server, and MAE_SMS gateway. The smart phones with in-built SQLite database are the clients. The MAE_SMS gateway is responsible to manage the schedule messages, send and receive SMS messages via the modem, and interact with the database to retrieve and insert new records. It also provides a Graphical User Interface (GUI) for the system administrator to monitor the system performance. Currently, the clients are Android smart phones that communicate to the server via the Internet or any network such as 3G or Wi-Fi. The smart phones have the capabilities to store the data into its own SQLite Database which could serve as temporary storage.

III. The Client (Android Smart Phone)

For our case, the client is an Android smart phone which is carried by the elderly. The Android smart phone must be GPS or Wi-Fi enabled and access to Internet. The phone is installed with our software that uses the GPS or W-Fi to find the exact location using satellites when any triggered event is initiated from a tracking device which can be another smart phone (Section 5.1) like any Android smart phone, iPhone or a server (Section 5.2). Upon successfully locating the position, the data will be registered in the phone's SQLite database for the caregiver to extract anytime from a server or another smart phone like iPhone or Android .The client can receive SMS reminder services from our server so that the elderly would not miss any important appointments or forget to take medications.

In our project, the push service is implemented in the Android smart phone which is the client. The service runs at the background and is transparent to the elderly. In order to be tracked, prior consent must be obtained from the elderly.

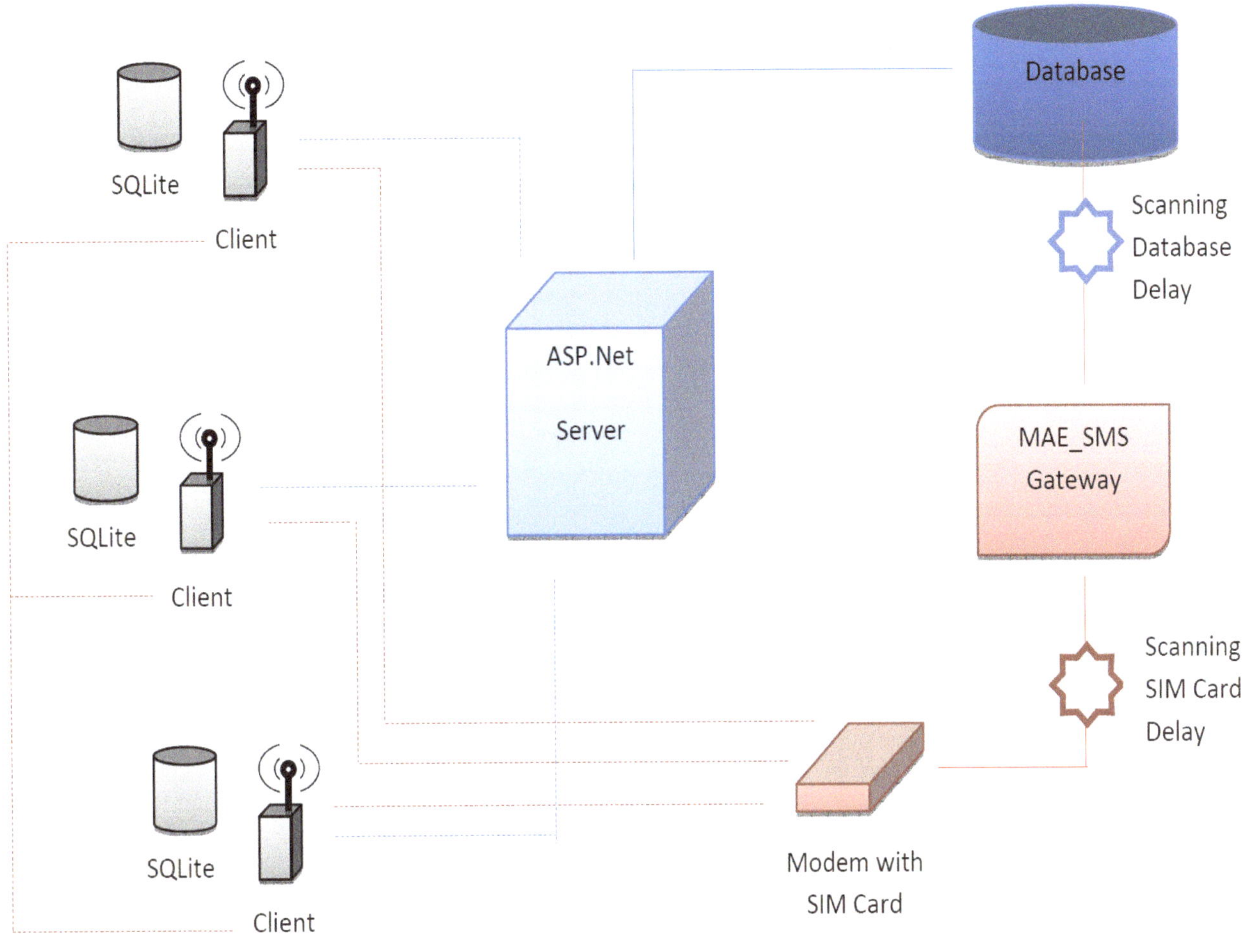

Figure 1. Overall system architecture

There are two ways of triggering the software in the Android smart phone. The first method is an interval counter that acts as an own signal of the triggering point. The interval can be set according to the user's preference. The second method of triggering would be the incoming messages like an SMS of preset command such as a special code. Both methods will trigger the software to activate the GPS or Wi-Fi function of the device. When the GPS or Wi-Fi program is activated by any of the above two events, it will begin to pull the location parameters.

The GPS technology is accurate in open areas but perform poorly indoors or between tall buildings. The Wi-Fi technology is used inside a building and gives more accurate positioning when the elderly is inside a building compared with GPS. Our system is able to switch automatically from GPS to W-Fi or vice-versa. Manual overrides are also implemented.

IV. The Server

There are two main functional modules of the server – SMS module and tracking module. The tracking module is briefly described in Section 5.2.

A. *SMS Module*

This module is written in C# programming language. We wrote a Window Application GUI with threading to monitor the system status and manage SMS sending and receiving messages. Figure 2 depicts the three main threads that handle the interconnection and synchronization among SQL database, messaging queue and the modem. Figure 3 describes how the algorithm works of the three threads in Figure 2. When the system starts, the program will connect the SQL database, retrieving all the due tasks and push the tasks into to the Microsoft Message Queue Server (MSMQ) with its standard message format. MSMQ is a fast store and forward service from Microsoft server that enables applications running at different times to communicate across different systems. Application sends messages to MSMQ and MSMQ uses queues of messages to ensure that the messages eventually reach their destinations. The system uses the mechanism of pulling data from the database in every time interval that was set in the 'scanning database delay' time. The program will then use the current date and time information to extract tasks from the server to schedule for the outgoing messages.

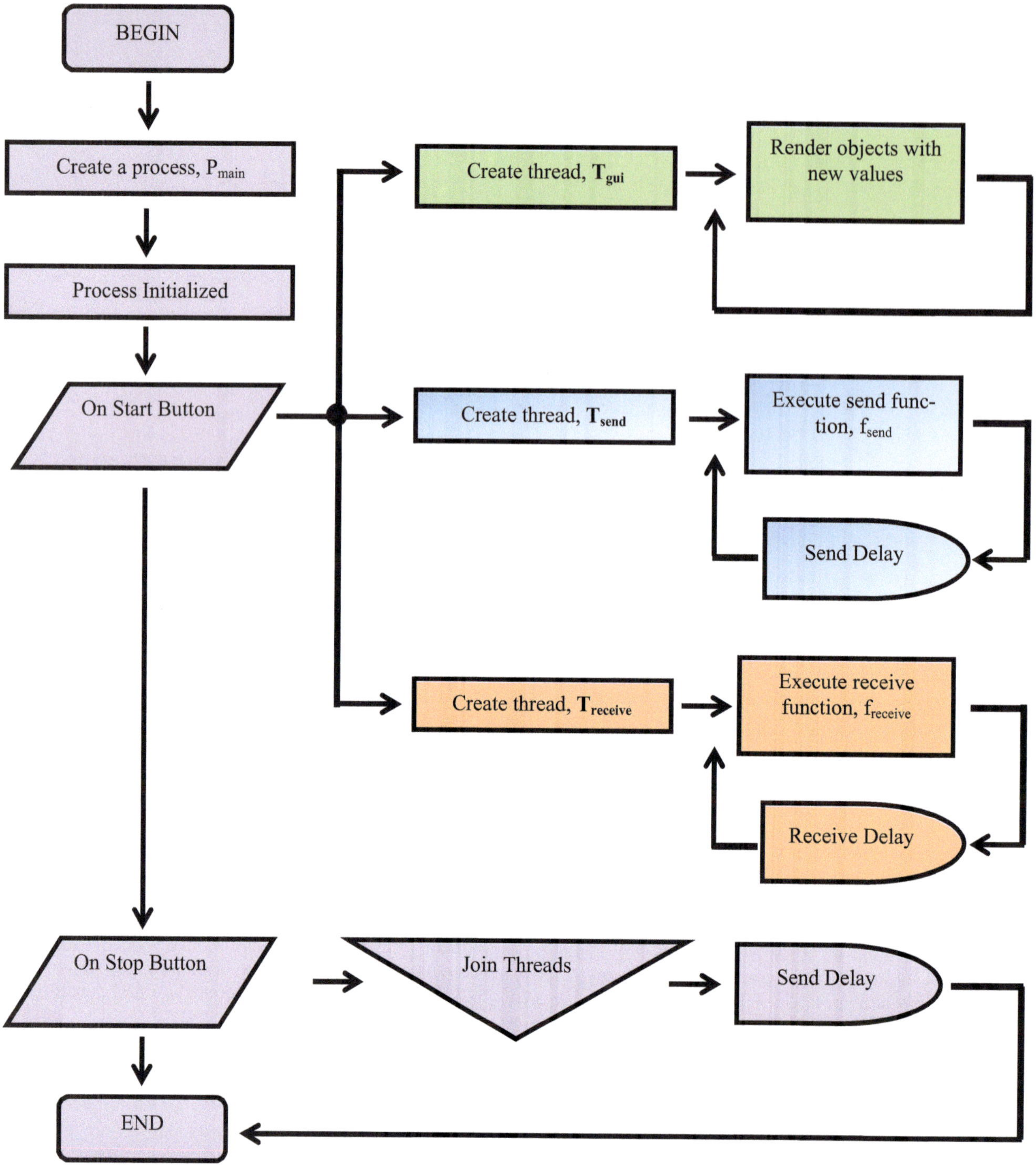

Figure 2. High level system flowchart of server

Algorithm

Step 1: Creation of Process (P_{main})

Create a process, P_{main} in the MAE Server in order to handle the SMS Scheduling, Sending & Receiving between Database, DB and the Message Queue Server (MSMQ).

Step 2: Initialization (P_{main})

P_{main} is fully initialized with required components such as serial port, interfaces. It also loads config data from text file.

Step 3: On Start Button Event ($E_{on-start}$)

The process, P_{main} will create 3 worker threads, T_{gui}, T_{send} and $T_{receive}$ in its process.

❖ ***GUI Thread***

The GUI of the system is handled by the worker thread, T_{gui}. The reason of using thread for GUI is because of the control buttons in the GUI. The system needs to be responsive even when it is running other functions, such as sending messages or receiving messages

❖ ***Sending Thread***

This worker thread, T_{send} will perform the function (f_{send}) over the period of scanning database delay, t=1,2,...,10 where delay time, t is predefined by user & measureable in minute.

$$Y = f_{send}(X, t * i) \quad i = 1,2,3,\ldots n \text{ (interval - measured in minutes)}$$

Whereas Y is the result of sending function (f_{send}) with the input of X (tasks due) over the scanning time interval, i. This function starts with t=1 and is repeating until the On Stop Button ($E_{on-stop}$) is invoked.

❖ ***Receiving Thread***

The worker thread, $T_{receive}$ will execute function ($f_{receive}$) over the period of scanning SIM delay, t=1,2,...,10 where delay time, t is pre-defined by user & measurable in minute.

$$Y = f_{receive}(X, t * i) \quad i = 1,2,3,\ldots n \text{ (interval - measured in minutes)}$$

Whereas Y is the result of receiving function ($f_{receive}$) with the input of X (incoming messages) over the scanning time interval, i. This function starts with t=1 and is repeating until the On Stop Button ($E_{on-stop}$) is invoked.

Step 4: On Stop Event ($E_{on-stop}$)

On Stop event occurred when the stop button is clicked. Upon the event occurrence, the program will attempt to join the threads to complete its task with the allowance of sending delay interval.

Figure 3. High-level algorithm of the threads in Figure 2

V. Mobile Tracking

We employed two techniques to illustrate mobile tracking. The first technique is mobile-to-mobile tracking. The second technique is server-to-mobiles tracking.

A. *Mobile-to-mobile tracking*

This tracking is based on one-to-one relationship of the communication pattern. For example, the caregiver iPhone can be used to track one elderly Android phone. The caregiver and the elderly are communicating with each other and the tracking is based on one elderly at one time. The one-to-one technique is suitable for the individual since the caregiver's intention is to track only one elderly at a time. The algorithm of mobile-to-mobile tracking is similar to the algorithm presented in Figure 4 except the variation where the replying/monitoring takes place in a server.

B. *Mobiles to Server Tracking*

This type of tracking is used in organization or care-center where the number of elderly exceeds the number of caregivers. It is known as one-to-many relationship. For instance, a number of elderly Android phones will send their individual location coordinates to the server periodically. In this relationship, the location data are stored and monitored in a server instead of a smart phone. Instead of relying on the device SQLite Database, the location data will be also stored in the server's SQL database with higher capabilities. The server is set to track and present graphically the last 10 locations of each registered elderly. Figure 4 depicts the algorithm for mobile-to-server tracking.

VI. Performance Results

A. *Collect Location Data*

We have conducted the testing for a period of 10 working days across Singapore with six Android smart phones with the following results in percentage of success rate of each working day.

Working day 1: 70%, Working day 2: 10%, Working day 3: 18%, Working day 4: 16%, Working day 5: 42%, Working day 6: 78%, Working day 7: 99%, Working day 8: 99%, Working day 9: 75%, Working day 10: 79%

In day one to five, the average percentages of locations collected is below 40%. The reason is due to the timeout error occurred at the mobile phone side because the public server, where the location coordinates are obtained, is physically located in United States (US).

In order to solve this problem, we have relocated the server from US to Singapore. Thus in day six to 10, the average performance of location data collection has increased to an average 86% . In day seven and eight, the percentages of collectable data are close to 100%.

B. *Server SMS Performance*

Our performance measurement is based on two factors: (1) Message size in characters and (2) Delay timer for scanning database and SIM card. With the sending 20 messages, the performance is measured in 'total sending time (sending)' and 'total receiving time (receiving)' for all messages. The definition of 'total sending time' is the time taken from the sending of first message to the last message when the whole system is functioning well. The timestamp is taken only when the sending of message is successful. The 'total receiving time' is defined as the time taken from the receiving of first message to the last message when the whole system is functioning well. The timestamp is recorded when it is registered in the database.

The matrix is the representation of timer set for scanning database (sending) delay and scanning SIM (receiving) delay respectively. The settings for delay timing are 1 minute and 2 minutes. Four different ranges of randomly chosen messages were used, with the average of 16, 39, 77 and 134 characters. Figure 5 shows the results of average message size versus the total sending time and total receiving time for 1x1, 1x2, 2x1, 2x2 matrixes.

In Figure 5, as the average message size increases, the sending time also increases. The reason is that the system requires more times to push the characters into the network when the number of characters increases. Although the delay time of scanning database (sending time) is increased, the total sending time does not change much. The reason could be the schedule time of the messages. The messages are scheduled to send at one time and the system is very quick in sending out all messages. It results in little impact on total sending time. The change of scanning database delay is actually affected on the schedule time. For example, if the system is running at 5 minutes interval and the last scanning were carried out at 00:00:00; the next scanning will carry out at 00:05:00. If the task were scheduled on 00:02:00, instead of carrying out in its schedule time, the task will be delayed to next schedule time.

The receiving time (red line) is hugely affected by two factors – time taken for scanning SIM card and time taken in the 3rd party network like the public telecommunication network provider. When there is an increase in scanning time delay, the receiving time from 1st to last messages increases. With the increase in messages time, the two factors above were taking longer time to receive the messages, and it results in increase on receiving time. The 3rd party network delay has potential be more severe.

Figure 6 depicts the sending and receiving time versus matrixes with average fixed message size. As shown in Figure 6, the sending time (blue line) does not vary much even the timing were changed. This is why the sending time has almost no impact, except for failure to send the message from the device.

The receiving time (red line), on the other hand, has much impact on system performance. The measurements were taken in 4 points – 1x1, 1x2, 2x1 and 2x2. In general, when the delay time in scanning SIM card for receiving is increased, the receiving time also has increased. This result is what we expect to happen because longer time is needed to register the received messages.

***Collection*: Creation of Gathering Data Process (P_{locate}) in Client**

Create a process, P_{locate} in the application in order to activate the location manager for the period of t, 60 seconds. With the valid t period, P_{locate} will collect the location data sets, X with the accuracy of z.

For t=60, $Y_i + Z = f[X_i]$ for $i=1$ to ∞

where t is measurable in seconds, f is the location retrieving function with technologies (X) which is both GPS and cell triangulation. The function will output location data set (Y) which contains longitude and latitude values along with the accuracy (Z).

The function, f can be defined as

Activate Global Positioning System (GPS) Receiver
Retrieve Longitude(x_1), Latitude(y_1) & Accuracy(z_1) using Satellite
Retrieve Longitude(x_2), Latitude(y_2) & Accuracy(z_2) using Wi-Fi
Save Longitude(x), Latitude(y) based on best Accuracy (z)

***Storing*: Save the location Data in SQL in Server**

After the collection of t=60, P_{locate} will choose the best accurate location data set Y based on Z value. Using data connection, It is then sent to in Web Service function (f_{write}) that will insert into the SQL database (DB) in the Server.

For New DB record in the server, $r_{new} = f_{write}[Y_i]$

The f_{write} function will

Do
Establish a new connection to SQL Database Table "Location"
Get result from set of Longitude(x), Latitude(y) and Accuracy(z)
Select the best accuracy (minimum value of z) location data set
Attempt to save into the SQL Database
Close SQL Database connection
Reply Attempt status to the Updating Device
While (No of Attempt<3) & (Not Successful)

***Viewing*: Monitoring the Locations in Server**

The viewing or monitoring is done in the server's web interface. The location result will be plotted in the Google Map API in the browser.

Response, $R = \max_T[\text{records in DB}]$ where T = recorded time in milliseconds

The response R is the result of following steps:
Establish a new connection to SQLite Database "Location"
Get latest result on the database (T is maximum)
Save into the SQLite Database
Close SQLite Database connection
Render on Map Object on Browser

Figure 4. Server-to-mobiles tracking algorithm

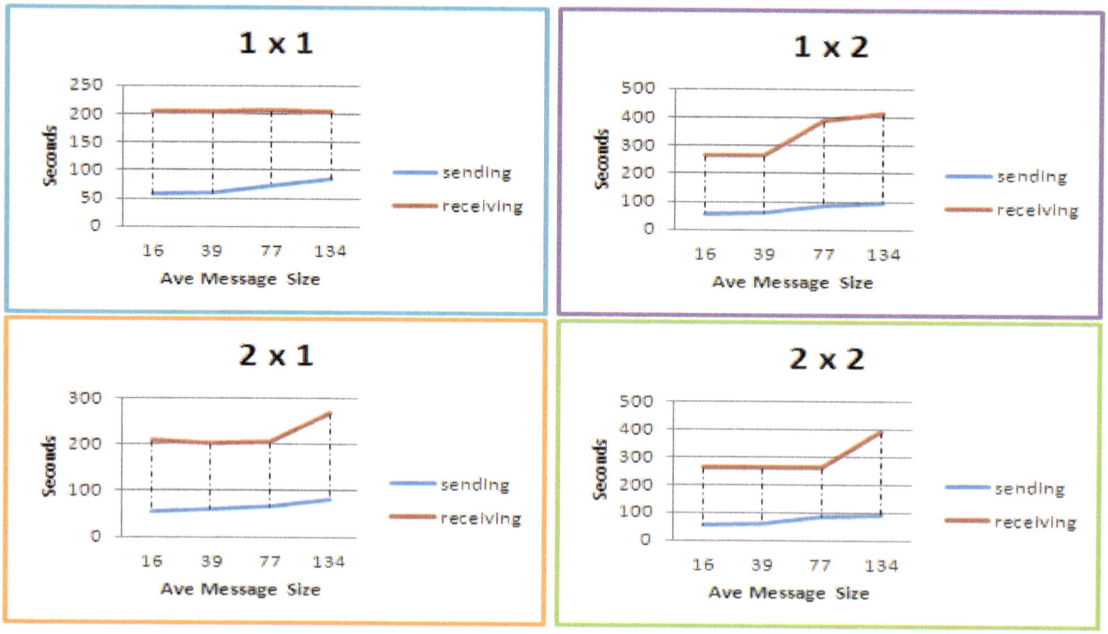

Figure 5. Average message size versus send and receiving time for matrixes

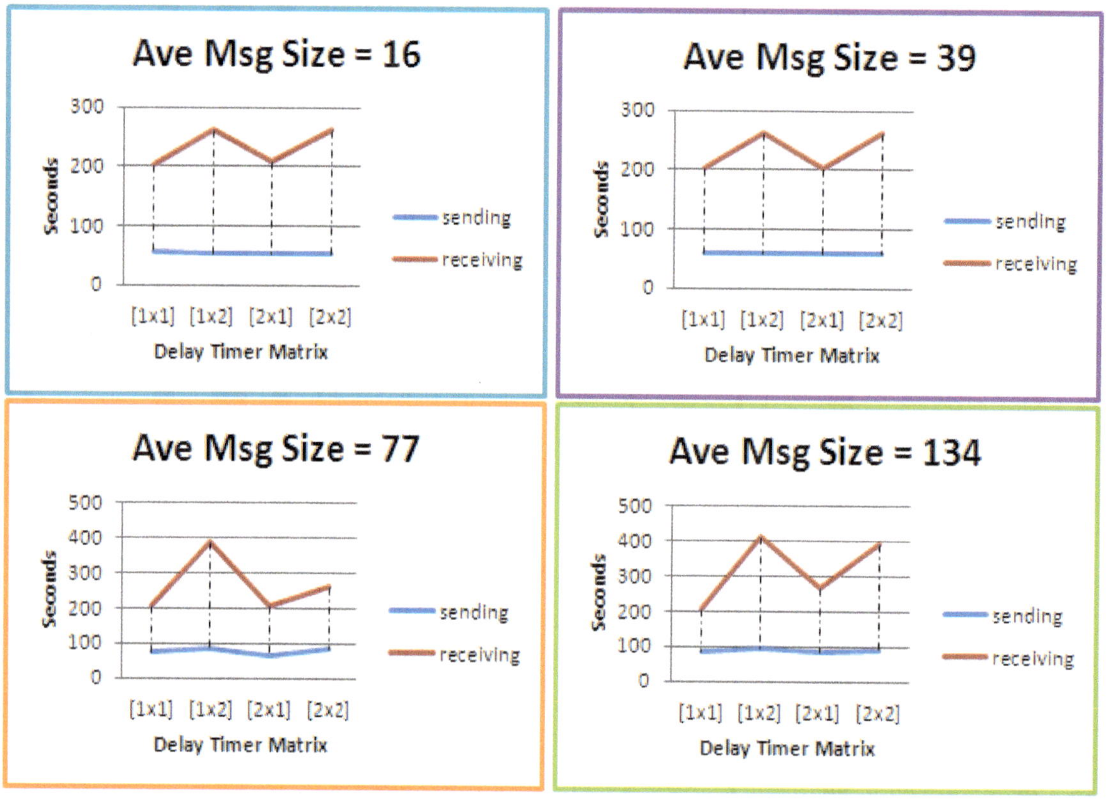

Figure 6. Sending and receiving time versus matrixes with average fixed message sizes

VII. Feedback of Pilot Test

The pilot test for a period of two weeks with the real users was conducted with the elderly volunteers in Singapore. The MAE program was pre-installed in the testing mobile handsets that were later distributed to the volunteers. This is the first time testing the system in the public. As the targeted audiences are retirees and elderly who are still active and healthy, they are still mobile and we aim to make the system to be user-friendly and useful to their daily activities. After the testing, we conducted one-to-one session feedback with the volunteers about the purpose of the system, user-friendliness, overall experience and expectations on future developments. The questions and results posted for the feedback is as follows:

Question 1: The concept is to allow the caregiver to find out the location of elderly when required. It is also to support a reminder services for elderly, such as appointment or medication. Do you think this concept is excellent, good, still okay, not useful, or poor? We have 40% of the respondent chose excellent and 60% chose good. The first question is aimed as receiving the feedback on the idea of MAE system. Most volunteers agree that this idea of connecting caregivers to elderly while allowing the elderly to have their own freedom of movement is good for them.

Question 2: While you are volunteering for this pilot-test, you feel that the test is very easy, easy, still okay, complicated or troublesome. 100% of the respondent chose easy. The second question is about user-friendliness. As The target audiences are either non-technology person or little knowledge about technology, making the test easy and no trouble is the top priority. We have achieved that since most volunteers felt the test was just to carry the mobile phone as usual and it was easy.

Question 3: What is your overall experience (including technical and administrative support)? 20% of the respondent chose excellent. 40% of the respondent chose good. 40% of the respondent chose acceptable and 20% of the respondent chose need improvement. The third question is about overall experience. This experience differs since the expectations are found to be different among the volunteers. Some volunteers felt that multiple incoming messages are nuisance and they are very much bothered since there could be confusion between testing messages and meaningful messages.

Question 4: Please suggest improvements of the test.

Our respondents gave the suggestions below. More time required to get used to the capabilities of smart phone

- The weight is a little heavy – should be lighter smart phone
- Function to interact with other users
- System should be more reliable.
- Long battery consumption

Some volunteers felt that they are new to smart phones and therefore, needs more times to get familiarize with the functions in the phones. Many applications are by default installed in the phones and that could create confusion for elderly audience. The weight of the smart phone is heavier compared to the small and compact conventional hand phones.

VIII. Conclusion

We presented the MAE system: Overall system architecture, client, server, and performance results. User feedback has shown to some extent that the MAE system enhances the elderly experienced in a mobile platform. To our best of knowledge, there is no similar enterprise mobile tracking and reminder system for multi-users being deployed in nursing homes. The elderly are moving more independently and confidently while the caregivers are given authority to find their locations as and when necessary. The study consists of latest mobile and information technologies to fulfill the core function of this project.

With the application installed on the mobile phone, the elderly can be monitored from either a computer terminal or another mobile phone. The location parameters were generated from the GPS that could trigger by a timer or SMS messages to the target mobile phones.

During the real life testing for 14 days, we have found many unexpected issues such as the timeout or page reset due to slow connection, the unnecessary location updates, device switching off from the users and the network coverage weakness. The interactive user interface will be required that make the application more attractive to users since some users prefer on the active side. Another factor that requires the mass deployment would be the application's compatibility on different devices and operating systems such iPhone, Blackberry and Windows mobile. Currently, a small group of elderly are using smart phones, while some are still hanging on the older and simpler type mobile phones. More works from manufacturers are required in order to suit the needs of the elderly. With technology advancement, applications will be more reliable since current situation is depending on many factors, such as data speed, network coverage and user-friendliness experience.

Figure 7. Functions of MAE server including social networks

IX. Future Work

We are implementing the software tracking program in other smart phone platforms [31] like iPhone and Windows Mobile and the social networking capabilities in smart phones and MAE server (Figure 7) as well.

References

[1] A. Anand, C. Manikopoulos, Q. Jones, C. Borcea, "A quantitative analysis of power consumption for location-aware applications on smart phones," IEEE Industrial Electronics, 4-7 June 2007, pp. 1986-1991.

[2] S. J. Barbeau, et. al, "A location-aware framework for intelligent real-time mobile applications," IEEE Pervasive Computing, Issue 3, March 2011, pp. 58-67.

[3] J. Liu, W. Lan, "Retrieving the GPS information of telephone based on Symbian OS," 3rd IEEE Conference on Computer Science and Information technology (ICCSIT), vol. 8, 2010, pp. 145-147.

[4] S. J. Fusco, "Exploring the social implications of location based social networking: an inquiry into the perceived positive and negative impacts of using LBSN between friends," 9th International Conference on Mobile Business and Global Mobility Roundtable (ICMB-GMR), 13-15 June 2010, pp. 230-237.

[5] A. Beach, et. al "WhozThat? Evolving an ecosystem for context-aware mobile social networks," IEEE Network, vol. 22, July-August 2008, pp. 50-55.

[6] N. Li, G. Chen, "Sharing location in online social networks," IEEE Network, vol. 24, September-October 2010, pp. 20-25.

[7] R. Cheng, Z. Yang, F. Xia, "iZone: a location-based mobile social networking system," 3rd International Symposium Parallel Architectures, Algorithms and Programming (PAAP), 2010, pp. 33-38.

[8] X. Shu, Z. Du, R. Chen, "Research on mobile location service design based on Android," 5th International Conference on Wireless Communications Networking and Mobile Computing (WICOM), 24-26 Sept 2009, pp. 1-4.

[9] C. Lehrer, I. Constantiou, T. Mess, "Exploring use patterns and perceived value of location-based services," 9th International Conference on Mobile Business and 2010 Ninth Global Mobility Roundtable (ICMB-GMR), 13-15 June 2010, pp. 107-115.

[10] K. Virrantaus, et. al "Developing GIS-supported location-based services," 2nd International Conference on Web Information Systems Engineering, 3-6 Dec 2001, vol. 2, pp. 66-75.

[11] R. Meier and V. Cahill, "On event-based middleware for location-aware mobile applications," IEEE Transactions on Software Engineering, Issue 3, May-June 2010, pp. 409-430.

[12] F. Chang, "A trusted smart phone and its applications in electronic payment," Information Technology and Applications (IFITA), 16-18 July 2010, pp 405-408.

[13] S. Shin, et. al, "Person wide web: active location based web service architecture using wireless infrastructure," IEEE Tencon 2010, 21-24 Nov 2010, pp. 2024-2029.

[14] B. Unhelkar, S. Murugesan, "The enterprise mobile applications development framework," IT Professional, Issue 3, May-June 2010, pp. 33-39.

[15] P. Leijdekkers, V. Gay, "Personal heart monitoring and rehabilitation system using smart phones," International Conference on Mobile Business (ICMB), 26-27 June 2006.

[16] J. K. Zao, W. Mei-Ying, T. Peihsuan, J. W. S. Liu, "Smart phone based medicine in-take scheduler, reminder, and monitor," IEEE International Conference on e-health Networking Applications and Services (Healthcom) 1-3 July 2010, pp 162-168.

[17] O. Postolache, "Indoor monitoring of respiratory distress triggering factors using a wireless sensing network and a smart phone," IEEE Instrumentation and Measurement Technology Conference, 5-7 May 2009, pp. 451-456.

[18] I. E. Ivanov, V. Gueorguiev, V. Bodurski, V. Trifonov, "Telemedicine and smart phones as medical peripheral devices (computational approaches)," Developments in e-systems engineering (DESE), 6-8 September 2010, pp. 3-6.

[19] F. Hu, Y. Xiao, Q. Hao, "Congestion-aware, loss-resilient biomonitoring sensor networking for mobile health applications, Issue 4, May 2009, pp. 450-465.

[20] Y. Jaeyeol, et. al, "Android platform for English tutoring," 5th FTRA International Multimedia and Ubiquitous Engineering (MUE), 28-30 June 2011, pp. 302-305.

[21] A. P. Massey, V. Ramesh, V. Khatri, "Design, development, and assessment of mobile applications: the case for problem-based learning," IEEE Transactions on Education, Issue 2, May 2006, pp. 183-192.

[22] N. Yanrong, F. Feiya, Z. Yu, Y. Xiaozhou, "Smart phone for the perishable goods distribution management," International Forum on Computer Science-Technology and Applications, 25-27 December 2009, pp 131-134.

[23] S. S. Hyong, Y. M. Geum, J. Jeong, "A study on the development of disaster information reporting and status transmission system based on smart phone," International Conference on ICT Convergence (ICTC), 28-30 September 2011, pp. 722-726.

[24] Y-W Moon, H-S Jung, C-S Jeong, "Context-awareness in battlefield using ubiquitous computing: network centric warfare," 10th IEEE Conference on Computer and Information (CIT), 29 June-1 July 2010, pp. 2873-2877.

[25] S. Murugaraj, T. Srinath, K. Assadarat, M. Ilias, "Enabling secure mobile access for electronic health care applications," Pervasive Health Conference and Workshop 2006, 29 Nov – 1 Dec 2006, pp. 1-8.

[26] S. Zhang, P. McCuullagh, C. Nugent, H. Zheng, "Activity monitoring using a smart phone's accelerometer with hierarchical classification," 6th International Conference on Intelligent Environments (IE), 19-21 July 2010, pp. 158-163.

[27] J. Park, H. yu, K. Chung, E. Lee, "Markov chain based monitoring service for fault tolerance in mobile cloud computing," IEEE Workshops of International Conference on Advanced Information Networking and Applications (WAINA), 22-25 March 2011, pp. 520-525.

[28] K. Schelfhout, D. Weyns, T. Holvoet, "Middleware for protocol-based coordination in mobile applications," IEEE Distributed Systems Online, Issue 8, August 2006.

[29] N. Kayastha, D. Niyato, P. Wang, E. Hossain, "Applications, architectures, and protocol design issues for mobile social networks: a suvery, "IEEE of the Proceedings, Issue 12, Dec 2011, pp. 2130-2158.

[30] Q. Han, N. Venkatasubramaniam, "Information collection services for QoS-aware mobile applications," IEEE Transactions on Mobile Computing, Issue 5, May 2006, pp. 518-535.

[31] D. Gavalas, D. Economou, "Development platforms for mobile applications: status and trends, Issue 1, IEEE Software, Jan-Feb 2011, pp. 77-86.

AUTHORS

Yoong Cheah Huei, Thet Swe, Lee Zheng Xian, and Neo Shou Yee are with School of InfoComm Technology, Ngee Ann Polytechnic, Singapore 599489 (ych1@np.edu.sg, tsw6@np.edu.sg, s10068683@np.edu.sg, s10075235@np.edu.sg)

8

Interactive Prototypes for Mobile Collaborative Learning (MCL) to Substantiate Pedagogical Activities

Abdul Razaque, Khaled Elleithy
University of Bridgeport, Bridgeport, CT, USA

Abstract—**Deployment of new emerging technologies have attracted and motivated a great deal of attention from researchers and scientist in various disciplines to develop and design the learning management systems, virtual learning environments and conference systems with support of MCL for education and other organizations. This technological revolution of connecting large numbers of potential learners through mobile telephony creates new challenges for developing the new architectures for education and business. Some of the MCL paradigms are proposed and implemented in various fields such as university, corporate, health and military but they lack completeness. This paper introduces new server and client side interactive prototypes with support of various working components, which help the users in obtaining the contents from server to meet the pedagogical requirements. This paper also proposes and integrates the features of content-server with cache-server to provide the faster delivery of contents for MCL. Furthermore, the paper proposes and implements some features of a novel "group application" to support asynchronous and synchronous and multimodal features to facilitate the students for MCL. Finally, the paper validates the features of interactive architecture and applications. This contribution will encourage and motivate the students to pursue their education again because the students will be able to get the course contents at anytime and anywhere.**

Index Terms—**Design; Development; Theory, Mobile collaborative learning (MCL); mobile devices; Server side architecture; client side protocol; group application; cache server.**

I. Introduction

Latest studies show that MCL is highly focusing paradigm for research particularly in distance and online education. The idea of mobile-based learning is different from traditional classroom-based learning. The mobile-based learning pedagogical method provides many possibilities. It helps the people to bring together working in same or different organizations for achievement of any specific goal. MCL exhibits intellectual synergy of various combined minds coming together to handle the problem and stimulate the social activity of mutual understanding. According to survey, MCL provides some limited features and functionalities to support the knowledge-sharing [1]. Deployment of new emerging technologies and its fast growing trend encourages the people to develop supporting MCL environment for real learning management systems, virtual learning environments and conference systems. These environments lack the most promising supportive architectures and softwares. From other side, The embedded emerging technologies onto mobile devices; such as blue tooth, motion sensors, blackberry, Infra-Red, cameras, Global positioning System (GPS) , GSM and faster broadband connections will make the dream into reality. Mobile devices are also more suitable to incorporate the existing services by employing web-based interface. The developed applications for mobile devices provide motion recognition, geo-location, academic profile of user and history of activities [6] & [15]. All of these rich features can provide a promising platfarm to foster healthy MCL environment.

We believe that MCL contributes a larger pedagogical agenda. This paradigm is being highly acknowledged and implemented in schools, colleges, universities and different organizations across the world. From other side; MCL lacks the Interactive architecture and intellectual applications to substantiate the pedagogical requirements. Many MCL theories handles some pedagogical issues explained in [16]. However, all of these MCL theories face some major challenging issues in open, large scale, dynamic and heterogeneous environments. The most performance affecting issues are knowledge sharing, faster delivery of contents, request for modified contents, delivery of large rich multimedia contents (video-on-demand), asynchronous and Synchronous Collaboration, support for multimodal, provision for archive updating, user friendly interface, middleware support, and virtual support.

To handle these issues, The contribution proposes and implements innovative client and server based architectural prototypes to improve pedagogical activities. These prototypes will handle almost all of the major challenging issues and provide an access to all users in obtaining the contents on mobile devices at anytime and anywhere.

II. Related Work

In this section, we discuss the salient features of related published work. Kwang lee & Abdul Razaque [1] have introduced the idea of server based architectural prototype at the abstract level to support MCL in education. On the basis of usability testing and ranking discovered the features necessary for MCL. Xiaoyong Su, et al.[14] proposed the four layer framework and prototype for multimedia mobile collaborative system. The proposed framework supports user, device and session management. The authors claim that mobile collaborative environment could possible be made by upgrading the devices and network technologies. The contribution is interesting and providing the abstract level of concept for introducing the framework for MCL. Vladimir Zanev and Rodney Clark [2] have discussed the concept of prototype for wireless course management system. They handle login and authentication, wireless calendar,wireless syllabus, and wireless testing. The paper also presents contents of course and guides the teacher to interact with students by using HTML based interface. Lahner F. and Nosekabel H. [3] have implemented a program to display e-learning contents on mobile devices in the University of Regensburg, Germany. Mildard Marcelo et al. [4] have designed C-note to provide an opportunity for collaboration in the uuniversity of Vaxjo, Sweden. They store the notes and information in database. They use an specialized C-Pen to scan the material of research project with support of C-note application. In addition scanned information is restored in database and the teacher can check the activities of each member of group during collaboration.

Jorge Bardosa et al.[7] have proposed prototype for mobile and ubiquitous devices in undergraduate course of reference (nicknamed GRefe). The prototype consists of user profile that stores the information regarding learning process. Allision Druin et al.[5] have discussed the prototype for their ongoing participatory design project with intergenerational design group to create mobile application and integrated onto iP Phone and iPod touch platforms. They claimed that designed application could provide the opportunities to bring the children and grand parents together by reading and editing the books. Ch. Bouras et al. [12] have introduced INVITE architecture and discussed the user's requirements to meet the demand of e-learning in collaborative virtual environment. They have also discussed the technology and standards required for designing the INVITE. Mobile Learning community (MLC) in Department of Computer Science, University of Illinois [13] has launched the project “exploring social trust in mobile educational environments" funded by NSF, Google Inc and Vodafone Inc. The project has developed four applications Recorder, Announcement, Quiz and Message. The main purpose of the project is to introduce the mobile learning material for course "cs241" for undergraduate students. All of these studies in MCL environment show that neither any contribution particularly worked on architecture based design nor on supporting systems and specific applications running on mobile. They mostly develop the courses and few contributions touch the technology but not implemented. This paper introduces rich MCL client-server based architecture that supports to management systems, virtual learning environments and conference systems within educational institutions and several organizations. The paper also develops an interactive application for MCL running on this architecture. It meets the basic requirements such as exchange for the delivery of communication contents for group discussion, short message service (SMS), Emails, audio, video and offer on-line information to collaborative group. The challenging issues motivating this contribution are: does this contribution provide innovative and interactive opportunities for synchronous and asynchronous online communications? to what degree architecture is more effective and result oriented? does architecture develop multicultural standards for MCL? Does paper handle all the important features necessary for effective MCL? Are group application and other light weight software threats sufficient to meet pedagogical needs and other collaborative requirements? How the findings are used to engineer the requirements of Students, teachers and other administrative staff of University of Bridgeport (UB) and other institutions? What are the socio-economic impacts of the research on USA and beyond the world? The goal of this paper is to obtain the learning materials to meet pedagogical needs and other collaborative requirements for educational and other organizations through mobile devices at any time and everywhere.

III. Prototype Design of MCL at Server Side

The server side protocol for MCL is envisioned as promising platform that substantiates latest technologies and mobile applications to meet pedagogical requirements and other collaborative activities within educational institutions and beyond. This integrates various functional components to cover all necessary features for MCL from sending SMS to large size of videos. It supports to content generation, content fragmentation, content buffering, content modification, content integrating, content diagnosing, content retrieving , content refinement, content visualization and ultimately to dissemination of results. Content server is a centeral processing unit at server side protocol shown in figure 1 with architectural design, which is based on content server engine (CSE). Content server continuously senses the traffic to receive the request from client side. If new request is arrived then it is dealt with requirement of user. CSE identifies the mobile users on basis of mobile information device profile (MIDF), status

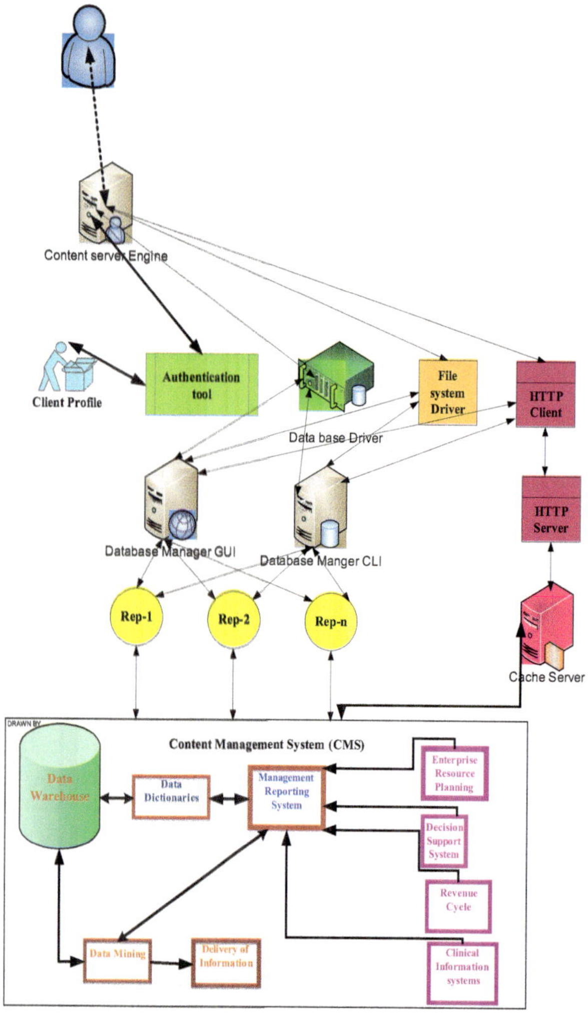

Figure 1. Architectural Prototype for MCL at server side

of previous network condition and requested URL. Another promising feature of server side protocol is to satisfy the authorization and authentication process in order to provide the access to legitimate users only. This job is done with help of CSE to verify the status of the users. If illegitimate user sends the request for getting the contents for collaborative learning that request is declined. CSE is implemented on internet information server (Web server) that also provides the access to legitimate users. CSE deals with three types of services normal, low priority and high priority. If request is about normal service that is handled with file system manager. It supports to text, graphs and small size of videos. Requests about large size of videos is dealt with database manager that is considered as low priority service provider. High priority covers all type of data services including text, graph, images and voice. This task is performed with integration of cache server, which sets its own hyper text transfer protocol (HTTP) connection. Integration of cache server process provides the faster delivery of learning material. With introduction of cache server, time can be saved of backups and log monitoring because the substantial time is spent on these activities every day. It improves the performance of health, military and law enforcement departments because they need high priority services to achieve their targets at anytime and anywhere.

CSE can save and get data from either database driver or file system driver. Both database and file system drivers forward the request to the database manager GUI and database manager CLI. Single CSE can support to many storage mediums.To obtain the contents faster, we propose HTTP client and integrate with CSE. At the server side, we also introduce the cache server, which has feature of setting up its own hypertext transfer protocol (HTTP) connection. The cache server gets the request from HTTP client and delivers the requested contents immediately; if they are already cached on server, otherwise forwards the client's incoming request to content management system because cache server has direct access for getting the requested contents. There is another advantage of the cache server because it does not need backups and log monitoring. Database manager CLI is functional component at server side. It creates programming interface, documentation and module. The beauty of database manager CLI is to provide the faster delivery of contents. From other side, it monitors, generates, restores and manages backups. CLI provides the limited services but supports for priority services. It fosters interactive and background operations. We access the database manager CLI tool by using C++, Java, XML and other high level programming languages. To overcome the limitation of CLI, The server side also introduces database manger GUI tool that works like database manager CLI with same but more extra features. GUI involves more extra features, which are diagnosing the contents, displaying information, several interactive options and updating database software. The database manager GUI also supports several featured mobile devices. The most promising feature of GUI is provision of friendly user-based interface to access the remote mobile devices to start successful MCL process.

GUI and CLI collectively provide the access to connect with different repositories for storing and extracting the contents for mobile devices. Both GUI and CLI make decision to which repository should be selected for obtaining the requested contents. The repository plays a role as bridge connecting GUI and CLI with EDW for obtaining the requested contents. The rapid increase in consumption of different medias cause of mixing all the contents and resulting loss of data occurs. To control this issue; various repositories are introduced at MSCS for each media that handle the scalability issues and loss of data. Each repository serves as table to obtain and store different kind of contents. The multiple use of repositories capture and preserve communication process and serve as tangible indicators for improving the quality of mobile learning [11]. The repositories are supported with content management system (CMS). The CMS provides logical storage for different devices. Hence we can switch from Rep1 to Repn in single table for finding the requested contents easily for collaboration.

IV. Client Side Prototype for MCL

The Feed Demon and iOS version of net news wire support iPhone and iPod Touch. They help mobile devices to get the contents from sever. The latest version of net news wire lite 4.0 was introduced to substantiate the apple macintosh store (AMS) with limited features. The AMS provides the digital distribution platform for Macintosh OS [17] and [18]. Despite of some missing features, net news wire 4.0 helps the users to obtain and share the contents with other mobile users. To collaborate with other users, the supporting software Really Simple Syndication (RSS) 2.0 is used. The RSS creates RSS feed (online resources) for members of group. The RSS feed is storage area used for asynchronous collaboration. The handheld devices store the contents obtained from server side onto their respective RSS feed. The RSS aggregator is another enhancement with integration of video aggregator. It is highly influential feature to support several on-line sources (RSS feed) and helping to store the contensts. The video aggregator collects and sorts out on-line video for RSS feed. To make successful MCL, Apple iPhone can be used with Lite feed RSS reader. Lite feed involves several utilities such as compressing files, cache and fast access to obtain the feeds. Lite feed also helps the handheld devices to view the whole article and share an information among other users. The passage of article is shown in figure 2.

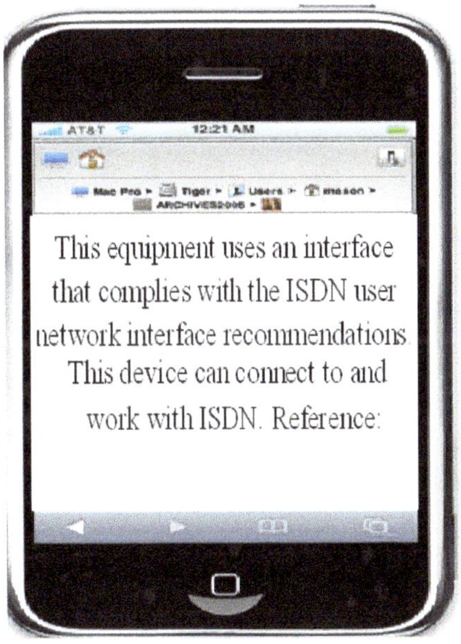

Figure 2. Received contents from server side

Another application is also bit-Torrent based on peer-to-peer that supports mobile RSS 2.0. The application allows a client to download the files automatically. The RSS 2.0 is the main source of sharing and downloading the information for MCL. The specific RSS aggregators can resolve the pedagogical issue because they are very helpful in building the blog [8]. Mobile RSS provides audio and video updating [1]. RSS reader is another supporting software that can check the subscribed feeds of the clients. It also builds user friendly interface. Each participant is assigned separate feed to store and read the information. When any mobile device wants to collaborate with rest of group. It obtains the contents from server and stores on its RSS feed. It provides an opportunity to other users to use an information for MCL. The interested user in collaboration takes one more responsibility to send multicasting message to invite the group-members to start MCL. The multicasting message is broadcasted but based on the size of group. The users can join and leave the group and recognized with class ID. The track group membership and route datagram are required to send multicast messages to the members of group. The group membership management is used to control the local delivery at at local level. Global is used to handle the remote multicasting [9]. Internet Group management protocol is used to track the location of clients. When participants send the response with reports they are recognized as members, taking part during MCL. Similarly, the global is inter-domain multicast protocol that is used for multicasting the message. The beauty of the protocol is to control the scalability and network convergence issues. It also helps the users to download the contents by using RSS 2.0 software from feed of interested participant who wants collaboration. All of the users get the contents from RSS fed then start the collaboration. The process of multicasting and storing the contents is shown in figure 3.

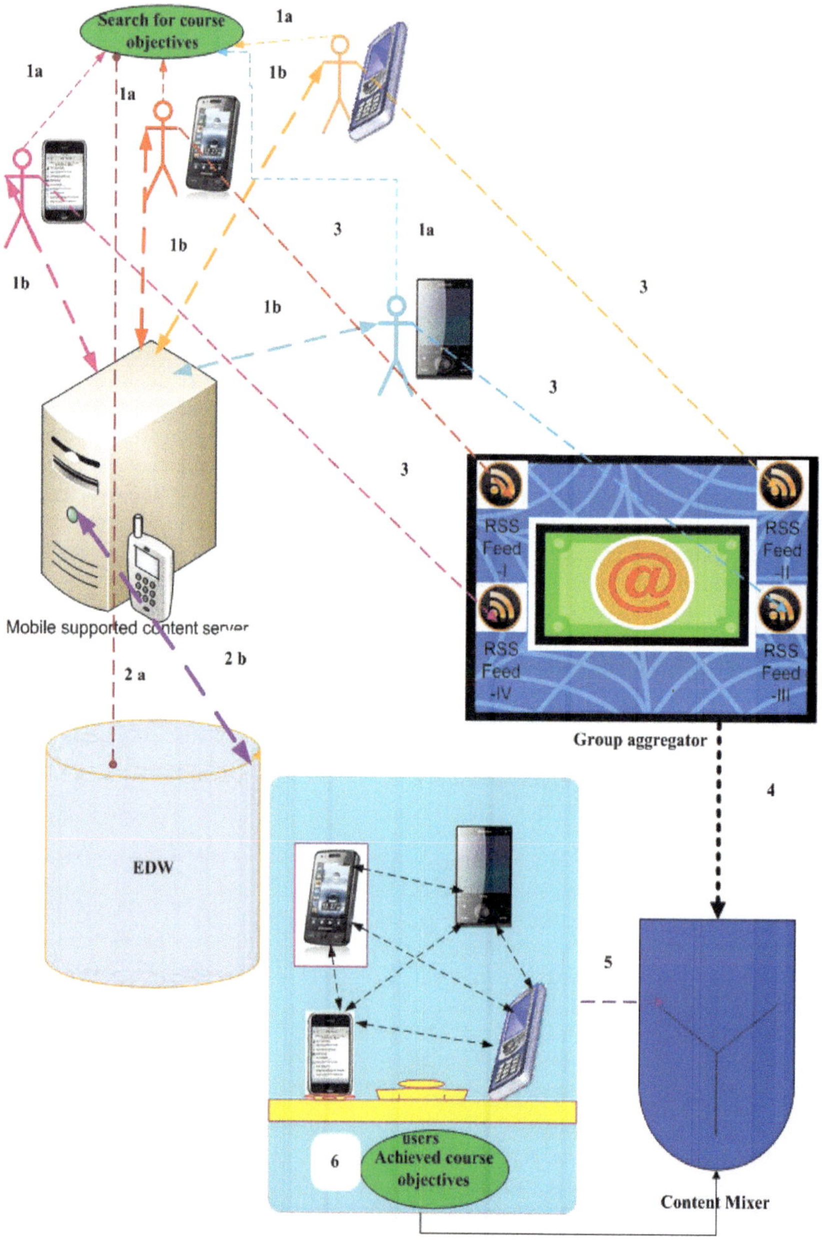

Figure 3. Content storing and extracting process from client side

V. Group Application and Case Study Prototype for MCL

MCL is a emerging concept in education since last several years. It allows the users to gain computer-based information through mobile devices. MCL provides various advantages such as portability, context awareness, connectivity and social interaction [6]. Mobile devices can be more successful tool for collaboration, allowing the students to obtain and share the information to meet the pedagogical objectives. From one side, mobile creates bridge of several opportunities, and from other side, limitations affect the deployment in MCL environment. The highly performance affecting factors are small size of screen, mobility, navigation issue, bandwidth, low resolution, and limited memory [1] and [6]. With deployment of emerging technologies and its best use make the task easy and accessible to all. The paper proposes and implements user side group application with some features at an initial state. The prototype of application is used as case study that resolves the barriers being faced mobile users to obtain the contents at anytime and anywhere. The prototype of group application helps the students to obtain the required contents to meet the course requirements. Assume the students want to obtain the contents of course "mobile and wireless communication" As these devices get the course contents from the server of UB to make MCL. The sufficient information is stored in EDW to meet the standard of course explained in [2]. Information of courses cover textbook information, course ID course name and course description. The main focus of case study is around the course description that comprises of course contents. The objective is completion of course contents during the semester to fulfill course requirements. The Course objectives are met with support of quizzes, class tests assignments, discussion, research projects, labs etc. These all of the requirements are obtained through MCL. Figure 4 supports the process of case study prototype for MCL.

Suppose Integrated Services Digital Network (ISDN) and its components are the contents of the first chapter of mobile and wireless communication course. To meet the course requirements, we want to create MCL environment but due to limited capability of mobile devices and installed software threads cannot provide easy access to achieve the objectives. To meet the requirements, we recommend to use feed demon or net news wire software with support of (RSS) 2.0 software. Furthermore, we suggest to incorporate android mobile operating system, which consists of software stack for mobile devices. The following recommendations will be implemented to add new application.

1. add new application with name " group" in application section by using Java programming and eclipse.
2. extend resource manager and activity manager and provide extra responsibilities to control different features of "group" application.
3. modify and extend libraries section to facilitate structured data storage.
4. enhance display Driver of Linux Kernel to support MCL activities.

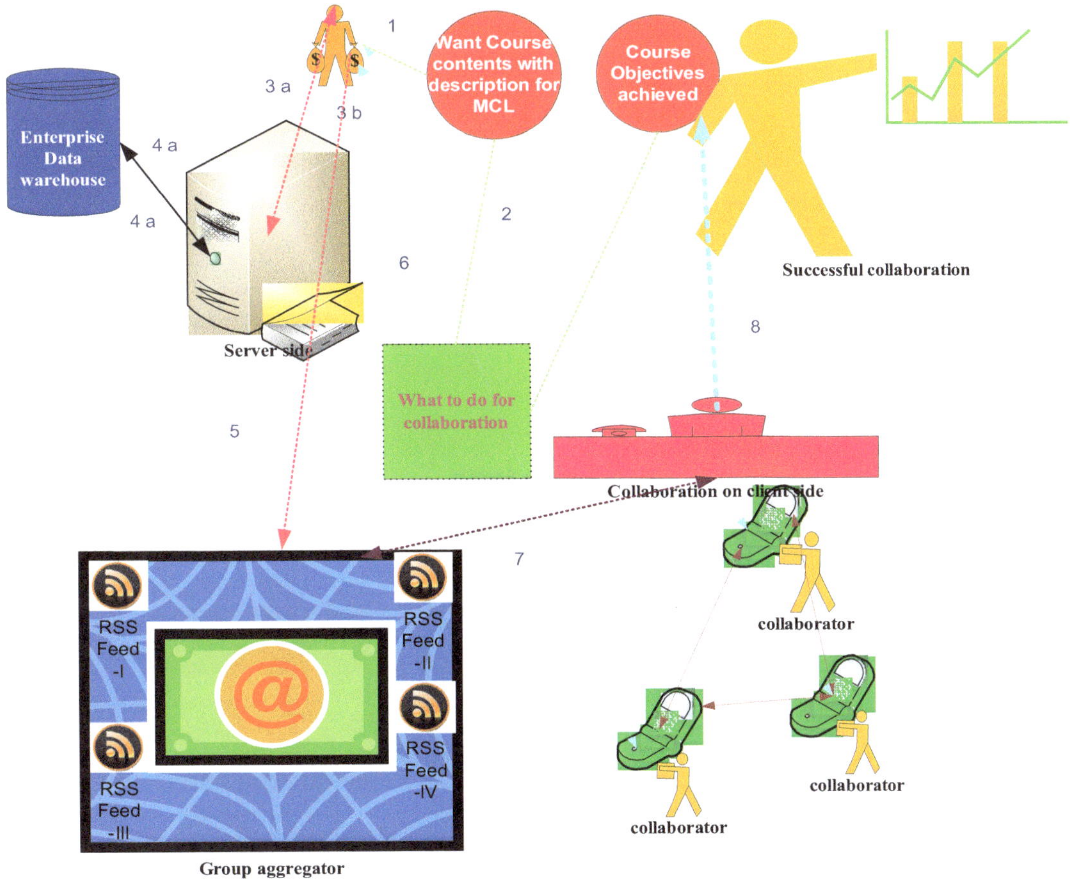

Figure 4. Process for obtaining the course contents through MCL

If these recommendations are implemented in android, new application "group" with new features will support for obtaining the contents from server side. group application is shown on android with other applications in figure 5.

The "group" application consists of two basic options Control and Delivery. The control option performs several functions including add new contact, edit contact, delete contact, existing collaborative group (C-G) and make new C-G. Delivery Option consists of two sub basic functions "Receive" and "send". When user wants to obtain the contents from server side, delivery option is used and contents are saved by using the store option. The store option controls and handles all type of data. Store involves file, audio and video. The store has download option. The participating users use this option to download the data and stored onto RSS 2.O for collaboration. Each section of store option manages the several files, audio and video. When contents are obtained from server that are saved onto store. The uploaded option of application is used to upload the contents to RSS Feed. The upload option helps the users to put any type of contents in RSS feed. The contents to be stored are used by member of the group for MCL. The requested user makes group call after storing the contents in RSS feed. The user also informs the collaborative group to join for MCL by making group call as shown in figure 6.

When participants of group receive an invitation message for collaboration, they start to download the contents from RSS feed of requested user. When group finish the process of downloading the required contents from RSS feed of inviting user. Finally they start to follow the process for MCL given in figure 7.

If members need file for MCL, they open to read an information of the contents. If an information is not sufficient for clarification of the topic, they can play the video of related information obtained in figure 8.

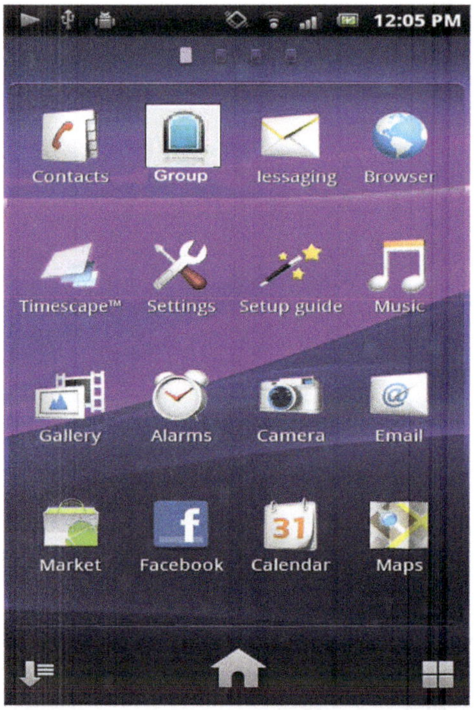

Figure 5. Group application with other applications

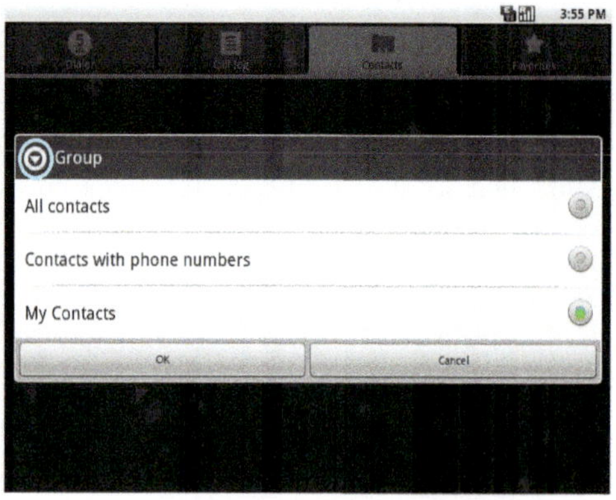

Figure 6. Making group call to all participants

Figure 7. Sharing of contents for Mobile Collaborative Learning (MCL)

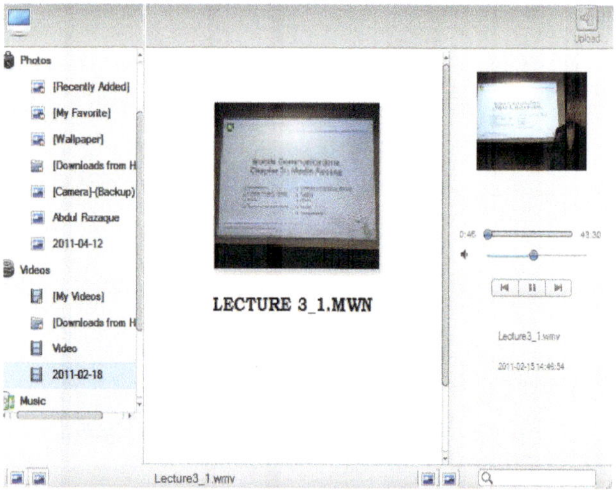

Figure 8. Sharing of Video for Mobile Collaborative Learning (MCL)

The "group" application provides utility to manage the contents by using separate options for each media . The file option is used to get text-based contents otherwise audio and video options are used. When all of the members get sound knowledge of the topic, subsequently initiate the process of collaboration by using existing collaborative group (Existing CG) option. The process is supported with TCAP, ISUP, Q.931, H.323 and other underlying protocols. These protocols collectively support real-time audio and video communication [1] & [10].

VI. Analysis & Validation of Architecture and Applications

The most important and critical task is validation of application and architecture. The paper uses two methodologies applied in validity testing supported with five likert formula. The methodology involves laboratory and field testing. The laboratory testing is conducted for architecture and application including light weight software threads. The field testing covers application and few light weight software threads. The benefits of laboratory testing involve human interaction and heuristic evaluation to accomplish the task. The paper does not depend on single testing procedure in laboratory. It comprises of four testing phases including unit testing, integration testing, performance analysis requirement or beta testing and user acceptance testing. The unit testing is initiated after finishing each task to check the functional performance of each component or feature. If any bug is detected that is removed at the initial stage. The unit testing is more helpful for introducing cost-effective and high-quality product explained in [19] & [21]. Integration testing (IT) makes the task easier because it combines all integrated aggregate into one unit to identify the problems to be occurred. IT reduces number of possible errors to make successful analysis [20]. Requirements analysis is applied when establishing the goals for conducting the validation. It combines the features of usability attributes and applies nine generic usability attributes explained in [22] & [23]. IT helps to find out the reflection and attitude of users when using mobile application for accomplishment of task. The users try to understand the presented contents for improving their performance. The quality of menus, navigation design of application, accuracy and completeness is identified to achieve certain goals. User acceptance testing (UAT) is recognition of party's work done by other party. It is final testing process to screen out any issues if users identify. The laboratory testing method is handy to focus on comparing multiple interface designs and mechanisms for input data of mobile devices. UAT makes easier to deploy video device to capture the emotions and reactions of participants when using mobile application. The dynamic mobile features are hard to simulate in laboratory and they need real environment to identify the valid findings about the performance of features explained tin the studies [24] $ [25]. This statement motivates to measure the performance of group application and including few light software threads outside the laboratory. The field testing faces three basic problems reported in (Palen & Salzman, 2002). To control these issues, we introduce the scope of mobile features prior to conducting field testing. The scope highlights movement and position of testers and environment scenario. It is more convenient to deploy effective methods for collection of data in field. We use real mobile devices in both testing methodologies, which help the testers to collect realistic facts and figure in static, dynamic and heterogeneous environments. The basic data collection methods w involve the questionnaire, interview, verbal protocol and regular meetings to get feedback from participants and testers to improve the quality of Interactive architecture and applications. In this section, we discuss the procedure of testing for group application based on heuristics and questionnaires. We apply both testing methodologies with 08 participants to be conducted at the University of Bridgeport. The participants analyze and evaluate the some basic features of server side Interactive protocol and group application. With this testing procedure, the users transfer files between laptops and the mobiles. The testing procedure invites the participants belonging to different background. Some are familiar and belonging to mobile and wireless communication field and fewer possess less expertise in this field but know how to use mobile devices to make MCL. The group leader who is much familiar with this field records the performance for the features of Interactive architecture and applications on the basis of activities performed during the testing. The performance analysis requirement or beta testing method also involves three steps. First, we introduce the testing procedure from design phase to conducting the test. Second, make all the related operations of architecture and applications and finally, we give the questionnaire to all the participants based on 5-level Likert method, so that testers give the feedback and answers after completing the appropriate tasks shown in figure 9.

When questionnaire procedure is finished, we arrange the meeting in which all the participants come and share their experiences and give their impression regarding the test. We also get suggestions regarding the redesign and modification of architecture and application. When participants leave the room, we compile the results and arrange in form of table for better understanding shown in table I.

Figure 9. Testing procedure for some features of group application

TABLE I.
Showing the comments of participants after analysis and validation

Post Usability Testing of fields	Comments of participants
Should be easy to use	Strong Positive=08; Positive=00;Neutral=00; Negative=00;Strong Negative=00
Should be User friendly interface	Strong Positive=08;Positive=00;Neutral=00;Negative=00;Strong Negative=00
To be flexible to collect and extract the data.	Strong Positive=06;Positive=01;Neutral=01;Negative=00;Strong Negative=00
To provide text, graphs, images, audio and video services to meet the requirements of related course of study	Strong Positive=06;Positive=01;Neutral=01;Negative=00;Strong Negative=00
To support multimodal MCL	Strong Positive=05;Positive=01;Neutral=02;Negative=00;Strong Negative=00
The Students should have alternative choices for selecting any topic to discuss.	Strong Positive=03;Positive=02;Neutral=02;Negative=01;Strong Negative=00
To provide the facility to contact and invite the members for collaboration	Strong Positive=02;Positive=03;Neutral=03;Negative=00;Strong Negative=00

VII. Conclusion and Future Work

The main objectives of introducing the architecture-based prototypes are to obtain the learning materials on handheld devices particularly on mobile devices. The deployment of these prototypes in educational institutions will foster the pedagogical activities. The students can get course-contents at anytime and anywhere. The paper introduces the novel architecture for server to support MCL in education. The design and development process of server-based architecture provides faster delivery of contents to users. The content server is also optimized by integrating it with cache server at server side. This integration saves time of users to collaborate quickly. The client-based architectural prototype gives complete directions to researchers and market oriented persons how to make the mobile device efficient to obtain the contents from server and store onto RSS 2.0 for MCL. The paper also targets to implement some basic features of the new group application and recommends some valuable suggestions in android operating system of mobile devices. Furthermore, the paper discusses the development process of group application including the prototype based case study for the University of Bridgeport to meet the course requirements. Finally, the paper validates the features of interactive architecture and applications to prove the claim. In future, we will use marking and ranking method to find out the basic user requirements. In light of basic user requirements, group application will be extended and also validated by using usability testing method and heuristic evaluation.

References

[1] Kwang Lee and Abdul Razaque" Suggested Collaborative Learning Applications for Mobile Devices",HCI International, (Book Chapter), Hilton Orlando Bonnet Creek, Orlando, Florida, USA,Volume 6769/2011, pp. 611–620, 9-14 July 2011.

[2] Valadimir Zanev and Rodney Clark,"Wirelss Course Management System", 43rd ACM Southeast Conference, Kennesaw, GA, USA, pp-2-118-2-123, 18-20 march 2005.

[3] Lahner F,Nosekabel H, " The Role of Mobile Devices in E-Learning First experiences with a wireless E-Learning environment", Proceedings of the IEEE International Workshop on wireless and mobile technologies in Education, Vaxjo, Sweden,August 29-30, 2002.

[4] Mildard Marcelo, Perez Juan and Hope Ulrich, C-Notes, "Designing a Mobile and Wirelss Application to support Collaborative Knowledge Building, Proceedings of IEEE International Workshop on Wireless and Mobile Technologies in Education, Vaxjo, Sweden, August 29-30,2002.

[5] Allison Druin, Benjamin B.Benderson and Alex Quinn, "Designing Intergenerational Mobile Storytelling" IDC 2009-Workshops, Como, Italy, 3-5 June, 2009.

[6] Abdul Razaque, Khaleed Eliethy and Nyembo salama" Design and development of hilological prototypes for mobile collaborative learning (MCL)",ASEE Northeast Section, West Hartford, USA, 29 April ,2011.

[7] Jorge Barbosa, Rodrigo Hahn, Debora N.F.Barbosa and Claudio F.R.Geyer," Mobile and Ubiquitous Computing in an Innovative Undergraduate Course", SIGCSE07, Covington, Kentucky, USA, March 7-10, 2007.

[8] Yueh-Min Huang, Yu-Lin Jeng and Tien-Chi Huang,"An Educational Mobile Blogging System for Supporting Collaborative Learning", International Forum of Educational Technology & Society (IFETS), ISSN 1436-4522, 2009.

[9] S. Deering, ÒHost extensions for IP multicasting,Ó Internet Request For Comments, RFC 1112, August 1989.

[10] Paul E. Jones, "H.323 Protocol Overview", October 2007: from website:www.packetizer.com

[11] kerstin, "How to handle multiple repositories in Eclipse",metafora-project: A platform for learning to learn together, Apr 8, 2011

[12] Ch.Bouras, G.horing, V. Horing, V.Triantafillou and Th.Tsiatsos," Architectures Supporinge-Learning through Collaborative Virtual Environments:the Case of INVITE", 2000.

[13] https://agora.cs.illinois.edu/display/mlc/members

[14] Xiaoyong Su, B.S.Prabhu, Chi-Cheng Chu and Rajit Gadh,"middleware for multimedia mobile collaborative system", IEEE third annual wireless telecommunications symposium (WTS, 2004), Calpoly Pomona, California, USA, May 14-15, 2004.

[15] Sergio Martin, Ivica Boticki, George Jacobs, Manuel Castro and Juan Peire," Work in Progress - Support for Mobile Collaborative Learning Applications, 40th ASEE/IEEE Frontiers in Education Conference, Washington, DC, 2010.

[16] Nilgun Ozdamar keskin and David Metcalf," The current perspectives, theories and practices of mobile learning",TOJET: The Turkish Online Journal of Educational Technology – volume 10 Issue 2, April 2011.

[17] Brent Simmons, "The return of net news wire Lite" from website: www.inessential.com, retrieved, 12 June, 2011.

[18] Brent Simmons,"Net news wire acquired by Black Pixel", from website: www.inessential.com , retrieved, 12 June, 2011.

[19] .B. George and L. Williams, "An Initial Investigation of Test-Driven Development in Industry," presented at ACM Symposium on Applied Computing (SAC),Melbourne, FL, 2004.

[20] Paul Grenyer,"Integration Testing A Java Enterprise Application Data Access Layer: Enterprise Application Development in Java, article,February 2010.

[21] Patrick J. Schroeder, Darrin Rothe,"Teaching Unit Testing using Test-Driven Development", Workshop on Teaching Software Testing 2005.

[22] Öquist, G., & Goldstein, M., "Towards an Improved Readability on Mobile Devices: Evaluating Adaptive Rapid Serial Visual Presentation".Interacting with Computers, 15(4), 539-558, 2002.

[23] Ziefle, M. "The influence of user expertise and phone complexity on performance, ease of user and learnability of different mobile phones", Behaviour &Information Technology, Vol:21(5), pp- 303-311, 2002.

[24] Palen, L., & Salzman, M.," Beyond the handset: designing for wireless communications usability",ACM Transactions on Computer-Human Interaction (TOCHI), 9(2), PP-125 - 151, 2002.

[25] Kjeldskov, J., & Stage, J.,"New Techniques for Usability Evaluation of Mobile Systems", International Journal of Human-Computer Studies, 60, 599-620, 2003.

[26] Beck, E. T., Christiansen, M. K., & Kjeldskov, J," Experimental Evaluation of Techniques for Usability Testing of Mobile Systems in a Laboratory Setting", Proceedings of OzCHI 2003, Brisbane, Australia,2003.

Authors

Mr. **Abdul Razaque** is Phd candidate of computer science and Engineering department in University of Bridgeport. His current research interests include the design and development of learning environment to support the learning about heterogamous domain, collaborative discovery learning and the development of mobile applications to support mobile collaborative learning (MCL), The congestion mechanism of transmission of control protocol including various existing variants, delivery of multimedia applications. He has published over 20 research contributions in refereed conferences, international journals and books. He has also presented his work more than 10 countries. During the last two years he has been working as a program committee member in IEEE, IET, ICCAIE, ICOS, ISIEA and Mosharka International conference. Abdul Razaque is member of the IEEE, ACM and Springer Abdul Razaque served as Assistant Professor at federal Directorate of Education, Islamabad. He completed his Bachelor and Master degree in computer science from university of Sindh in 2002. He obtained another Master degree with specialization of multimedia and communication (MC) from Mohammed Ali Jinnah University, Pakistan in 2008. Abdul Razaque has been directly involved in design and development of mobile applications to support learning environments to meet pedagogical needs of schools, colleges, universities and various organizations. E-mail: arazaque@bridgeport.edu.

Dr. **Khaled Elleithy** is the Associate Dean for Graduate Studies in the School of Engineering at the University of Bridgeport. His research interests are in the areas of mobile wireless communications, network security and formal approaches for design and verification. And Mobile collaborative learning. He has published more than one hundred twenty research papers in international journals and conferences in his areas of expertise. Dr. Elleithy is the co-chair of International Joint Conferences on Computer, Information, and Systems Sciences, and Engineering (CISSE).CISSE is the first Engineering/Computing and Systems Research E-Conference in the world to be completely conducted online in real-time via the internet and was successfully running for four years. Dr. Elleithy is the editor or co-editor of 10 books published by Springer for advances on Innovations and Advanced Techniques in Systems, Computing Sciences and Software. Dr. Elleithy received the B.Sc. degree in computer science and automatic control from Alexandria University in 1983, the MS Degree in computer networks from the same university in 1986, and the MS and Ph.D. degrees in computer science from The Center for Advanced Computer Studies in the University of Louisiana at Lafayette in 1988 and 1990, respectively. He received the award of "Distinguished Professor of the Year", University of Bridgeport, during the academic year 2006-2007. e-mail: elleithy@bridgeport.edu.

Learning Confidence Intervals with Mobile Devices

F.J. Tapia-Moreno, H.A.Villa-Martinez
University of Sonora, Hermosillo, México

Abstract— **Mobile learning (m-learning) enhances learning skills in some students. Mobile phones, Tablets, PDAs, Pocket PCs, and Internet can be used jointly in order to encourage and motivate learning wherever and whenever students want to learn. In this work, we show learning objects for teaching and learning inferential statistics using mobile devices. With these learning objects, students can calculate confidence intervals based in either a large or a small data sample obtained from a normal or a non-normal population. These objects have been designed for devices with Android operating system.**

Index Terms—**m-learning, inferential statistics, learning objects.**

I. INTRODUCTION

In April of 2012 we presented a paper about the main results obtained in the project *Statistics-to-Go* [1] in which we developed learning objects for descriptive statistics by means of three Java-ME suites named *Module 1*, *Visual Summary Measures*, and *Skewness Measures*. We wrote the corresponding user manuals. These learning objects are currently used by the students who take the statistics courses offered in the University of Sonora. The success in the use of these learning objects has motivated us to build a module for inferential statistics.

In this paper, we present learning objects for calculate confidence intervals based in either a large or a small sample of data obtained from a normal or a non-normal population.

The remainder of this paper is composed as follows: Section II presents a short review about the state of art in mobile learning. Section III shows details about Android operating system and what can be created on mobile devices which use this operating system. Section IV presents basic confidence intervals theory. Section V shows the latest advances in the project *Statistics-to-Go*. Finally, Section VI presents our conclusions and future work.

II. MOBILE LEARNING

Mobile learning (also known as m-learning) is the acquisition of knowledge by means of some mobile device, for example a cellular phone or a tablet computer [2]. m-learning is extremely effective in classroom environments where modern technology may not be available to all students, or in schools where there are not sufficient computational resources for all students. In such situations, using a mobile device for learning, a device that can be shared among a group of students, is a great way to ensure that pupils are still able to benefit from the diverse opportunities presented by educational technology.

Some of the reported uses of m-learning are: to enhance the user experience in visitors of an art museum, to take advantage of the learning context [3], to allow students to participate on an academic discussion forum [4], in training for personnel in disaster or emergency situations [5], to suggest didactic methods that are adequate to a specific teaching situation [6], and in the learning of science [7], languages [8], and medicine [9].

The differences between m-learning and other types of learning can be studied by considering both the technology involved and the educational experience. Regarding technology, m-learning differs in the use of portable equipment that allows students to access learning objects anytime and anywhere. Regarding the educational experience, Traxler [10] compares m-learning and traditional e-learning using keywords. This way, m-learning is "personal", "spontaneous", "opportunistic", "informal", "pervasive", "private", "context-aware", "bite-sized", and "portable", whereas e-learning is "structured", "media-rich", "broadband", "interactive", "intelligent", and "usable". The same author notes that some of these distinctions can disappear as mobile technology advance, but properties as informality, mobility, and context will remain.

III. ANDROID OPERATING SYSTEM

Android is an operating system designed primarily for mobile devices such as cellular phones and tablet computers. Initially developed by Android, Inc., which Google bought in 2005, Android was unveiled in 2007 along with the founding of the Open Handset Alliance, a consortium of hardware, software, and telecommunication companies devoted to advancing open standards for mobile devices. The first Android-powered phone was sold in October 2008 [11].

Android-based cellular phones lead the market. According to Tech-Thoughts [12], Android has a share of almost 70% in the global smartphone market, while iOS, made by Apple, accounts to about 14%. In fact, Android has the largest share in every developed country, except in the United States and Japan where iOS is the dominant mobile operating system [13, 14].

In the tablet world the story is similar. According to BGR [15], Android worldwide market share grew from 39.8% in 2011 to 42.7% in 2012, while iOS share diminished from 56.3% to 53.8%.

IV. CONFIDENCE INTERVALS

In statistical inference, one wishes to estimate population parameters (for example average, variance, standard deviation, proportions, and differences between means and/or proportions) using observed sample data. A confidence interval gives an estimated range of values which is likely to include the unknown population parameter. This estimated range is calculated from a given set of sample data. Each confidence interval has associated a measure of the certainty that the interval confidence contains the population's parameter. This measure is called *confidence level* and is indicated by $1-\alpha$, where α, a number between 0 and 1, is called the significance grade. The most used confidence levels are 90%, 95%, and 99%. Each one of these numbers represents the probability that the confidence interval contains the real valor of the population parameter. Figure 1 shows a confidence interval for the population's average μ based in an observed sample data. $\mu_{\bar{x}}$ is the sample's average.

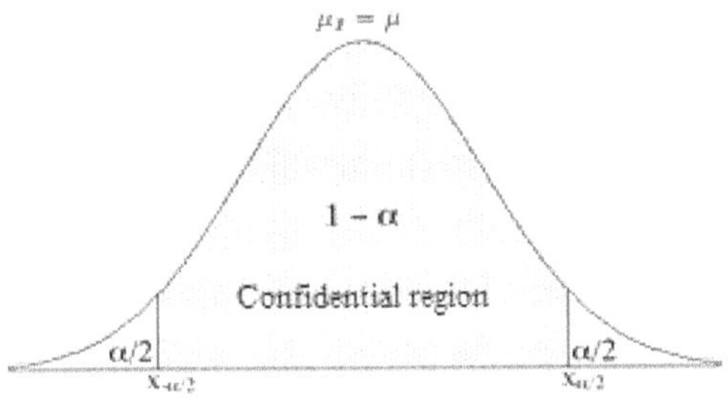

Figure 1.

According to Triola [16], from a population sample a confidence interval can be constructed for the population's mean using the normal distribution when: 1) the size of the sample is larger than 30 ($n \geq 30$) (this is due to the central limit theorem), or 2) when the sample size is less than 30 ($n < 30$), but the population has a normal distribution and its standard deviation is known. For this case, the confidence interval is given by Equation (1).

$$\bar{X} \pm z_{1-\alpha/2} \cdot \frac{\sigma}{\sqrt{n}} \quad (1)$$

Note that if the population's standard deviation is unknown, then Equation (2) should be used for determining the confidence interval.

$$\bar{X} \pm z_{1-\alpha/2} \cdot \frac{S}{\sqrt{n}} \quad (2)$$

where $\bar{X}$ is the sample's mean, S is the sample's standard deviation, n is the size of the sample, and $z_{1-\alpha/2}$ is the critical value of the normal standardized distribution [16].

If the size of the sample is less than 30 ($n < 30$), the population a has normal distribution, and the standard deviation is unknown, then the Student's distribution is preferred for constructing the confidence interval for population's mean [2]. In this case, the confidence interval is given by Equation (3)

$$\bar{X} \pm t_{n-1,1-\alpha/2} \cdot \frac{S}{\sqrt{n}} \quad (3)$$

where $\bar{X}$ is the sample's mean, S is the sample's standard deviation, n is the size of the sample, $n-1$ are the degrees of freedom, and $t_{n-1,1-\alpha/2}$ is the corresponding critical value for the confidence grade $(1-\alpha/2)$ of the Student's distribution.

When the size of the sample is less than 30 ($n < 30$) and the population has a non-normal distribution, then the Chebyshev's Theorem can be used to build a confidence interval for the population's mean [16]. Equations (4) and (5) show the confidence interval for this case when the population's standard deviation is known and unknown, respectively.

$$\bar{X} \pm k \cdot \frac{\sigma}{\sqrt{n}} \quad (4)$$

$$\bar{X} \pm k \cdot \frac{S}{\sqrt{n}} \quad (5)$$

where $k = \sqrt{1/\alpha}$.

In a similar way, confidence intervals for a population's proportion can be build using the normal distribution as an approximation to the binomial distribution if $n \geq 30$, and either $n \cdot p \geq 5$ or $n \cdot (1-p) \geq 5$. Here, n is the sample size and p is the probability of success in the experiment [16]. The confidence interval for this case is presented in Equation (6).

$$\bar{X} \pm z_{1-\alpha/2} \cdot S_{\bar{p}} \quad (6)$$

where $S_{\bar{p}} = \sqrt{\frac{\bar{p}\cdot(1-\bar{p})}{n}}$ or $S_{\bar{p}} = \sqrt{\frac{\bar{p}\cdot(1-\bar{p})}{n}} \cdot \sqrt{\frac{N-n}{N-1}}$ if $n < 0.05 \cdot N$. Here N is the population size, and $\bar{p}$ is the probability of success in the experiment.

Similarly, confidence intervals can be constructed for the difference between the means of two populations and for the difference between the proportions of two populations. The confidence interval for each one of these cases is shown in Equations (7), (8), (9), (10), (11), and (12).

$$(\bar{X}_1 - \bar{X}_2) \pm z_{1-\alpha/2} \cdot \sigma_{\bar{X}_1-\bar{X}_2} \quad (7)$$

$$(\bar{X}_1 - \bar{X}_2) \pm z_{1-\alpha/2} \cdot S_{\bar{X}_1-\bar{X}_2} \quad (8)$$

$$(\bar{X}_1 - \bar{X}_2) \pm t_{(n_1+n_2-2),1-\alpha/2} \cdot S_{\bar{X}_1-\bar{X}_2} \quad (9)$$

$$(\bar{X}_1 - \bar{X}_2) \pm k \cdot \sigma_{\bar{X}_1-\bar{X}_2} \quad (10)$$

$$(\bar{X}_1 - \bar{X}_2) \pm k \cdot S_{\bar{X}_1-\bar{X}_2} \quad (11)$$

$$(\bar{p}_1 - \bar{p}_2) \pm z_{1-\alpha/2} \cdot S_{\bar{p}_1-\bar{p}_2} \quad (12)$$

where $\sigma_{\bar{X}_1-\bar{X}_2} = \sqrt{\frac{\sigma_1^2}{n_1} + \frac{\sigma_2^2}{n_2}}$; $S_{\bar{X}_1-\bar{X}_2} = \sqrt{\frac{S_1^2}{n_1} + \frac{S_2^2}{n_2}}$; $S_{\bar{p}_1-\bar{p}_2} = \sqrt{\frac{\bar{p}_1\cdot(1-\bar{p}_1)}{n_1} + \frac{\bar{p}_2\cdot(1-\bar{p}_2)}{n_2}}$, and $\bar{X}_1$, $\bar{X}_2$, σ_1, σ_2, S_1, S_2, $\bar{p}_1$, $\bar{p}_2$, n_1, and n_2 are respectively, the means, standard deviations of the populations, standard deviations of the samples, proportions of the samples, and the samples sizes.

If a sample is taken from a normal distribution, confidence intervals can be built for both population's variance and standard deviation using the chi square distribution [17]. Both confidence intervals are given in Equations (13) and (14), respectively.

$$\frac{(n-1)\cdot S^2}{\chi^2_{n-1,\ \text{upper}}} \le \sigma^2 \le \frac{(n-1)\cdot S^2}{\chi^2_{n-1,\ \text{lower}}} \qquad (13)$$

$$\sqrt{\frac{(n-1)\cdot S^2}{\chi^2_{n-1,\ \text{upper}}}} \le \sigma \le \sqrt{\frac{(n-1)\cdot S^2}{\chi^2_{n-1,\ \text{lower}}}} \qquad (14)$$

where the upper and lower sub-indices identify the percentiles points of the chi-square distribution that was used for constructing the confidence interval.

Finally, we can build confidence interval for the ratio of variances of two normally distributed populations using the *F* distribution [16] employing Equation (15).

$$\frac{S_1^2/S_2^2}{F_{1-\alpha/2}} < \frac{\sigma_1^2}{\sigma_2^2} < \frac{S_1^2/S_2^2}{F_{\alpha/2}} \qquad (15)$$

One last remark is that a person who wants to calculate confidence intervals needs statistics tables, like those in [17], in order to obtain the proportion of the area under the normal curve, the Student *t*, the chi-square, and the *F* factor. On the other hand, our system obviates this requirement.

IV. RESULTS

In this second stage of our project, we have developed learning and teaching objects for Inferential Statistics. Each one of the objects calculates a confidence interval for a distinct situation. The first object computes confidence intervals for the population's mean. The application uses Equations (1) or Equation (2) depending if the value of the population's standard deviation is known or not. Figure 2 shows the output of this object for a sample with size $n = 57$, a mean of 7.5, a standard deviation of 2.1, and a confidence level of 95%. That is, $\alpha = 0.05$.

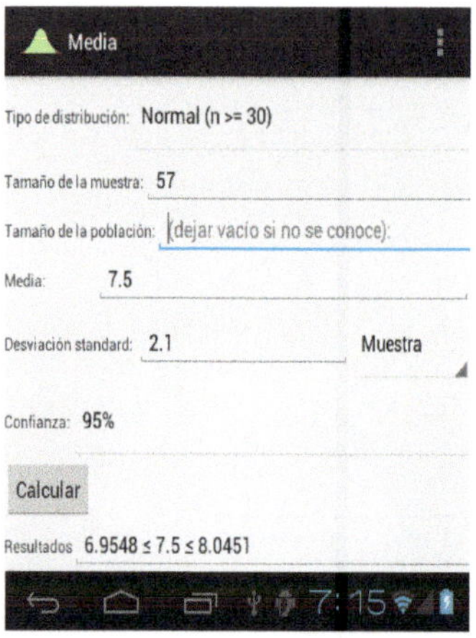

Figure 2.

The second object computes confidence intervals for the population's average when the size of the sample is less than 30 ($n < 30$), the population has a normal distribution, and the standard deviation is unknown. For this case, the Student's distribution is the preferred way for constructing the confidence interval for the population's mean. In this case, the system uses Equation (3) in order to calculate it. Figure 3 shows the output of this object for a sample with size $n = 21$, a mean of 7.5, a standard deviation of 2.1, and a confidence level of 95%.

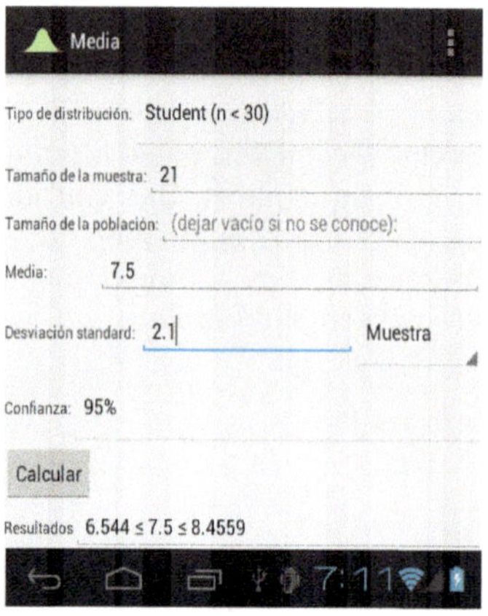

Figure 3.

The third object computes confidence intervals for the population's average when the population has a non-normal distribution. The Chebyshev's theorem is used to build this confidence interval. The application uses Equation (4) or Equation (5) depending if the value of the population's standard deviation is known or not. Figure 4 shows the output of this object for a sample with $n = 57$, a mean of 7.5, a standard deviation of 2.1, and a confidence level of 95%.

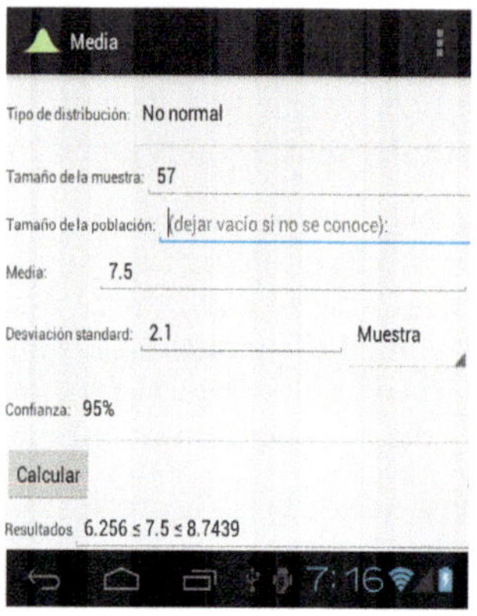

Figure 4.

The fourth object calculates confidence intervals for a population's proportion using Equation (6). This application can be used only when both conditions mentioned in section IV are satisfied. Figure 5 presents the output of this object for a population with size $N = 400$, and a sample of size $n = 57$ with a proportion of $\hat{p} = 0.58$ and a confidence level of 95%.

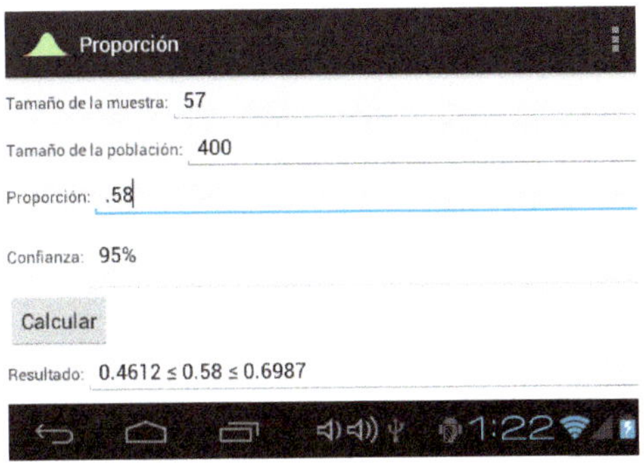

Figure 5.

The fifth object computes confidence intervals for the difference between the means of two populations using Equation (7) or Equation (8) depending if both population's standard deviations are known or not. Figure 6 shows the output of this object for two samples with sizes $n_1 = 71$ and $n_2 = 55$, sample's means of $\bar{X}_1 = 17.58$ and $\bar{X}_2 = 11.6$, and sample's standard deviations of $s_1 = 4.8$ and $s_2 = 1.256$, with confidence level of 95%.

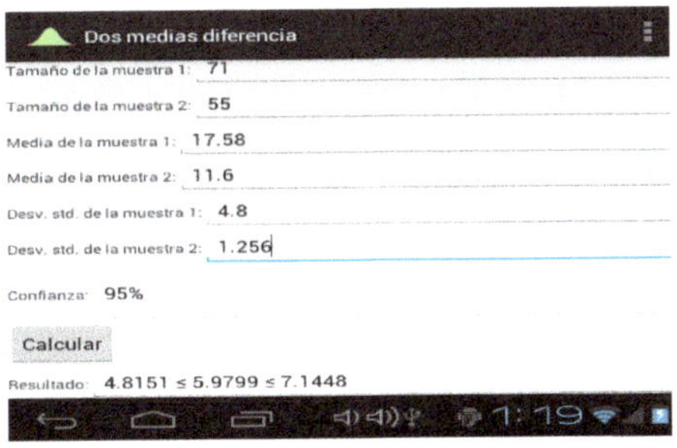

Figure 6.

The sixth object computes confidence intervals for the difference between the means of two populations when both two samples are little (n_1, n_2<30), and both populations have approximately normal distributions. Equation (9) is utilized by the object to calculate these intervals.

The seventh object calculates confidence intervals for the difference between the means of two populations when these populations have non-normal distributions. The object uses Equation (10) or Equation (11) depending if both population's standard deviations are known or unknown. The input data, and the output is similar to the output of the fifth object, shown in Figure 6, with the difference that the seventh object uses Student's distribution in order to estimate the confidence interval.

The eighth object estimates confidence intervals for the difference between the proportions of two populations when the conditions mentioned in the section IV are satisfied. The object uses Equation (12) to calculate the confidence intervals. Figure 7 shows the output for two samples with sizes $n_1 = 45$, and $n_2 = 65$ with proportions of $\hat{p}_1 = 0.70$, and $\hat{p}_2 = 0.55$, with a confidence level of 95%.

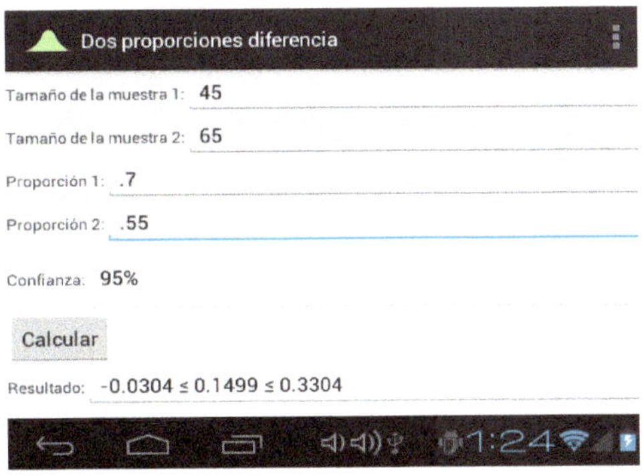

Figure 7.

The ninth object computes confidence intervals for the population's variance and standard deviation using the chi square distribution. The object uses Equation (13) or Equation (14) to estimate these confidence intervals. Figure 8 shows the result output for a sample with size $n = 70$, a standard deviation of 7.5, and a confidence level of 95%.

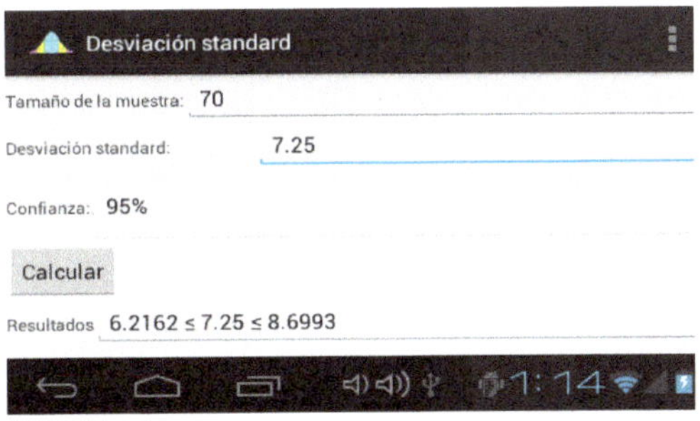

Figure 8.

The tenth object calculates confidence intervals for the ratio of the variances of two samples employing Equation (15). Figure 10 shows the output of this object for two samples with sizes $n_1 = 54$ and $n_2 = 84$, variances of $s_1 = 8.6$ and $s_2 = 5.25$, with a confidence level of 95%.

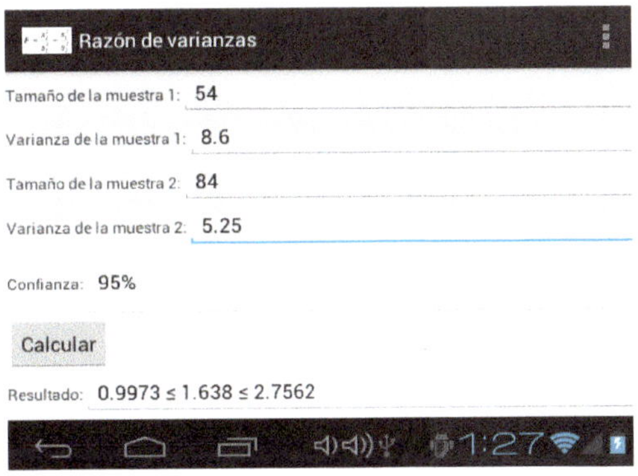

Figure 9.

VI. CONCLUSIONS

We have presented the main achievements obtained after completing the second stage of the project *Statistics-to-Go*. We have designed learning objects that enhance learning of statistical concepts exclusively using a cellular phone and taking advantage of the ubiquity of personal technology for educational purposes. We are exploiting the possibilities of mobile phones for teaching and learning statistics helping by the way those students who do not have possibilities for buying a computer or paying for Internet service. These learning objects cover most of the syllabus of the introductory inferential statistics course offered in the first semester at the University of Sonora.

In the third stage of this project we will design learning and teaching objects to perform **statistical hypothesis tests, which is** a method of making decisions using data from a scientific study, and that the students learn in a second course of inferential statistics.

Also we will test the efficacy of these objects on the students' mobile devices and we will gather their opinion about the usability of our applications.

REFERENCES

[1] Tapia et al. *Elaboration of Statistics Learning Objects for Mobile Devices*. iJIM, 6, 2, p. 51-54. April 2012

[2] Molnár, Györgi, J.: *New Learning Spaces? M-Learning's, in Particular the iPad's Potentials in Education*. iJIM, 7, 1, p. 56-60. January 2013.

[3] Bhaskar, N. Uday.; P. Govindarajulu *Advanced and Effective Learning in Context Aware and Adaptive Mobile Learning Scenarios*. iJIM, 4, 1, p. 9-13. January 2010.

[4] Motiwalla, Luvai F. *Mobile learning: A framework and evaluation*. Computers & Education 49, p. 581-596. 2005.

[5] Colley, Jo; Claire Bradley; Geoff Stead; Jessica Wakelin. *Global MedAid – evolution of an mlearning app for international work-based learners*. mLearn 2012. Helsinki, Finland. 2012.

[6] Möller, Andreas; Stefan Diewald; Luis Roalter; Barbara Beege; Matthias Kranz. *MobiDics: Cooperative Mobile E-Learning for Teachers*. mLearn 2012. Helsinki, Finland. 2012.

[7] Sohaib, Ahmed; David Parsons. *Evaluating 'ThinknLearn': A Mobile Science Inquiry Based Learning Application in Practice*. mLearn 2012. Helsinki, Finland. 2012.

[8] Mmaki, Jantjies; Mike Joy. *Multilingual mobile learning – A case study of four South African high schools*. mLearn 2012. Helsinki, Finland. 2012.

[9] Hardy, Ernst; John Harrison; David Griffin. *Iterative design and delivery of high impact, multiple platform, scenario-based interactive mobile learning activities in the health sciences*. mLearn 2012. Helsinki, Finland. 2012.

[10] Traxler, J.: *Defining, Discussing, and Evaluating Mobile Learning: The moving finger writes and having writ....* International Review of Research in Open and Distance Learning, 8, 2, 1-12. 2007.

[11] Kerfs, Jeremy: *Beginning Android Tablet Games Programming*. p. 1-2. APress. 2011.

[12] http://www.tech-thoughts.net/2012/12/smartphone-market-share-trends-by-country.html#.UQRSaL-mz2Y. Last accessed Jan 26, 2013.

[13] http://www.kantarworldpanel.com/dwl.php?sn=news_downloads&id=113. Last accessed Jan 26, 2013.

[14] http://www.kantarworldpanel.com/global/News/news-articles/Windows-sees-strong-European-growth. Last accessed Jan 26, 2013.

[15] http://bgr.com/2012/12/05/tablet-market-share-2012/. Last accessed Jan 26, 2013.

[16] Triola, Mario F. *Elementary Statistics with Multimedia Study/Guide*, 10th Edition. Addison-Wesley Longman. 2007.

[17] Neave, Henry. *Statistics Tables for Mathematicians, Engineers, Economists. and the Behavioural and Management Sciences*. College Audience. 2011.

AUTHORS

F. J. Tapia-Moreno is with the Department of Mathematics, University of Sonora, Hermosillo, Sonora 83000 Mexico (e-mail: ftapia@gauss.mat.uson.mx).

H. A. Villa-Martínez is with the Department of Mathematics, University of Sonora, Hermosillo, Sonora 83000 Mexico (e-mail: hvilla@gauss.mat.uson.mx).

Development of eStudent iOS Mobile Application

M. Antic, S. Jovanovic and S. Cvetanovic
Metropolitan University, Belgrade, Serbia

Abstract—**iOS is the Apple mobile operating system, for Apple mobile devices. Initially developed for iPhone, and later for iPod Touch, iPad, AppleTV. The Metropolitan University has a Web application titled eStudent, which enables students by Web to get information about their marks for all subjects, their financies, exam scheduling, professors and assistents, and send exam registration and feedback about teaching, etc. This paper explains the development of the mobile application eStudent on the iOS platform. This application enables students whenever they want, by using their iPhone mobile phone, to get access to the information from the eStudent Web application, and to present it on their iPhone User Interface (UI). This paper explains in details software requirements analysis, system architecture, system modelling, and UI of the eStudent iOS mobile application.**

Index Terms—**Apple, API, Ruby On Rails, Objective C**

I. Introduction

iOS (iPhone OS) is the Apple mobile operating system (operating system for Apple mobile devices). Initially developed for iPhone, and later for iPod Touch, iPad, AppleTV. Regarding statistics from Oct.2011, AppStore (Apple shop of iOS applications) had more than 500,000 iOS applications, downloaded more than 18 billion times! Couple of arguments why to use iOS platform [1,2]?

- Nonsegmented market
- Excellent user satisfaction (experience) with iOS devices
- Precise and clear advices for developers
- Many students use iOS devices (iPhone, iPad, …)

The Metropolitan University in Belgrade uses the eStudent Web application, which enables students by using Web to get information about their marks, financies, to send exam registrations, and send feedback about teaching quality, etc. However, since more and more students have a smartphone in their pockets, often iPhone, there is a need for eStudent mobile application, to enable students to use eStudent Web application via their iPhone mobile phone, whenever they want.

This paper explains the development of the eStudent iOS mobile application. It consists of:

- Software requirements analysis (application functionality analysis)
- Defining system architecture
- Models of eStudent iOS mobile application
- API (Application Programming Interface) development
- Development of User Interface

II. Software Requirements Analysis

During the analysis, the software requirements of iOS mobile application eStudent are groupped in separate parts, and the main purpose is to explain how to adapt eStudent Web application to be presented on iPhone mobile device. The eStudent Web application has five separate parts, so consequently, the iOS eStudent mobile application is also divided in five parts (five tabs).

<u>(1) Reports:</u> "Reports" show student marks for all current teaching subjects. The main goal here is to adapt as better as possible the presentation of information to the size of the mobile phone screen and to the "native" controls of iOS mobile device (designed by Apple). During this process, the experience of developer is an important factor, and guidelines from Apple Human Interface Guidelines, which advices designers and devopers how to present data on the small screen. Fig. 1 shows a "Report" of the eStudent Web application for one of teaching subjects.

<u>(2) Financies</u>**:** This page shows the financies data (financial obligations, debts, and payments during the school year, etc.). Here, it is important to have an easy access to the financies data both from the current and from the previous year. So, in the navigation bar, it will be enabled to access easily the previous year, not just the current year. Also, it is convenient to separate payments and obligations into two separate tables, so some data filtering patterns can be applied. Fig. 2 presents the student financies page in the eStudent Web application.

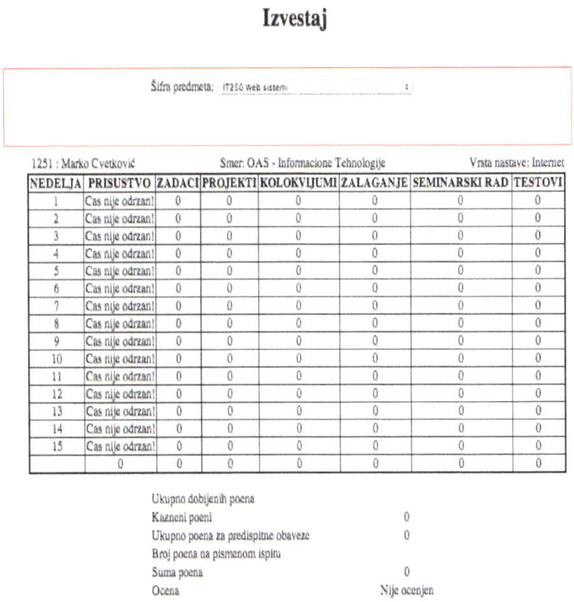

Izvestaj

Šifra predmeta: IT250 Web sistemi

1251 : Marko Cvetković — Smer: OAS - Informacione Tehnologije — Vrsta nastave: Internet

NEDELJA	PRISUSTVO	ZADACI	PROJEKTI	KOLOKVIJUMI	ZALAGANJE	SEMINARSKI RAD	TESTOVI
1	Cas nije odrzan!	0	0	0	0	0	0
2	Cas nije odrzan!	0	0	0	0	0	0
3	Cas nije odrzan!	0	0	0	0	0	0
4	Cas nije odrzan!	0	0	0	0	0	0
5	Cas nije odrzan!	0	0	0	0	0	0
6	Cas nije odrzan!	0	0	0	0	0	0
7	Cas nije odrzan!	0	0	0	0	0	0
8	Cas nije odrzan!	0	0	0	0	0	0
9	Cas nije odrzan!	0	0	0	0	0	0
10	Cas nije odrzan!	0	0	0	0	0	0
11	Cas nije odrzan!	0	0	0	0	0	0
12	Cas nije odrzan!	0	0	0	0	0	0
13	Cas nije odrzan!	0	0	0	0	0	0
14	Cas nije odrzan!	0	0	0	0	0	0
15	Cas nije odrzan!	0	0	0	0	0	0
	0	0	0	0	0	0	0

Ukupno dobijenih poena	
Kazneni poeni	0
Ukupno poena za predispitne obaveze	0
Broj poena na pismenom ispitu	
Suma poena	0
Ocena	Nije ocenjen

Figure 1. Report page in Web eStudent

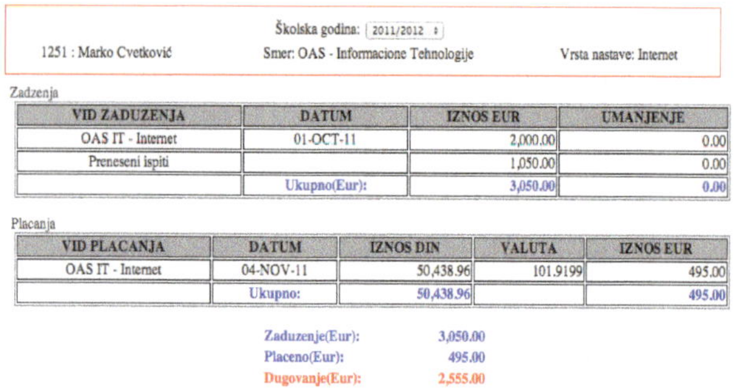

Školska godina: 2011/2012

1251 : Marko Cvetković | Smer: OAS - Informacione Tehnologije | Vrsta nastave: Internet

Zadzenja

VID ZADUZENJA	DATUM	IZNOS EUR	UMANJENJE
OAS IT - Internet	01-OCT-11	2,000.00	0.00
Preneseni ispiti		1,050.00	0.00
	Ukupno(Eur):	3,050.00	0.00

Placanja

VID PLACANJA	DATUM	IZNOS DIN	VALUTA	IZNOS EUR
OAS IT - Internet	04-NOV-11	50,438.96	101.9199	495.00
	Ukupno:	50,438.96		495.00

Zaduzenje(Eur): 3,050.00
Placeno(Eur): 495.00
Dugovanje(Eur): 2,555.00

Figure 2. Financies page in Web eStudent

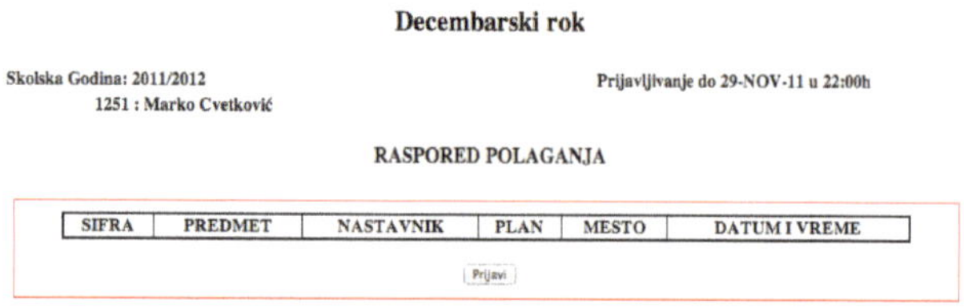

Decembarski rok

Skolska Godina: 2011/2012
1251 : Marko Cvetković

Prijavljivanje do 29-NOV-11 u 22:00h

RASPORED POLAGANJA

SIFRA	PREDMET	NASTAVNIK	PLAN	MESTO	DATUM I VREME

Prijavi

Figure 3. Exam registration page in Web eStudent

(3) <u>Exam registration</u>**:** The exam registration page is very important for students, because it is one of the most important functions of eStudent application. It must show all relevant information, eg. confirmation about exam registrations, registration deadline, data about dates and times of scheduled exams, etc. After the requirements analysis, it was decided to present the whole exam scheduling list on the same screen, with the possibility to reset the exam date and time by chosing the relevant date from the list. Also, it is decided to enable the exam registration to be performed easily by a single tap (pressing a single button). And, the server response will be showed in a corresponding notifying window. Fig. 3 shows the present look of the exam registration page in the eStudent Web application.

(4)<u>Professsors</u>**:** The page "Professors" gives relevant information about lecturers and teaching assistents for relevant teaching subjects for every student. It was found that all relevant data can be presented in a single table.

(5) <u>Student's feedback</u>**:** Students send their feedback about teaching quality (their satisfaction and impression about teaching quality for every teaching subject) twice a year (once during a tea\xching semester). After the requirements analysis, it was decided to present relevant data on the mobile screen as a HTML document, created by the University Web server and sent to the mobile device after a request sent by student's mobile phone.

Besides the previous five requirements regarding the data presentation, we have another three requirements:

(6) <u>Analysis of API</u>**:** API - *Application programming interface*, has the goal to provide input data for eStudent mobile application. The input data will be taken from the Web server. For each Web server page, a HTTP request for getting data (from the Web server) will be sent, and the Web server will sent the required data (to the iOS eStudent mobile application) in the JSON format. However, since the Information system of the Metropoltan university (ISUM) is in the process of upgrading and renovating, and that the final version of API is not ready in the moment , a prototype API was created. This prototype API

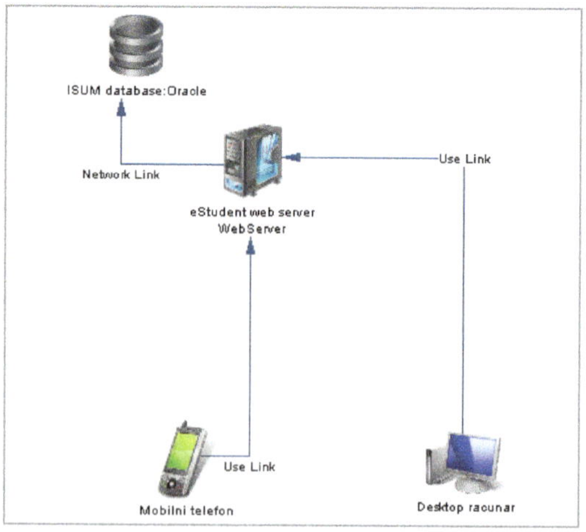

Figure 4. System architecture

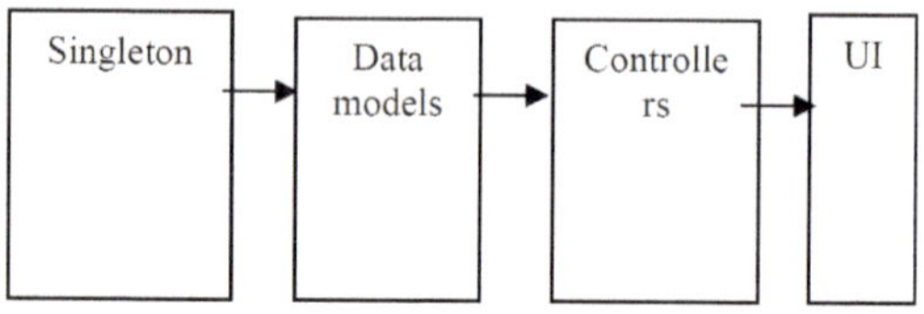

Figure 5. Structure of iOS eStudent

will be used as a good representative, during the process of finalizing the future API in the final ISUM environment, which will enable iOS eStudent to work online in the real environment.

(7) Number of API requests: In order to maximize the user satisfaction, for every server page (for every of five basic screens**),** data will be sent separately. In this way, we minimize the size of JSON files, for 3G (or *Edge)* connection, and the user will not wait much during the process of inputting data sent from the the Web server.

(8) <u>Secure autentification and authorization</u>: This requirement is resolved simply by requesting from the user to input his/her name, family name and a password. After this, the user can access the Web application. During every HTTP request sent to the Web server, it is necessary to do the basic HTTP autentification, to block any unauthorized use of API. Besides this baic HTTP autentification, for every request, the student ID number is requested in order to get data for the corresponding student.

III. System Architecture

The system architecture is given in the Fig. 4, resulting from system requirements. It is a distributed mobile application, where the iOS eStudent mobile application communicates with the eStudent Web server by using the HTTP ptotocol (GET, POST), while the data are in th JSON format. Native applications for iOS are written using the Apple framework Cocoa Touch (for mobile interfaces). The structure of the iOS eStudent mobile application is divided in four parts (Fig. 5):

- Data models, describing data
- Presentation layer - .xib files, describing user interface UI
- Controllers – Business logic which controls UI and manipulates data
- Singleton object, responsible for communicating with API

IV. PROCESS MODEL AND OBJECT MODEL

On the highest abstraction level, all processes are shown in a form of a tree, where we can see also subrocesses presented. Five separate processes: report viewing, financies viewing, exam registration, professors viewing, and sending feedback. Fig. 6 shows this process model.

Fig.7 i 8 present object models of the eStudent iOS application. Based on defining system Use Cases, set of program classes has been identified. In Fig.7 the following classes are included: Singleton, Subject, Week, Semester, Course, CourseRegistration, Professor.

In Fig.8 classes are: ReportViewController, DetailedReportViewController, FinanciesViewController, FinanciesYearViewController, RegistrationViewController, RegistrationDetailedViewController, ProfessorsViewController, FeedbackViewController, LoginViewController.

V. USER INTERFACE

Login View Controller **(Fig. 9)** appears after starting (openning) the application, in order to enable the access of the user to the system, and to send necessary data to the Web server. After the tap of the Login button, the iOS eStudent mobile application sends a HTTP POST request to the Web server, to check entered data via eStudent Web application and API. If entered data are correct, Login View is removed and further steps are allowed. If entered data are incorrect, in Login View, the view is unchanged, and the user can try to enter data again.

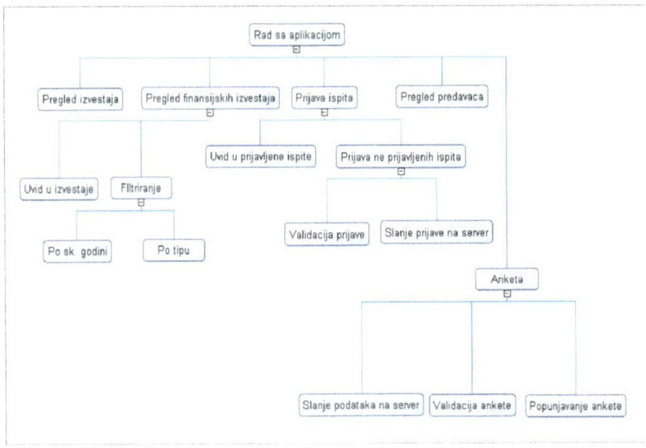

Figure 6. Process model in eStudent iOS

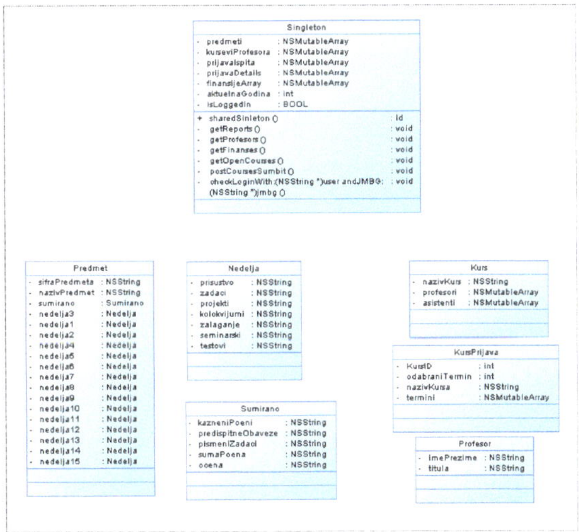

Figure 7. Data object models and Singleton class

Apple's iOS UI is very good, and other big IT companies are trying to copy it. One of UI controls on iOS, very practical, and used in all applications with several views (more than 2-3 views) is TabBar. TabBar is the control which enables quick change between views, it is always visable and is placed on the bottom of the screen. TabBar controller provides five different view controller*s*:

- Report View Controller
- Financies View Controller
- Application View Controller
- Professor View Controller
- Feedback View Controller

Report ("Izveštaj") View Controller at its first appearance takes fresh data from eStudent Web application using GET API request. After parsing the JSON file (obtained from Web eStudent), the table is filled in with the list of all teaching subjects relevant for the particular student. Title and ID for every relevant teaching subject is shown in the single table. Every box in the table has an arrow, indicating to the user to tap the box in order to get more information about the particular subject (Fig. 10 and 11). This new set of data are shown immediately, they are already in the mobile phone, taken from Web eStudent in the current JSON file. There are 16 sections, the first section is summing up, showing data for the whole semester for the chosen teaching subject, and the rest of 15 sections represent 15 weeks of the teaching semester, and the student marks for each week.

Exam registration ("Prijava ispita") View Controller**:** After a tapping tab "Prijava", the iOS application contacts theWeb application usung HTTP GET method, and collects all the information needed for this view:

- List of subjects which are enabled to the student to try to pass in the current examination period
- list of dates and times for every subject
- Deadline information

Fig. 12 shows a *view* with a list of subjects and corresponding examination dates/times. I f any subject is already registrated for examination, this is then clearly marked and the examination date/time is given. Also, if the student changes his mind about exam registrations, he can make changes in the examination registration.

Figure 8. View controllers object model

Feedback ("Anketa") View Controller (Fig.13): For each semester, students have to send a feedback about their satisfaction with teaching (lectures, professors, teaching msaterials). Since it is more practical to send this data in HTML format, there is a WebView embedded in Feedback View Controller, enabling creating HTML code, and this HTML code will be sent by Web server (ie. by Web eStudent application). Feedback is filled in as a multiple choice document. During this process, students can use the Next button, located in the navigation bar PickerViewControl, enabling a quick movements through the feedback document. This small but useful functionality significantly speeds up the process.

Financies ("Finansijski Izveštaj") View Controller (Fig.14): Financies View shows student's financial obligations, payments and debts, to the Metropolitan university. Different data are presented in separate boxes, enabling a good visibility on a small space. There are three sections: Sum, Oblgations, and Payments. Using segmented control, students can filter presented data, depending what is interesting for them in the moment, to choose either a summing report, or a detailed report with obligations and payments.

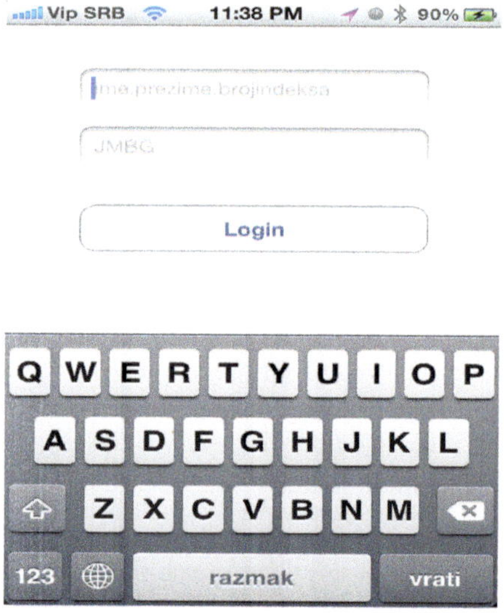

Figure 9. Login View Controller

Figure 10. Report View Controller

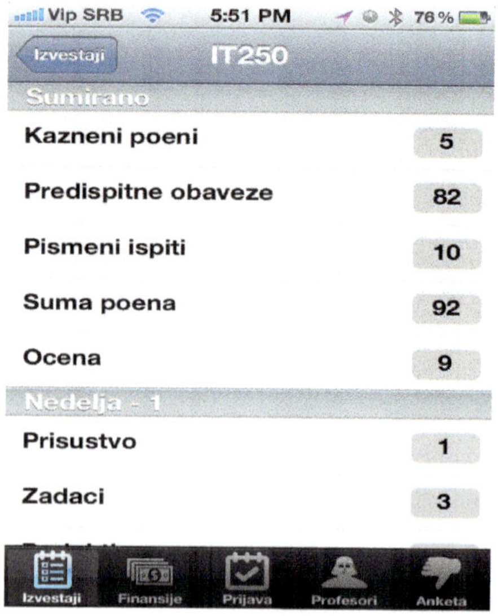

Figure 11. Report for IT 250

Figure 12. Exam registration View Controller

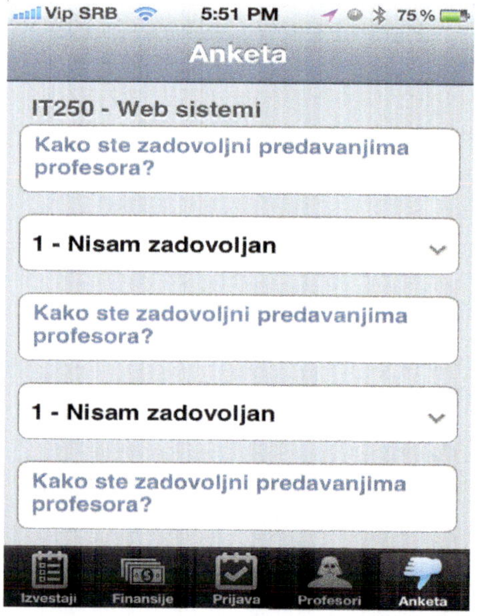

Figure 13. Feedback View Controller

Professors view is a simple view. After a tab "Professors", a request is sent to the Web server, for fresh data. Then, a table is formed with several sections, each section for an exam, and within the section are data about lecturers and assistents for corresponding subject. Fig. 15 shows this view.

Every view controller works independently between each other. Fig. 16 presents view control hierarchy in the eStudent iOS mobile application.

Figure 14. Financies report

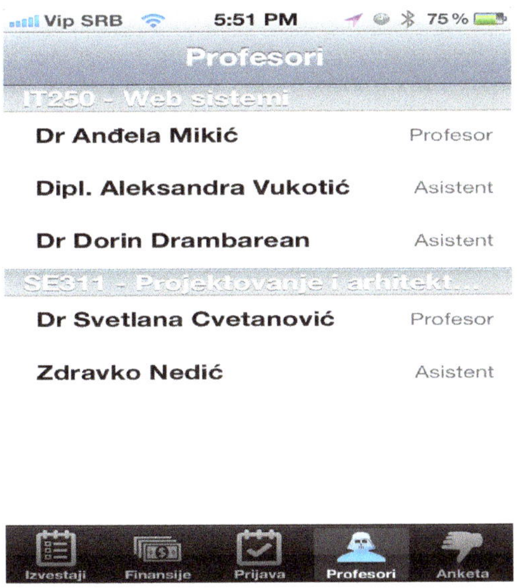

Figure 15. Professors View Controller

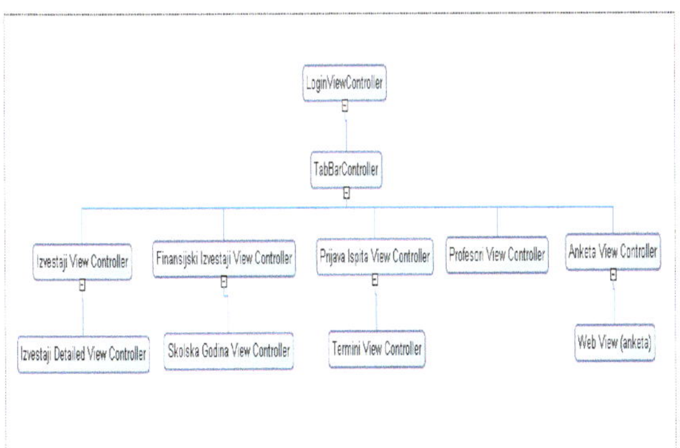

Figure 16. View Controllers hieararchy

VI. API

Many mobile applications communicate with Web servers, making it interactive and interesting. Eg. Twitter, Facebook, etc. communcate with Web services using HTTP protocols. But, the question is how such mobile applications communicate with Web applications? Answer is, they do it using Application Programming Interface (API) [3,4].

API enables developers of mobile applications (or any other application) to connect it easily with a Web application, and enable using all functions defined by API. In most cases, API is the only open point of an application tpwards the external world, so it is important to make API safe against any potential security risks. Figure 17 presents the connection between the iOS eStudent mobile application, API and the Web eStudent application, and iOS operating system.

During the development of the iOS eStudent mobile application, it was decided to make a prototype API layer first ("Demo API"), which enables the application to work and to be fully developed, and the final (real) API can be developed in some later stage, after finalizing the ISUM information system. This Demo API will enable an easy development of the final API, as an integral part of ISUM (Information system of University Metropolitan), which will operate on a production data base. Demo API is produced in Ruby On Rails, a popular Web framework for Ruby, and for every API request it was neccessary to define details regarding input parameters and return information. API was tested by RestClient application (http://code.google.com/p/rest-client/). API layer is protected by BASIC HTTP Autentification, and, concretely, in this Demo API, the access data are: username: mladjan, password: mladjan.

This is the lst of API methods:

- login_client
- get_reports
- get_profesors
- get_open_courses
- getFinance
- makeCoursesSignup

API program listing has approximately around 10 pages. For example, the method **get_profesors**. This method returns data about professors and corresponding teaching subjects, in the JSON format. Expected parameters are: name, surname, student ID number. These parameters must be sent as parameters of a POST method! If the request is successful, this method will return data in the JSON format. In case of any error, the method will return ERROR (in text format).

Format of the JSON file returned by the method is:


```
[ {
"profesori" : [ "Dr An•ela Miki•" ],
"asistenti" : [ "Dipl. Aleksandra Vu-
koti•", "Dr Dorin Drambarean" ],
 "predmet" : "IT250 - Web sistemi"}, {
"profesori" : [ "Dr Svetlana Cvetano-
vi•" ],"asistenti" : [ "Zdravko Nedi•"
],
"predmet" : "SE311 - Projektovanje i
arhitektura softvera"} ]
```

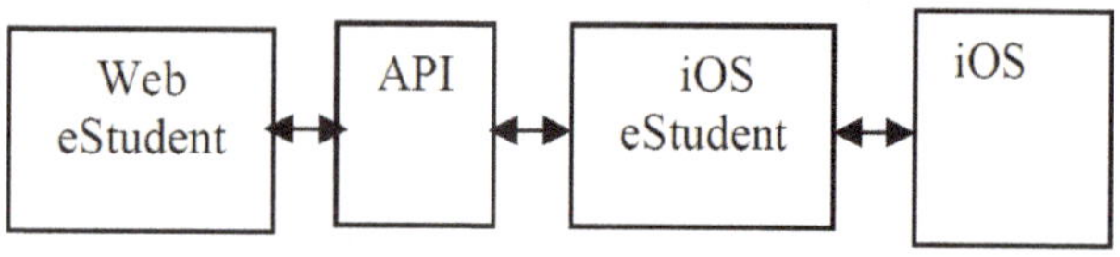

Figure 17. API connection

VII. OBJECTIVE-C PROGRAM LISTING

Objective-C is an object oriented programming language, made by combining the sintax of C and *Smalltalk* [5, 6]. Today, Objective-C is well known as the language of OSX and iOS applications, and supported by Apple Inc. Like C/ C++, the program code in Objective-C is separated in two separate files, for every program class(the header file, and implementation fajl). The iOS eStudent application is written in Objective-C. The program listing in Objective-C has around 60 pages.

Here, we have only a small section of this program listing, as an illustration. This is one of simple program classes, titled Kurs:

```
// Kurs.h
// eStudent
//
// Created by Mladjan Antic on 1/3/12.
//        Copyright        (c)        2012
__MyCompanyName__.All rights reserved.
//
#import <Foundation/Foundation.h>
#import "Profesor.h"
/**Model for Course*/
@interface Kurs : NSObject
/**Course name*/
@property nonatomic  retain  NSString
*nazivKursa;
/**Professors*/
@property nonatomic  retain  NSMutab-
leArray *profesori;
/**Assistents*/
@property nonatomic  retain  NSMutab-
leArray *asistenti;
@end
```

And, implementing the class Kurs looks like this:

```
// Kurs.m
// eStudent
//
// Created by Mladjan Antic on 1/3/12.
//        Copyright        (c)        2012
__MyCompanyName__.All rights reserved.
//
#import "Kurs.h"
@implementation Kurs
@synthesize  nazivKursa,  profesori,
asistenti;
@end
```

VIII. CONCLUSION

This paper explains the development of the eStudent iOS mobile application, developed to enable the students of the Metropolitan University, via their Apple mobile phones, to access data from the eStudent Web application. The paper explains some important aspects of design, analysis, modeling, solved during the development of the eStudent iOS mobile application. The focus of the paper is the design and organization of the mobile application, and its UI. Design simplicity of the iOS eStudent, intuitive UI, and its speed, are its advntages. eStudent iOS is dependent of the corresponding Web servis, ie. Web application eStudent, which means that any new function of the eStudent iOS must be followed by the development of the Web eStudent.

ACKNOWLEDGMENT

REFERENCES

[1] http://www.asktog.com/columns/070iPhoneFirstLook.html

[2] http://developer.apple.com/appstore/guidelines.html

[3] **CS193P iPhone Application development**, Stanford University, http://www.stanford.edu/class/cs193p/cgi-bin/drupal/

[4] **Vandad Nahavandipoor**, iOS 5 Programming Cookbook, O'Reilly Media 2011, http://oreilly.com/iphone/

[5] **Stephen G. Kochan**, Programming in Objective-C, Sams Publishing, 2011, ISBN 0672325861, www.amazon.com

[6] M.Antic,S.Jovanovic, S.Cvetanovic, eStudent iOS mobile application, 3rd Int. Conf. eLearning-2012, Belgrade, Sept. 27-28.

AUTHORS

Mladjan Antic was with the Belgrade Metropolitan University until 2011, as a software developer. He is now with his own company, which produces mobile applications software (e-mail: mladjan.antic@me.com).

Slobodan Jovanovic, was with Strathclyde University, Glasgow, Scotland during 1993-2008. Since 2008 he is with the Faculty of information technology, Belgrade Metropolitan University, as a professor in Computer science (e-mail: Slobodan.jovanovic@metropolitan.ac.rs).

Svetlana Cvetanovic has been with the Faculty of information technology, Belgrade Metropolitan University, since 2007, as a lecturer in Software Engineering and a director of Master studies (e-mail: svetlana.cvetanovic@metropolitan.ac.rs).

11

Mobile Tourist Guide – An Intelligent Wireless System to Improve Tourism, using Semantic Web

Hosam F. El-Sofany[1], and Samir A. El-Seoud[2]

[1]Cairo Higher Institute for Engineering, Computer Science, and Management, Cairo, Egypt

[2]Technische Universität Darmstadt (TU Darmstadt), Darmstadt, Germany

Abstract—**With the recent advances in Internet and mobile technologies, there are increasing demands for electronic access to tourist information systems for service coordination and process integration. Mobile computing and mobile devices are used to implement various tourist services (e.g. electronic tourist guides, digital interactive maps, and tourist e-commerce transactions). However, due to disparate tourist information and service resources such as airlines, hotels, tour operators, it is still difficult for tourists to use them effectively during their trips or even in the planning stage. Neither can current tourist portals assist tourists proactively. To overcome this problem, we propose the analysis, design, and implementation of the "Mobile tourist guide" system, that access through wireless devices and use Semantic Web technologies for effective organization of information resources and service processes. The proposed system provides the users with various services such as: 1) displaying the shortest path between the sources and destinations the visitors specify, 2) displaying general information of shops, newest events of the plaza and shops, 3) provides service of hotel, restaurant and cinema-ticket reservations, 4) provides user-friendly administration service. The Admin can manage the position, blocking path details, general information of hotel, restaurant, shops and plaza, and reservation details via web browser without changing the framework of the system. The system prototype has been developed on the top of Java 2 Micro Edition which offers an ideal platform for the development of full-fledged, interactive and portable applications tailored for resource constrained mobile devices. The paper presents our development experiences and highlights its main advantages and limitations in relation to the implementation of such kind of applications.**

Index Terms—**Electronic tourist guide, Mobile tourist guide, Semantic Web technology, Tourist Information System, Tourist application.**

I. Introduction

Tourism is a leading industry in the e-business. There are several reasons behind this fact. Tourism is an information-intensive industry with a long value chain. Since 1960s this industry represents a classical field of information and communication technology applications. Characteristics of electronic tourism marketplace lead to many challenges. One of them is growing need for interoperability between information systems, allowing seamless information exchange between tourism organizations. The traditional solution for interoperability issues was to develop interfaces between each pair of communication systems, but development and maintenance of such programs are too expensive and not flexible enough for the electronic tourism marketplace. There have been many attempts to standardize data exchange within tourism electronic markets in order to reach global interoperability. As a result there are some global standards used only by large companies and some standards developed for the needs of small companies, but used only on national level [1].

The World Tourism Organization predicts that by 2020, tourist arrivals around the world will increase over 200%. Tourism has become a highly competitive business all over the world. Competitive advantage is increasingly driven by the advancement of information technology and innovation. Currently, the Internet is the primary source of tourist destination information for travelers.

With the recent advances in hardware and software technologies, the Internet is quickly evolving towards wireless adoption. New mobile applications running on these devices provide users with easy access to remote services available anytime and anywhere, and will soon take advantage of the ubiquity of wireless networking in order to create new virtual worlds. Besides, intelligent software agents can run on these devices and can provide personalized assistance to tourists during their trip. Together with traditional information agents such as hotel broker agents, tour planning agents, and other disparate tourist resources, they form a *Multi-Agent Information System,* for collaborative and intelligent assistance to tourists.

At the same time, Semantic Web technologies have been maturing to make e-commerce interactions more flexible and automated. Ontology defines the terms used to present a domain of knowledge that is shared by people, databases, and applications. In particular, ontology encodes knowledge, possibly spanning different domains as well as describes the relationships among them. Currently, ontology is actively being developed in various business domains. The Semantic Web provides explicit meaning to the information available on the Web for automated processing and information integration based on the underlying ontology. Our aim in this project is to describe characteristics of the electronic tourism as well as challenges related to the tourism electronic marketplace and to discuss how semantic web technologies can help to resolve these challenges.

In this research, we intend to expand tourist coordination and integration towards complete support by employing all the above-mentioned technologies. The main challenge of our system is to provide an effective coordination and integration of disparate information and service resources anytime, anywhere, as well as the provision of personalized assistance and automation to the tourists, each having different preferences and support requirements that often being changed during the trip. With the help of ontology, our system can help tourists better understand and guide them to specify their needs and preferences collaboratively, so that the appropriate information and services resources could be located from the Semantic Web.

Because scalability and flexibility, tourists cannot be flexibly assisted in a centralized manner. The assistance of increasingly powerful mobile devices becomes the enabling technologies. Under individual's instructions and preferences, intelligent software agents within the system can be delegated to help recommend, plan, and negotiate personalized activities and schedules, thereby augmenting the user's decisions collaboratively. We introduce a scalable, flexible, intelligent, and multi-agent "*Mobile tourist guide*" system, with agent clusters for tourist service coordination and integration. Each agent cluster comprises several types of agents to achieve the goals of the major tasks of a tourist's trip, such as, information gathering, preference matchmaking, planning, service brokering, commuting, and mobile servicing. The agents also make use of ontology from the Semantic Web to search information and make recommendations to the tourists. Further, we detail how this can be effectively implemented with Web service and Semantic Web technologies, integrating disparate Internet tourist resources

The research goal is to analysis, design, and implement the "*tourist guide*" system, that access through wireless devices (such as mobile telephones, Personal Digital Assistants - PDAs, and laptops). The following are some services and functionality provides by the proposed research study:

- The system provides general information of hotel, restaurant, shops, hospitals, and companies. As well as the newest events of the plaza and shops.
- The system provides service of hotel, restaurant and cinema-ticket reservations.
- The system provides service of displaying the shortest path between the sources and destinations the visitors specify.
- The system provides user-friendly administration service. The Admin can manage the position, blocking path details, general information of shops and plaza, and reservation details via web browser without changing the framework of the system.
- The system tackles the difficulties of displaying the routes on different floors and many problems in mobile web development.

II. System Overview

The Semantic Web as the next generation web is the vision of having background knowledge about the meaning of web sources stored in a machine-processable and – interpretable way. The area of tourism is highly dynamic area that currently already extensively uses the available Internet technologies. However, the shortcomings of the existing technology are that information finding and extraction as well as the interpretation of the information contained in the web sources is left to the human user. The management and interoperation of semantically diverse tourism information systems are facilitated by Semantic Web technology that provides methods and standards, which allow accurate access to information as well as flexibility to comply with needs of tourism information system users and administrators. In this research project we introduce a new Semantic Web framework that can enable knowledge sharing and reuse. The Semantic Web uses agent technologies to formally model information represented in web resources. This makes it accessible to humans and computers working together, perhaps in conjunction with intelligent network services such as search agents.

The "Mobil tourist guide" system represents a relatively new trend in the field of tourism and involves the use of Semantic Web technology and mobile devices as electronic tourist guides. While much of the underlying technology is already available, there are still open challenges with respect to design, usability, portability, functionality, and implementation aspects. This research study presents the analysis, design and implementation of the "Mobil tourist guide" system, that allows the users to download these personalized applications either directly to their mobile device or first to a PC and then to a mobile terminal (through infrared or Bluetooth). Thereafter, network coverage is not further required as the applications execute in standalone mode and may be updated when the user returns online. The system prototype will use Java 2ME, which offers an ideal platform for the development of full-fledged, interactive and portable applications tailored for resource constrained mobile devices.

In our research study we consider number of important issues related to the Semantic Web that belongs to four categories: *Semantic Web languages*, *ontologies*, *semantic markup of Web pages*, and *Semantic Web services* [2].

Semantic Web Languages: In order to represent information on the Semantic Web and simultaneously make that information both syntactically and semantically interoperable across applications, it is necessary to use specific languages. There are a lot of such languages around, and most of them are based on XML, XML Schemas, RDF, and RDF Schemas, all four developed under the auspices of W3C and using XML syntax.

Ontologies: Ontology comprises a set of knowledge terms, including the vocabulary, the semantic interconnections, and some simple rules of inference and logic for some particular topic. Ontologies applied to the Web are creating the Semantic Web. Ontologies provide the necessary armature around which knowledge bases should be built, and set grounds for developing reusable Web-contents, Web-services, and application. Ontologies facilitate knowledge sharing and reuse, i.e. a common understanding of various contents that reaches across people and applications.

Semantic Markup: Ontologies merely serve to standardize and provide interpretations for Web content, but are not enough to build the Semantic Web. To make Web content machine-understandable, Web pages and documents themselves must contain semantic markup, i.e. annotations which use the terminology that one or more ontologies define and contain pointers to the network of ontologies.

Semantic Web Services: Intelligent, high-level services like information brokers, search agents, information filters, intelligent information integration, and knowledge management, are what the users want from the Semantic Web. They are possible only if a number of ontologies populate the Web, enabling semantic interoperation between the agents and the applications on the Semantic Web, i.e. semantic mappings between terms within the data, which requires content analysis.

A. Semantic Web and Tourism Information Systems

The Semantic Web appears as a promising technology for implementing E-tourism information system. The Semantic Web constitutes an environment in which human and machine agents will communicate on a semantic basis [8]. One of its primary characteristics, viz. shared understanding, is based on ontologies as its key backbone. Ontologies enable the organization of tourism objects around small pieces of semantically annotated tourism objects [9]. Items can be easily organized into customized tourism services and delivered on demand to the user, according to her/his profile and business needs.

The term "Semantic Web" encompasses efforts to build a new WWW architecture that enhances content with formal semantics. That means, content is made suitable for machine consumption, as opposed to content that is only intended for human consumption. This will enable automated agents to reason about Web content, and produce an intelligent response to unforeseen situations.

"*Expressing meaning*" is the main task of the Semantic Web. In order to achieve that objective several layers of representational structures are needed. They are presented in the Figure 1, among which the following layers are the basic ones [8]:

- The XML layer, which represents the structure of data,
- The RDF layer, which represents the meaning of data,
- The Ontology layer, which represents the formal common agreement about meaning of data,
- The Logic layer, which enables intelligent reasoning with meaningful data.

The important property of the Semantic Web architecture (i.e., common-shared-meaning and machine-processable metadata), enabled by a set of suitable agents, establishes a powerful approach to satisfy the E-tourism requirements. The process is based on semantic querying and navigation through tourist services, enabled by the ontological background. The Semantic Web can be exploited as a very suitable platform for implementing an E-tourism information system, because it provides all means for electronic tourism: ontology development, ontology-based annotation of tourism objects, their composition in tourism recourses and active delivery of the tourism services through E-tourism portals.

The main style related to our research in the Semantic Web is the layered architecture. The layered architecture consists of a number of layers organized hierarchically as in Figure 1. Each layer provides a service to its upper layer and serves as a client to the layer below. This kind of architecture depends on the sequence of incremental steps, which means the increasing level of abstraction; if the function of one layer changes, this means that two other layers may be effected.

In the tourism industry, the Semantic Web technologies offer considerable benefits in terms of travel management,

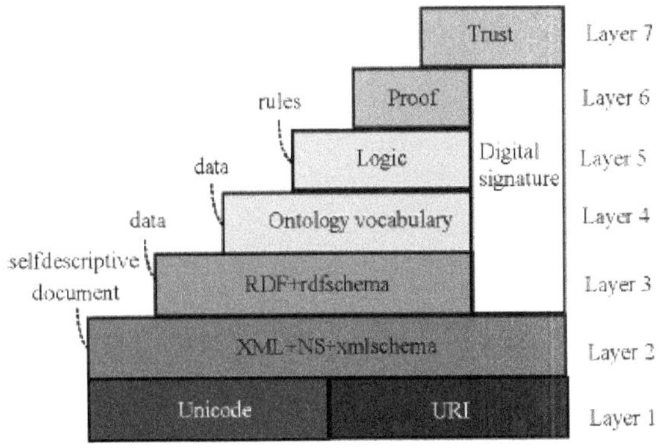

Figure 1. Layers of the Semantic Web

content and document management, knowledge management, supply chain management and integration of distributed applications and services [16,17]. The Semantic Web technologies help in the realization of the vision of the intelligent wireless Web in the following ways: Deeper understanding of the semantics of document content and travel task structure using ontologies, will help the mobile tourist in intelligent information retrieval, extraction and processing, thereby helping him to accomplish elements of a travel plan. Through the introduction of ontological reasoning, Semantic Web techniques are suitable for flexibly discovering abilities in using information that was not specifically designed or intended for a particular use case. Given the dynamic nature of travel processes, this would help in on-the fly resource discovery and integration. Mobile tourists will be able to dynamically locate highly specific data and services on an as-needed basis [18].

Travel enterprises very often undertake their processes in different ways. Differences in the meanings of terms make collaboration difficult. The use of shared ontologies and semantic standards will ensure increased interoperability across devices, platforms and applications [19].

Separation of presentation and data, as ensured by the Semantic Web technologies, will ensure the use of the same middleware tier for both mobile and fixed network clients. The Semantic Web technologies can provide a standardized way to interpreting context, enabling both human and software agents to infer new context knowledge and consequently take intelligent actions.

B. System Architecture

Our basic Semantic Web framework illustrated in Figure 2, provides the tourist with different types of tourism services. In this research study we focus on the implementation of these services that provide the tourist with the following services:

- General information of hotel, restaurant, shops, hospitals, and companies
- The newest events of the plaza and shops.
- Hotel, restaurant and cinema-ticket reservations.
- displaying the shortest path between the sources and destinations the visitors specify, and
- Semantic search.

On the other hand, the system allows the administrators to manage the position, blocking path details, to purplish the general information of shops and plaza, and reservation details via web browser.

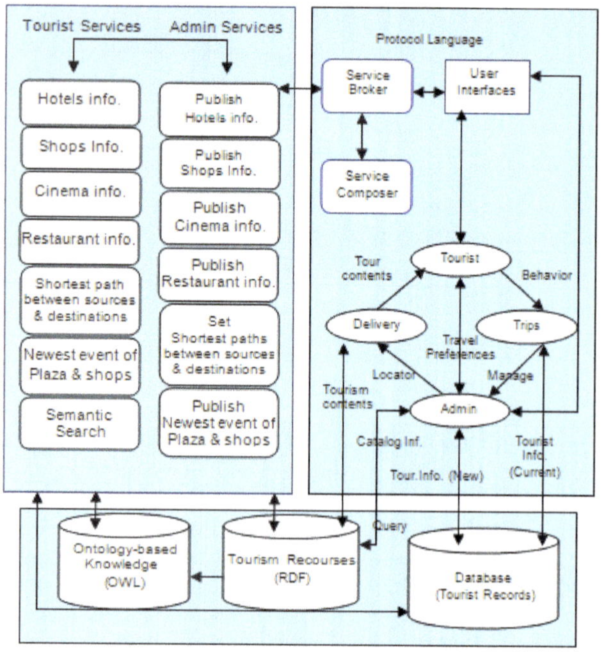

Figure 2. Basic Semantic Web Model for Tourist Guide System

C. *Using Mobile technology for E-Tourism*

M-tourism means the provision of tourism services on wireless devices: portable computers, PDA's, mobile telephones and Tablet PC. Our mobile tourist guide system has functions to enter and to organize the tourism objects in the database and to make them available for tourists and administrators using the E-tourism management information systems (TMIS) portals in which the tourism process and its activities have been structured. The M-tourism object management systems are now compatible for all kind of user devices, the conventional desktop PC as well as the mobile devices laptop, PDA and mobile telephone.

- **The advantages:** The advantages of mobile tourism refer mostly to the mobile user: a great flexibility, an improved tourist schedule is possible, increased productivity during dead moments and just in time tourism.
- **Difficulties and limitations:** Nevertheless they have very small screens, limited memory capacity and the large diversity of mobile devices obstruct a good tourism experience. In the M-tourism field we notice that the tourism services must answer to specific conditions. We will have to get around the technical restrictions so we can create a good tourism services. The usage of video, audio, clear interfaces and divided tourism services must contribute to this. Furthermore we will have to adapt the content to the needs of the mobile user. Because the user has a fragmentized time schedule, we will have to be sure that the tourism objects are not to long. Dividing the knowledge in smaller modules offers a solution.

III. Analysis and Design Methods

We build first the general architecture of the system, set the graphical user interface and design the structure of the database that stores the required data and information. In order to achieve our goals we defined the following tasks:

A. *System Analysis*

The analysis phase is the main phase in which the assessment system requirements are identified in more details. The goal of this phase in the system development is to represent the system goals into defined functions and operations of the intended applications. The tourist guide system requirements are documented in a complementary set of artifacts (flowcharts, use case diagrams, system sequence diagrams…). Each artifact provides a different perspective of the system under design and contains distinct requirements. The combination of these perspectives establishes the tasks that are to be accomplished by the system.

- **Information Gathering Techniques:** By assisting of some specialist employees from different tourism companies, all the system services and materials are checked. The following are some information gathering techniques that used in the tourist guide system analysis.
  - *Interviews*: The most common technique for gathering requirements is to sit down with the clients (tourists, tourism companies, and administrators) and ask them what they need. Our discussion would help us plan out ahead of time, based on the type of requirements that we are looking for.
  - *Prototyping*: We use this technique for gathering requirements. In this approach, we gather preliminary data and requirements that we use to build the initial version of the system.
  - *UML* (Universal Modeling Language) use-cases is used also in the analysis of the tourist guide system as a model tool, in which the functional requirements are extracted and described along with a scenario of the flow of events.

The system requirements are classified in two types: *functional* and *non-functional* requirements.

- **Functional requirements:** The tourist guide system allows different types of users to interact with the system according to given privileges. There are three main users of the system: *administrator*, *tourism company*, and *tourist*, as shown in Figure 3. The system provides the following applications (services) to each user:

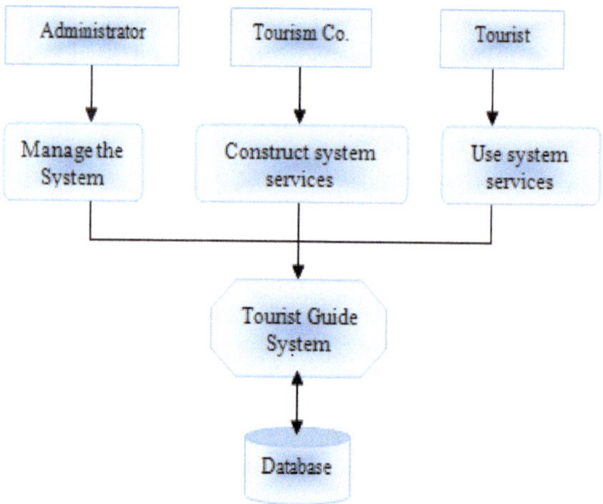

Figure 3. The main users of the system

1. *The administrator application*: This application provides the tourism administrator with simple tools to manage the information of the tourist guide system. The following are some features provided by the administrator application:
   - Activate/inactivate the system.
   - Import and export the system' database
   - Send automatic email immediately containing the new password whenever the user changes his/her password.
   - Administer the basic data and information of the system: hotels, restaurant, shops, hospitals, companies, and etc.
2. *The tourism company application*: This application provides the tourism companies with various tools to construct the system services. The following are some features provided by the tourism company application:
   - Change account password
   - Manage (add, update, delete and display) all the tourist guide system services.
3. *The tourist application*: This application provides the tourist with interactive tools (web-based and mobile-based) to use the system services. The following are some features provided by the tourist application:
   - Create (and change) account password
   - Search for information of hotel, restaurant, shops, hospitals, and companies.
   - Search for the newest events of the plaza and shops.
   - Hotel, restaurant and cinema-ticket reservations.
   - displaying the shortest path between the sources and destinations the visitors specify.
   - Access the tourist guide system through wireless devices (such as, Mobile phone, PDA and Tablet PC) and take quick tour.

- **Non-functional requirements:** In this part, we have verified and proved the system' aspects and interfaces that are not directly related to the functional behavior, such as: *usability, reliability, supportability, performance, and security*:
  - **Usability:** Is the system providing a coherent and flexible user interface that can be used easily with all users? Has the system an obvious component which allows the user to navigate easily? Are the system applications easy to use by tourism companies and tourists? Is the system adaptive to user needs? Is the tracking component working properly and produce the expected results
  - **Supportability**: Can the system be easily modified or extended easily? Can tourism companies and administrators add materials easily?
  - **Contents and Design**: Are the design and contents of the system satisfactory to tourists, tourism companies, and system administrators? What are their feedbacks? Does the system cover well the topics? Are they useful and meet their goals? Quality of texts, images, sounds, and clips? We will compare the progress of tourists.
  - **Performance**: Are the performance requirement concerned with quantifiable attributes of the system, such as response time, throughput availability and accuracy?
  - **Flexibility**: Is the system flexible, since it will be used regularly with a new set of initial data and can accommodate changes in database such as modifying the contents?
  - **Security**: Is the system preventing unauthorized users to access the system?

- **Hardware and software resource:** We have used some software during our first prototype of the system development such as: Windows XP professional, Windows Server O/S, Microsoft office Excel, Microsoft Word, SQL Server database system, MS Visual Studio.Net, and Photo Shop.

  To extend our research work, by using Semantic Web and Mobile technologies, we used other new languages and tools including: XML- eXtensible Markup Language, XML Schemas, RDF- Resource Definition Framework, RDF Schemas, OWL- Web Ontology Language, and Java 2ME.

B. System Design

Any successful web-based system should have an easy to use graphical user interface (GUI) that facilitates its use and flexibility. Thus the proposed system should possess an attractive GUI that is suitable to the potential users. The interface should be tested by its users and updated whenever necessary.

The tourist guide system interface is entirely web-based, and doesn't necessitate any technical skills from the potential users. The system is appropriate for web-based search, with access to the Internet, or mobile-based search using the wireless devices.

- **Database design:** Database is an important component in our system. We have used Microsoft SQL Server database as a database management system (DBMS). The DBMS stores the following entities:
  - Tourist (name, id, nationality, tel, email, …)
  - accompany (name, id, relation, …)
  - Hotel (name, tel, add, …)
  - Hospital (name, tel, add, …)
  - Shops (name, tel, email, add, …)
  - Restaurants (name, tel, email, add, …)
  - Companies (name, tel, email, add, …)
  - Conferences (name, type, location, start_date, end-date, …)
  - Services (Sno, type,…)
  - Travel_agent (id, name, tel, fax, email, add, …)
  - Tours (location, arrive_time, pickup_time, location, …)
  - Admin (id, name, tel, email, …).

We have chosen the entity-relationship model to represent the database design. As shown in Figure 4, we present the initial entity-relationship diagram of the tourist guide system, that contains the main entities (tables) used for creating the system database.

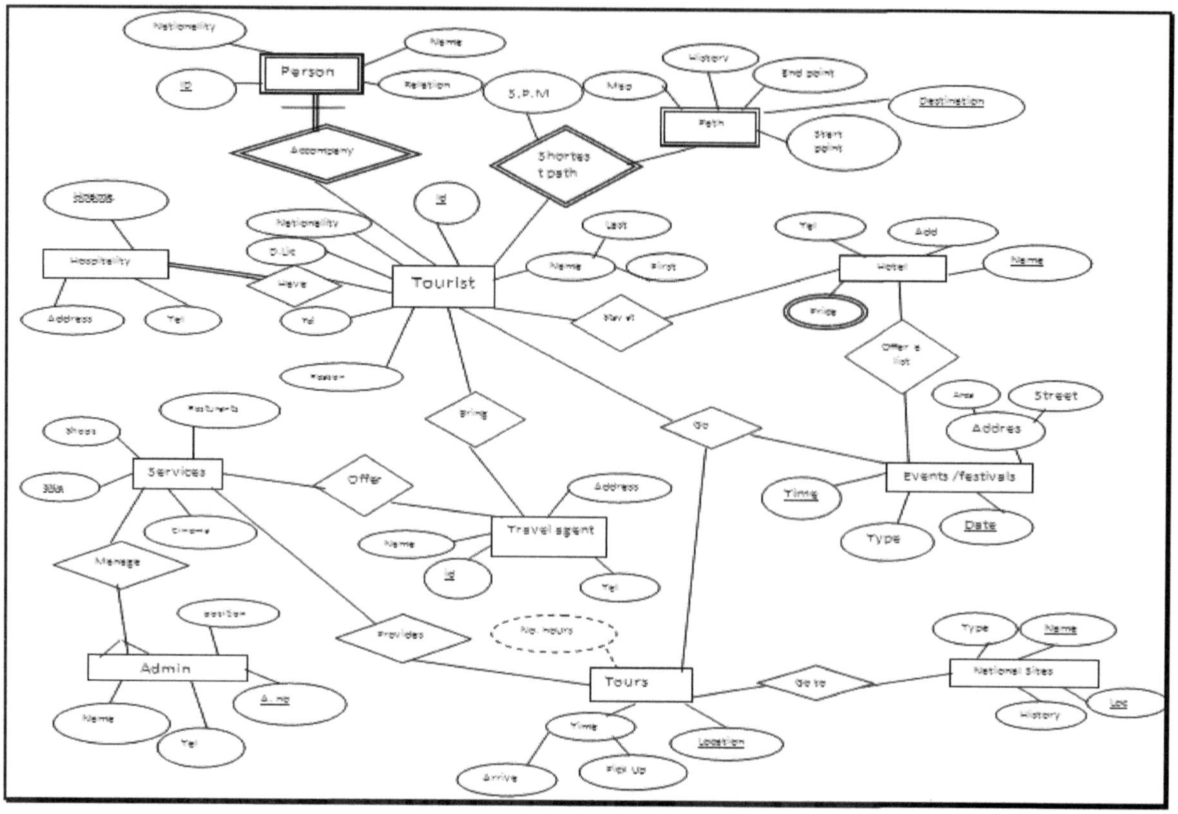

Figure 4. Basic entity relationship diagram

A. Advantages of the System

The proposed research study is of considerable significance due to the following facts:

- The new generation of the Web, the so-called Semantic Web, appears as a promising technology for implementing E-tourism systems.
- E-tourism and M-tourism (Electronic and Mobile tourism) research areas can benefit from Semantic Web technologies. The Semantic Web technology has enabled by a set of suitable agents, which seems to be powerful enough to satisfy the E-tourism requirements fast, and just-in-time.
- The Semantic Web has opened new horizons for internet applications in general and for E-tourism in particular.
- It is extremely important for Tourism Authority to improve the tourism services, thus promoting significant changes among tourism institutions.
- The study is significant to the researcher who will be working on open, hot and recent research areas of the next generation Web.
- The study is significant to researchers who will be working on building Web-based system using new techniques (such as; Semantic Web, and Mobile technology) and new languages and tools (such as; *XML*-eXtensible Markup Language, *XML Schemas*, *RDF*-Resource Definition Framework, *RDF Schemas*, *OWL*- Web Ontology Language, and Java 2ME).
- The study allows the (new) employees and workers to use his/her wireless devices (PDA, Mobile telephone, and Smart phones) anytime and anywhere to know the places that they are looking for easily and quickly.
- Tourists, especially first time tourists, will feel comfortable. This is because all what they need to know about national sights; restaurants and other interesting places will be in front of them by pressing a few buttons.
- Tourists will be able to save time by knowing the information they need without having to call or spend time on the internet searching.
- First time drivers, or drivers who came from abroad will be able to find their way quickly and smoothly.
- The hotels, hospitals, companies and other important sectors will not receive unnecessary calls like calls that ask about the building's site or departments' phone numbers, etc. This is because this kind of information will be available on the proposed system.
- The system guides the researchers and guests that came for conferences to spend an easy tour in the country.
- The system guides the visitors that came from outside for festivals and Eid period, for places where celebration is, without wasting their time.

IV. Conclusion and Future Work

In this paper, we have presented the architectural details and design for "Mobile tourist guide" system, that used Semantic Web technologies for effective organization of information resources and service processes. The proposed system allows the users to: display general information of hotels, restaurant, shops, hospitals, companies, etc; display the shortest path between the sources and destinations the visitors specify; the newest events of the plaza and shops; provide service of hotel, restaurant and cinema-ticket reservations; and allow the administrators to manage the position, blocking path details, general information of hotel, restaurant, shops and plaza, and reservation details via web browser without changing the framework of the system. The system prototype has been developed on the top of Java 2 Micro Edition which offers an ideal platform for the development of full-fledged, interactive and portable applications tailored for resource constrained mobile devices. The paper presented our vision of the Semantic Web and its potential benefits to tourism information systems. In tourism much data is only loosely structured or given in texts – a mode that is too weak for rich querying. The Semantic Web provides additional mechanisms adequate for dealing with these structures, enabling new applications for commerce and communications, especially. We have proposed a Semantic web framework that supports multiple platforms (in particular wireless mobile ones) and a methodology for the analysis and design of Mobile tourist guide system.

As for future research, our key focus is about thr implementation of "Mobile tourist guide" ontologies. Despite the considerable efforts toward the construction our system ontologies for agent interoperation, some difficulties and challenges are still open. Throughout the paper, we assume ontologies to be consistent and available to the agents. In reality, many ontologies are redundant in the sense that they present the same domain with little or no interoperation between different ontologies. The ontologies play a main role in the system implementation, as they provide a shared representation of the domain and of the concept that the agents need to use.

REFERENCES

[1] Egypt Authority Web site.

[2] Fayed Ghaleb, Sameh Daoud, Ahmad Hasna, Jihad Jaam, Samir A. El-Seoud, and Hosam El-Sofany, "E-Learning Model Based On Semantic Web Technology". International Journal of Computing & Information Sciences (IJCIS), Vol. 4, No. 2, P63-71, August 2006.

[3] Berners-Lee, T., "Semantic Web on XML," Dec.2000.

[4] C Cody Burleson. Introduction to the Semantic Web Vision and Technologies.

[5] Matthews, B.,"Semantic Web Technologies. JISC Technology and Standards Watch.

[6] Davis, M., Moving to Unicode 5.1.

[7] Berners-Lee, T., Uniform Resource Identifiers, URI Generic Syntax. IETF.

[8] Patel-Schneider, P.F. and D. Fensel. Layering the Semantic Web: Problems and Directions. In the Proceeding of 1st International Semantic Web Conference (ISWC 2002). Sardinia, Italy, 9-12 June, pp: 476, 2002.

[9] Buraga, S. and G. Ciobanu. A RDF- based model for expressing spatio-temporal relation between web sites. In The 3rd International Conference on Information Systems Engineering. IEEE Computer Society. pp: 355. IEEE Computer Society Washington, DC, USA, 2002.

[10] Miller, E., "An Introduction to the Resource Description Framework", in D-Lib Magazine, May.1998.

[11] Nejd, W., D. Olmedillal and M. Winslett. Peer Trust: Automated Trust Negotiation for Peers on the Semantic Web. Lecture Notes in Computer Science: Secure Data Management. Springer Berlin/Heidelberg, vol.3178/2004. pp: 118-132., 2004.

[12] Matthew Richardson , Rakesh Agrawal and Pedro Domingos, et al. Trust management for the semantic web. Lecture Notes Comput. Sci., 2870: 351-368, 2003.

[13] Brian Matthews and Theo Dimitrakos., "Deploying Trust Policies on the Semantic Web.

[14] Russell Cloran and Barry Irwin. XML Digital Signature and RDF.

[15] Mark Bartel, John Boyer, Barb Fox and Brain La Macchia, et al. XML Signature Syntax and Processing.

[16] Kanellopoulos, D., Kotsiantis, S., Pintelas P. Intelligent Knowledge Management for the Travel Domain. GESTS International Transactions on Computer Science and Engineering, Vol. 30, No.1, pp. 95-106, 2006.

[17] Dimitris K. and Sotiris K., "Towards Intelligent Wireless Web Services for Tourism". IJCSNS International Journal of Computer Science and Network Security, VOL.6 No.7B, July 2006.

[18] Lassila, O., Adler M. Semantic gadgets: ubiquitous computing meets the semantic web. In: D. Fensel, J.A. Hendler, H. Lieberman, W. Wahlster (Eds.) Spinning the Semantic Web: Bringing the World Wide Web to Its Full Potential, MIT Press, 2003, pp. 363– 376, 2003.

[19] Missikoff, M., Werthner, H. Höpken, W., Dell'Ebra, M., Fodor, O. Formica, A., Francesco, T. HARMONISE: Towards Interoperability in the Tourism Domain. Proceedings of Information and Communication Technologies in Tourism 2003, ENTER 2003, pp. 58-66, Helsinki: Springer, 2003.

AUTHORS

Hosam F. El-Sofany is with Cairo Higher Institute for Engineering, Computer Science, and Management, Cairo, Egypt

Samir A. El-Seoud is with Technische Universität Darmstadt (TU Darmstadt), Darmstadt, Germany.

Digital Social Media: An Interactive Technology Incorporated as a Competitive Advantage for Business

P. A. Correia Pereira[1] and I. García Medina[2]
[1] University of Vic, Barcelona, Spain
[2] Glasgow Caledonian University, Glasgow, UK

Abstract—**In a more transparent and dynamic world, in which consumers trust other consumers more for advice and recommendations on products and services, the continuity of organizations appears to be associated with socialization, the sharing of interests and the interaction with the audience. This is associated with the incorporation of digital technologies to business, specifically the use of social media.**

Consequently, it is timely and interesting to explore the phenomenon of virtual socialization, although it is a little-studied field and what is needed is an innovative and theoretical approach based upon theories of marketing and communication. Expertise in these areas is present in all organizations and their performance is important for appropriate development of them.

This work is a qualitative analysis about the behavior, reactions and attitudes of individuals to organizations, in order to understand the social factors that contribute to sustainable competitive advantages of organizations which can support strategic and future actions.

We conclude that relevant factors associated with the tacit knowledge of the organization, specifically to learning and social interaction of the organization and their knowledge of virtual communities. The higher the coexistence of factors, the more difficult is the replication and greater will be the hypothesis of sustainable competitive advantage.

Index Terms- **Social Networks, Digital Communication, eMarketing.**

I. THE IMPORTANCE OF SOCIAL MEDIA FOR DIGITAL MARKETING COMMUNICATION

The reinvention of society and the modern organizations due to the digital influences have been called by several authors with different designations like Wikinomics [21], said:, Groundswell [11], Tribes [6], among others. Call to the whole movement mass collaborations, social networks, social networking or social media, and are mainly characterized by the speed, transparency, collaboration, trust, ubiquitous, and socially. The idea is to bring people together in a single environment, a large brain with direct connections (producer-consumer).

Social media have revolutionized the way people communicate and share information between each other. The consolidation of social media on a daily basis is increasingly intense, although some cases have been highlighted, in particular Facebook.

According Reference [7], "The advantage of the competitiveness comes from a technological introduction and the constant adaptation along the time to the technological evolution inserted in the strategy of the organization."

But technology alone can hardly provide a sustainable competitive advantage to business, the value comes from the sphere of innovation and business structural changes (new business models, process changes,...) and the product / service [14].

Reference [17] said: "the phenomenon has changed social behaviors and captivated new users, especially in female and senior citizens, who did not use information technologies as much before pre social media when users were overwhelmingly male." With the emergence of social networks these audiences have developed a greater affinity with digital technologies, having a regular and dynamic coexistence in the virtual world, occupying a prominent place in virtual social interactivity. The participation on a massive scale of thousands of people throughout the world to the phenomenon of social media has contributed to the exponential growth of the various social networks, allowing them to achieve the critical mass necessary for the social media to become an appreciated channel of communication for many organizations.

The use of the online phenomenon by organizations involves developing a communication program of integrated marketing, maximizing the individual and collective contribution of all activities of communication, defining for the virtual channels a more personal program of communication in the field of direct and interactive marketing [19].

This will make it possible to perceive the role of the communicational marketing in the creation of content and interaction with customers in the various social platforms such as Facebook, Twitter, Linkedin, Youtube, MySpace, Orkut, blogs and other social networks [19].

Reference [22] state that digital social networks are the amplifier of internal capabilities of the organization at the level of behavior and technology. To use "outsourcing" for the management of social networks is a strategy ready to fail because it transposes to third parties what should be the essence of the organization in the creation of competitive advantages to perceive and relate to its target audience. Embarking on interactive marketing in social networks, is not just one more option for communication with customers by copying the actions of the competition. It consists of a deep change in the behavior in the organization. Only the experience and the corresponding temporal allocation will lead to interactions of a high level that allow the individual consumer to create unique experiences with the company and it will become the key to new sources of competitiveness.

An approach to transparency and trust concentrates the whole organization, and is an integral part of the activities of management, in particular in the perception of the internal and external environment, the vision and mission, defining and implementing strategies and monitoring. The growing importance of social media as a tool of marketing-mix, has encouraged the expansion of similar technologies, new and powerful on-line applications are to be launched especially in the mobile channel, which will be the main form of access to social networks in the near future, mobility will be driven by devices such as smartphone, e- readers and tablets.

According reference [17]: "the ubiquity of the technology, inherent characteristic of the new tools, the communication between devices, connectivity and the convergence of technologies of information associated with georeferenced location, will contribute to a most enriching experience of the user."

Reference [3] state that "the visibility and convergence are moving to a new era of greater transparency and commitment in the relationships individual - individual and individual – organization". However, the majority of them are still linked to the ranking of their profiles, "how many friends I have in my profile", "who said I like? ", as if that resembled the popularity or fame, how important is this information that is worth as much as the hits of the site in the process of compromising between both sides.

II. SOCIAL MEDIA MARKETING COMMUNICATION STRATEGIES

While the practice of management may not be evolving as fast as it once did, the environment that faces 21st century businesses is more volatile than ever. This new century may still be young, but it has already spawned a sizable brood of daunting managers' challenges that are markedly different from the ones that taxes our forebears; the new reality calls for new organizational and managerial capabilities. To thrive in an increasingly disruptive world, companies must become as strategically adaptable as they are operationally efficient. To safeguard their margins, they must become gushers of rule-breaking innovation [7].

The strategic basis for social media is centered in real relationship with people and on the need to eliminate the noise created by the rankings, find out what is important in long-term relationships between the organization and its fans and focus on genuine relationships, honest, lasting and value to both understand who is interested in maintaining a relationship with us and is truly on and react to our presence and actions.

Important contacts are those with whom we have a relationship of proximity and interactivity that allows the exchange of views and discussions on common interests, people who actively participate with their opinion on the subject, people who talk about winter sports because they have passion, compliment or criticize you because they live in this community. People are connected by real things and matters of mutual interest rather than rankings.

The difference is about people feeling essential to the relationship. Consequently, "networking" is important when there is a real link for everyone and is a useless distraction when the real connection does not exist, although we are constantly seduced by the image of useless popularity based on the numbers, it is important to know who follows my path and what I would follow them and go after these people and get the privilege of one day be able to gain value from this relationship [19].

Therefore, marketing in social media is not "speak loud" ads to people, but to lead, to do something that captivate, connect people who are interested in a connection with a mutual order. People want to follow someone who is genuine and authentic.

For reference [11] the strategic plan of social media establishes four procedural steps: people (who are the customers and for what they are prepared?), goals (interact in the media in the marketing area or sales grow, internal participation), strategy (what is the relationship with customers? use customers to spread the company message in the market?), technology (what are the applications to use, blogs, wikis, social networking?).

The objectives of the strategic plan are to enable the consumer to identify the existence of the company (company and products / services), differentiate from the competition, being a reference in the labor market and have the preference of consumers (be valuable throughout time), [19].

Create and implement a social strategy is difficult, mainly because there are few precedents and models to follow, so it is necessary to be constantly aware of the challenges and prepare to solve them [11]. Without a comparative basis, experimental models are predominant; however, make multiple errors in the field of social media is by far desirable than be inactive, is the world of the social economy [17], [21].

Certainly, the implementation of a strategy by an organization for social media will drastically alter the way business is conducted [1]. This is the adaptation to a new world, where the influence of relationships and search engine optimization are important and news are spread to all the word very quickly [19].

III. SOCIAL MEDIA MANAGEMENT – THE ROLE OF COMMUNITY MANAGER

The bidirectional communication management in online communities and the marketing activities of the organization are maximized by allocating specific resources to these functions, however, the allocation of responsibilities for managing social media to the Community Manager, is an issue that is not consensual and that is being debated, so there are different views and positions and various denominations. Its role is to build, manage and grow online communities around a brand.

In summary the role of the Community Manager resembles to a television debate mediator which manages the speaking time of the participants, the issues under discussion and the argument avoiding conflicts. In social networks this role is more comprehensive, it is assumed to be an advocate for community members, sponsor / supporter of the brand, making sure that the discussion is productive and positive and avoiding negative / prejudice discussions.

The Community Manager provides customer support, responding to questions and doubts, monitors the web to find out what is being said about the company and its products / services and acts as an ambassador for the company.

Following this perspective, the manager must try to understand the participants in online communities, how they use social media, how to look for information about products or topics of interest and answers to their questions. The frequent social networking monitoring allow the Community Manager to observe these attitudes, perceived regularity, with the intention of predict personalities and behaviors of the participants in the social network about the company, its products and services and lead members community in the right direction, keeping a positive conversation about a common interest, so as to ensure that members have an interaction valued by both.

Therefore, the Community Manager must ensure bidirectional communication, ensuring that members provide feedback while the brand keeps them updated, ensures that both members and managers learn the most of each other, and that they pay attention to their role in the company, its responsibilities and objectives and be connected to the various departments of the company, especially to marketing, to customer service and administration. The definition is not accurate, due to the multiple tasks involved, sometimes be defending the client but cannot deny their brand loyalty [13].

IV. BE SUSTAINABLE

The effective use of social media can be a source of competitive advantage for organizations, difficult to replicate by others, due to the connection architecture to the company itself, occupying a strategic role at the level of differentiation or cost, instead of a merely presence role. Answering to what have company to offer that is unique and valuable to current and potential customers?

The strategic sustainability of organizations for social networks lies in an ongoing analysis of the behavior of individuals and their needs and desires.

The suitability of products and services to the needs and desires of consumers are a constant challenge for any organization. The current media ecosystem transformations are aspects that should be analyzed from the synchronization perspective between the organization and consumers.

To obtain a competitive advantage, it is necessary to carry out business activities better than competitors, then, the competitive advantage is synonymous of a superior performance. The differences in how firms choose or execute strategic activities are the basis of competitive advantage [2].

A company gets a sustainable competitive advantage when implementing a major strategy followed and implemented by few opponents simultaneously, and they have difficulties in acquiring the necessary resources to implement it. A company gets a temporary competitive advantage when implementing a major strategy, followed by some opponents, and they do not have significant difficulties in acquiring the necessary resources to implement that strategy. A company faces constant competition, when implementing an important strategy that is simultaneously implemented by several opponents. A company is in a competitive disadvantage when implementing a strategy that is not important, so it does not reduce costs or increase profits [12].

V. UNDERSTAND THE LOCAL SOCIAL MEDIA MOVEMENT

Although we live in a worldwide economy it is interesting to know how local economies are adjusting to the global social media phenomenon.

The overall studies of marketing communication on the subject of social media are relatively recent, but this does not mean that are not relevant, as Kelsey said: “Internet is a really powerful and inexpensive channel for communication, distribution and sales”. Also reference [18] explains that before the web, companies used the media only for promotion, but with the Internet, companies use the media to develop relationships with customers.

The problem is that it is necessary to know how to use digital social media and how to manage the customer relationship in order to take advantage of them and before to arrive to this stage it is necessary to know the publics.

During the four months period between October 2011 and January 2012, 150 individuals and 15 companies were analyzed in Funchal and Lisbon (Portugal). To have the perception of knowledge degree about social networks in particular on Facebook and why people participate in the network and adhere to brand pages.

Understand the strategic vision of those responsible for company´s marketing and communication about Facebook. Analyze the role of Facebook in marketing activities and interactive communication (users and companies).The contribution of Facebook for marketing and communication actions. (Competitive advantages and contribution to brand loyalty).Which in this case, are related with the integration of social networking on marketing communication to obtain sustainable competitive advantages.

Although the evolution of information technology is a catalyst for a more intense online social experience, it is important to understand how to live the virtual experience and how evolve the communication between individuals and companies (how to adapt to this new consumer audience), face the current short term models based on offline actions, reactive strategic actions, miss perception of users, lack of information on social network consumers’ life.

Despite the recognized importance of social media for marketers, few companies are prepared to manage this change.

In most cases there is not a social network integration in the company activities in all departments and workers, as well as viral marketing planned actions, models and platforms for analysis and monitoring of companies participation in the network (except the monitoring actions allowed by Facebook), which enable to measure the impact of social media strategies in business, creating new assets as "crowdsourcing" (to use volunteers and collective knowledge of the Internet to solve problems, create content and solutions).

Today the focus continues to be very connected to sales and promotions and to traditional communication channels when it should move to create interactions with meaning for the audience through content before focusing on sales.

VI. CONSIDERATIONS AND FINDINGS

The reinvention of societies and organizations is shown to be connected to the social essence of human motivation and the evolution of the specie, which is a reflection of the agglomeration of individual behavior in society.

Meeting the needs of social connection has allowed technological tools related to social networks to have a very high degree of adhesion in a short period of time.

Social networks are a new marketing tool that are more specific than digital marketing and which go beyond communication with a philosophy shared by all organizations in their actions and reactions, which connect all departments and employees.

Understanding which factors are relevant for the communication between companies and consumers in the social networking phenomenon is essential to define the company future.

The main reason for participation in social networking is social communication but there are other interests, related with the companies, their products and services. Users become "fans" of theirs pages, read, comment and interact, it further points out with "likes" the published content, and have a predilection for frequent communication. For users, the company's presence in the social network is an important factor in reputation.

The use of social networks as a means of communication to integrate multiple communication tools are confirmed by individual users but only a minority of companies representatives users feel that Facebook integrates with other tools.

The emergence of social networks and the concentration of various communication tools into a single platform have captivated individual users and have revolutionized the way people communicate and share information. However, this has not been identified as a purely technological change but above all as a cultural one. It is a process of social transformation where technology is an inseparable element of social and cultural trends. For businesses, the adoption of social networks has not been as significant as it still has a traditional view of business involvement in social networks. They see it as a push communication in detriment of the interactivity that this channel requires its participants.

The strategic vision of those responsible for marketing and communication on the social network Facebook is a new way of communication is enforced through the real-time electronic dialogue. Companies consider social media as a key element of their marketing strategy and not just as a channel of their traditional communications strategies (although there are some differences in the social network participation which are related with the company size). In addition, sharing information interactively with clients could hardly be done through traditional channels. This bidirectional communication management is a key success factor of this type of communication strategies. Before the advent of social networks, organizational communication was traditionally characterized as unidirectional, the company transmitted the message and the audience was limited to hear it without having the option to respond, allowing the organization to control the message.

The bidirectionality between the company and fans is also a way to segment marketing activities and communication, a source of information about needs and desires. The fact that the company maintains a stable relationship with the "fans" results in increased brand loyalty.

The evidence continues to show us the fact that companies will pass through the biggest change of their short history. The models of network collaboration and innovation can provide to managers with new possibilities to arouse the human potential, whose success is in challenging the knowledge / experience that companies accumulated over several years of thought.

In recent years, the practice of marketing an expansion of his coverage has been verified, thus amplifying the transaction connected with the interactive relations. The basic objective of trade relations goes beyond the transaction also includes the development of stable and lasting relationships with users, which are beneficial to both parties, leading to increased confidence and consequently the willingness to help each other, reflecting directly in reducing the costs and the realization time of transactions.

Organizational survival increasingly depends upon socialization, sharing interests and activities with the audience linked to the incorporation of digital technologies in their activities, especially those related to social networks. Technology emerges as a support for the satisfaction of social connection, transforming communication between people and companies, making it much more dynamic and transparent.

Facebook stands out for its size and orientation to relationships, within social networks Facebook is dominant. It is an emerging area of convergence of the pre-existing communication media, heterogeneous collaborative cultures and participatory audiences.

However, the adoption of Facebook technology is not sufficient for the integration in the online social movement, as the technology of Facebook and digital marketing themselves do not constitute a competitive advantage, it depends on the ability of the company to look for opportunities in your market, how they integrate them into routines and management capacity of the organization. It implies a strategic approach to monitoring consumer behavior and detection of their needs and desires by the company.

Non-Facebook participants who observe this phenomenon and its view of future participation indicate that most have no intention of using Facebook in the near future because they are not interested. This means that companies have to simultaneously maintain the traditional channels because there is a part of the public/consumer still remains outside of social networks.

As we have seen, there are many positive factors associated with the participation in the social network. Interactivity, the detection of customer needs, adjustment of supply, transmission of content without geographical boundaries and ease of implementation of viral marketing campaigns are the most import of these factors.

Concerning the less positive factors of social network participation, we could mention the importance of the lack of credibility of some information, the excess of information, the lack of response from some brands and delay in responding to customer requests.

Finally, it is important to note that monitoring and learning through experience in the network as part of culture and organizational routines are specific factors that may become a competitive advantage by being relevant factors associated with tacit knowledge of the organization.

VII. CONCLUSIONS

The physical and digital worlds are becoming increasingly linked, socialization and human relations decreases their physical dependence on connecting people around the world from different ethnicities and beliefs, a result of increased connectivity and convergence of information and mobile technologies, allowing to be online regardless of geographic location and therefore to act and to have access to information on social networks in real time.

Many companies start realizing that the number of fans on their pages is not synonymous with success in social media, the value is not the number of fans in their communities but the compromise between people and the company increased by social networks. Consumers are no longer passive, they are taking an active role in organizational communication, so organizations need to adopt a culture of inter-relational customers, based on trust and transparency, the numbers will come as a result of this work, but if companies have no idea of this relationship, they do not know what they are doing in social networks.

There is a progression in the development of standards of transparency on-line. Consumers will manage and share their data with trading partners, making it a viable business model.

The major social players (Amazon, Google, Facebook) accumulate individualized information about each of their users in four groups of variables of marketing segmentation: demographic, geographic, psychographic and behavioral. In order to optimize the selection and provision of information customized to the individual profile, this customization can be extended to other areas of business and other businesses cut superfluous information received by consumers, which increase the efficiency and effectiveness of traditional marketing to social media.

New applications for the various social and mobile platforms will tend to a high customization, increasing the link between organizations and customers, this will provide increased communication effectiveness of on-line advertising with a social and interactive nature of the online marketing, viral and "worth of mouth" with a growing consumer participation in organizational communication. Obviously, collaborators need to get used to working in a wider and more dynamic community.

REFERENCES

[1] Brown, E., (2010). Working the crowd, social media marketing for business. British Informatics Society Limited. Swindon.

[2] Davis, L., Dehning, B. &Stratopoulos, T., (2003). Does the market recognize IT-enabled competitive advantage? Information & Management 40, 705-716.

[3] Dunay, P & Krueger, R. (2010), Facebook marketing for dummies, Wiley Publishing, Inc.

[4] Godin, S., (2007). Permission marketing: turning strangers into friends and friends into customers. Pocket Books, Simon & Schuster UK Ltd.

[5] Godin, S., (2009). All marketers are liars: The power of telling authentic stories in a low-trust world. Penguin Group.

[6] Godin, S., (2010). Linchpin: Are you indispensable? How to drive your career and create a remarkable future. Piatkus Books.

[7] Hamel, G., (2007). The future of management. Harvard Business School Press.

[8] Kelsey, T., (2010). Social networking spaces, from Facebook to twitter and everything in between, Apress.

[9] Kotler, P., Kartajaya, H. &Setiawan, I., (2010). Marketing 3.0: from products to customers to the human spirit. John Wiley & Sons.

[10] Kotler, P., Keller, K., Brady, M., Goodman, M. & Hansen, T. (2009). Marketing management. Pearson Education Limited.

[11] Li, C. &Bernoff, J., (2011) Groundswell: Winning in a world transformed by social technologies. Harvard Business School Press.

[12] Mata, F., Fuerst, W. & Barney, J., (1995). Information technology and sustained competitive advantage: A Resource-Based analysis. MIS Quarterly/December, 487-505.

[13] Ng, De., (2011). Online community management for dummies (For Dummies (Computers)). John Wiley & Sons. New Jersey.

[14] Peppard, J. & Ward, J., (2004). Beyond strategic information systems: Towards an IS capability. Journal of Strategic Information Systems, 13, 167–194.

[15] Peppard, J. & Ward, J., (2005). Unlocking sustained business value from IT investments. California Management Review, 48, 52-70.

[16] Peppard, J., Lambert, R. & Edwards, C., (2000).Whose job is it anyway? Organizational information competencies for value creation. Information Systems Journal, 10, 291-322.

[17] Qualman, E., (2009). Socialnomics: How social media transforms the way we live and do business. Wiley Publishing, Inc.

[18] Schubert, P. & Leimstoll, U., (2007). Importance and use of information technology in small and medium-sized companies, Electronic Markets, 17, 38 -55.

[19] Scott, D. (2010). The new rules of marketing and PR. John Wiley & Sons, Inc..

[20] Scott, D. (2011). The new rules of marketing & PR: How to use social media, online video, mobile applications, blogs, news releases, and viral marketing to reach buyers & PR: how to use social media, blogs. John Wiley & Sons; 3rd Edition.

[21] Tapscott, D., & Williams, A., (2008). Wikinomics: How mass collaboration changes everything. Atlantic Books.

[22] Wilson, H., Guinan, P., Parise, S. & Weinberg, B., What's Your Social Media Strategy? USA: Harvard Business Review. Vol.89, n. 7/8, 2011.

AUTHORS

P. A. Correia Pereira, is with the University of Vic, Sagrada Familia, 7, 08500 Barcelona, Spain (e-mail: pacorreia4@gmail.com).

I. García Medina is with the Glasgow Caledonian University, Cowcaddens Rd, Glasgow, Lanarkshire G4 0BA, UK (e-mail: irene.Garcia2@gcu.ac.uk)

13

Mobile Remote Control for Home Automation

M.M. Ramses, E. D. Laura Victoria and A. S. Raúl
University of Colima, Colima, México

Abstract— **Domotic: Existing home automation technologies permit persons to control elements of the home using a computer connected to a local network. However, the real challenge is to remotely control objects, permitting users to exercise greater freedom and versatility. This work presents an inexpensive controller device developed for high density constructions typical of many developing countries throughout the world that is designed to remotely control home luminescence, air conditioner and water pumps through mobile devices such as a smart phone or tablet. Results show the packet delivery was acceptable within a range of 15 meters in a concrete, two-story dwelling, thus demonstrating its applicability in small homes in countries which share the same construction features as those of Mexico.**

Index Terms—**domotic; home; automation; mobile**

I. Introduction

A simple definition of home automation is "a system that automates different things in the house". Home automation, also known as domotics, is a field of study that has generated much interest because of potential benefits related to health, security, comfort and ecology, among others [1].

Today, research is being carried out worldwide, but particularly in Japan, Europe and the United States, because of the potential commercial benefits of products that employ wireless technologies to control systems and remotely monitor dwellings [2].

Mobile devices are handheld devices such as smart phones or tablets that can have touch or non-touch display screens, or even mini keyboards. These 4G compatible devices currently provide sufficient computing capacity, programming and transmission range to make controlling household appliances or functions possible [3].

The current revolution in smart phone and tablet technologies has significantly impacted the lifestyle of people throughout the world and the number of users has increased to over 6.8 billion [4]. A study of over 5,000 people carried out in The United States, The United Kingdom, China, India, South Korea, Indonesia and Brazil shows that 84% of the respondents report that mobile technologies have improved their lives as it allows them to contact family and friends, as well as allows them to carry out business communications and online operations. [5].

Because the initial cost of a smart phone or tablet can be up 650 dollars, the need to provide added value for the consumer market is imperative. This additional value for mobile or tablet users can be provided either through additional applications or by expanding the convenience, range, and use of mobile technologies [6]. One area that is increasingly attractive is home and office automation. The ability to remotely control devices will inevitably grow and evolve as the processing power of mobile devices increases, the size and efficiency of control devices decreases, and the effective transmission range of mobile devices expands.

This work focuses on using of mobile devices inside and outside the home to communicate with the controller developed to control lights, water pumps and air conditioner through smart phones and/or tablets using three different web browsers (Google Chrome, Safari, Opera) with the goal of making these two functions more accessible to the general public and fully functional in high density constructions of at least 30 countries [7].

II. State of the Art

The relevance of traditional cable networks has decreased considerably due to their greater costs and limitations. Traditional cabled networks, which require considerable physical infrastructure, are increasingly being replaced by wireless technologies because they are more flexible and require far less infrastructure [8].

The first communication protocol used in a home automation system was named X10, which was developed by a Scottish company called Pico Electronics in 1975 [9].

The X10 protocol works through household electrical systems, transmitting pulse bursts and managing communication among compatible devices.

A few months later, the first X10 controller/actuators to handle electrical currents were developed to receive instructions they received via communication protocols [10].

Figure 1. X10 Switch Actuator [10].

After creating this protocol, more integrated projects that incorporated these new technologies to control home devices started to appear as the area of home automation began to grow.

One of the projects published using X10 was developed in 2004 entitled "Architectures for Ubiquitous Environments" [11]. This project consists of a central computer used as a server within the local network to receive requests from a PDA device (Figure 2) to turn on or off any device connected to an X10 controller. The major drawback of this technology is that the controller entirely depended on the home's electrical wiring, which in many cases proved to be inadequate.

Figure 2. PDA domotic application [11].

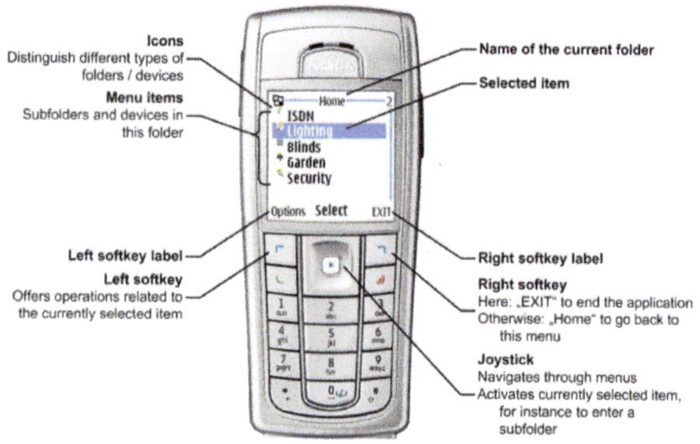

Figure 4. Cell Phone application [15].

Three decades after the first cabled networks were introduced, a series of high-level protocols for low-power wireless communication called Zigbee, designed by Zigbee Alliance, greatly facilitated indoor wireless networking and communications [12]. In 2005 the Xbee wireless radio was introduced to work with Zigbee to provide connectivity between various points [13].

Figure 3. Xbee Wireless RF Module [13].

Once wireless networks for automation environments started to become popular, several proposals were put forth to develop a "Domotic multipurpose platform based on wireless technology with distributed processing capabilities" [14].

These projects consisted of independent controllers that were installed in each room and communicate over the wireless network and the actual monitoring and control was carried out by an interconnected network consisting of cameras, sensors and actuators. This permitted persons to control the home network through a PC, but required persons to be present to actually monitor and control the household.

Mobile networks began to migrate to mobile technologies through; GPRS, which is available in most areas, covered and offers the advantages of packet-oriented data communications and use a cell phone to control the devices (Figure 4) [15].

A project was developed to control different household devices such as television or lamps are described in [16]. This proposal uses cameras installed in strategic positions of the house that provide real-time visual monitoring of the devices to be controlled.

All of the cameras are connected to a central computer that contains the necessary software to recognize the elements that are targeted with a laser. Consequently, line of sight was a vital concern. In this scheme, the computer takes care of sending the command to the actuator to turn on or off the targeted device.

The term smart buildings refers to a relatively new concept in which multiple home appliances communicate with each other using short range wireless communicated technologies such as ZigBee [13] and WiFi [17].

The University of Greenwich performed experiments with these 2 wireless technologies and the results show that WiFI on channel 6 (2426 MHz to 2448 MHz) and ZigBee-Pro on channel 12 (2409 Mhz to 2411 MHz) operate without any serious difficulties [17].

The Hebei University of Science and Technology examined the possibility of employing a WSN and service robots for smart home applications. In this case, the service robots can be considered mobile nodes that provide additional sensor information, improve and repair the connectivity, and collect information from surrounding wireless sensor nodes [18].

Currently, there are several manufactures that wish to enter the field of home automation systems and Belkin is one of the most important companies involved in this field [19].

This company has developed a device that connects to the network and functions as a controller/actuator. Its drawback is that it is limited to lamps and each element operates independently, so the devices cannot work together, thus making it is necessary to work them individually [19].

III. Overview of the System

The presented architecture is designed to be implemented in homes constructed with high-density materials, which commonly share the following features in many parts of the world:

Wall-to-wall construction: There are no lateral spaces between houses as the houses are built with adjoining walls [7].

Interior wall construction: Interior walls are made of 20-cm. thick baked brick wall with between 4 and 6 cm. of cement/sand finish. The actual space for wiring, plumbing, etc. is actually hollowed out of the brick itself. In short, the walls have no air spaces and the materials are much denser than their American or European counterparts.

Ceiling construction: Ceilings are traditionally made of a 10 cm solid concrete slab, although more modern constructions combine concrete and cinder blocks [7].

Due to the thickness and density of the house internal and external walls, the controller/actuator must be at a distance no greater than 15 meters from another controller/actuator or the communication server to be completely secure.

The controller/actuator can turn on and off any water pump, including 220 V pumps [20].

It is possible to turn lights, water pumps and the air conditioner on and off from any part of the world from a computer or mobile device that is connected to the Internet.

If electrical power should be turned off or fail, the server and drivers return to normal operation 7 seconds after the power is restored [21].

Figure 5 shows the proposed scenario where the developed controller was used to control lights, air conditioner and water pumps in Mexican households.

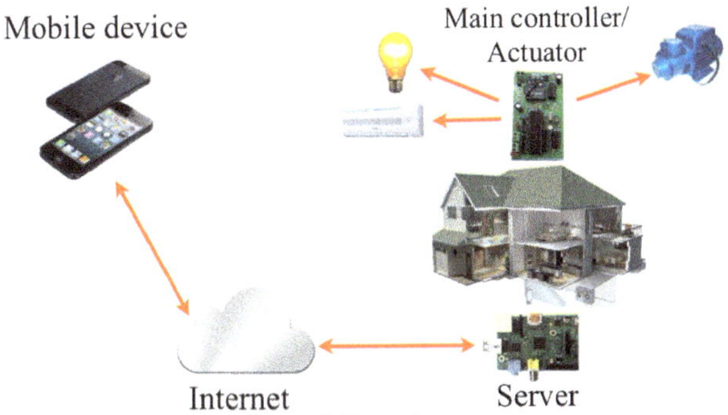

Figure 5. Domotic system.

- Main controller/actuator: Necessary to meet the basic needs of home automation systems and connects to the server via Xbee S1 radius using the IEEE 802.15.4 protocol. This controller acts as an actuator to connect directly to the lamps and water pumps to turn them on or off, depending on the instructions received by the server.
- Server: Essential to provide connectivity between the radios and the Internet. It consists of a Raspberry PI that has a Web server, PHP and MySQL. The server has an Xbee radio that is wirelessly connected with the Main Controller / Actuator. It connects to the Internet via the local network (Ethernet) or can function as access point using a wireless card. This computer was chosen because of its low power (5V, 700mA) and small size [22].
- Internet: Needs to ensure that the server and the mobile device are connected to the Internet so that they can remotely control the functions of turning the lights, air conditioning and water pumps on and off. Importantly, if an Internet connection is not available, persons can still control the automation system within the physical confines of the home.
- Mobile device: Provides communications to the home automation network remotely using 802.11/3G/LTE. The device should function regardless of the operating system used, although the Apple IOS operating system for the iphone and iPad was used in this work.

Importantly, the domotic system also requires a main computer; therefore, a Raspberry PI was also used in this work. This element of the home automation system is very important because it consumes relatively little power, possess a stable operating system and can act as a Web Server (PHP/ MySQL) [23].

Figure 6. Main controller/actuator

The Main Controller/ Actuator shown in Figure 4 was developed to be connected directly to the domestic electrical installation. This device includes an Xbee S1 radio that is synchronized with a server to receive on/off requests for lights and water pumps. The microcontroller allows control of up to 16 elements (lights or water pumps) that can be distributed throughout a household using one relay per electronic element [24]. One of the main features of the Main Controller / actuator is its size, because it is smaller than a regular pen.

The controller is low consumption and operating at 5V. It is necessary to have one per room with a maximum distance of 15 meters between each to ensure proper operation. An important feature of this device is that costs only 22 dollars with the hardware needed to control lights, water pumps and air conditioners by infrared transmission.

To avoid clogging the communication between the controller and air conditioners, will be located at the site of the luminaire replaces the traditional focus for Compact Fluorescent Lamp or LED Lamp, depending on the lumens needed in space (Figure 7).

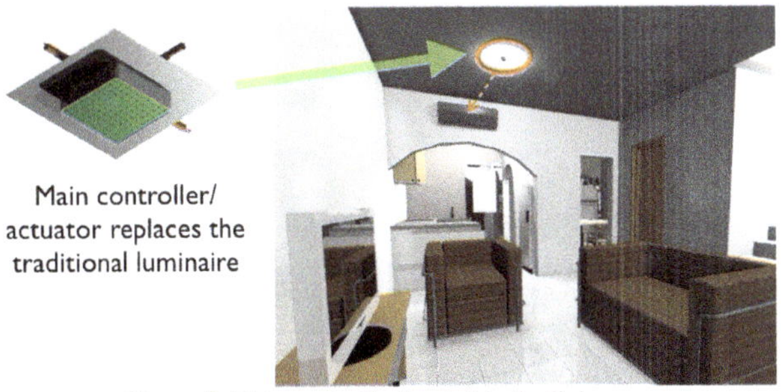

Figure 7. Main controller located in the living room.

The last element of this system is the mobile device. For this case study, an iPhone 5 and an iPad 4 with iOS 6.1 were used however it can use any device with a Web browser.

An interface Web and an iOS application were developed to communicate the mobile device with the Raspberry through the local network and the Internet. The IEEE 830 standard was employed to develop the software and IEEE 829 was used for software testing as it permits home electronics users to turn apparatuses on and off by touching a digital switch with a finger (Figure 8).

This iOS application makes it possible to control the air conditioner and lighting by means of the interphase shown in Figure 9.

This application makes saving energy possible by automating the state of the devices depending on the real-time home demands (i.e. at home, at work, on vacation, etc.), which users can program (Figure 10).

This architecture permits users to remotely monitor and control lights, air conditioning and water pumps wirelessly through the Internet from any mobile communications device (mobile phone, i-pad, tablet, etc.).

IV. Results

Binding and packet loss rate tests were performed in a typical Mexican home built with high density materials (Figure 7) to identify the coverage of the Xbee radios that conform this architecture.

Figure 8. Lights and Water Pumps

Figure 9. Lights and Air Conditioner

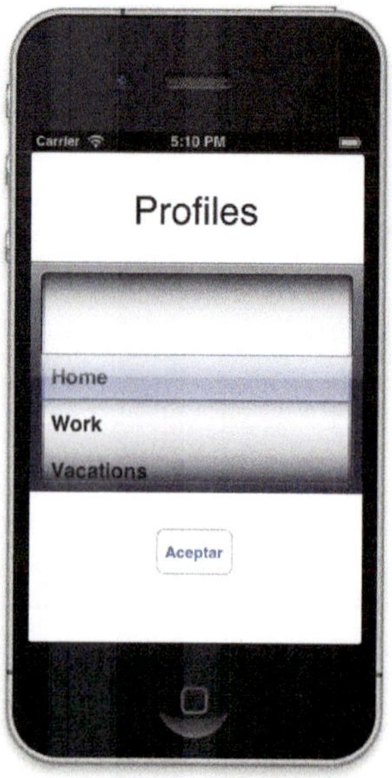

Figure 10. Profiles

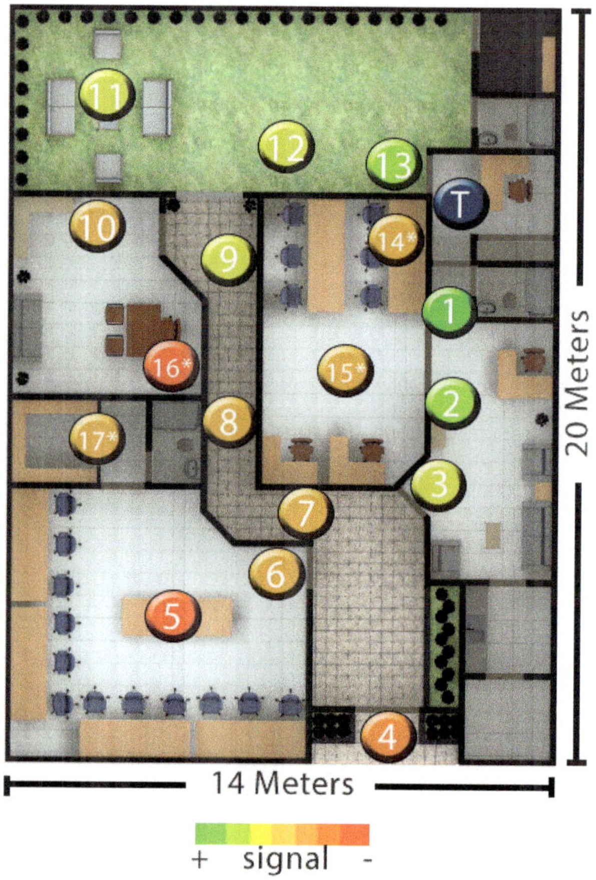

Figure 11. Home

As seen in Figure 11, nodes were placed in different sections of the house to determine the received signal strength from the node T (transmitter) to any node listed. The circles marked with the symbol (*) correspond to nodes placed on the roof of that position.

Table 1 shows the values of RSSI (Received Signal Strength Indication) obtained from each node placed in the home:

TABLE I.
RECEIVED SIGNAL STRENGTH INDICATION

ID	Received Signal Strength Indication		
	Meters	*Xbee Indoors*	*Experiment*
1	3	-56.69604549	-43
2	6	-65.12488537	-59
3	9	-70.05544063	-62
4	15.13	-87.35517543	-88
5	14.31	-86.50820013	-88
6	13.32	-85.41846082	-75
7	10.7	-82.08924517	-80
8	10.05	-81.13662511	-80
9	6	-73.29610672	-68
10	7.5	-76.68795717	-85
11	9.86	-80.84650498	-68
12	9.03	-79.50988421	-76
13	2.82	-61.81953174	-66
14*	3	-62.76005687	-73
15*	6.7	-74.97343105	-80
16*	8.07	-77.80138667	-88
17*	10.05	-81.13662511	-85

It was observed that wireless communications are continuously attenuated by multiple elements that affect the transmission / reception of information, so that this type of infrastructure cannot be used in all the scenarios [25].

If a distant transmitter is moved closer to a receiver, the strength of the transmitted signal at the receiving antenna will increase. Measure the strength of the signal at the receiving antenna is one method of determining the quality of the communication link [26]. It is for this that when designing a wireless network, it is important to consider how the signal propagates. In order to get reasonably reliable estimates of the behavior of an electromagnetic signal is necessary to make use of electromagnetic propagation models.

Electromagnetic propagation models aim to predict to what extent affects the propagation medium to electromagnetic energy carried by between two antennas that are communicating [27].

Radios were placed in different areas of the house to check the extent of the radio signal and perform calculations using the formula for the propagation model exposed by Farahani [28]:

$$P_d = P_0 - 10 * n * log_{10}(f) - 10 * n * log_{10}(d) + 30 * n - 32.44,$$

Where:

- Pd = signal power (dBm) at a distance d
- P0 = signal power (dBm) at zero distance from the antenna
- f = signal frequency in MHz
- d = distance (meters) from the antenna
- n = path loss exponent, and is determined experimentally

Table II illustrates the different values that the factor "n" can take in different environments. Given the experimental stage, we consider a factor n=2.8 and n=3.5 corresponding to indoors office building, without line of sight.

TABLE II.
"N" FA TOR FOR DEFAULT SETTINGS [28]

n	Environment
2.0	Free Space
1.6 -1.8	Indoors, line of sight
1.8	Supermarket
1.8	Factory
2.09	Conference room
2.2	Store
2 – 3	Factory indoors, without line of sight
2.8	Residence indoors
2.7 -4.3	Indoors office building, without line of sight

The figure 12 shows the comparison between the data obtained through Farahani model and the results of this experiment.

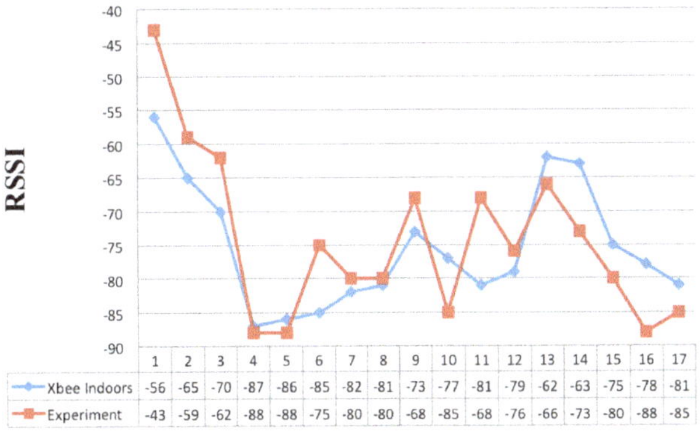

Figure 12. Comparison Chart of Xbee Indoors and Experimente

V. CONCLUSIONS

The experiment helped us to verify that can control and monitor lights, air conditioner and water pumps remotely using the inexpensive controller developed for use in high density constructions typical of many developing countries.

Future work will include expanding the applications of the controller we developed to implement new infrared communication protocols with the goal of applying them to home appliances and multimedia devices.

REFERENCES

[1] G. Eason, B. Noble, and I. N. Sneddon, "On certain integrals of Lipschitz-Hankel type involving products of Bessel functions," Phil. Trans. Roy. Soc. London, vol. A247, pp. 529–551, April 1955. *(references)*

[2] H.M.P. David Muñoz, Héctor Souto., "La domotica" 2008.

[3] P. Viswanathan, "What is a Mobile Device" 2008.

[4] mobiThinking, "Mobile subscribers worldwide," 2013.

[5] N. Gibbs, "Your life is fully mobile" 2012.

[6] Amazon, "Mobile devices"" 2013.

[7] Wester Forms, "Viviendas de concreto", 2013.

[8] T. Gaspar, H. Rodrigues, S. Odedra, M. Costa, J.C. Metrolho, A. Bardill, and S. Prior, "Handheld devices as actors in domotic monitoring system," Industrial Informatics, 2004. INDIN '04. 2004 2nd IEEE International Conference on, pp. 547-551.

[9] A. Home, "A brief history of X10" 2004.

[10] X10, "Actuator X10," 2012.

[11] Z. Salvador, R. Jimeno, A. Lafuente, M. Larrea, and J. Abascal, "Architectures for ubiquitous environments," Wireless And Mobile Computing, Networking And Communications, 2005. (WiMob'2005), IEEE International Conference on, pp. 90-97 Vol. 94.

[12] Zigbee, "Zigbee protocol," 2012.

[13] Digi, "Xbee," 2013.

[14] A. Araujo, D. Fraga, J.M. Moya, and O. Nieto-Taladriz, "Domotic platform based on multipurpose wireless technology with distributed processing capabilities," Personal, Indoor and Mobile Radio Communications, 2004. PIMRC 2004. 15th IEEE International Symposium on, pp. 3003-3007 Vol.3004.

[15] A. Rosendahl, and G. Botterweck, "Mobile Home Automation - Merging Mobile Value Added Services and Home Automation Technologies," Management of Mobile Business, 2007. ICMB 2007. International Conference on the, pp. 31-31.

[16] F. Chavez, F. Fernandez, R. Alcala, J. Alcala-Fdez, and F. Herrera, "Evolutionary learning of a laser pointer detection fuzzy system for an environment control system," Fuzzy Systems (FUZZ), 2011 IEEE International Conference on, pp. 256-263.

[17] M.A. Hannan, and K. Arshad, "An experimental study of interference in Smart Buildings," Communications, Signal Processing, and their Applications (ICCSPA), 2013 1st International Conference on, pp. 1-5.

[18] H. Wang, J. Wang, and M. Huang, "Building a Smart Home System with WSN and Service Robot," Measuring Technology and Mechatronics Automation (ICMTMA), 2013 Fifth International Conference on, pp. 353-356.

[19] Belkin, "Wemo Switch,"2013.

[20] I.G. Martín, "Relevadores," 2008.

[21] E. Alex, "Raspberry reboot," 2012.

[22] R.P. Foundation,"raspberry pi,"2012.

[23] Coyán, "Raspberry Pi Server", 2012.

[24] Microchip, "PIC18f4550 DataSheet," 2006.

[25] B.B. Luu, B.J. O'Brien, and D.G. Baran, "A Soldier-Robot Ad Hoc network," in Proceedings of the Fifth Annual IEEE International Conference on Pervasive Computing and Communications Workshops(PerCom'07), White Plains, new York, USA., 2007.

[26] MaxStream, "Received Signal Strength Indication," 2004.

[27] R.C.P. Vega an A.C. López, Sistemas de telecomunicación vol. 7: Ed. Universidad de Cantabria, 2007.

[28] S. Farahani, ZigBee Wireless Networks and Transceivers: Newnes, 2008.

AUTHORS

M. M. Ramses received the Engineering degree in Telematic's Engineer at the University of Colima - Colima, México, in 2012, and is a Master student in Computer Science in the same institute. His current research interests are mobile applications for home automation and Energy Saving in homes.

E. D. Laura Victoria received the Engineering degree in Telematic's Engineer at the University of Colima - Colima, México, in 2010, and is a Master student in Computer Science in the same institute. His current research interests are projects related to control system in domotics environments and wireless communications networks.

A. S. Raúl realized a Doctorate in Mobile Computing at the University of Sheffield, United Kingdom. In 2007, Dr. Aquino realized a postdoctoral stay in Electronic and Telecommunications at the Research Center of Higher Education in Ensenada, México, and in 2008, Dr. Aquino realized a postdoctoral stay in Telecommunications at the National Autonomous University of México. Currently, his areas of interest are design, development, simulation and implementation of Telecommunication networks: Wireless Body Area Networks (WBANs), Wireless Sensor Networks (WSNs), Wireless Local Area Networks (WLANs), and Wireless Metropolitan Area Networks (WMANs).

Bringing Remote Labs and Mobile Learning together

D. May, C. Terkowsky, T. Haertel and C. Pleul
TU Dortmund University, Dortmund, Germany

Abstract - **Within (remote) laboratories in Engineering Education students have the chance to do own experiments and by that gain own experiences in their learning processes. Apart from technical questions, one of the most intriguing aspects in this context is how students can document their learning process and show to others (teachers and/or other students) what they have achieved. Another aspect concerns the question of learner's mobility during the learning process. If the laboratory can be accessed remotely, why do we constrain learners in their level of liberty by forcing them to sit in front of a fixed computer to use a location-independent environment for experimentation? Therefore, rendering this environment available for mobile devices is the logical consequence. Furthermore, integrating mobile devices into the course's technical environment means to take a whole new approach to the teaching and learning process itself. It is especially a question of embedding mobile devices into the users' workflow (or better "learn flow") rather than a simple question of accessibility. The following article features an example of how remote laboratories can be linked with mobile devices and e-portfolios, thus creating a unique learning environment helping learners to document their personal learning processes and to exchange them with others while at the same time being flexible in means of time and place. This combination of topics has been realized within one subtask of the project "ELLI – Excellent Teaching and Learning in Engineering Education" at TU Dortmund University.**

Index Terms—**engineering education, e-portfolios, mobile learning, remote laboratories, tele-operated laboratories**

I. INTRODUCTION TO TELE OPERATED LABORATORIES AS A PLACE FOR LEARNING

Once having graduated, former engineering students will work on solving real problems creatively and they will work with real technical equipment- regardless if they head for a career in a company or in the academic sector. But do these people get into contact with this equipment during their studies? Most of their time engineering students are sitting in the lecture hall following the presentation in which the professor explains to them the course's content. In other words: The students try to understand and memorize what they have to know in order to pass the course's exam. In "classical" lectures there is only little space and time for the students to understand the big picture of the subject and the inherent research process with its questions, research activities and result interpretation. Hence, not few lectures feature results of research activities without providing their greater context or the research questions which were important at the beginning of the research process. Even if a professor would like to do so, in many cases there is simply not enough time for it. One possibility to change this fact could be the use of laboratories in teaching and to implement experiential [1] and research based learning in the teaching and learning process [2]. To bring the students in contact with laboratory equipment means bringing them in contact with the technical equipment of their future profession and giving them the chance to develop central technical competences for the technical part of their future career. In addition to the technical competences, for us the students' work in laboratories offers the opportunity to add aspects of

- systemic thinking
- problem definition
- responsibility
- innovation

The work presented in this paper will base on the achievements of the PeTEX project, will deploy its technological infrastructure and will optimize it. By this we will extend the possibilities innovate the existing concept. The main conception of the further development is the combination of the topics virtual learning environment, mobile learning and creativity. All this work will be carried out as a subtask of the new project **ELLI–Excellent Teaching and Learning in Engineering Education** which is funded by the German Ministry of Research and Education until 2016.

A. Constraints and solutions for the use of laboratories in education

A very important factor that hinders the use of laboratories by students in teaching is the cost of such equipment and the organizational effort of co-locating students, equipment and supervisors. Especially small universities often face the situation that they neither can afford all the laboratory equipment nor can allow the students to use it by themselves as it might get damaged. This means in many cases that lab experiments, if the professor tries to integrate them into the lecture, are either only shown via video or that the faculty's staff shows the equipment during guided tours through the laboratory. This means a real dilemma for modern engineering education.

One possible way out if this dilemma—wanting the students to develop technical competences on the one hand and having them done experiments but not being able to use the equipment on the other hand—are tele-operated (called "remote") and virtual laboratories. With them the laboratory equipment can be used by different universities from different places or very risky experiments can be done completely virtually.

B. PeTEX – Platform for eLearning and Telemetric Experimentation

Important research on the use of remote laboratories in teaching engineering aspects was done by the universities of Dortmund (Germany), Palermo (Italy), and Stockholm (Sweden) within the project *PeTEX* – Platform for e-learning and Tele-operative Experimentation [3], [4]. The technical part of PeTEX was carried out at TU Dortmund University by its Institute of Forming Technology and Lightweight Construction (IUL, Prof. Tekkaya) and integrated in close co-operation with the Center for Higher Education (former Center for Research on Higher Education & Faculty Development, Prof. Wildt) [5], [6]. Within this project comprehensive research on using remote laboratories in teaching was carried out. Therefore a network of three prototypes in the field of manufacturing technology was developed [7], [8].

Our work's overall context is the implementation of research-based and experiential learning by using laboratories in higher engineering education at universities. As explained above, our aims are that students get into contact with real technical equipment, understand the greater context of research and gain technical competences for their future work. In this context we will proceed as follows:

- Considering that laboratory equipment is expensive and some experiments even can be dangerous for students, we will explain why and how remote laboratories should be integrated in teaching. (chapter II)
- Up until today a weakness of such teaching approaches is the need for an open designed learning environment in which the students can act independently while at the same time can be guided through the learning process. We will work out how e-portfolios can help in this context and how they can be used to document and reflect on the learning process. (chapter III)
- In a final step we will change over to the topic mobile learning. In order to support the students' learning process as much as possible and to leave them the choice of using the software virtually every time and from everywhere we will open the software for the use from mobile devices such as smart phones and tablet computers. By showing different scenarios we explain how the use of mobile devices can support the learning process significantly and how they can help to promote creativity. (chapter IV)
- At the end of the paper we will explain shortly what our future steps will be in order to put our plan into action. (chapter V)

II. Learning with Remote Laboratories

Learning through experiments in general has become a central part in modern higher engineering education [9]. Implementing experiential learning and research-based learning by the active use of laboratories in higher engineering education by students is a teaching and learning concept which supports the constructivist approach. The learning arrangement is designed from the learners' point of view because the user is the one to design his learning process and "walk" through the learning objects while constructing his knowledge inside an active process.

As mentioned in the introduction, the students have the opportunity to get into contact with the physical equipment of their future professional life as well as to make practical and theoretical experiences with equipment, methods and processes of empirical research by the use of laboratory equipment in teaching. That is why doing technical experiments in a laboratory is an adequate way of applying, enhancing and testing knowledge the students have acquired during the lecture and developing central competences by doing so.

A. Kolb's Experiential Learning Cycle

The use of laboratories in teaching and learning environments can basically be traced back to understanding of learning explained by Kolb: "Learning is the process whereby knowledge is created through the transformation of experience" [1]. Kolb states that learning involves the acquisition of abstract concepts which can be applied flexibly in a range of situations. According to Kolb's theory, the impetus for the development of new concepts is provided by new experiences. Kolb's concept of experience is defined in his experiential learning theory consisting of a four-phase cycle in which the learner traces all the foundations of his learning process:

- *Concrete Experience*: A new experience of situation is faced or a reinterpretation of an existing experience takes place.
- *Reflective Observation*: The new experience is analyzed, evaluated and interpreted. Of particular importance are any inconsistencies between the experience and the understanding of it.
- *Abstract Conceptualization*: Reflection gives rise to a new idea or a modification of an existing abstract concept.
- *Active Experimentation*: Transforming the new abstract concept into operation, the learner interacts to the world around him to check what emerges.

In his four-step learning cycle Kolb explains that at the beginning of each learning process there is a real learner's experience (step 1) which is followed by a reflective observation (step 2). From that point on the learner tries to conceptualize what he has experienced (step 3), starts to experiment actively (step 4) and generates new experiences. This is the start of a new cycle. With every loop—from the simple to the complex—the student enhances his experiences. Thus, the learning cycle transforms learning activities into a helix of experience-based knowledge, skills and competences.

B. The learning process in the light of research processes

Since its beginnings university always has been a place not only for learning but for research, too. In order to be more concrete, at university these two processes always were thought and implemented in unison, hence they inspired be each other. For both learning and researching, explicit steps can be defined which describe the process' sequences. For the learning process we already explained the sequences as mentioned above by showing Kolb's learning cycle. For the research process a first approach can be to define the following steps: Make a practical experience, define the research question, implement research activities and interpret the results. If you now look

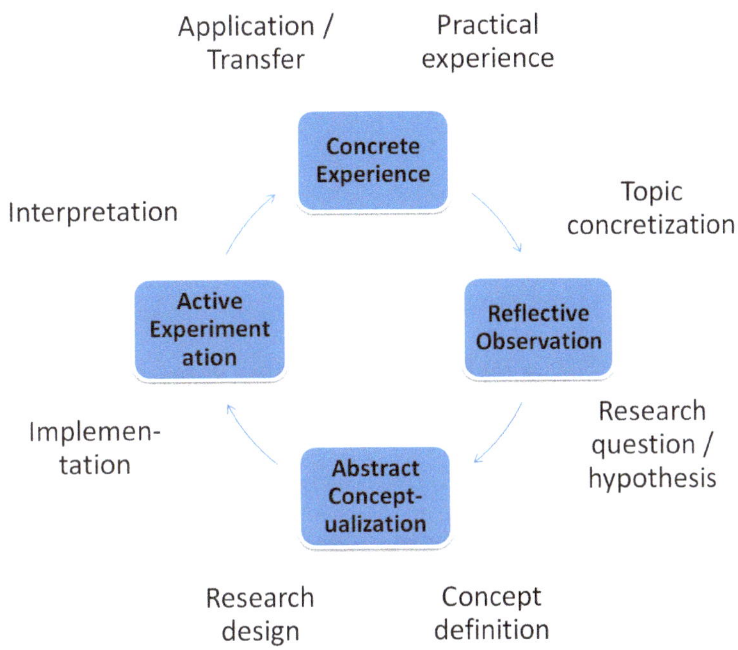

Figure 1. Synchronized learning and research process [16]

at the learning process as well as the research process and define both as a circle, you will find discover that both can be synchronized. Wildt [10] did this and showed clearly that very similar steps can be identified (see Fig. 1).

Surprisingly this fact has never had a severe impact on the way teaching is done at universities. As mentioned above, especially in engineering studies classical lectures are the most often used teaching method. This is the fact although there are some alternatives, which would permit a combination of learning- as well as research processes.

C. Research based learning

Research based or experiential teaching and learning in higher education is one adequate way of implementing learner centered teaching. In addition to that Herrington and Oliver worked out the importance of an authentic learning environment for a successful learning process [11]. According to them, this authentic learning environment can be achieved by teaching and learning activities in laboratories in which students can face a real context and carry out real activities. By connecting the actions in laboratories in a next step to real problems—e.g. from current research or from the industry—the students are able to go the whole way from the question at the beginning of an experiment to the final use of the results and can experience the relevance of their work.

D. Active experimentation using tele-operated equipment

Employing remote and virtual laboratories in teaching provides a vast range of opportunities to implement experiential learning in the field of mechanical engineering following the path of research-based learning [5]. In the following we will explain how a remote laboratory was put into practice. We noticed a discussion within the community whether remote labs can and/or should replace real laboratories. This may not be our concern in this paper as we do not want to advocate for or against one or the other laboratory solution without looking at its circumstances. There are and there always will be situations in which the use of real or remote laboratories makes more or less sense than the other.

One example in the context of manufacturing technology, namely forming technology, can be to use the remote lab concept for the purpose of material characterization. This could be offered in addition to a conventional lecture or in order to enhance traditional hands-on labs during the phase in which students prepare themselves for the lab or when they would like to rework some of the test steps while writing the lab report. Following the approach based on Kolb's experiential learning cycle, students can deal with basic concepts of metal forming during the lecture to test and see what they discussed in class by doing experiments on their own in order to create their own knowledge as well as using the remote experiential equipment. Another opportunity could be that students are given a real engineering problem related to material behavior. They are asked to work on this problem in small groups by planning and carrying out experiments using tele-operated equipment. Finally they have to present their explorations and and suggestions on how they would deal with the problem [5]. In order to support this entire process and especially the step of "active experimentation" an appropriate level of clear interaction and feedback needs to be integrated to the tele-operated experimental setup. In the PeTEX project a complete experimental setup (Fig. 2) has been moved to a new level using innovative engineering design, modern concepts of automation, measurement technology and robotics as shown in Fig. 3.

All aspects have been brought together by developing a clear and interactive user interface providing real time feedback of the running experiment. In Fig. 4 the screen for the uniaxial tensile test is shown. While using the live camera stream (1), users can investigate the surrounding test apparatus, e. g. sensors or clamping devices. Afterwards the learner initiates the preparation of the experiment (2), using the integrated 6-axis robot to select and check an appropriate specimen. To freely configure the experiment, relevant test parameters (3) can be filled in. When the test is started (4) the robot positions the specimen to the fully automated clamping device. The developed innovative concept of the fully automated clamping process and parallel measuring of relevant values is mentioned in [3].

Also during the test, a high level of interaction is provided to the user by manipulating the camera view or pausing and continuing the test. Pausing the test—which means the load is not further increased since that moment—causes a reaction by the material. This phenomenon is graphically visible in the real time diagram (6) and also in the real time test data at the header bar (5). Comparisons with prior test data are available within the data base (7)

Figure 2. Testing machine

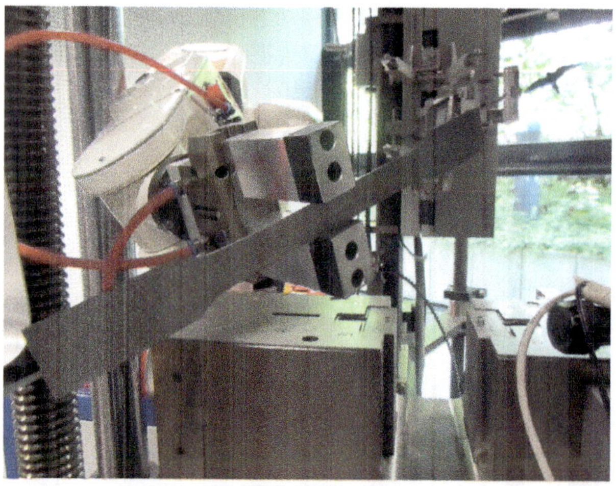

Figure 3. Robot positioning a specimen

and the graph (6). After the experiment is finished, learners are provided with a data package including all results for further analysis and investigation.

Additionally the entire tele-operated experimental environment was made available with the learning content management system Moodle. Within Moodle, we conducted the alignment of four, for us elementary, areas for this kind of socio-technical system. This socio-technical alignment for tele-operated laboratory learning consists of the adjustment of the technical, didactical, media and social level. By integrating the tensile test environment into Moodle, this socio-technical alignment was put into a usable as well as flexible environment.

An often quoted challenge to such open designed learning concepts is that it turns out that a very sophisticated concept is needed for teachers to enable them to document and evaluate the learners' behavior and achievements during the learning process using a virtual laboratory. It is obvious that such a concept requires different systems for the instructor to accompany the learner through the learning process and—above all—to evaluate the achieved learning outcome. The following passages present the future thoughts concerning a concept for the learning process' documentation in context with the use of remote laboratories in combination with e-portfolios. This will be followed by explanations on the use of mobile devices in this context.

III. E-PORTFOLIOS AND THEIR USE IN EXPERIENTIAL LEARNING

In addition to the open learning concept which is supported by the use of laboratory equipment in general, the use of remote laboratories within the PeTEX project was designed for the usage by a very heterogeneous learner group composed of students and professionals [12]. This means that the software for the learning process' documentation as well must be designed very open in order to prevent system inherent barriers for different learner types. In this context it should be mentioned that the PeTEX system intends to merge higher education and the workplace as well as to create an international learning community not being limited to one institute. These 3 aspects—documenting the learning process , building a

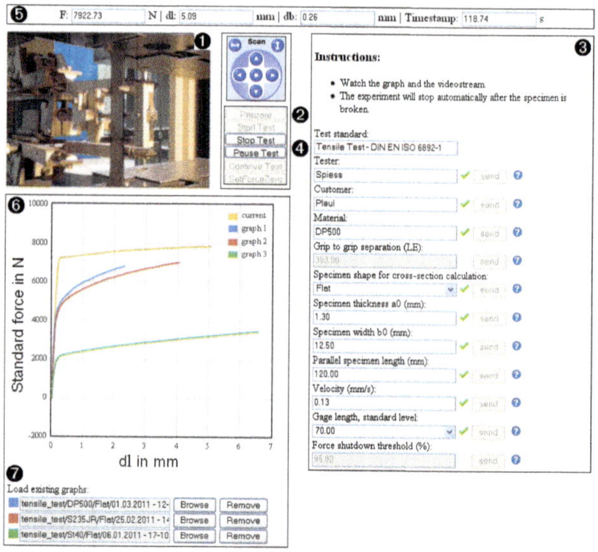

Figure 4. Interface to the tele-operated experiment [4]

learners' community and connecting the students' work with their future professional work—are requirements the software has to address and accomplish (see Fig. 5).

Software which seems to be adequate and which is frequently discussed in similar contexts is the e-portfolio [13]. E-portfolios are based on the general idea of portfolios, referring to the idea of collecting different kinds of documents in a folder in order to reflect personal learning processes and to exchange them with others [14]. E-portfolios support the same, but they are made online and provide the collection of different kinds of data like texts, tables, photos, videos and audio [15]. E-portfolio software could be added technically to the Moodle environment—which is already used in the PeTEX context— very easily because an online e-portfolio application especially designed for Moodle already exists by the name of Mahoodle. In the following we will explain why e-portfolios fulfill the three main requirements in the new PeTEX context [6].

A. E-Portfolio as a learning process documentation

The user, regardless whether in higher or in professional further education, can arrange all the data he wishes to document or to show in different ways in order to create his own portfolio just like his personal page in any social network. He can present experiments and its results, show photos from the test set-up, explain his thoughts on the research and so on. Additionally, he can permit other users, learners or teachers, to view his e-portfolio. By creating such an e-portfolio the learner can document his own learning and research process and start reflecting on the experiments he does during his research-based learning process [16]. This reflection is an important aspect as it corresponds to his personal learning circle. Especially for students the e-portfolio can serve as a means of orientation or checkpoint in the own field of research [1], [16], [17]. By the same way the teacher can evaluate the learner's action by regarding a learners' portfolio. Because other persons are able to see the collection in the portfolio it can be said that it is not only a way of documenting the learning process but as well a way to communicate it so that a collaborative learning process can be achieved. This leads to the next use of E-Portfolios in the PeTEX context.

B. E-Portfolio as a learning community software

Considering the e-portfolio as software for documentation and evaluation is just one application of the system. A constructive enrichment by using e-portfolios is the community building. Every author of an e-portfolio is able to allow other users to view various parts of his portfolio as well as to view others' portfolios. This means that learners who are doing experiments in the PeTEX system and filling their e-portfolios have the chance to get into contact with each other via the portfolio software. They can see what others are especially interested in, start discussing about it, give comments and help each other in case of an emerging problem during the experiential learning process. Along the course of this interaction, a specialized community on remote laboratories evolves within the PeTEX context.

C. E-Portfolio as a bridge between the university and the workplace

The PeTEX system is designed for the usage in higher education and workplace learning. This means in a first step that both user groups can use the e-portfolios in the explained way of use. A further future thought is to use the e-portfolio as a livelong system to document own competencies from university level on and during the whole professional life. This should be explained by an example in three steps:

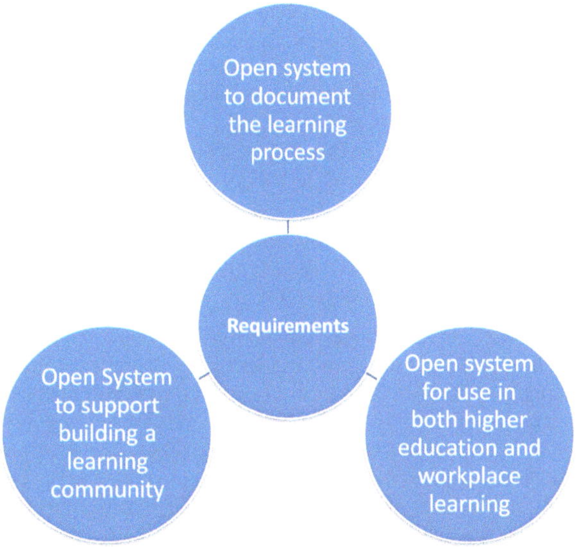

Figure 5. Requirements for a software system documenting the learning process in remote laboratories

Step 1 - An engineering student starts working with the PeTEX system at the university. He uses the system in order to document his experiments. During his studies he conducts different experiments, compares them and collects all research results and data in his e-portfolio in order to scientifically describe a certain material behavior which was observed (e.g. while pausing the test for a couple of seconds), and reflects on his own way of learning. The teacher is able to evaluate his learning behavior. This can be regarded as the main use of e-portfolios at university.

Step 2 - Because the PeTEX system as well addresses workplace learning the e-portfolios can be viewed as a bridge from university to professional life. Depending on the specific use of e-portfolios by the student he can take his portfolios to present himself to potential employers. Hence, they can see what the students did in this field of his studies and whether he accords with the company's needs. In this context the e-portfolios can support application processes.

Step 3 - Once the former student pursues his career as an employee, there is no need to stop working with his portfolio. He still can work on his collection by documenting new experiments as well as gained knowledge and competencies from his new occupation. By doing so, the employee does not stop reflecting on his learning process. His e-portfolio grows with every year and successively becomes a better representation of his professional life and his competences. Especially the last aspect matches perfectly with the advantage of the PeTEX system as small and medium sized companies use it to enhance their technological skills by doing research with the PeTEX hardware. In addition to that they can use the e-portfolios to implement a system for the documentation and measurement of the employees' skills and competences. This could be supported by the lifelong use of e-portfolios.

Summing up all these aspects it can be concluded that the use of e-portfolios in the PeTEX context can support the idea of experiential and research-based learning even if there are a couple of challenges to meet [16]. The portfolios can be used to document and present the research and learning process, to build up a specially focused learning community as well as to merge university and workplace learning.

IV. SCENARIOS OF USING MOBILE DEVICES IN COMBINATION WITH E-PORTFOLIOS

Another frequently mentioned new concept in context with higher education is mobile learning. From a technical perspective mobile learning means to integrate mobile devices like cell phones, smart phones or tablet computers in the learning process [17]. Among others, one of the advantages of mobile learning is that previously unplanned time periods can be utilized for learning and that learning processes can be initiated virtually everywhere [ibid.].

In addition to that using mobile devices can support the creativity process because new ideas mainly come spontaneously and the fact of carrying a mobile device throughout the day makes it easily possible to put down a note with an idea and to work on it later or to work on it immediately as we will explain in the scenarios. Bringing e-portfolios on tablet devices for example could be an opportunity to combine the concepts presented in this text with mobile learning. In the following we will present different scenarios how the use of mobile devices can enrich the concept of remote laboratories in higher engineering education. These scenarios differ mainly in terms of individual or collaborative learning processes and self-directed or teacher-directed learning processes. These four aspects in different combinations lead to the scenarios, so that it becomes obvious how flexible the use of e-portfolios is in our learning concept. As we wanted an open learning environment for working with the laboratory equipment remotely it seems that the e-portfolio software fulfills this requirement perfectly. Fig. 6 shows the four scenarios that are explained in the following.

A. Possible scenarios

Scenario 1 "Using the software in creative moments"- A first scenario could be that a student is thinking about his experiments while sitting at home and watching TV or while he goes out with his friends. He is really struggled by his research work, thinks about his parameters, his results and why his experiments offered the results that showed up. Suddenly he has an idea on a hypothesis and wants to check it by rereading his last experiments in the portfolio or doing a new experiment. Because he can use the software for connecting with the experiential environment by using his tablet computer there is no need to wait until the next day for doing the experiment over again at the university but he can stay where he is and even can stay sitting on the sofa for checking his hypothesis. The new result can immediately be put in his portfolio so that he documents his new step within his research process. Using a simulation in a virtual laboratory instead of the remote experiment can be a method to (pre-)check the hypothesis first and then to carry out the real experiment remotely [18].

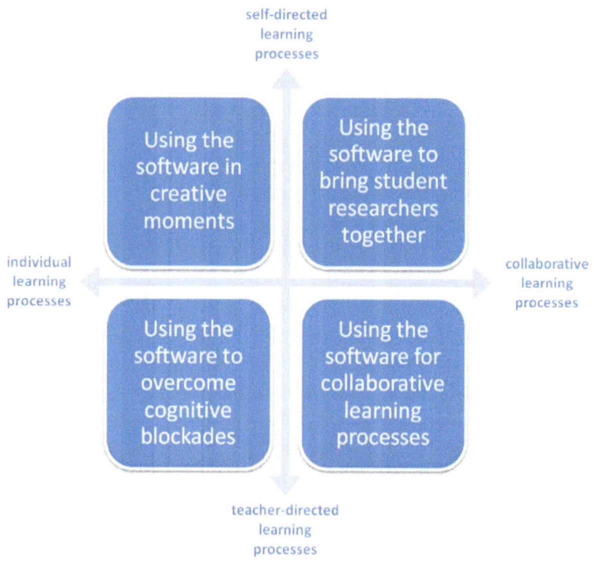

Figure 6. Four scenarios for using mobile devices in combination with remote laboratories

Scenario 2 "Using the software to bring student researchers together"- With his mobile device (regardless if smart phone or tablet PC) the student can access his personal e-portfolio in which he documents his experiments and personal competency development in this sector from wherever he wants to. Sitting in the train on the way to or on the way back from university he could flip through his experiments and look what he found out as different results. At the same moment another student looks on the first student's portfolio. He finds out that his own research had quite similar results even though he used different parameters or—even more challenging—he used the same parameters and material but had different results. Knowing this he contacts the first student via a chat or e-mail, as well using his mobile device, and they can communicate about their common results at this very moment and work together on future experiments [19].

Scenario 3 "Using the software to overcome cognitive blockades" - A third scenario could foster the students' ability to consider different perspectives about their questions: After having performed an experiment that was given to him by the teacher, the student possibly does not know why he did not get the expected results or does not know how to interpret the results. He asks himself why the experiment did not work as it should have, but he cannot find the answer. While writing his e-portfolio as documentation for the teacher's evaluation, he could start the "creative-help app", which helps him to use different perspectives on the problem: Firstly, he is asked to make a (mental) headstand following the question "What else could I do to get the wrong results from experimenting?" If that does not help to find the answer, he secondly will be asked to describe his experiential design and his assumptions in a way that a ten-year-old could understand it. If those methods, which are rather close to the problem, still cannot help him, the "creative-help app" will suggest a force-fit technique by showing a picture that does not have anything to do with a problem (for example a lady beetle, a daisy chain, a bottle of wine) and asking the student to find relationships between the picture and his experiment. This method helps to leave the well-trodden paths and forces the students to look from completely other perspectives on their problem. It often results in very unconventional or provoking ideas, but rethinking the obviously unsuitable solutions sometimes leads to the one really good idea that would not have appeared to his mind without having made the detour [20].

Scenario 4 "Using the software for collaborative learning processes" - The fourth scenario could be evoked by the teacher. He can give the students—as a kind of homework—the exercise to check an explicit hypothesis by implementing adequate experiments. Using the e-portfolios the students can stay in contact without the necessity to meet at the university and in combination with mobile devices they can be virtually be anywhere going through a collaborative learning process. Because the e-portfolios software has a connection to the experiential environment the students as a group can do the experiment and discuss the results with regards to the homework's hypothesis in one pass and without changing the learning environment [21].

B. *Proof of concept*

In order to put the plans explained above into practice different research has been carried out. Our work on remote laboratories and creativity bases on finished research projects at TU Dortmund University and several other European universities. Based on this research we developed a first proof of concept for the mobile devices running with Android. The software permits the user to do a remotely run experiment by checking the parameters, starting the experimentation process and following the results. The next technical step is to bring the software from the proof of concept status to a level on which it can be tested and improved with students on a larger scale. As a consequence thereof it would have to be further connected with the e-portfolio software and it has to be worked out for iOs devices. These are the steps for the coming year.

V. CONCLUSION, DISCUSSION AND FUTURE PLANS

With this paper we explained four different but in our project newly connected aspects:

- We discussed why laboratories are a place for conducting experiments and why they can be important for modern engineering education. The central idea is to engage the students in teaching and learning environments which are connected closely to their future working environment.
- In addition to that, we discussed the aspect of student centered learning environments in general and why they are vitally important in higher education. They are essential for having the students do the things they have to learn by themselves. In our context we underlined this statement by the example of the synchronized learning and research process. This is the only way for students to develop competences and to reach a high level of learning outcomes.
- The learning environment was the next aspect we focused on. If students are learning in laboratories or with the help of laboratory work it is obvious that they need a special learning environment in order to reflect on their learning process. Questions like "How can I document my learning progress for other learners or the teachers?" or "How can I communicate with my classmates during the learning process?" are becoming more and more important. We want to address these questions with the use of e-portfolios as a place to document and to communicate learning processes as well as to merge studies at university with the later workplace environment.
- Finally we explained our plans in context with the use of mobile devices. There are several scenarios thinkable (we concentrated on four of them) in which the use of smart phones and/or tablet PCs extend the opportunities of learning environment substantially. With the use of mobile devices previously impossible learning scenarios become feasible. The fact that these devices are highly portable of course plays a central role in this context. We finished with a technical proof of concept for Android devices.

All the aforementioned aspects are combined in our work. In order to innovate the teaching and learning in engineering education we will design the "Mobile Lab Portfolio". This environment integrates the use of remote labs, e-portfolios and mobile devices. See in the following the central advantages of the presented concept:

- As the equipment of laboratories is either very expensive to purchase and maintain at every university or not always accessible for students the use of remote or virtual laboratories is a good alternative to face this dilemma.
- Using the equipment virtually in simulations or remotely from wherever they can help the students to do experiments just as a pre-check on personal hypothesis or even when they are not physically in the laboratory.
- Learning processes that are achieved by the usage of the laboratories can be documented in e-portfolios.
- These portfolios are a good opportunity in order to document the experiments for personal use or for evaluation by an instructor. By looking at his students' portfolios the instructor can either see what kind of experiments the students have done and what they learned from it.
- If the portfolios are not kept hidden from other students but are open for other users to look at them and comment on the achievements, there is the opportunity for a community to evolve working together on the experiments. The e-portfolio software should be made accessible for mobile devices, too.
- This paves the way to mobile learning, which means that the learning process is not bound to any location. From virtually everywhere and at every time the user can work on their portfolios and communicate with each other.

As this is a work-in-progress paper the upcoming step for this year will be to implement the e-portfolio software in the PeTEX system and to make it accessible for the students' mobile devices. Once this has been achieved, first tests with students can be carried out and the system can be evaluated and improved.

Of course technical problems will arise during the implementation and they may be even difficult to solve. But in the end most problems stay simply technical and it became obvious in the PeTEX project that every technical problem will be solved sooner or later. At this moment we focus on the concept and first technical steps for our future work. During the whole work we want to concentrate on the didactic background for this. The question if any of the explained aspects will help the students to learn more and gain real competencies during their studies is and will stay our main focus. Not everything which is possible from the technical point of view or even can be technically designed does make sense for the learning process and higher education. Therefore we look at teaching and learning from the students' perspective. That leads us to our concepts explained above first, without asking in detail about how everything can be implemented yet. Our team strives to support a better engineering education- that is what our focus lies on.

VI. References

[1] D.A. Kolb, "Experiential learning. Experience as the source of learning and development", Englewood Cliffs, N.J.: Prentice-Hall, 1984.

[2] L. Huber, J. Hellmer, and F. Schneider, „Forschendes Lernen im Studium - Aktuelle Konzepte und Erfahrungen", Bielefeld, Universitätsverlag Webler, 2009.

[3] C. Pleul, C. Terkowsky, I. Jahnke, and A.E. Tekkaya, "Platform for e-learning and tele-operative experimentation (PeTEX) - Hollistically integrated laboratory experiments for manufacturing technology in Engineering Education", Proceedings of SEFE Annual Conference, 1st World Engineering Education Flash Week. Lissabon, Portugal, Bernardino, J. and Quadrado, J.C., 2011, pp. 578-585.

[4] C. Terkowsky, C. Pleul, I. Jahnke & A. E. Tekkaya, "Tele-operated Laboratories for Production Engineering Education - Platform for E-Learning and Telemetric Experimentation (PeTEX)", International Journal of Online Engineering (iJOE) 7, Special Issue EDUCON 2011, 2011, pp. 37-43.

[5] C. Pleul, C. Terkowsky, I. Jahnke & A.E. Tekkaya, "Tele-operated laboratory experiments in engineering education – The uniaxial tensile test for material characterization in forming technology", in "Using Remote Labs in Education. Two Little Ducks in Remote Experimentation", Ed. by Garcıa-Zubıa, J. and Alves, R. G. Deusto Publicaciones, 2011, Chap. 16, pp. 323-348.

[6] C. Terkowsky, I. Jahnke, C. Pleul, D. May, T. Jungmann & A.E. Tekkaya, "PeTEX@Work. Designing CSCL@Work for Online Engineering Education ", in "Computer-Supported Collaborative Learning at the Workplace - CSCL@Work.", ed. by S. P. Goggins, I. Jahnke, and V. Wulf, Springer, 2013, pp. 269-292.

[7] C. Terkowsky, I. Jahnke, C. Pleul, R. Licari, P. Johannssen, G. Buffa, M. Heiner, L. Fratini, E. Lo Valvo, M. Nicolescu, J. Wildt & A.E. Tekkaya, "Developing Tele-Operated Laboratories for Manufacturing Engineering Education. Platform for E-Learning and Telemetric Experimentation (PeTEX)", In International Journal of Online Engineering (iJOE). IAOE, Vienna, Vol.6 Special Issue: REV2010, 2010, pp. 60-70.

[8] I. Jahnke, C. Terkowsky, C. Pleul & A.E. Tekkaya , "Online Learning with Remote-Configured Experiments", In Kerres, M., Ojstersek, N., Schroeder, U. & Hoppe, U. (Eds.): Interaktive Kulturen, DeLFI 2010 – 8. Tagung der Fachgruppe E-Learning der Gesellschaft für Informatik e.V., 2010, 265-277.

[9] L.D. Feisel, and A.J. Rosa, "The Role of the Laboratory in Undergraduate Engineering Education", Journal of Engineering Education, 2005, 121-130

[10] J. Wildt, „Forschendes Lernen: Lernen im „Format" der Forschung", Journal Hochschuldidaktik, Jg. 20 Heft 2, 2009, 4-7

[11] J. Herrington & R. Oliver, "An instructional design framework for authentic learning environments", Educational Technology Research and Development, 48 (3), 2000, 23-48

[12] C. Terkowsky, I. Jahnke, C. Pleul & A.E. Tekkaya, "Platform for E-Learning and Telemetric Experimentation (PeTEX) - Tele-Operated Laboratories for Production Engineering Education", In Auer, M.E. , Al-Zoubi, Y & Tovar, E. (Eds.), "Proceedings of the 2011 IEEE Global Engineering Education Conference (EDUCON) – Learning Environments and Ecosystems in Engineering Education". IAOE, Vienna, 2011, 491-497.K. Himpsl and P. Baumgartner, "Evaluation von E-Portfolio-Software", Teil III des BMWF-Abschlussberichts, Einsatz von E-Portfolios an (österreichischen) Hochschulen." www.bildungstechnologie.net. 2009.

[13] A.C. Breuer, „Das Portfolio im Unterricht – Theorie und Praxis im Spiegel des Konstruktivismus", Münster, Waxman Verlag GmbH, 2009

[14] R. Reichert, „Das E-Portfolio - Eine mediale Technologie zur Herstellung von Kontrolle und Selbstkontrolle", in „Kontrolle und Selbstkontrolle – Zur Ambivalenz von E-Portfolios in Bildungsprozessen", Ed. by Thorsten Meyer et al. Wiesbaden, VS Verlag für Sozialwissenschaften – Springer Fachmedien Wiesbaden GmbH, 2011

[15] G. Reinmann & S. Sippel "Königsweg oder Sackgasse? – E-Portfolios für das forschende Lernen", in „Kontrolle und Selbstkontrolle – Zur Ambivalenz von E-Portfolios in Bildungsprozessen", ed. by Thorsten Meyer et al. Wiesbaden: VS Verlag für Sozialwissenschaften – Springer Fachmedien Wiesbaden GmbH, 2011

[16] H. Blom, „ Der Dozent als Coach", Neuwied/Kriftel, Luchterhand, 2000

[17] A. Kukulska-Hulme & J. Traxler, "Mobile learning: A handbook for educators and trainers", Milton Park and New York, Routlegde and Taylor & Francis Group, 2005

[18] D. May, C. Terkowsky, T. Haertel & Christian Pleul, "Using E-Portfolios to support experiential learning and open the use of tele-operated laboratories for mobile devices," in: REV2012 - Remote Engineering & Virtual Instrumentation, Bilbao, Spain, Conference Proceedings. IEEE, 2012, 172-180.

[19] C. Terkowsky, D. May, T. Haertel & C. Pleul, "Experiential remote lab learning with E-Portfolios - Integrating tele-operated experiments into environments for reflective learning," in: 15th International Conference on Interactive Collaborative Learning and 41st International Conference on Engineering Pedagogy in Villach, Proceedings, IEEE, 2012.

[20] C. Terkowsky, D. May, T. Haertel & C. Pleul, "Experiential Learning with Remote Labs and E-Portfolios - Integrating tele-operated experiments into personal learning environments," in: International Journal of Online Engineering (iJOE). IAOE, Vienna, Vol 9, No 1, 2013, 12-20.

[21] D. May, C. Terkowsky, T. Haertel & C. Pleul, "The laboratory in your hand - Making remote laboratories accessible through mobile devices," in: Proceedings of the 2013 IEEE Global Engineering Education Conference (EDUCON),"Synergy from Classic and Future Engineering Education", Technische Universität Berlin, Berlin, Germany, March 13-15, IEEE, 2013, 335-344.

Acknowledgment

The authors sincerely thank: Prof Dr.-Ing A. Erman Tekkaya, Dr.-Ing. habil. S. Chatti, Prof. Dr. Dr. J. Wildt, Dominik May, (TU Dortmund University), Prof. F. Micari, Prof. E. Lo Valvo, Prof. L. Fratini, PhD R. Licari, PhD G. Buffa (University of Palermo, Italy), Prof. M. Nicolescu, P. Johansson (KTH–Kungliga Tekniska Högskolan, Stockholm, Sweden) for their great cooperation during the PeTEX project.

Last but not least the authors sincerely thank Prof. Dr.-Ing. Thorsten Jungmann (FOM Hochschule Düsseldorf), Emanuel Bielski (TU Dortmund University) and all friends and colleagues at TeachING-LearnING.EU—the first subject center for engineering education in Germany—for many inspiring discussions and the productive view exchange.

Authors

D. May is managing director of Teaching-Learning.EU and with the Center for Higher Education, TU Dortmund University, Dortmund, Germany (e-mail: may@teachning-learning.eu).

C. Terkowsky is head of the Engineering Education Research Group at the Center for Higher Education, TU Dortmund University, Dortmund, Germany (e-mail: claudius.terkowsky@tu-dortmund.de)

T. Haertel is head of the ELLI project at the Center for Higher Education, TU Dortmund University, Dortmund, Germany (e-mail: tobias.haertel@tu-dortmund.de)

C. Pleul is head of the Research Group on Laboratories in Engineering Education (LEED) at the Institute of Forming Technology and Lightweight Construction, TU Dortmund University, Dortmund, Germany (e-mail: christian.pleul@iul.tu-dortmund.de)

Lunar Observation Support System Using Smartphone AR for Astronomy Education

K. Tian[1], M. Endo[2], M. Urata[1], K. Mouri[3] and T. Yasuda[1]
[1]Nagoya University, Nagoya, Japan
[2] Chukyo University, Toyota, Japan
[3] Nagoya City Science Museum, Nagoya, Japan

Abstract—**It is possible to improve learners' understanding of the concept of lunar phases by using Observation based Learning (OBL). In this research, Smartphone Augmented Reality (AR) technology was used to develop a multi-viewpoint AR-based mobile learning (M-VARML) system for moon observation that can be used in the real world environment. An experiment was conducted to evaluate the usefulness of our system. Participants were given five moon observation tasks, which they had to perform manually and using the M-VARML system. After each task, they were given follow-up questionnaires on the usefulness of the functions in the system. Finally, they were asked about the operational usability and the effect our system had in motivating them to pursue lunar observation. The results show that the M-VARML system is effective in improving the observation and learning of students and in enhancing their motivation to pursue lunar learning.**

Index Terms—**Smartphone Augmented Reality, Lunar Observation, Lunar Phase, Astronomy Education.**

I. Introduction

Lunar phases are astronomical phenomena that we observe in our daily life. Yet, it may be one of the most difficult concepts to teach in astronomical education [1]. In a survey conducted in Japan, over 50% of pupils had difficulty in understanding lunar phases [2]. In addition, it is reported that only 30% undergraduate students could explain the mechanism of lunar phases [3]. In order to help learners grasp this abstract concept, research has been conducted to study the development and usefulness of educational materials for lunar phase learning using several different approaches. One type is the virtual learning environment. For example, a learning system is available that allows the user to manipulate astronomical models of the sun, earth, and moon which exist as visible, tangible bodies [4]. The tangible learning system, titled "The Phase of the Moon," was implemented in a six-grade elementary school science class unit. A 3D animation moon phase concept system has been applied to science learning in the fourth grade [5]. Several issues, however, arise with a virtual learning environment. For example, much of the research is confined to restricted locations such as computer rooms or laboratory. Lack of outdoor learning seriously limits students from clearly understanding the position and movement of moon phases in the real sky. A virtual learning system is abstract and difficult to apply to the real world environment. For example, in one study indicated that after students obtained elevation angle of the moon information from a virtual learning system, it was difficult for them to measure the actual elevation angle of moon in the real world [6]. Thus, the positive effects of virtual learning environments are limited in motivating and learning about lunar phases.

Planetarium curators believe that the most important element in astronomy education is observing real stars outdoors [7]. It is important to observe real objects and real phenomena to learn science; this helps the learners experience and understand the target objects or phenomena in their real world with its surrounding conditions and sceneries [8]. Observation-based learning (OBL) is one such kind of learning approach that meets the requirements of experiential learning; it can be employed in the real world to circumvent issues that exist in the virtual learning environment. Suzuki confirmed that on-going observation and conversation could enhance college students' ability to understand the phases of the moon [9]. An activity that could greatly aid a student in lunar phase learning is observing the moon in sky. Through observations, a student can understand the rising and setting of the moon. Further, by daily observation of lunar phases, they can comprehend the periodic waxing and waning of the moon and understand the relationship among the sun, earth, and moon. An approach to OBL in lunar education includes using an application on mobile device. A lunar observation support system that can be used with mobile phones was developed [10]. This system enables students to observe the moon in open air and send observational data through their mobile phones to the server. However, this system had issues in the real world environment. Observation of the moon is often restricted by time, place, and weather. On cloudy or rainy days, the observer failed to see the moon. Furthermore, in cities, it is often inconvenient for observers to view the lunar phases due to high rises or light pollution. Learners could not find time on thirty days to completely observe a lunar cycle because of their busy schedules. One typical solution is to use Augmented Reality (AR) technology to address these issues. Augmented Reality (AR) is a variation of Virtual Reality (VR); it adds virtual information to the real world and interacts with it. Kikuo and Tomotsugu [11] consider AR as a new pattern that could be applied in teaching that has great potential for future development. This new and unique teaching method and strategy can help learners with little computing experience to interact easily. Interest in mobile AR applications has grown with the dramatic increase in the usage of smart phones in the past few years. Present day smart phones and tablets integrate fast processors with graphics hardware, large touch screens, and onboard sensors (GPS, compass, and accelerometers) to create gadgets that are ideal for indoor as well as outdoor AR experiences, such as, CityViewAR [12].

There are a number of popular AR astronomy applica tions, such as Google Sky Map, Star Walk, and Stellarium Mobile. These applications utilize the Smartphone's GPS and sensors to locate the stars. These applications name the star that is being viewed through the phone. However, the purpose of these applications are not designed for lunar learning and teaching and cannot meet the learning goal of lunar phases; for example, they have not dynamic latitude and longitude line to help students to study the azimuth and elevation angle of moon in the sky. In addition, there are no functions that show one-month moon orbit, shape, and age in the AR view. Most application AR views do not have a camera view, and thus a visual view in the real world environment cannot be obtained. Further, these applications do not provide a function that explains the mechanism of lunar phases, as it relates to everyday observation of lunar phases.

Setozakai [13] proposed that it is necessary to help learners to understand the astronomy topics from multi-viewpoints, namely the geocentric model and the heliocentric model viewpoint. It states that it is important to switch between the two viewpoints to understand astronomy subjects. Based on Setozakai's theory, Kawasaki [14] indicated that it is important to observe the appearance of moon from the earth viewpoint, and, at the same time, to understand the relationship between the sun, earth, and moon from the viewpoint with sun as the center of the universe to study the mechanism of lunar phases. Therefore, it is optimal to construct a multi-viewpoint observation environment for students to observe lunar phase in the real world environment.

In this research, Smartphone AR technology was applied to develop a multi-viewpoint AR-based mobile learning system for supporting moon observation that can be used anytime and anywhere, irrespective of the weather conditions. In order to assess the learning effect, usability and motivational effect of this system, a task-based experiment was conducted. Participants were given five moon observation tasks to perform manually and using M-VARML system. After each task, they were given follow-up questionnaire about the usefulness of the functions in the system. Finally, they were asked about the operational usability and the learning attitude of system. The results show that the M-VARML system enhances a learner's learning ability for lunar observation. It has high motivational effect on lunar observation and operational usability.

II. Research Method

This section explains the learning goal of the system, the system design, the system interface, and its functionality.

A. Learning Goal of Lunar Observation

The proposed system was developed as an educational tool for lunar phase observation. The learning functions of the system are based on learning indicators in the student's science course unit, titled "Lunar Phase." This course set the learning goals of moon observation of M-VARML as follows:

(1) Observe the moon, its azimuth and elevation angle in the sky every day.

(2) Observe the moon every day to witness it rising in the east and setting in the west.

(3) Observe the moon every day to understand the periodic waxing and waning of moon in 30 days.

(4) Observe the azimuth and altitude, changing of the shape of the moon by observing the moon at the same time in 30 days.

(5) Understand the positional relationship among the sun, moon, and observer.

(6) Understand the positional relationship among the sun, earth, and moon from the viewpoint with sun as the center of the universe.

(7) Understand the relationship between moon age and shape of lunar phases.

B. System Outline

The M-VARML System allows learners to observe the appearance of moon from the earth viewpoint, at the same time, understand the relationship of sun, earth and moon from the sun-center of universe viewpoint to study the mechanism of lunar phase. The outline of system is as shown in the Fig.1. The learners could switch the two viewpoints to observe moon and achieve the learning goal of moon that we have stated.

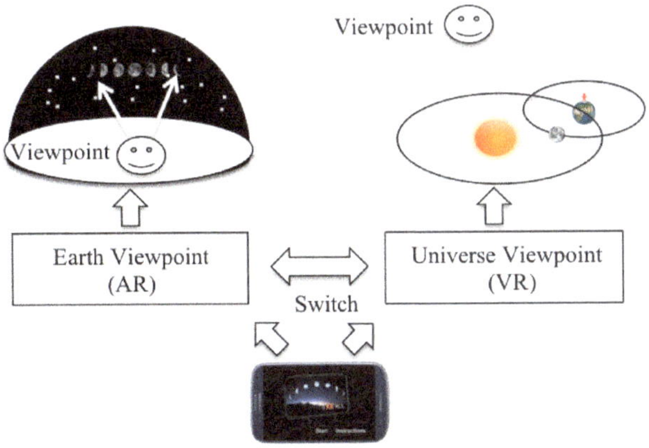

Figure 1. Outline of M-VARML

The M-VARML system has two main views: the AR view and the Universe View. In order to achieve the learning goals of lunar phase observation (see Section II-A), we set the following specifications of the learning environment in M-VARML system:

(a) The azimuth, elevation angle, shape, moon age of moon can be observed in the real sky through the AR view function.

(b) The azimuth, altitude of sun can be observed in the real sky through the AR view function.

(c) The orbit of moon between 1:00 AM and 12:00 PM can be observed in the real sky using the "One Day" function.

(d) The periodic waxing and waning of the moon in the sky can be observed through the "SetDay" function.

(e) The relationship between moon age and the shape of moon can be observed through the "SetDay" function.

(f) The positional relationship among the sun, observer, and moon can be understood through the AR view.

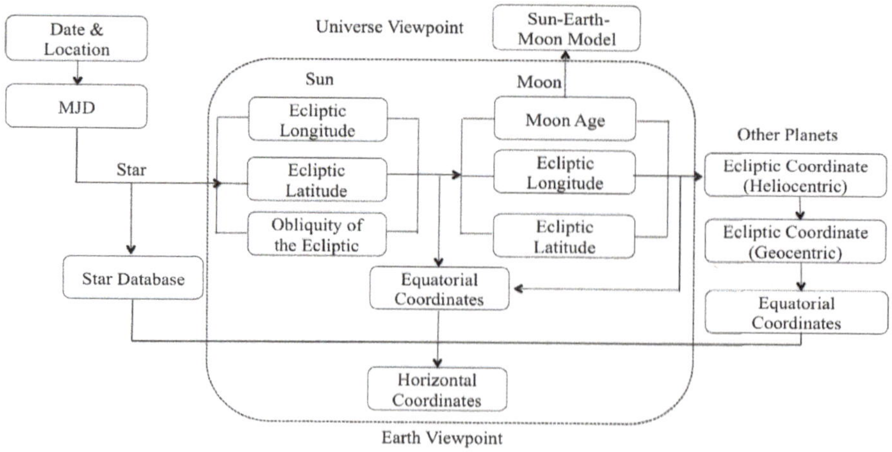

Figure 2. Astronomy simulation model

(g) The relationship among the sun, earth, and moon with respect to the universe can be understood through the universe view function.

In the above specifications, the specification (a) helps to achieve the learning goal (1). The specification(c) helps to achieve the learning goal (2). The specification (d) helps to achieve the learning goal (3) and (4). The specification (e) helps to achieve the learning goal (7). The specification (b) and (f) help to achieve the learning goal (5). The specification (g) helps to achieve the learning goal (6).

C. Astronomy Simulation Model

The M-VARML system is based on the astronomy model that the earth revolves around the sun, and moon revolves around the earth while rotating on its own axis. Our research group has been focusing on the simulation of astronomical phenomenon for a long time. Based on the astronomy simulation model, which is as shown in Fig.2, several astronomy education system [15] & [16] have been developed. In this research, we employ the astronomy simulation model to develop the M-VARML system for supporting moon observation from multi-viewpoints.

D. AR View Function

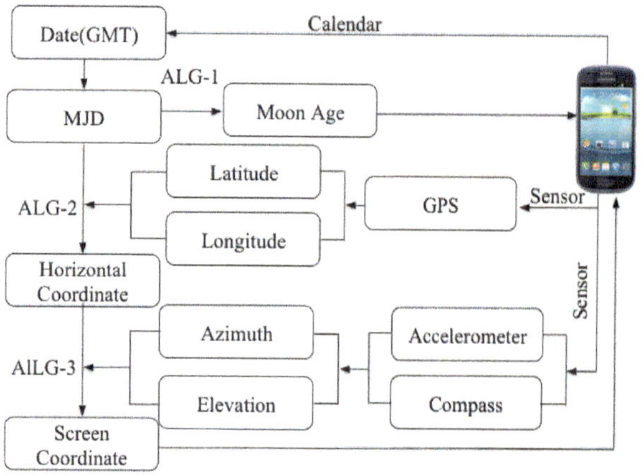

Figure 3. Procedure for calculating moon position in AR view

The M-VARML system shows virtual moon images overlaid on a live video camera background, making the virtual moon image appear in the real world. To visualize the moon correctly in the AR view, it is necessary to get real-time data from the Smartphone through built-in sensors. This is then combined with the astronomy simulation model (see Fig.2) to calculate the screen coordinates of the visual moon in AR view correctly registered to the real sky, which users observe in the sky.

The procedure for calculating the position of the moon is showed in Fig.3. Firstly, it is necessary to obtain real-time date and time from the calendar of the Smartphone and to calculate the Modified Julian Day (MJD) [17] based on the acquired date and time. Then the longitude and latitude data is obtained using GPS sensor, which is responsible for tracking the geographical position of the Smartphone. Combining these data with the horizontal coordinates algorithm (ALG-2), the moon age, azimuth, elevation of the real-time moon and sun can be computed. The azimuth and elevation data is the horizontal coordinate data of the moon in the sky. This is then used to derive the data of the screen coordinate of the moon and register the moon in the AR view, corresponding to the actual moon in the sky. The accelerometer sensor and compass sensor are used to measure the viewing direction and elevation angle of the Smartphone device that the user is using to observe the moon in the sky. Based on this tracking information, we use the screen projection algorithm (ALG-3) to convert the horizontal coordinates to a screen coordinates to locate the moon in the AR view. After this procedure, the virtual moon is shown in the AR view appropriately maps the moon from the real sky.

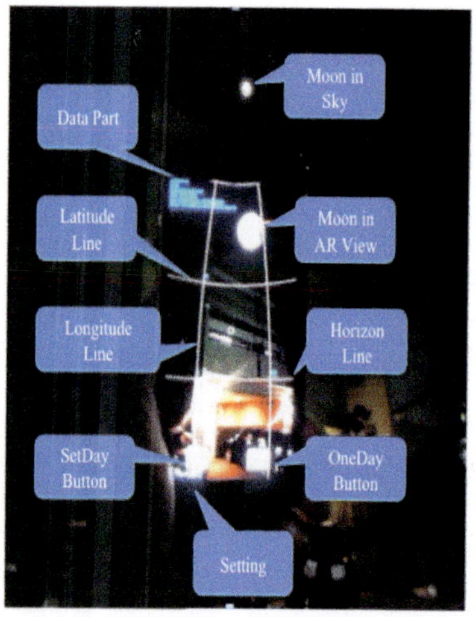

Figure 4. AR view interface

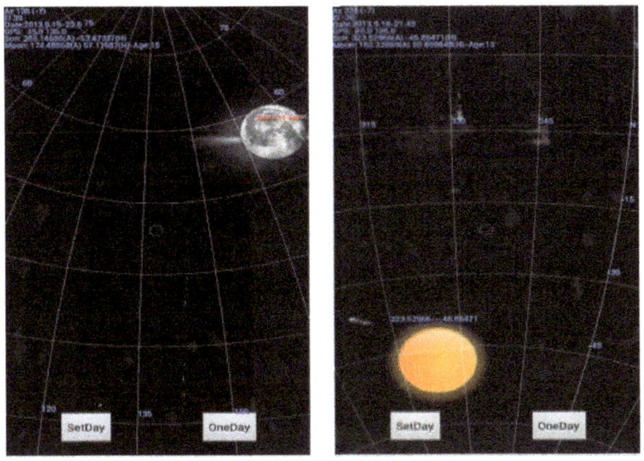

(a) Moon (b) Sun

Figure 5. Moon and Sun in AR view

The complete AR view interface of M-VARML is shown in Fig.4. The main visual components overlaid in the AR view include the data part, visual moon image, latitude line, longitude line, horizon line, and two buttons. The data part shows the data of orientation, elevation angle of mobile device, GPS, date, time, azimuth (sun and moon), elevation (sun and moon), and moon age. The horizontal line is the zero degree on the latitude line, indicating the horizon in the real world. N, E, S, W indicates the direction that the Smartphone device is pointing to in real world, corresponding to the north, east, south and west directions, which, on combining with latitude and longitude line helps learners search and observe the moon and sun as shown in Fig.5. In addition, learners could modify the setting including the degrees of latitude and longitude, size of the text, and color, etc., from the setting interface.

The "OneDay" function meets the specification of learning environment (c). After the users touch the "OneDay" button, as shown in Fig.6 (a), the one-day moon orbit will be displayed in the AR view as shown in Fig.6 (a). The learner rotates the mobile device to observe moon's orbit from 1:00 am to 12:00 pm. It takes the learners a few minutes to observe the moon rising in the east and setting in the west, thereby assisting learners to achieve the learning goal (2).

The "SetDay" function in the AR view meets the specification-learning environment (d) & (e), and the interface is as shown in Fig.7. Users touch the "SetDay" button to step into the "SetDay" interface as shown in Fig.7 (a). Users can choose the number of days of the moon from 1 to 30 days. For example, if a learner selects 30 days of the lunar phase, the lunar phases of the next 30 days will be overlaid in the AR view. In this view, the learners can observe the periodic waxing and waning of the moon, the orbit, azimuth, elevation angle, age, and shape of consecutive moons in the sky in a particular location at once. This merely takes a few minutes to complete the lunar cycle observation. This function helps learners to achieve the learning goal (3), (4), and (7).

Through the AR view function, learners not only could construct a visual feeling for lunar observation in the sky, but also could complete the learning goal of lunar observation from earth viewpoint at anywhere, anytime irrespective of the weather restriction in a short time.

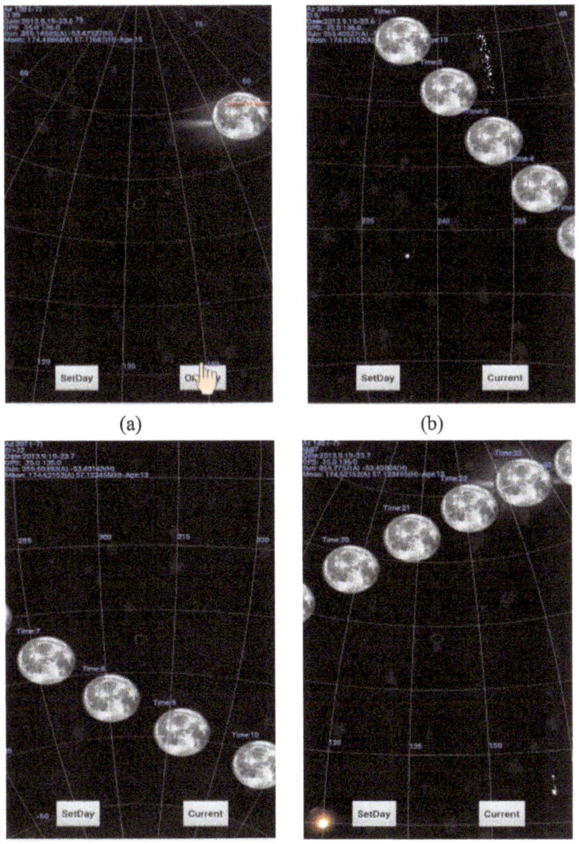

(a) (b) (c) (d)

Figure 6. Observe one-day moon orbit in AR view

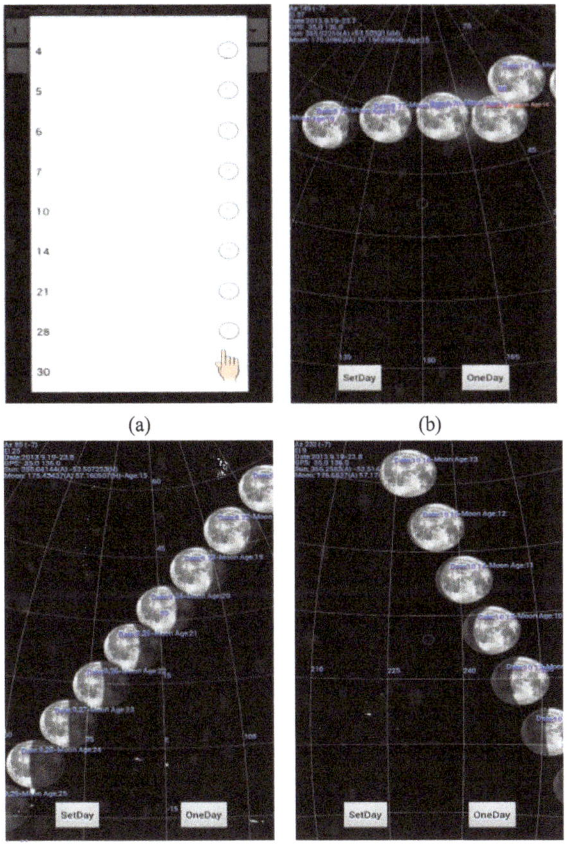

(a) (b) (c) (d)

Figure 7. Observe waxing and waning of moon in AR view

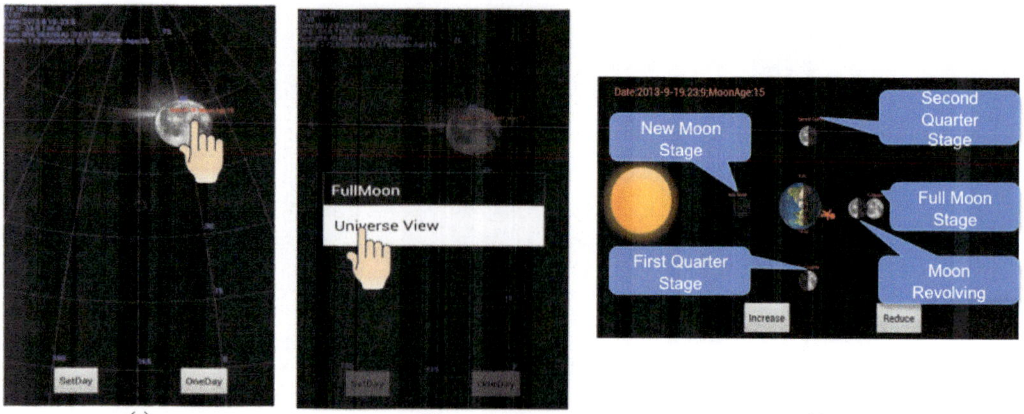

(a) (b) (c)

Figure 8. Step into the universe view

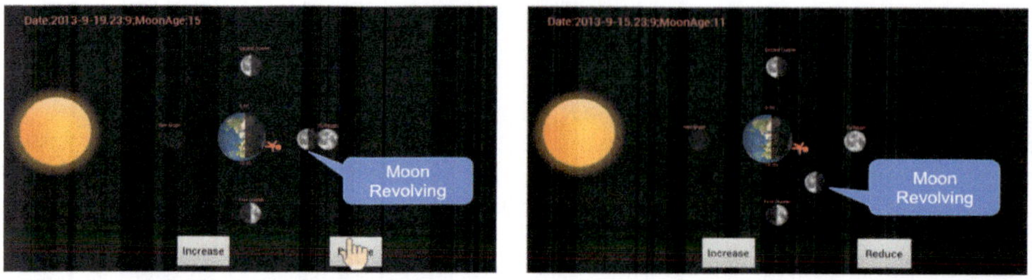

(a) Operating the increase button

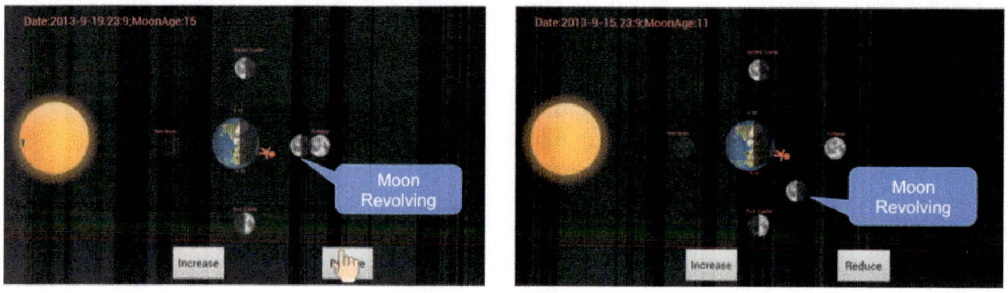

(b) Operating the reduce button

Figure 9. Interactive interface of universe view

E. Universe View Function

Learners observing the phases of the moon often wonder what causes this moon phase. The best way to understand the phases of the moon is to examine an earth-moon-sun view. We used the sun-earth-moon orbit astronomy model to develop the Universe view function. When the user touches the current moon image in the AR view, the user can see an item named Mechanism. Onselecting this item, the user steps into the Universe view interface as showed in Fig.8.

The lunar phases are due to how the sun illuminates the moon and the relative positioning of the earth, moon and sun. Fig.8 shows that the moon revolves around the earth while the earth is rotating on it axis to give us day and night. The person icon on the earth indicates a user standing on the earth observing the moon. It takes the moon about 30 days to revolve around the earth. During those 30 days, as the moon orbits the earth, it grows in size from a new moon, to first quarter, to full moon. It then decreases in size in reverse order to last quarter and then backs to the new moon. Thus, in the Universe View, we visualize four key moon phases of the moon orbit. As shown in Fig.8(c), Thus, in the Universe View, we visualize four key moon phases of the moon orbit.

As shown in Fig.8(c), when users observe the actual moon in the sky, if they know that the moon age is 15 that particular day, and they see the proportion of the visible moon from the earth viewpoint, they can identify the relationship among the sun, earth, and moon through the Universe view function. Users discern that the moon is at the full moon stage.

In the AR view function, users can observe the moon at the same location and time in 30 days. We have also provided users with an interactive function to see the relationship among the sun, earth, and moon at the same location and time during 30 days in the Universe view. Learners can use the increase (see Fig.9 (a)) and reduce buttons (see Fig.9 (b)) to experience it. The Universe view meets the stated specification of the learning environment (g). This improves their knowledge regarding the mechanism of lunar phases.

III. SYSTEM IMPLEMENTATION

The M-VARML system is developed on the Google Android operating system and software development kit. The developing code toolset is Eclipse, ADT, and Android 2.1 SDK. The programming language is Java. The application is supported by a range of smart phones and tablet devices that run on Android 2.1 or higher, and have the required sensors (i.e., camera, GPS, electronic compass, and accelerometer). Our M-VARML was implemented on an Android Smart phone. The mobile device was a Samsung Galaxy S3.

IV. Experiment Design

For this research, two experiments were conducted to confirm the usefulness and usability of the system; the number of participants in each experiment is 10 (18 males, and 2 females, most of whom were in their twenties). They were mainly undergraduate and graduate students from the Nagoya University participated in the two experiments. It was verified that they had an interest in lunar phase observation. The first experiment [18] was learning goal experiment; the other one was task-based experiment in which participants were given specific tasks. In the first experiment, the learning goals of system (See Section II-a) were explained to the learners, and they were instructed on how to use the M-VARML system can be used to observe lunar phases in the real world environment. After the subjects used the system, they answered a questionnaire. The results of questionnaire showed that the system was helpful to learners in observing the lunar phase. However, interviews with participants indicated that they found the various functions of system perplexing and they were unable to grasp the usefulness of each function in the system.

Therefore in the second experiment of the present study, participants were given five tasks, which were based on the aforementioned learning goals of lunar phase observation (see Section II-a), the tasks are listed in Table I. and the purpose of these five tasks is to evaluate the usefulness of each function in system. Fig.11 shows the relationship between the tasks and the questions. Consider task 2 as an example; this task was to evaluate the usefulness of the "OneDay" function, which helps learners to achieve the learning goal (2).

(a) (b)

Figure 11. Experimental situations

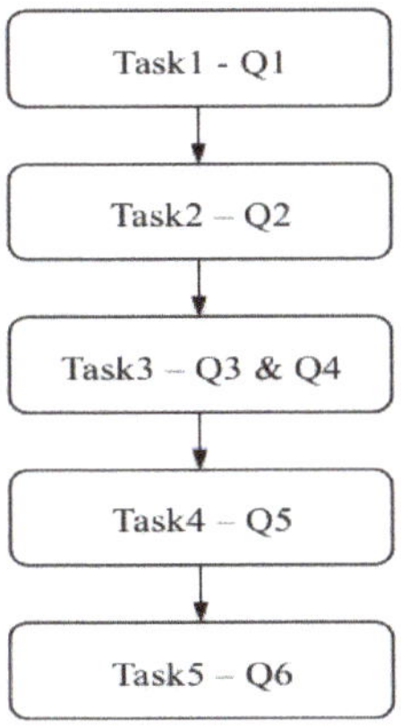

Figure 11. The relationship of the task and questions about usefulness.

TABLE I.
EXPERIMENTAL TASKS FOR LIST

No. Items	Contents
Task1	Firstly, participants use the M-VARML system to search and observe real time moon to understand its azimuth and elevation angle without the latitude and longitude line in the AR view. Then users choose the function that shows latitude and longitude line in AR view. Lastly, the participants use the system again to search the moon and observe the azimuth and elevation angle of real time moon in the sky.
Task2	Initially, the participants guess the one-day moon orbit and use their fingers to draw the line of one-day moon orbit in the sky. In order to confirm their drawings, they use the "OneDay" function to observe the one-day moon orbit in the AR view.
Task3	Firstly, the users guess the one-week moon orbit, the shape and age of the moon. Then, they use their fingers to draw the line of one-week moon orbit in the sky and draw the shape of after one-week moon on paper with a pen. In order to confirm their drawing, they use the "SetDay" function and choose seven days moon phase to observe the future one- week moon orbit, shape, and age in the AR view. Then the users guess thirty days of the moons orbit and its shape, and then they use their fingers to draw the line of 30 days moon orbit. Finally, they choose 30 days lunar phase to observe the 30 days moon orbit, shape, and age in the AR view.
Task4	Firstly, the participants guess the position of sun in the sky by observing the shape of the moon. They manually show the positional relationship among the sun, observer, and moon. In order to confirm the positional relationship, users use the system to search the sun and moon in the AR view.
Task5	Initially, the participants guess the relationship among the sun, earth, and moon in the universe view and draw the relationship of these objects on paper with a pen, in accordance with the shape of the moon. Then, they click the moon image to step into the Universe view to verify the relationship among the sun, earth, and moon. Then, they can click on the increase and reduce buttons to explore the relationship between the sun-earth-moon in one month.

TABLE II.
QUESTIONNAIRE ABOUTY USEFULNESS

No. Items	Contents
Q1	The function of latitude and longitude line in system is more helpful to me in understanding the azimuth and elevation angle of real time moon in the starry sky.
Q2	The function of "OneDay" in system is helpful to me in observing the movement orbit of one-day moon in the starry sky.
Q3	The function of "SetDay" in system is helpful to me in observing the periodic waxing and waning of moon in 30 days.
Q4	The function of "SetDay" in system is helpful to me in understanding the relationship between moon age and shape of moon.
Q5	The function of AR view in system is helpful to me in understanding the positional relationship among sun-observer-moon in the real-world environment.
Q6	The function of Universe view in system is helpful to me in understanding the positional relationship among sun, earth, and moon from the universe viewpoint.

TABLE III.
QUESTIONNAIRE ABOUT USABILITY AND SATISFACTION

No. Items	Contents
Q7	It is easy for me to operate the system.
Q8	It is interesting for me to use the system to observe lunar phase.
Q9	The system is helpful to me in enhancing my motivation on moon observation.

These participants were asked to perform moon observation tasks in the real world environment. Some examples that were selected are shown in Fig.11. The tasks are listed in Table I. After each task was completed, the participant was given follow-up questions seeking the usefulness of the M-VARML system corresponding to the task. These questions can be found in Table II.

After completing all the five tasks and the corresponding questions about the usefulness, participants were give three questions and a free-response item on usability and satisfaction of the system. These questions can be found in Table III. All surveys used a five-point Likert-scale, the responses to which were coded as 1 = strongly disagree through 5 = strongly agree.

V. EXPERIMENT RESULT

A. Usfulness of System

The results of the survey pertaining to the usefulness of system during the task stage is as shown in Table IV. The mean of the questionnaire item Q1 was 4.60, showing that the users deemed the latitude and longitude line in AR view was helpful to them in observing the azimuth and elevation angle of the moon and sun in sky. During task 1, users agreed that the latitude and longitude line is helpful to them in constructing a visual feeling on azimuth and elevation angle in real sky. Therefore, for the usefulness of "OneDay" moon function (Q2), the participants confirmed that this function is very useful in observing the moon rise and set in a day since only one participant was able to draw the correct orbit line of one-day moon in the sky while performing task 2. Similarly, only one user could complete task 3. Some participants even lacked any remote perception of the relationship between the moon age and its shape in the sky. Therefore, the participants conclusively agreed that the function of "SetDay" was helpful to them in observing the periodic waxing and waning of moon, and in understanding the relationship between the moon age and the shape of the moon. The mean of for Q3 and Q4 was 4.60 and 4.40, proving the same. After completing task 4, most of the participants thought that the AR view was helpful in understanding the relationship between the sun, observer, and moon from the earth viewpoint, as is evident from the mean of 4.20 suggests for Q5. As all the questionnaire items (Q1, Q2, Q3, Q4, Q5) regarding the usefulness of the AR view, have a mean above 4.2, it indicates that the AR view is very helpful to the learners in observing the lunar phase and is able to provide a visual learning experience in the real world. In addition, four participants succeeded in completing task 5; users thought that the Universe view function was helpful in understanding the relationship among the sun, earth, and moon from the universe viewpoint. However, as compared to the aforementioned AR view function, the mean of Q6 was just 3.60 since

TABLE IV.
RESULT OF THE QUESTIONNAIRE ABOUT USEFULNESS

No. Items	1	2	3	4	5	Mean
Q1	0	0	0	4	6	4.60
Q2	0	0	1	2	7	4.60
Q3	0	0	1	2	7	4.60
Q4	0	0	1	4	5	4.40
Q5	0	1	0	5	4	4.20
Q6	1	2	1	2	4	3.60

TABLE V.
RESULT OF THE QUESTIONNAIRE ABOUT USABILITY AND SATISFACTION

No. Items	1	2	3	4	5	Mean
Q7	0	1	3	3	3	3.80
Q8	0	0	2	5	3	4.10
Q9	0	0	3	3	4	4.10

most of participants considered that it was better to use the 3D space to show the relationship of sun, earth, and moon. Some of participants suggested that it would be better for new users if additional explanatory text were added in the Universe view to aid them in understanding the mechanism of lunar phases. From this analysis and the experimental results, we can draw the conclusion that the functions of our system are very useful for learners for lunar observation.

B. Usability of System

In term of interface designing, most of the users agreed that the system was easy to operate, with the mean of Q7 being 3.80, which is shown in Table V. They were able to operate the system to complete the moon observation tasks. The participants did report that the data in the AR view was a little hard to understand due to the usage of professional terms in that view. According to the interviews, it was reported that the size of the text should be larger or that a different set of colors would make the interface easier to understand.

C. Satifsaction of System

In addition, the participants appeared to be highly interested in using the system to observe lunar phase, with the mean of Q8 being 4.10. Participants agreed that the system is helpful to them in enhancing their motivation on moon observation in their daily life, which the mean of Q9 is 4.10.

In the questionnaire we also asked question that required participants to freely write down their opinions. 2 participants mentioned it would be better if we highlighted the horizon line and 1 user suggested that we use the arrow guidance to find the moon. We also asked participants what new feature they would like to see in the system. 3 participants suggested that it would be better to have a full moon reminder function. Three 3 users asked if it would be possible to include tidal information related to the lunar phase.

VI. CONCLUSION

In this study, we utilized the Smartphone AR technology and 2D content to construct multi-viewpoint observation environment for learners to observe lunar phases in the real-world environment without restrictions of weather, time, or place. The system has an AR view (to study the orbit, azimuth, elevation angle and circle of moon phases) and a Universe view (to help understand the mechanism of moon phases from the universe viewpoint) to support the learners with moon observation.

The results of the experiment revealed that each function in system was very usable and useful, these functions assisted learners in achieving the defined lunar observation learning goals using the M-VARML system. Therefore the system is helpful to learners to perform moon observation in the real world environment. Furthermore, the M-VARML system enhanced the learners' interest and motivated the learners to observe moon phases in their daily lives.

For future work, we plan to improve the interface design of the system so that even the learners with limited knowledge can use it and add more features, such as providing astronomy news about lunar phase. Further, our research group works in collaboration with Nagoya City Science Museum. Astronomy Curator plays a very important role in astronomy education. An important part of the future work includes discovery of ways to make best use of the knowledge of the curator in learning about the lunar phases. Our future challenges include the implementation of the M-VARML in Nagoya City Science Museum to test the learning effects of system.

ACKNOWLEDGMENT

The M-VARML system for lunar observation that was developed as result of this research is an important part of a project named "Astronomy Education Platform for Use in Smartphone Augmented Reality Technology" which cooperated with the Nagoya City Science Museum and Chukyo University in Japan. This project was supported by the Grants-in-Aid for Scientific Research Japan (No. 25280131). In addition, thanks to Mr. Tom Kabara of Mei Writing in Nagoya University for patiently guiding us in writing our paper.

REFERENCES

[1] R. Zheng and J. Olsen, "Developing the Lunar Phases Concept Inventory," *Proceedings of the 2002 Physics Education Research Conference*, 2002.

[2] H. Agata, "Science Education Collapse - A Survey about the Astronomy Education in Elementary School," *The Astronomical Herald*, vol.97, no. 12, pp.726-736, 2004.

[3] A. Ito, M. Chida, H. Tabara, "Variation of Astronomical Knowledge of University Students Revealed from Investigations Carried out in 1976 and 2006," *Bulletin of Integrated Research Center for Educational Practice Faculty of Education, Utsunomiya University*, vol.30, no.7, pp. 473-482, 2007.

[4] Y. Morita and N. Setozaki, "Practical Evaluation of Tangible Learning System: Lunar Phase Class Case Study," *Proceedings of Society for Information Technology and Teacher Education International Conference*, pp. 3718-3722, Mar, 2012.

[5] C. Ma, "Teaching Effects of 3D Animation Applied to Science Learning in the Fourth Graders-The Case of Moon-Phase Conception," M.S. thesis, National Taitung University, Taiwan, 2008.

[6] Y. Lin, "A study of AR-based Moon Observation System for Learning," M.S thesis, National Hsinchu University of Education, Taiwan, 2012.

[7] M. Kondo, M. Mouri, and Y. Yausda, "A Multiplatform Content Management System for Curators to Provide Teaching Materials in Astronomy Education," *Journal of Theoretical and Applied Information Technology*, vol. 4, no. 9, pp. 874-881, 2005.

[8] M. Soga, M. Ohama, Y. Ehara and M. Miwa, "Real-World Oriented Mobile Constellation Learning Environment Using Gaze Pointing," *IEICE Transactions on Information and Systems*, vol.94, no.4, pp. 763-771, Apr, 2011

[9] M. Suzuki, "Conversations about the Moon with Prospective Teachers in Japan," *Science Education*, vol.87, no.6, pp. 1-13, 2003.

[10] H. Miyata, M. Suziki, M. Fukahori, and T. Akamastu, "Development and Educational Practice of a Lunar Observation Support System by Using Mobile Phones for Science Education," *International Journal on Advances in Intelligent System*, vol. 1, no. 1, pp. 1-10, 2008.

[11] M. Kikuo and K. Tomotsugu, "Augmented Instructions a Fusion of Augmented Reality and Printed Learning Materials," *Proceedings of the fifth IEEE international Conference on Advanced Learning Technologies*, pp. 213-215, July, 2005.

[12] G. Lee, A. Andreas, and A. Kim, "CityViewAR: A Mobile Outdoor AR Application for City Visualization," *11th IEEEE International Symposium on Mixed and Augmented Reality and Communication Engineers Technical Report,* pp. 51-56, Dec, 2007.

[13] N. Setozakai, Y. MORITA, and T. TAKETA, "Study on Effective Use Scene of Multi Aspect Type VR Teaching Materials in Astronomical Education," *The Institute of Electronics Information and Communication Engineers Technical Report,* vol.107, no.391, pp. 51-56, 2007.

[14] T. Kawasaki, N. Iwane, and Y. Matsubara, "Development of a VR-based Astronomy Learning Environment using Intuitive Manipulation and Switching of the View corresponding to User's Body Movement," *Japan Society for Educational Technology*, vol. 34, no. 3, pp. 153-160, 2010.

[15] E. Akao, M. Endo, T. Naka, K, Iwazaki, K. Mouri and T. Yasuda, "Design and Development of Tablet-based Digital Teaching Material for Annular Eclipse Lectures based on Cooperation with a Museum," *2012 The Society of Socio-Informatics*, pp.301-304, Sep, 2012.

[16] M. Ura, M. Endo, M. Yamada, S. Miyazaki, K. Iwazaki, K. Mouri and T. Yasuda, "A Smartphone Application for Searching Constellations Interactivity," *Workshop on Advanced Image Technology Technical Program,* pp.283-288, Jan, 2013.

[17] M. Jean, "Astronomical Algorithm," pp. 59-60, 1998, *Willmann-Bell.*

[18] K. Tian, M. Endo, M. Urata, K. Mouri and T. Yasuda, "Ubiquitous Augmented Reality Mobile Learning System for Supporting Moon Observation," *International Conference on Information and Social Science*, pp. 223-242, Sep, 2013.

AUTHORS

K. Tian is PhD Candidate in the Graduate School of Information Science at Nagoya University, Japan. He can be reached by tianke0711@gamil.com.

M. Endo is an associate professor in the School of Engineer at Chukyo University, Japan. He can be reached by endo@om.sist.chukyo-u.ac.jp.

M. Urata is an assistant professor in the Graduate School of International at Nagoya University, Japan. She can be reached by mayu@nagoya-u.jp.

K. Mouri is a curator of the Planetarium in Nagoya City Science Museum, Japan. He can be reached by mouri@nagoya-p.jp.

T. Yasuda is a professor in thc Graduate School of Information Science at Nagoya University, Japan. He can be reached by yasuda@is.nagoya-u.ac.jp.

Global Telecommunications Security: Effects of Geomagnetic Disturbances

D.J. McManus[1], H.H. Carr[2], and B.M. Adams[1]
[1]The University of Alabama, Tuscaloosa, AL U.S.A.
[2]Auburn University, Auburn, AL U.S.A.

***Abstract*—Global information and communication technologies permeate organizational structures, while questions of security pervade strategic plans of corporations worldwide. From the spectacular to the sublime, the effects of geomagnetic disturbances (i.e., electrical current produced by solar storms) can be as devastating to an organization's telecommunications systems as a hacker breaching a firewall. Using a dataset spanning 31 years (1978-2009) with 580,000 solar activity records, we investigate the effects and relationships of natural anomalies, specifically solar storms, on the security of corporate telecommunications. The ionosphere is a natural barrier around the earth to protect it from the sun and serve as a shield, but some electrical currents break this barrier causing significant telecommunications outages and security breaches within corporations. In this exploratory empirical study, we present the initial evidence that tracking geomagnetic disturbances can provide vital cautions for business continuity planning. The results of the study should help organizations with strategic planning efforts with respect to their overall security, especially as it relates to telecommunications.**

***Index Terms*—Communication Systems, Information Technology, Risk Analysis, Solar Radiation, Organizational Security**

I. Introduction

The adoption of global information and communications technologies is vital to the strategic plans of corporations worldwide, while the issue of secure data transmissions continues to plague network administrators. These administrators must consider all potential threats to their systems' vulnerabilities. One such threat comes from geomagnetic disturbances, as evidenced by solar storms (i.e., interactions of charged particles from the sun with the earth's atmosphere). These disturbances (or storms) are a direct result of solar flares or bursts of energy from the sun, which produce electromagnetic radiation across the electromagnetic spectrum at all wavelengths from long-wave radio to the shortest wavelength gamma rays [1]. These geomagnetic storms have been reported to cause problems for pipelines, electric power systems, and telecommunications networks. The costs associated with these storms can be significant. For example, an estimated $500 million increase in the wholesale price of electricity resulted from solar storms, while government satellite disruptions average $100 million per year [2, 3]. Computer and network administrators consider solar flares a double threat to telecommunications security due to the induction of electricity into wires and equipment, as well as, the interference with radio/wireless communications.

"Telephone and power lines continue to be strung on poles where they are susceptible to the inductive noise and even equipment damage from the solar flares. As early as 1859, telegraph systems were disabled by flares; nowadays, the induced energy might well disable a telephone central switch or computer modem. Areas of the country especially vulnerable are those with long-distance distribution lines as found in the northwest US and Canada. For the wireless environment, flares will add significant noise in the spectrum, making communications difficult and even disabling receivers"[1, p.234].

Although natural anomalies have affected communications systems over the past two centuries, the effects of electrical currents of solar storms, as seen in sunspots, can be as devastating to the security of an organization's telecommunications systems as a hacker breaching a firewall. In this research project, we conduct an exploratory empirical analysis of geomagnetic disturbances and their impact on communications systems.

II. Geomagnetic Disturbances

The electrical "zaps" of the geomagnetic disturbances date back to the telegraph. Since the telegraph used long electrical conductors, it was the first system to be affected by solar storms. Interruptions caused by solar storms date back to 1848, with an electrical telegraph anomaly between Florence and Pisa, Italy. It was not uncommon for the telegraph system to be unusable following such storms. Today, telegraph concerns are rare but many other wire-based systems are of vital interest. "On August 4, 1972, an outage of the L4 coaxial cable system in the midwestern US occurred during a major geomagnetic disturbance. An examination of this disturbance showed that, at the time of the outage, the Earth's magnetic field had been severely compressed by the impact of high-speed particles from the Sun" [4, p. 2]. While the shielding of coaxial cables typically avoid such interferences, on this occasion, it was to no avail.

Solar storms continue to have an impact on continental cable systems, submarine cable systems and fiber optic cable systems, due to the induced voltages produced by geomagnetic disturbances. Although the fiber optic cabling is being used for transatlantic communication, there is still a conductor in the cable to carry power to the repeaters; hence, the electrical voltages from a solar event can still affect the communications signal. This event was duly noted in March 1989, when there was a geomagnetic storm that generated high voltage levels that affected the power supply cables. "Future cables, because of improvements in the fiber optics, may use fewer repeaters and require a lower driving voltage. However, downsizing the power feed equipment without taking account of the induced voltages may leave future systems more vulnerable to geomagnetic effects" [4, p. 2].

One of the best-documented solar disturbances occurred in China during June, 2000. China's Telecommunication Services experienced several geomagnetic disturbances over a two week period, with June 2nd, 6th, and 9th being the most notable. These solar storms have been credited with the interference of telecommunications services, i.e., shortwave radio services, communication satellites, and navigational systems, throughout the country for approximately 17 hours [5]. "Radio waves suffer lots of disturbances as a result of the irregular behaviour of ionosphere which is caused by erratic solar radiation from the sun"[6, p.114].

"During solar flare eruptions high-energy radiation, such as x-ray, and charged particles are hurled out of the Sun... when these particles and energetic radiation impact Earth, they may cause spectacular displays of aurora. However, the impact can create geomagnetic storms and deposit excess amounts of energy on satellite components and power grids, causing these systems to fail. Solar flares can also disrupt telecommunications, corrode pipelines, upset navigation systems, and pose radiation hazard to orbiting yuhangyuan (e.g,; astronauts in Chinese)"[5, p. 2].

These high energy charged particles in the solar winds can disturb the earth's barrier, which is important because the ionosphere reflects the High Frequency (HF) signals used by satellite communications and GPS navigation systems that support U.S. Department of Defense, emergency services, broadcasters, marine and aviation operators [4].

The allure of solar storms, i.e., Aurora Borealis or northern lights in the Northern Hemisphere, has been pondered for centuries. Benjamin Franklin first brought attention to the "mystery of the Northern Lights" in 1779, by theorizing that overcharged electrical charges illuminated the sky [7]. As presented in Table I, there is evidence that allows researchers to postulate that the aurora (i.e., the result of solar storms) indirectly affects electrical communication systems, that is, they are the evidence of charged particles.

TABLE I.
HISTORICAL ACCOUNTS OF SOLAR EVENTS

Event Date	GEOMAGNETIC DISTURBANCES
June 15, 2000	China Telecom Services – Beginning June 2, 2000 solar storms are believed to be responsible for interfering with telecommunications: June 9 - shortwave radio services affected throughout China for 17 hours. The solar disturbance interfered and interrupted a communication satellite and some navigation related systems [5].
March 1989	Fiber Optic Cable - New submarine cables are using optical fibers to carry the signals, but a conductor carries the power to the repeaters. At the time of the storm, a new transatlantic telecommunications fiber-optic cable was in use and large induced voltages were observed on the power supply cables [4].
Aug. 4, 1972	Coaxial Cable - An outage of the L4 coaxial cable system in the mid-western US occurred during a major geomagnetic disturbance [4].
Feb. 10, 1958	Submarine Cable System - Transatlantic communication from Clarenville, Newfoundland, to Oban, Scotland proceeded as alternately loud squawks and faint whispers as the naturally induced voltage acted with or against the cable supply voltage [4].
Mar. 24, 1940	Early phone system - Phone communications in the US were disrupted and voltages in excess of 500 V were thought to have occurred. In Sweden, several magnetic storms produced voltages large enough to start arcing in the carbon protectors, ultimately resulting in fire [4].
Feb. 19, 1852	Chemical Telegraph - Bain's "chemical telegraph" used specially prepared paper: current from a stylus caused a chemical reaction leaving a coloured mark on the paper. Electrical current increased so much that a "flame of fire" followed the pen and set fire to the paper [4].
Sept., 1851	New England, United States – A solar storm occurred and took complete possession of all the telegraph lines in New England and prevented any business from being transacted during its duration [8].
Nov. 17, 1848	Electric Telegraph in Italy - A man named Matteucci had the opportunity to observe the magnetic influence of geomagnetic disturbances. The soft iron armatures employed in the electric telegraph between Florence and Pisa remained attached to their electro-magnets, as if the latter were powerfully magnetized, without the apparatus being in action and without the currents in the battery being set in action [9].
Nov. 17, 1848	Electric Telegraph in England, a man named Highton witnessed a strange event - the magnetized needle of the electric telegraph was always driven toward the same side, even with much force [9].
April 14, 1779	Benjamin Franklin's interest in the mystery of Solar Storms (i.e., Aurora Borealis or Northern Lights) begun on his voyages across the North Atlantic to England. He ascribed the shifting lights to a concentration of electrical charges in the polar regions intensified by the snow and other moisture. He reasoned that this overcharging caused a release of electrical illumination into the air. In this essay, which he wrote in English and French, Franklin analyzed the causes of these geomagnetic disturbances. It was read at the French Académie des Sciences on April 14, 1779 [7].

I. GLOBAL TELECOMMUNICATIONS SECURITY

Scientists believe that electromagnetic space storms will have a negative impact on telephone lines, television signals, aircraft navigation systems, and create widespread power outages [6, 10]. Models have been developed to forecast average trends of solar storms [11, 12, 13]; however, as organizations conduct business continuity planning, they must understand the significant impact of all levels of solar storms to their telecommunications systems. Organizations certainly understand the effects of lightning from a thunderstorm; therefore, it is equally important for them to prepare for the potential degradation resulting from solar flares.

Geomagnetic disturbances, even when infrequent, can breach a corporation's telecommunications system, thus, putting the company at risk for a security breach. "The effects [of solar storms] may be particularly strong in high latitude(s), and trigger powerful currents in telephone and electrical equipments"[6, p.116]. It is evident that modern technologies are adversely affected by solar storms, thus, organizations must recognize the need for a business continuity plan that can address wired and wireless environments. As companies review their global locations, consideration must be given to the impact of solar disturbances in a specific region of the world during a specificseason, time of day, and/or magnetic latitude. Organizations that have worldwide locations have increased vulnerability to solar activities as global divisions will be affected as the earth turns. Business continuity plans must be developed to include the necessary risk assessment and appropriate response to geomagnetic disturbances that result in telecommunications outages.

II. Research Methodology

The National Oceanic and Atmospheric Administration (NOAA) Polar Operational Environmental Satellites (POES) track the daily activity levels of electromagnetic activity with specific focus on our Sun. This empirical data includes approximately 580,000 observations (i.e., activity records) for the years 1978-2009. POES monitors the power flux of the sun-producing protons and electrons, thus, using the power flux measurements to determine the total power input in a specific polar region [14]. These measurements are based upon an activity index, results of 100,000 satellite passes, ranging from 1 to 10, where 10 is the highest; the total power dissipation is measured in gigawatts. We extrapolated the activity data from the NOAA POES to determine trends based upon historical events.

For each instance of data, the power and corresponding activity level are tracked, as well as the date, center time, hemisphere, satellite, and normalization factor. The center time is the time when the satellite is halfway complete in the pass over the pole; the hemisphere distinguishes whether the data was taken from the northern or southern hemisphere; the satellite is determined by the satellite number recording the event; and the normalization factor (n) represents the level of confidence in the estimate of power. A normalization factor value of 2.0 or less for n signifies strong confidence, while a value of 2.0 or greater means there should be less confidence in the power and activity level measurements [14].

During our preliminary investigation, we evaluated data relating to historical geomagnetic disturbances that have impacted telecommunications systems throughout the world. Our purpose was to analyze individual known events, such as those reported in Table 1. In this stage of the research project, statistical analyses were conducted in an attempt to evaluate the relationships of historical events and geomagnetic disturbances occurring during reported outages, which are attributed to solar storms. We focus our findings, in this preliminary stage, on the events of the China Telecommunications services outages in June, 2000. The geomagnetic disturbances beginning June 2, with additional storms on June 6th and June 9th are clear indicators that solar storms can significantly impact telecommunications equipment and services as well as shortwave radio service. The most serious of these three events, interfering with the China telecommunications systems, occurred on June 9th, resulting in an outage of 17 hours, with interruptions to communication satellites and navigational systems. These geomagnetic disturbances were well documented by NOAA POES, including timeline and activity level of the solar storms.

III. Discussion

In our preliminary investigation, the activity level of the solar storms for June, 2000 was evaluated by conducting a time series analysis using the measurement of gigawatt power. Figure 1 reports the findings of this statistical process for June 1-4, 2000 in the Northern Hemisphere. The disturbance in China was reported approximately 3:30pm (local time). This timeline would correspond to approximately 2:30 a.m. in North America and 9:30 a.m. in Europe. As indicated in Figure 1, there is clear evidence of a significant magnetic disturbance, (i.e., during early morning of June 2nd). In addition, the solar storm is stronger during the summer months; thus, the reported disruptions in China were at maximum strength, resulting in significant outages.

Based upon the results in Figure 1, a closer examination was warranted, so a time series plot is provided for June 1-2, 2000, as shown in Figure 2. Observations 29 and 37 become more evident; however, one should question the impact of observation 37, with a value of 195. Based upon the criteria presented by NOAA POES, observation 37 has a quality reading of $n > 4$ indicating that there is less confidence in the reliability of the measurement. Observation 29, with a value of 101.2, has a quality measure of $n=1.2$ indicating a very reliable measurement. Although there is less confidence in the reliability of observation 37, there is clear evidence that a geomagnetic disturbance occurred that had a significant impact on the country of China.

Furthermore, Figure 3 presents a histogram of gigawatt power readings by hemisphere for all data collected in the year 2000. The overall distributions of the power measurements are skewed; however, these readings, and any resulting actions, must be tempered by the quality of the measurements as determined by n. This is the juncture where thresholds for threat levels need to be established and perhaps establishing boundaries for Level 1, 2, or 3 emergency recovery plans.

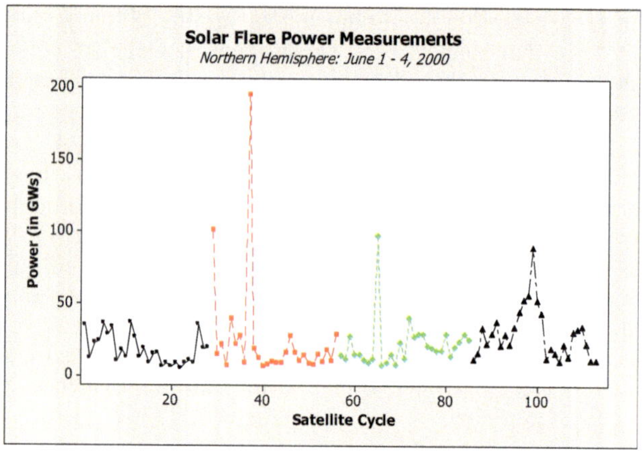

Figure 1. Solar Flare Power Measurements – Northern Hemisphere: June 1- June 4, 2000

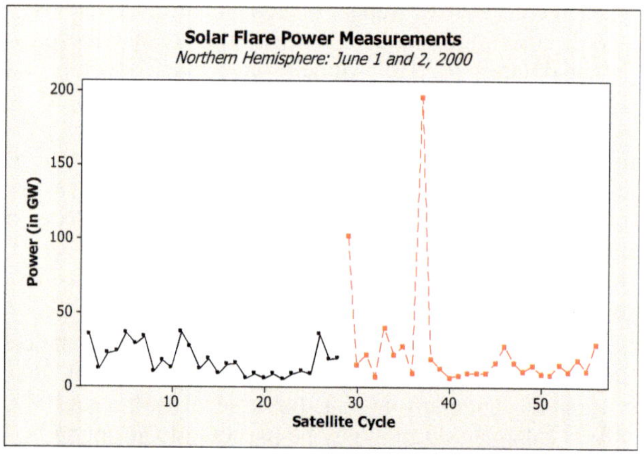

Figure 2. Solar Flare Power Measurements – Northern Hemisphere: June 1 and June 2, 2000

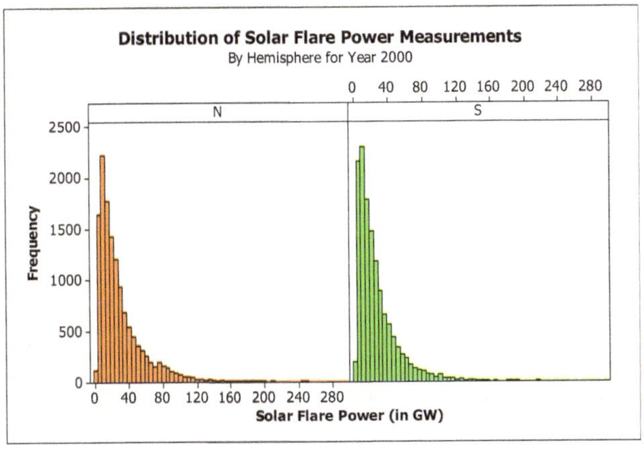

Figure 3. Distribution of Solar Flare Power Measurements – By Hemisphere (Northern and Southern) for Year 2000

Power readings for June 6th and June 7th, 2000 are reported, by hemisphere, in Figures 4a and 4b. On both of these days major solar storm activity was experienced and significantly impacted Chinese telecommunications services. It should be noted that the impact of solar storms may potentially be influenced by a number of factors such as the season (i.e., tilt of the earth's axis), time of day (i.e., earth's orientation to the sun), altitude, or latitude. In both cases, China is protected from solar events by the earth's rotation, which could explain the lack of disruption in the Chinese infrastructure. As noted in Figures 4a and 4b, solar disturbances may impact different hemispheres with varying degrees; hence, this should be considered in evaluating business continuity planning.

The power measurements reported in Figure 5 certainly correspond with the short wave radio, communication satellite and navigational system disruptions, experienced in China on June 9th and 10th, 2000. There is clear evidence that the Northern Hemisphere experienced extremely strong geomagnetic disturbances that coincided with outages of 17 hours. In a comparison of the solar storm activity levels on these two days with June 2nd, 6th, and 9th, it is important to note the relative severity of the power measurements and severity of the outage.

NOAA POES recorded the eruptions on the Sun during this reported outage. These events caused by an M-class flare at 3:00 p.m. on June 2nd affected the southern region of China, such as Guangzhou, Hong Kong and Haikou. The second geomagnetic disturbance event was recorded approximately midnight on June 6th, resulting from two X-class flares that produced large amounts of charged particles. Fortunately, since this event occurred at night, the country of China was facing in the opposite direction away from the high energy radiation. However, two days later on June 9, during a daylight event, charged particles slammed into the Earth's atmosphere and caused a severe storm in the ionosphere, resulting in wide spread disruption to shortwave radio services and other communications systems [5]. These coinciding events provided strong evidence that the geomagnetic disturbances can impact organizations worldwide. Therefore, the academicians of twenty research units associated with the Chinese Academy of Sciences joined together to develop a 15-year Space Weather Strategic Plan [5]. It is projected that more geomagnetic disturbances will occur, which necessitates the need for organizations to include in their business continuity plan recovery strategies for geomagnetic disturbances;

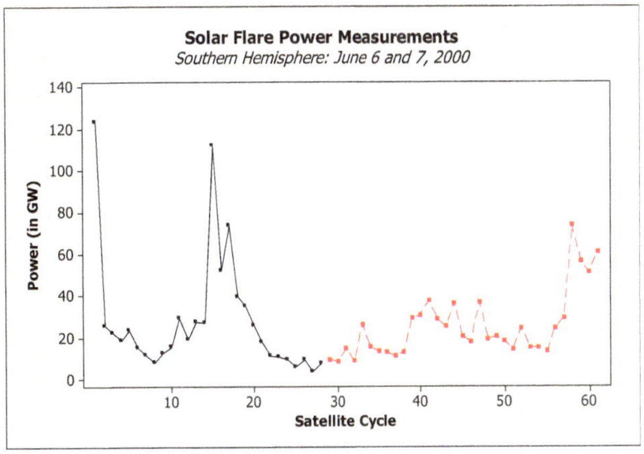

Figure 4a – Solar Flare Power Measurements – Southern Hemisphere: June 6 and June 7, 2000

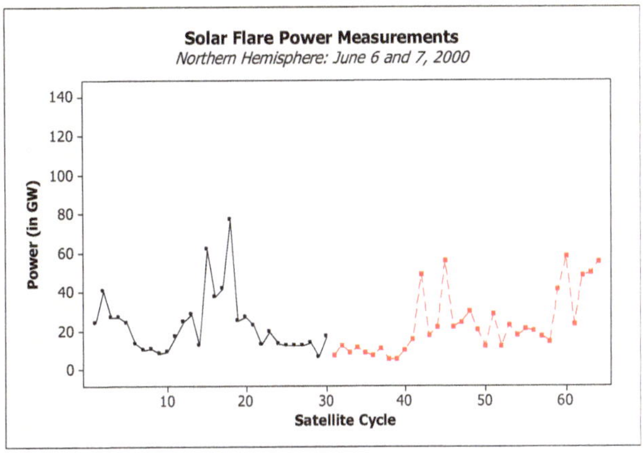

Figure 4b – Solar Flare Power Measurements – Northern Hemisphere: June 6 and June 7, 2000

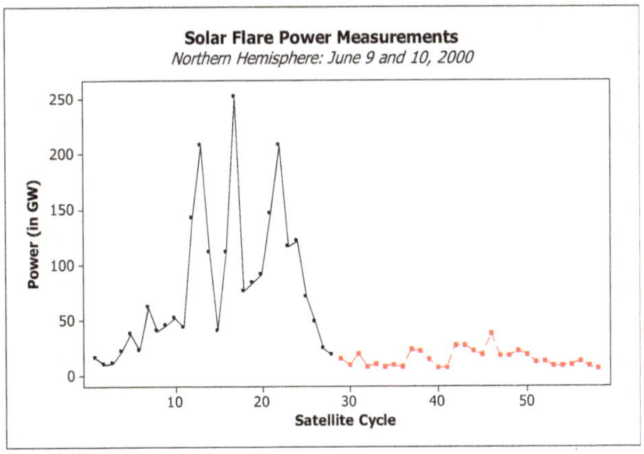

Figure 5. Solar Flares Power Measurements – Northern Hemisphere: June 9 and June 10, 2000

thus, protecting the integrity of their telecommunication systems.

The initial results from this research provide evidence that by using satellite readings of gigawatt power measurements, one can determine reasonable guidelines for Emergency Response Activity thresholds or threat levels. These thresholds or threat levels might be a function of location measures (e.g., magnetic latitudes, hemisphere), local time (e.g., am, pm), and altitude of city or region.

The development of predictive models to provide early warnings of threats would aid emergency response efforts. One approach is directly modeling the intensity of solar

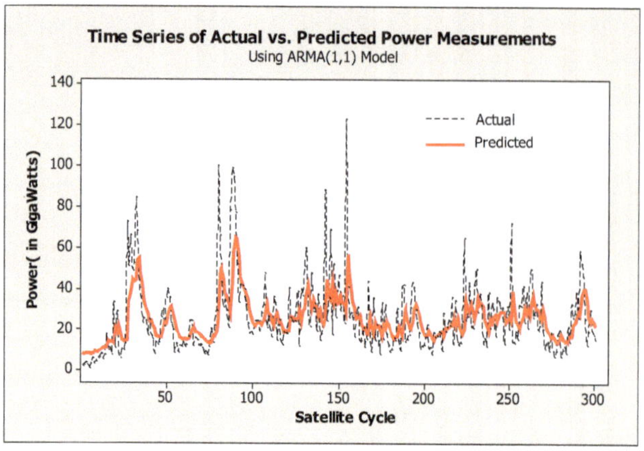

Figure 6. Solar Flares Power Measurements – Northern Hemisphere: Actual vs. Predicted

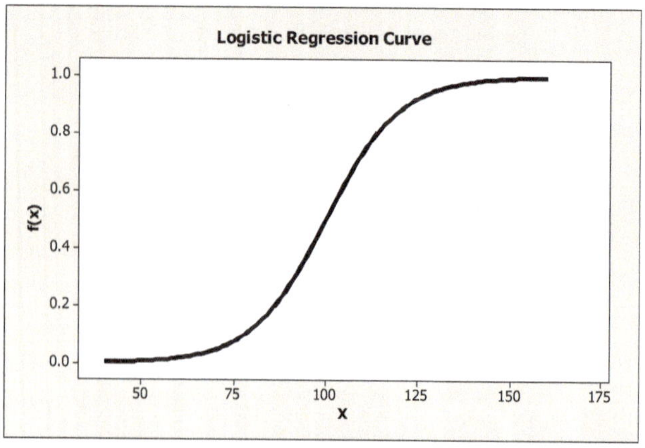

Figure 7. Illustrative Logistic Regression Curve for Modeling the Likelihood of Events

flares. As represented in Figure 6, a sequence of 800 consecutive satellite readings are modeled using an autoregressive moving average (ARMA) model of order (1,1). The actual observations are depicted by the smaller dashed lines while the predicted values are represented by the bold lines.

The ARMA (1,1) parameters were all statistically significant ($p<0.005$) as illustrated in Table II. While the predicted values reflect the average trend in solar flare activity, the extreme values are consistently underestimated. Predicting trends in solar flare activity is important but the extreme values are the source of threats to telecommunication infrastructure.

TABLE II.
STATISTICAL TEST OF ARMA(1,1)
FINAL PARAMETER ESTIMATES

Type	Coef	SE Coef	T	P
AR 1	0.8880	0.0259	34.22	0.000
MA 1	0.5846	0.0457	12.79	0.000
Constant	2.5839	0.2145	12.04	0.000
Mean	23.075	1.916		

A second approach is to model the threat level using a scale of 0 to 1 with zero (0) indicating no threat of disruption and one (1) indicating a significant disruption due to solar activity, such as the China solar event. Two potential predictive models are logistic regression models and neural networks. In the case of logistic regression, exogenous variables result in prediction values in the interval [0,1] which have nice probabilistic interpretations. The simplest form of the logistic regression model is given by

$$E(Y) = \frac{\exp(\beta_0+\beta_1 X)}{1+\exp(\beta_0+\beta_1 X)}$$

where E(Y) represents the expected value of the response variable (probability of a solar event) and X represents some exogenous variable. Figure 7 provides an illustrative graphical display of a logistic regression curve with $\beta_0 = -10$ and $\beta_1 = 0.1$. Note that as the value of X increase from 50 to 150, the estimated likelihood of a solar event increases from nearly zero to nearly one. In future research, we will attempt to develop practical prediction models using the findings of this initial investigation.

Additional input variables such as the presence of coronal mass ejections (CME), magnetic orientation of CME, and infrastructure location (longitude, latitude, elevation) are being evaluated as part of on-going research.

To date, the prediction of solar flares and geomagnetic disturbances are limited at best; however, by tracking the intensity of solar disruptions, we have the opportunity to predict potential dangerous outages. Companies may then assess potential costs of disruptions and reasonable counter measures for maintaining viable operations at the predicted threat level. As companies review their global locations, consideration must be given to the impact of geomagnetic disturbances in a specific region of the world during a specific season, time of day, and latitude.

IV. Conclusions

Globalized corporations depend upon information and communication technologies as a component of their strategic plans; hence, deciphering and mitigating risk to these technologies is vital to corporate business continuity plans. The International Telecommunication Union has reported in the year 2010, there are five billion mobile devices worldwide. Just imagine the 6.8 billion people on our planet without mobile devices. As companies review their global locations, consideration must be given to the impact of geomagnetic disturbances in a specific region of the world during a specific season, time of day, and latitude. Business continuity plans must be developed to include the necessary risk assessment and appropriate response to geomagnetic disturbances that result in information technology outages.

Acknowledgments

We would like to acknowledge and thank National Oceanic and Atmospheric Administration (NOAA) Polar Operational Environmental Satellites (POES) for their continued collection and sharing of solar activity and data sources. In addition, we would like to thank our student, Kristina Dueland for her efforts during the preliminary stages of our research. An earlier version of this research was presented at the 2010 Interdisciplinary Conference of AHLiST, Association of History, Literature, Science and Technology.

REFERENCES

[1] H. H. Carr, C. A. Snyder and B. N. Bailey, *The Management of Network Security: Technology, Design, and Management Control*. 1st ed., Pearson/Prentice/Hall, 2010.

[2] S.F. Oldenwald and J.L. Green, "Bracing the Satellite Infrastructure for a Solar Superstorm," *Scientific American Magazine*, August 2008.

[3] Data Users. NOAA Economics; *The Economic and Social Benefits of NOAA Data & Products. Weather and Water*, (2009), 1-2.

[4] Space Weather Canada, "Geomagnetic effects on communication cables," Government of Canada.

[5] W. Long, "Solarstorms impact China's telecom services." *Space-Daily*, 15 June 2000.

[6] B.A. Ezekoye and R. M. Obodo, "The effects of solar radiations on telecommunications", *Pacific Journal of Science and Technology*, 8(1), pp. 109-117, 2007.

[7] Library of Congress, "Benjamin Franklin in his own words", *Library of Congress, Exhibitions, American Treasures*, 2006.

[8] G. B. Prescott, "United States early radio history – telegraph", *History, Theory and Practice of the Electric Telegraph*, pp. 318-326, 1860.

[9] P. Norton, "The Aurora Borealis and the Telegraph", 2007.

[10] McManus, D.J., Carr. H.H., and Adams, B.A. Wireless on the Precipice: The Fourteenth Century Revisited. *Communications of the ACM*. June, 2011.

[11] Dikpati, M., deToma, G. and Gilman, P. Predicting the Strength of Solar Cycle 24 Using a Flux Transport Dynamo-based Tool. *Geophysical Research Letters*, 33, L05102, (Sept. 2006),

[12] Hathaway, D. and Wilson, R. Geomagnetic Activity Indicates Large Amplitude for Sunspot Cycle 24, *Geophysical Research Letters*, 33, L18101, (Sept., 2006),

[13] Hathaway, D. Solar Cycle Prediction. *NASA/Marshall Solar Physics*, (May, 2010).

[14] J. Green and L.Puga, "Introduction - auroral activity extrapolated from NOAA POES", 2007, *Space Weather Prediction Center.*

AUTHORS

D. J. McManus is with The University of Alabama, Tuscaloosa, Alabama, 35487 USA (e-mail: dmcmanus@cba.ua.edu).

H. H. Carr is with Auburn University. Auburn, Alabama 36849 USA (e-mail: Houston@business.auburn.edu).

B. M. Adams is with The University of Alabama, Tuscaloosa, Alabama, 35487 USA .

17

Innovations in Environmental Monitoring Using Mobile Phone Technology – A Review

M.J. Aitkenhead, D. Donnelly, M.C. Coull and E. Hastings
The James Hutton Institute, Craigiebuckler, Aberdeen, Scotland, UK

Abstract—**In recent years, the use of mobile phones and tablets for personal communication has increased dramatically, with over 1 billion smartphones out of a total of 5 billion mobile phones worldwide. The infrastructure and technology underlying these devices has improved to a level where it is now possible to integrate sensor technology directly and use them to acquire new data. Given the available resources and the number of technical challenges that have already been overcome, it would seem a natural progression to use mobile communication technology for field-based environmental monitoring. In this work, we review existing technology for acquiring, processing and reporting on environmental data in the field. The objective is to demonstrate whether or not it is possible to use off-the-shelf technology for environmental monitoring. We show several levels at which this challenge is being approached, and discuss examples of technology that have been produced.**

Index Terms—**Environmental monitoring, mobile technology, apps, sensors, mobile phones.**

I. Introduction

The potential of mobile phone technology to provide rapid, cost-effective environmental monitoring has expanded greatly in recent years. An important aspect of this is the fact that the millions of mobile phones currently being used around the world are the product of a mature technology that is mass-produced cheaply and reliably. These systems are designed to be flexible and act as miniature computers, to enable the user to carry out a number of different activities and also to allow developers to produce software, in the form of applications (apps) that can be installed easily and rapidly. This ready-to-use technological base, ubiquitous and purposefully designed to be extensible, provides fertile ground for environmental monitoring on a massive scale. Ref. [1] provides a broad survey of the technology being used for mobile phone sensing systems. Here we provide a more in-depth examination in relation to environmental sensing, focussing especially on the technological sophistication and user/device interaction level.

Environmental monitoring and healthcare are two areas that overlap in relation to sensing systems. Healthcare, particularly in developing countries or remote regions, can potentially benefit greatly from small, cheap sensors that can be plugged into mobile phones. Ref. [2] described a mobile phone add-on containing a sensor on a chip capable of detecting specific disease biomarkers rapidly and without requiring great expertise on the part of the user. Ref. [3] demonstrated a system that uses a mobile phone camera to detect and measure light levels in a fluorescing sensor chip. This kind of technology presents great potential for disease detection, biosecurity, food safety and environmental monitoring. Ref. [4] also demonstrated the use of mobile phone technology to relate activity and location (using built-in geolocation functionality of mobile phones to determine position and travel speed) to air pollution information, and showed that this technology could be used to provide an unobtrusive method of environmental monitoring for health purposes.

Many examples of technology in this area already exist. Ref. [5] discusses the potential for using mobile phones for environmental sensing, although they focussed on sensors embedded into the mobile phone rather than augmentation with additional devices. Work at The James Hutton Institute on linking statistical models to mobile phone apps with image analysis techniques has led to rapid field monitoring for soil [6].

Here we identify and review a number of examples over a wide range of technological sophistication and level of user/device interaction, in order to provide a summary (with notable examples identified) of what is currently available. We have focussed on instrumentation and software that uses mobile phone/tablet technology, rather than on small-scale computers such as the Raspberry Pi or Arduino or on large-scale automated monitoring systems that use mobile phone/GPRS networks for remote data transmission, as the focus is on technology that can be implemented ‘off the shelf’. This work is based upon an earlier report [7] developed for SEPA (Scottish Environmental Protection Agency) that focussed on aquatic monitoring, but we have broadened the scope to cover environmental monitoring in general.

II. Methods & Results

A. *Categorisation of technology*

There are many different solutions available within the field of environmental monitoring, with many aimed at human health, water quality and agriculture in particular. Grouping the different technological and computing developments into specific categories allows us to more easily compare these different solutions, and allows the reader to focus on the category that is of greatest interest to them. There are several possible frameworks of categorisation of mobile environmental monitoring solutions. We have chosen to group them in two ways: by technological sophistication (which is assumed to be closely related to cost) and by user/device interaction. The terms used in the listings below are our own, and are intended to reflect the category in each case while providing headings for a table in the Discussion.

1) Technological sophistication

- Observation-only apps – purely app-based, with no additional technology required and little or no interpretation of the information obtained. Commonly used to record imagery of events or specific features within the landscape, sometimes with the option to provide additional data from the user's visual interpretation (e.g. water quality, presence of a particular species, see later examples).
- Clever apps – again, purely app-based but with more sophisticated interpretation of information obtained by the user. An example could be where image analysis is integrated with additional knowledge available from knowing the user's geographical location and both are used to drive models or to provide information directly to the user from digital maps.
- Plug & play – off-the-shelf sensor devices that can be attached externally onto a mobile phone or tablet and used to augment the sensor technology of the device. These are ready to use and provide specific sensor readings.
- Clever kits – either for purchase in kit form, or recently developed and intended for testing and later development as a 'plug and play' device, this type of technology requires construction and a high degree of understanding from the user. Sensor devices of this kind are usually customisable and are designed to be used by researchers or professionals.

2) User/device interaction

- Crowd source – here, the user is not required to provide any interpretation of the results, but simply acts as a recorder of information (usually imagery, but sometimes audio recordings). The information is transmitted to a central location where it can be subjected to further interpretation by experts. This level of interaction can potentially take place at any of the levels of technological sophistication described above, although it tends to be seen more at the 'observation-only app' level.
- Personal use – the user is provided with readings, usually interpreted automatically to make them more useful, about their environment. There is no expertise required in handling the equipment or interpreting the information obtained, but some experience in using the equipment properly might be necessary.
- Decision support – information is taken from the user, either in the form of imagery for interpretation or numerical values (or both). This is integrated with information derived from knowledge of the user's position within the landscape, and is used to provide information that is of specific interest for land management etc. The outputs from this type of system tend to be less 'interpreted' than they would be for a member of the public.
- Mobile lab – the user of this technology is a specialist, often a scientist or someone looking to answer a specific question that requires sophisticated monitoring. Use of this technology requires an understanding of the system being monitored, training in the use of the technology and the ability to interpret sophisticated monitor readings. The use of the term 'tricorder' for this kind of multifunctional sensing device is increasingly common in the scientific and medical community in relation to smartphone use (e.g. [8]).

The following table shows how the above categorisation works, and lists the examples that we have been able to find that fall into each category, with links to a discussion of each example.

TABLE I.
CATEGORISATION OF SOLUTIONS, WITH EXAMPLES

	Observation-only apps	**Clever apps**	**Plug & play**	**Clever kits**
Crowd source	What's invasive NatureWatch	Water Quality Reporter		
Personal use	PEIR		Lapka Ghana air quality Thermodo Alcohoot	
Decision support	SoilWeb SoilMapp mySoil	SIFSS SOCiT		Platforma SINC MoboSens
Mobile lab		EMAP SEPA /Geofield	Tesla Micro Proscope	Sensordrone WiMoto smarTROLL

B. Presentation of relevant examples

We have presented below a list of examples that have been found of apps, mobile phone-related equipment and associated systems for environmental monitoring. The relevance of these examples varies, and where necessary we have identified the concepts or ideas that are relevant to environmental monitoring and that could be extracted for future systems. In each case, we have identified the design function, the potential of the system for environmental monitoring, the level of technological sophistication, the level and type of user/device interaction, any limitations on the use of the system in question and its suitability (i.e. the level to which it achieves the stated purpose).

Name: **What's Invasive**

Web address: http://whatsinvasive.com/

Designed function: Crowd-sourcing of invasive plant species.

Environmental monitoring potential: One of a number of community data collection apps, it is designed for use in collecting observations on the locations of invasive species of plants. It could equally be applied to animals, and is applicable in terrestrial or aquatic settings, and across a wide range of habitat types or ecosystems. Indirectly, this app and others similar to it can provide a template for capturing user observations of features or issues of interest e.g. pollution incidents, algal blooms, illegal dumping etc.

Technological sophistication: The app is relatively simple and acts more as a mechanism for capturing user observations than anything else. Photographs can be geotagged automatically, as can observations, allowing databases to be populated with user observations and images. However, there is no automated interpretation of the information sent, neither is there much in the way of moderation – the quality or relevance of an observation varies with the user.

User/device interaction: This app can be used with the minimum of training or experience, and requires no sophisticated technical knowledge on the part of the user. However, they are expected to be able to identify the species or other feature of interest that they are describing.

Limitations on use: This is very much a 'what goes in is what comes out' app – all information of use is entered by the users, whose level of expertise in the subject area may vary. It is difficult therefore to produce statistical or numerical evaluations of the topic of interest.

Suitability assessment: Potentially useful for capturing the spatial location and distribution of qualitative observations. Not capable of capturing quantitative observations unless used in parallel with monitoring equipment. In order to make use of this app for monitoring environmental quality more generally, it would have to be developed to allow different observations to be entered.

Name: **PEIR**

Web address: http://www.cs.cornell.edu/~destrin/resources/conferences/2009-jun-Mun-Sheddy-PEIR.pdf

Designed function: Developed at UCLA and demonstrated in 2009, which calculated exposure to particulate matter in cities based on positional information and mode of transportation.

Environmental monitoring potential: Assessing user risk exposure to airborne pollutants is useful by itself, and the methodology could be transferred to other types of hazard e.g. water-borne pollutant exposure assessment.

Technological sophistication: Records the user's position and movement automatically, and integrates this information with existing datasets. The results are interpreted for the user and presented in an understandable form.

User/device interaction: Relatively simple user interaction is involved – the app presents the user with a set of results about air pollution exposure and also sends this information to servers for later data interpretation.

Limitations on use: Only designed for airborne pollution currently, but could be enhanced to be applicable to water and other forms of environmental pollution relatively easily.

Suitability assessment: Designed to integrate information about the user's position with existing spatial datasets and expert knowledge about transport systems and air pollution. As such, it requires existing datasets to be available, so wider use would only be feasible when assessing exposure to pollutants that have already been mapped or can be modelled.

Name: **Water Quality Reporter**

Web address: http://www.bristol.ac.uk/aquatest/in-action/aquatest-system/wqr/

Designed function: Transmits information from a water testing device to water supply companies.

Environmental monitoring potential: This app is designed specifically to allow water monitoring results to be sent to a central system for processing, and has been extensively tested and used in the field. The measurements uploaded by the user are perhaps not the full range that could be applied.

Technological sophistication: Relatively easy to use and technologically unsophisticated – this app is designed to be used by anyone in the field who has access to the relevant water quality monitoring equipment and a mobile phone. The data recording side is quite well designed and provides a framework for later remediation and interpretation.

User/device interaction: Straightforward data input by the user – no connection between the monitoring device and the mobile phone.

Limitations on use: The user has to upload the measurements, rather than having a direct connection between the monitoring device and the mobile phone or tablet. This means that there is a time/effort constraint on uploading multiple measurements.

Suitability assessment: This app is suitable for water quality monitoring, but is fairly restrictive in that it requires user input rather than connection to the monitoring equipment, and limits the types of observations that can be entered. By connecting to different types of sensors it could have application in more general pollution monitoring, contaminated land assessment, and risk exposure studies.

Name: **Platforma SINC**

Web address: http://www.alphagalileo.org/ViewItem.aspx?ItemId=128267&CultureCode=en

Designed function: Equipment can be used to detect mercury concentrations in water, based on colour change in sensitive paper. Not fully implemented as a monitoring tool, but its use with mobile phones has been demonstrated.

Environmental monitoring potential: Good potential, needs to be more fully developed to be an integrated package. Some kind of standardised detection mechanism (i.e. standardized chemosensitive paper in a mechanism that allows easy imaging) would improve the functionality.

Technological sophistication: Relatively low-tech, although the image analysis software requirements for detection and processing of the chemosensitive paper colour change are not trivial.

User/device interaction: Some activities required on the part of the user, in relation to sample preparation and acquisition of appropriate imagery.

Limitations on use: The limitations on this kind of application relate more to the chemosensitive paper (or whatever sensing mechanism is used that changes colour). If there is a chemical of interest and a reliable colour-change chemistry that can be implemented in the field, then there are no limitations on application.

Suitability assessment: Currently most suitable for water pollution monitoring, but if appropriate chemical-colour change mechanisms can be implemented that are cheap, rapid and accurate, a wider range of pollution and contaminated land assessment would be possible.

Name: **Ghana Air Quality**

Web address: http://www.globalproblems-globalsolutions-files.org/unf_website/PDF/vodafone/tech_social_change/Environmental_Conservation_case3.pdf

Designed function: Carbon monoxide sensor linked to mobile phone technology, used as a pilot study.

Environmental monitoring potential: The sensor that links to the mobile phones was designed specifically to record atmospheric carbon monoxide levels. However, similar sensors could be developed with effectively identical electronic interfaces to the mobile phone technology.

Technological sophistication: The sensing device is the most sophisticated part of this system, with the recording app being relatively simple and low in 'options'.

User/device interaction: No user/device interaction, this is a portable recording device that the user (a taxi driver) transports in their vehicle and which records information about air quality over time. The recorded information is uploaded to servers for later processing.

Limitations on use: The limitations relate to the physical monitoring mechanisms – a water quality sensor that produces a translatable electronic signal could be linked to a mobile phone just as easily as a sensor that detects air pollution levels. However, such a sensor does need to exist in the first place.

Suitability assessment: Potentially suitable for environmental monitoring, if the monitoring equipment is available. An app for recording readings could be produced rapidly and cheaply.

Name: **Sensordrone**

Web address: http://www.kickstarter.com/projects/453951341/sensordrone-the-6th-sense-of-your-smartphoneand-be

Designed function: Multisensory device for mobile phones, for use in experimental environmental monitoring.

Environmental monitoring potential: This system has a lot of potential as it allows sensors that measure relevant parameters to be linked through Bluetooth technology to mobile phones, and provides a 'whole package' system for acquiring, storing and visualizing logged data.

Technological sophistication: This is a relatively sophisticated system, as it includes sensors, connectivity to mobile phones for download and storage of measurements, and apps for visualizing and controlling this information.

User/device interaction: While a lot of the information acquisition and storage is automated, there is still a degree of user/device interaction required in setting up the system for use and in ensuring that the observations are recorded properly.

Limitations on use: Currently, there are only a small number of sensors available, although this situation is changing rapidly. The user is therefore limited to using these specific sensors and cannot plug in a different sensor that may be of more relevance to their interests.

Suitability assessment: Highly suitable within a range of parameters. The system provides a complete package for rapid observation of the environment, although it does require quite a high level of understanding and implementation on the part of the user.

Name: **EMAP**

Web address: http://www.aecom.com/What+We+Do/Environment/Practice+Areas/Impact+Assesment+and+Permitting/Environmental+Mobile+Application+for+Projects

Designed function: Geographical Information System (GIS) package for mobile tablets, designed to allow information to be captured, edited and processed in the field. Designed as an all-in-one package that facilitates environmental monitoring at all stages from field data capture to reporting, all from a single mobile platform.

Environmental monitoring potential: Extremely useful as a mechanism for recording, processing and reporting. Not designed to allow monitoring devices to link directly to the tablet, so data entry is carried out manually.

Technological sophistication: A very sophisticated package, requiring a high level of user understanding, but with a lot of tools to allow field monitoring to be carried out more rapidly.

User/device interaction: The user interacts with the software package in a similar manner to that in which a GIS package would be used on a desktop. Additional functionality for reporting and filling in of technical sheets is also built in.

Limitations on use: The technology is proprietary to AECOM, a large American technical and management support company. Licensing may or may not be possible.

Suitability assessment: Extremely suitable for a wide range of environmental monitoring applications.

Name: **SEPA/Geofield diffuse pollution assessment**

Web address: http://www.sepa.org.uk/about_us/news/2013/sepa_wins_connect_ict_innovate.aspx?lang=en-gb

Designed function: Use of off-the-shelf tablets and sensor recording software in water monitoring. The application of this system is broad and has not been fully explored, as it allows users in the field to record a large number of different readings and to integrate these with existing spatial datasets.

Environmental monitoring potential: This system appears to have been designed to facilitate the recording and later interpretation of a wide range of environmental monitoring. Its potential lies therefore in assisting with the input, organisation and archiving of readings rather than the direct acquisition of the readings themselves.

Technological sophistication: High levels of technological sophistication are applied in the software that is used for the recording and organisation of sensor readings in the field. However, there is no mechanism for the user to integrate monitoring devices directly into the system and to record the readings without the human interface.

User/device interaction: The user must enter measurements into the system, which does not provide any feedback or processing but which does allow visualization of the data entered.

Limitations on use: The main limitation here is the user themselves – if a system could be developed that could accept data directly from a number of recording devices, then the user could act in a much more effective 'oversight' capacity rather than having to be directly involved in the measurements and their recording.

Suitability assessment: Highly suitable for monitoring of water and environmental pollution, and for a number of other monitoring requirements. The suitability lies more with the system's facilitation of data entry and recording, and with allowing later assessment, than with the direct recording of sensor readings or the integration of readings with existing data to provide 'added value'.

Name: **NatureWatch**

Web address: http://www.eea.europa.eu/mobile

Designed function: Allows citizen scientists to identify and report invasive species

Environmental monitoring potential: Similar to the "*What's Invasive*" app

Technological sophistication: Similar to "*What's Invasive*", but the user can also submit video and audio clips that could be of value in recording presence of invasive animals, birds, insects etc.

User/device interaction: This app can be used with the minimum of training or experience, and requires no sophisticated technical knowledge on the part of the user. Guidance is given on recognizing invasive species.

Limitations on use: This is very much a 'what goes in is what comes out' app – all information of use is entered by the users, whose level of expertise in the subject area may vary. It is difficult therefore to produce statistical or numerical evaluations of the topic of interest.

Suitability assessment: Potentially useful for capturing the spatial location and distribution of qualitative observations. Not capable of capturing quantitative observations unless used in parallel with monitoring equipment. Otherwise, similar limitations to "What's Invasive".

Name: **SIFSS**

Web address: http://sifss.hutton.ac.uk/

Designed function: Allows a user to find information about soil at their location (using the unit's GPS or interactive map) within Scotland and get a range of chemical and physical parameters such as pH, carbon content, texture, nitrogen, phosphorus, and base cations.

Environmental monitoring potential: A wide range of potential application e.g. assessment of nutrients and eutrophication – soil nitrogen & phosphorus are available in the app. Future developments are likely to include reporting on whether the user is in a Nitrate Vulnerable Zone (NVZ) and the HOST class (Hydrology of Soil Type) of the soil.

Technological sophistication: The most comprehensive of existing UK soils apps (albeit with information for Scotland only). Detailed information on up to 15 soil characteristics from the Soil Survey of Scotland are provided for each soil horizon, for both cultivated and semi-natural land covers (where they exist). The app accesses James Hutton databases, so always points to the most up to date data.

User/device interaction: This app can be used with the minimum of training or experience, and requires no sophisticated technical knowledge on the part of the user. Where several soil series are present at a location, the user is given a description of the appearance of the soil as a guide to selecting the correct series. No user information is incorporated in the app.

Limitations on use: Currently uses only the Scottish soil mapping information at 1:250,000, but future versions will incorporate higher resolution mapping where it is available. Could be adapted easily for other countries where relevant information is available.

Suitability assessment: Suitable for any field-based assessment that requires the type of information accessible through the app.

Name: **SOCiT**

Web address: http://www.hutton.ac.uk/research/groups/information-and-computational-sciences/esmart

Designed function: SOCiT provides the user with a near instantaneous field assessment of soil organic matter content using image analysis and modelling.

Environmental monitoring potential: Apart from determining soil carbon, potentially very useful conceptually – the integration of sensor information and environmental data sets in near-realtime.

Technological sophistication: From the perspective of the user this is quite a simple app to use, requiring only the device's camera (and a spade). However, there is a lot of server-side processing that integrates automated image analysis and neural network-based environmental data analysis.

User/device interaction: The user has to dig a hole, drop an image normalization card in it, take a photo and click a button for the soil organic matter to be calculated.

Limitations on use: Currently Scotland only (due to access to environmental data sets), potentially extending to rest of UK in the future.

Suitability assessment: At present this is a soil information tool but could provide a template for other apps integrating image and environmental data analysis.

Name: **SoilWeb**

Web address: http://casoilresource.lawr.ucdavis.edu/soilweb/

Designed function: Provide access to soil mapping information from the United States. This includes drainage class, water table depth and available water storage (in soil).

Environmental monitoring potential: Has potential in the US as detailed soil mapping is used in addition to the soil hydrology information.

Technological sophistication: This appears to be similar to SIFSS, but with more restricted user control of location selection – this app permits only input by the device GPS.

User/device interaction: Minimal "Get My Location" information required, which is automatically achieved.

Limitations on use: US only - blocked for Android distribution outside US and while available for Apple devices the only input is via the device's GPS.

Suitability assessment: Limited.

Name: **SoilMapp**

Web address: http://www.csiro.au/soilmapp

Designed function: Provides access to soil mapping information from Australia. This includes maps, photographs, satellite images, tables and graphs of data about nearby soils. Also includes soil's physical and chemical characteristics, including acidity (pH), soil carbon, available water storage, salinity and erodibility.

Environmental monitoring potential: Potentially useful due to the large number of attributes linked to soil hydrology, but only in Australia.

Technological sophistication: This is an interactive map, but gives access to very large amounts of data in a convenient manner.

User/device interaction: The user interacts with the soil map displayed on the device to obtain information. Information on site conditions can also be uploaded to CSIRO by users for subsequent use in map validation.

Limitations on use: Available for iPad only, data for Australia only.

Suitability assessment: Limited, but a good example of what can be done, and potentially useful in guiding future developments of existing apps.

Name: **mySoil**

Web address: http://www.bgs.ac.uk/mysoil/home.html

Designed function: To provide access to soil mapping information from the majority of Europe. This app provides extensive coverage of soil information, but with a limited number of attributes, and low precision (e.g. Organic Matter: "Medium" and pH: "Slightly Acidic")

Environmental monitoring potential: With current information available in the app quite low, however, it is likely that more data sets will be added in the future.

Technological sophistication: Fairly limited, a clickable interactive map.

User/device interaction: Uses a clickable interactive map with web mapping services (WMS). Users can also upload their observations which are subsequently displayed as pushpins.

Limitations on use: Europe only. While the extent of the cover within Europe is comprehensive, the depth of information being provided is quite low.

Suitability assessment: The app contains information from the UK's Countryside Survey and the Land Cover Map (2007). It is possible that other aspects of the data from these surveys, if incorporated in future versions (or new apps), could make this app more useful in monitoring or assessment.

Name: **Lapka**

Web address: https://mylapka.com/

Designed function: Lapka Personal Environment Monitor is a collection of sensors (which fit together in a wood and moulded plastic block) which interface with an iPhone via the associated app. The sensors are individual devices which plug into the audio jack of the phone and measure electrical conductivity (which they claim is highly correlated to nitrate concentration so is billed as a 'food quality' monitor), humidity and temperature, radiation, and electromagnetic fields.

Environmental monitoring potential: Depending on how sensitive the sensors are, having the capability of measuring electrical conductivity and temperature in one device which interfaces directly to the phone (thus allowing records to be associated with geotagged photographs) is appealing on the face of it. How the device would match up against purpose built sensors is unknown without testing. The package is sold for $220 (USD) which the makers claim compares favourably with the cost of a single humidity sensor at around $200, but puts it outside the range of most citizen scientists unless provided by a third party.

Technological sophistication: The various sensors are all available in other forms (e.g. Thermodo for temperature), but the package appears to be much more about style than function.

User/device interaction: Simply plug each sensor into the phone and the app records the appropriate measurement.

Limitations on use: Cost, sensors not especially sensitive or measuring different things to other devices.

Suitability assessment: Moderate/Low. Limited range of useable information, although having multiple sensors in one device is a plus.

Name: **Tesla Microscope**

Web address: http://tesla.dma.ucla.edu/TEST6/?q=research/fluorescent-microscope-cell-phone-attachment

Designed function: This device is a 'matchbox-sized attachment that converts a mobile phone's camera into a fluorescent microscope. The device utilises an inexpensive lens and battery-powered, light-emitting diodes to create a field of view some two orders of magnitude larger than previous cell-phone fluorescent microscopy technology. It is more than five times smaller than previous cell phone microscopes.

Environmental monitoring potential: The device is designed to be able to quickly and accurately analyze large sample sizes. The proposed use is in testing medical samples in remote areas, but the technology would easily lend itself to analysis of water and other fluids for pathogens. The inventors state that while it does not have the resolution of conventional microscopes, it is high enough to 'screen for labeled pathogens in drinking water or food and to image various body fluid samples to search for disease markers'.

Technological sophistication: Extremely high, although it appears to have been designed for size and robustness in the field. No indication if it being mass-produced or if it is still at the experimental design stage.

User/device interaction: Relies on a high degree of user input and interpretation.

Limitations on use: Some other equipment required (e.g. a colour filter), although this device is designed to be able to work with cheaper alternatives to the usual filters.

Suitability assessment: Potentially very suitable. Individual users would need training before it could be rolled out.

Name: **Proscope**

Web address: http://www.bodelin.com/proscope/proscope-micro-mobile

Designed function: A professional quality microscope which attaches to an iOS device. This relatively inexpensive device (c$150 (USD)) fits onto an iOS device and provides 20X to 80X magnification, with high quality optics. Includes stand, integrated LED lighting, and allows photographs of the subjects to be taking using the devices native camera.

Environmental monitoring potential: Potentially extremely useful for in the field monitoring of invertebrates and other biological specimens. Adding of geotagged photographs has potential for rapid biological field assessments.

Technological sophistication: High. Produced by a company with a track record for portable and/or internet enabled microscopes. Robust design specifically designed for field operation.

User/device interaction: Some familiarization with the equipment required (as well as the specific skills required to carry out the task in question).

Limitations on use: Interpretation skills of user

Suitability assessment: Wide range of potential applications

Name: **Thermodo**

Web address: http://thermodo.com/

Designed function: An external temperature sensor for a smartphone, tablet

Environmental monitoring potential: The device (which interfaces with an app from the same developer) is a very small temperature sensor which plugs into the audio jack of the phone. While it is designed to measure air temperature, there seems to be no reason why you could not plug an extension into the audio jack, and dangle the sensor itself in water or any semi-liquid medium such as slurry, food products etc. While it offers no real benefit over using a standard thermometer, as the app will (presumably) geotag the locations where the measurements are made, it would cut down on data processing time in the office plotting temperature readings spatially/temporally.

Technological sophistication: Medium/Low. The sensor itself is the most sophisticated part, although affordable (c$30 USD))

User/device interaction: Simply plug in the device, start the app and record.

Limitations on use: No better than a standard thermometer. Even if solution for working in liquids proposed above is workable, phones are not very robust so care would have to be taken in the field not to damage it.

Suitability assessment: Limited, but also shows the capability for developing small sensors which plug into smart device and can interface with a simple app.

Name: **Alcohoot**

Web address: http://www.getalcohoot.com/

Designed function: Measurement of blood alcohol concentration (BAC)

Environmental monitoring potential: This is a small sensor device which attaches to a smartphone (compatible with iPhone, iPad and Android devices currently) via a 3.5mm jack. Its intended purpose is to measure the user's blood alcohol content in the same way as a conventional breathalyzer. It uses a fuel cell sensor (similar to the type used in police breathalyzers) as opposed to a semi-conductor sensor, which allows it to be specific to alcohol (and unaffected by other solvents or chemicals such as hairspray, deodorant etc), as well as a patented air sensor to detect airflow into the device. If the sensor could be replaced with different kinds, then it might be possible to use as a relatively sophisticated air quality detector for measuring point source gas emissions which might help in tracing of chemical leaks, spillages etc.

Technological sophistication: The internal sensors are quite sophisticated, although the device itself retails for $75 (USD) so is still affordable. For the BAC measurements, the sensor needs to be periodically recalibrated which either would require return to base (RTB) or provision of some method for the user to do it in the field.

User/device interaction: The user simply connects the device to the phone and blows into the mouthpiece.

Limitations on use: Need for periodic calibration.

Suitability assessment: Limited to blood alcohol content currently, but shows the capability of designing and implanting add-on sensors for specific applications. The challenge would be to create small, accurate and reliable enough versions of other gas sensors for this to be a useable technology for pollution monitoring.

Name: **WiMoto**

Web address: http://www.wimoto.com/

Designed function: A variety of small wireless sensors which can measure ambient temperature, humidity, light, soil moisture, soil temperature, object temperature, human presence and movement.

Environmental monitoring potential: Unlike a number of the other sensor add-ons in this review, these can send their data wirelessly via Bluetooth. Assuming they are robust enough for deployment in the field, they could replace the need for manual download of temperature loggers and/or allow placement of the sensors in locations where accessing loggers frequently would be hazardous or time-consuming. The example use suggested for the temperature sensor on their crowd funding page is to monitor a swimming pool.

Technological sophistication: The sensors themselves appear no more complex than others mentioned here but the addition of battery powered Bluetooth is different.

User/device interaction: All data collected via a smartphone or tablet, which can then be uploaded to a cloud server. Potentially allows data to be collected more quickly and easily than traditional in-situ loggers.

Limitations on use: Battery powered, the makers claim that batteries last for about a year. Uses Bluetooth SMART technology so receiving device needs to be capable of acquiring signal.

Suitability assessment: Potentially very high, especially for light and temperature in water. Still at the crowd funding stage, but could be used to develop other sensors using similar infrastructure e.g. dissolved oxygen, specific chemicals.

Name: **MoboSens**

Web address: http://nanobionics.mntl.illinois.edu/mobosens/#

Designed function: A smartphone based sensor add-on which can accurately measure nitrate concentrations in water due to be released by end of 2013. Development of arsenic sensor underway, with plans in place to expand to sensors covering other pollutants (specifically heavy metals).

Environmental monitoring potential: Specifically designed to tackle a major diffuse pollution issue.

Technological sophistication: Sensor prototype has been developed at Illinois University. Further development and testing at the crowd funding stage.

User/device interaction: Not known – this is still at the testing stage with pilot citizen science group/crowd funders, but assumed to be simple given purpose.

Limitations on use: Not known. Some training presumed, and no information given about where data would be uploaded and analyzed beyond the pilot area.

Suitability assessment: Very high. If the technology can be expanded to detect a wider range of determinands, then it would be the ideal device for large scale, crowd sourced pollution data collection, particularly of water bodies.

Name: **smarTROLL**

Web address: http://www.in-situ.com/products/water-quality/handheld-systems/smartroll-multiparameter-handheld

Designed function: This app is designed to interface with the smarTROLL water quality probe produced by In-Situ Inc. The probe can measure dissolved oxygen, pH, ORP, conductivity (actual or specific), salinity, total dissolved solids, resistivity, and density, as well as physical parameters such as air and water temperature, barometric pressure, water level, and water pressure. The phone app replaces the need to have a handheld meter or logger attached to the probe, and allows data to be sent instantly to the office or uploaded to the cloud subject to having a phone signal.

Environmental monitoring potential: Specifically designed for this purpose.

Technological sophistication: Low for the app itself as it only records readings made by the probe.

User/device interaction: Very simple, no real user knowledge required as the probe (assuming correct calibration) can be put in water and readings taken without any assumed skill or technical expertise. Eliminates need to download separate logger and potential transcription errors from manually adding data.

Limitations on use: Specific to the sensor, which is expensive. No capability for expansion as it would require the probe itself to have extra sensors added.

Suitability assessment: Very high for the specific attributes measured.

III. Discussion

Most of the examples given above relate directly to the monitoring of one aspect of the environment, but in many cases the technology and software in question could be easily altered for use in a range of environmental monitoring tasks. In each case, a level of investment and redesign would be required. However, we emphasize that this investment is quite small in nearly every case – there are a lot of solutions available that would be very appropriate with some tweaking, and in each case most, if not all, of the technological barriers have been overcome through earlier development work.

Gaps in the above table show a trend that is not difficult to explain – the more sophisticated and/or expensive the technological application, the more likely it is to be used for scientific or industrial research. Complex equipment requiring experienced users is not used in citizen science, while purely app-based recording of simple observations does not occur with sophisticated monitoring of complex environmental conditions.

A notable example is the application of SEPA's existing Geofield framework for recording, organising and later processing of environmental monitoring readings using tablets and geospatial software developed by SigmaSeven (http://www.sigmaseven.com/geofield). What SEPA appear to have in this case is a system that solves half of the problem (we are not being disparaging here – this is half of a very big problem) and that facilitates a wide range of environmental monitoring. In order to take this system which is already in place and appears to work well a step further, it is necessary to remove the human element, or to at least reduce the amount of human data input required. Automated sensor acquisition using Bluetooth technology or Wi-Fi download would require a fair degree of further development in the data recording interface, but would enable measurements to be made more rapidly and would remove the risk of human error. GPRS (General Packet Radio Service) is another potential option here, although it may not always available in the field.

The systems that we have discovered in our search that may provide this 'missing element' are the Sensordrone, WiMoto and smarTROLL. While nowhere near as sophisticated in terms of presentation and immediate user ease of use as the Lapka, these systems are highly flexible and generalizable, and furthermore are to a certain degree 'programmable', i.e. the sensor readings can be controlled by the user in a manner that is not available through other systems. The developers of Sensordrone in particular have produced a system that is at this end of the development scale, with a wide range of potential applications that integrates well with mobile phone technology and produces useful information. However, the requirement for user expertise in achieving the potential of Sensordrone is a likely stumbling block. Comparison of the Sensordrone webpage with that produced by Lapka is the first sign that here we have something designed with scientists and 'early adopters' in mind, rather than members of the general public interested in knowing about potential contaminants in food that they have purchased. The other two systems, WiMoto and smarTROLL, lie somewhere between these two extremes.

Additional systems of note include the following:

- PEIR – a useful example of integrating automatically recorded sensor data with position and existing spatial datasets of relevance. The image below is extracted from the referenced link, as is the case with images relating to the other examples given.
- Water Quality Reporter – an example of a relatively simple, widely applicable app for uploading of observations. This concept could be readily and rapidly applied to almost any kind of water-related monitoring.
- Platforma SINC – cleverly applies colour interpretation from the mobile phone camera of sensor strips. A relatively low-tech but potentially rapid and simple tool.
- EMAP – this tablet-based GIS package is similar in concept to the system currently being used by SEPA and developed by Geofield. A very useful system for facilitating monitoring in the field, although the next obvious step (removing the human input requirements and connecting directly to sensors) needs to be made.
- Tesla microscope – a microscope attachment for mobile phones that could be used for monitoring waterborne bacteria, requires user expertise in identification but potentially extremely useful.
- Proscope – similar to the example above, with a few additional features including geotagging of images and easier recording of video.
- MoboSens – this system is not available yet and details are on the scant side, but appear to promise a wide range of specific sensors that can connect to mobile phones.

In many cases, the examples of mobile phone technology for environmental monitoring that we have reviewed in this work are still in the testing stage and require the user to 'assist'. This can involve either providing a mechanism for getting the sensor into the medium that is being tested (e.g. by dipping a colorimetric sensing material into water, [9]), or by providing interpretation of the raw recorded data (e.g. contextualising noise levels, [10]). In addition, for large-scale environmental monitoring using the millions of potential recording devices to actually take place, user (or in this case, device owner) participation must be achieved. Ref. [11] discusses potential incentive methods for accomplishing this. A step-change is required both in terms of automating data capture (either through augmenting the current sensor suite of a mobile phone, or finding ways to more cleverly turn data captured through existing sensors into useful information), and in engaging participants, such that they are willing and able to use their personal devices in this way. This last point will require consideration of personal privacy, data source anonymity and potential misuse or unwanted (by the people involved) use of the data that is captured. A final point, as discussed by [12], is that even if useful real-time environmental monitoring data is available, either from dedicated field researchers or crowd-sourced from the general population, we still lack ways of effectively communicating the appropriate information to people that might benefit from it. Ref. [13] provides some pointers towards how a model of 'participatory public health' might emerge.

Another consideration that is important for most of the systems discussed here is that of data roaming limitations. Mobile phone signals are still weak in many parts of the world, particularly in areas of low human population density. An important aspect of systems designed for environmental monitoring is that they should be able to record while off-line, and use periods of available signal strength to upload observations when possible. The same is true of systems based on SMS messaging, which can be used to record observations but which still require a signal for transmission. Intelligent design of controlling apps that make use of signal strength opportunities, rather than being forced to attempt transmission of observations as soon as they are made, will make field-based observations more successful.

IV. Conclusions

There are a number of different systems available on the market that are at least partially relevant for environmental monitoring, or that can inform developments in that direction. We have identified a number with particular promise, as well as categorizing the examples that we could find within a conceptual framework that enables comparison. Field-based environmental monitoring using off-the-shelf components is an area that is developing extremely rapidly, and trends in what is available would seem to indicate that more functionality and integration between sensors and mobile phone technology is highly likely in the near future. One aim of this paper was to demonstrate whether or not it is possible to use off-the-shelf technology for environmental monitoring. We feel that given careful selection of the technology within a restricted range of observation types, that it is indeed possible to do this, and that ease of use and breadth of application are improving rapidly.

Acknowledgments

The authors would like to thank Dr Peter Cock and Dr Paul Shaw of the James Hutton Institute for their advice in the preparation of this paper.

References

[1] W.Z. Khan, Y. Xiang, M.Y. Aalsalem, Q. Arshad, "Mobile phone sensing systems: a survey", *IEEE Communications Surveys and Tutorials 15(1), 402-427,* 2013.

[2] P.B. Lillehoj, M.-C. Huang, N. Truong, C.-M. Ho, "Rapid electrochemical detection on a mobile phone", *Lab on a Chip 13(15), 2950-2955*, 2013.

[3] S. O'Driscoll, B.D. MacCraith, C.S. Burke, "A novel camera phone-based platform for quantitative fluorescence sensing", *Analytical Methods 5(8), 1904-1908*, 2013.

[4] A. de Nazelle, E. Seto, D. Donaire-Gonzalez, M. Mendez, J. Matamala, M.J. Nieuwenhuijsen, M. Jerrett, "Improving estimates of air pollution exposure through ubiquitous sensing technologies", *Environmental Pollution 176, 92-99,* 2013.

[5] N.D. Lane, E. Miluzzo, H. Lu, D. Peebles, T. Choudhury, A.T. Campbell, "A survey of mobile phone sensing", *IEEE Communications Magazine 48(9), 140-150*, 2010.

[6] M.J. Aitkenhead, "The E-SMART project".

[7] M.J. Aitkenhead, D. Donnelly, M.C. Coull, "Innovations in aquatic monitoring", *CREW project number CD2013_04*, Available online at: crew.ac.uk/publications.

[8] S. Novella, "Will your smartphone become a tricorder?", *Science-based medicine.*

[9] H. El Kaoutit, P. Estevez, F.C. Garcia, F. Serna, J.M. Garcia, "Sub-ppm quantification of Hg(II) in aqueous media using both the naked eye and digital information from pictures of a colorimetric sensory polymer membrane taken with the digital camera of a conventional mobile phone", *Analytical Methods 5(1), 54-58*, 2013.

[10] E. Kanjo, "NoiseSPY: A Real-Time Mobile Phone Platform for Urban Noise Monitoring and Mapping", *Mobile Networks & Applications 15(4), 562-574*, 2010.

[11] X. Sheng, X. Xiao, J. Tang, G. Xue, "Sensing as a Service: A Cloud Computing System for Mobile Phone Sensing", *Proceedings 11th IEEE Sensors Conference, 2012, Taipei, Taiwan, 2239-2242*, 2012.

[12] F.J. Kelly, G.W. Fuller, H.A. Walton, J.C. Fussell, "Monitoring air pollution: Use of early warning systems for public health", *Respirology 17(1), 7-19,* 2012.

[13] D. Van Sickle, "The Next Generation of Public Health Approaches to Asthma in Asia and the Middle East", *Asia-Pacific Journal of Public Health 22, 229-235*, 2010.

Authors

M.J. Aitkenhead is with the James Hutton Institute, Aberdeen, Scotland, UK AB15 8QH (e-mail: matt.aitkenhead@hutton.ac.uk).

D. Donnelly is with the James Hutton Institute, Aberdeen, Scotland, UK AB15 8QH (e-mail: david.donnelly@hutton.ac.uk).

M.C. Coull is with the James Hutton Institute, Aberdeen, Scotland, UK AB15 8QH (e-mail: malcolm.coull@hutton.ac.uk).

E. Hastings is with the James Hutton Institute, Aberdeen, Scotland, UK AB15 8QH (e-mail: emily.hastings@hutton.ac.uk).

18

Problems in Context-Aware Semantic Computing

O.A. Nykänen and A. Rivero Rodriguez
Tampere University of Technology, Tampere, Finland

Abstract—**Acknowledging the user context, e.g., position and activity, provides a natural way to adapt applications according to the user needs. How to actually capture and exploit context, however, is not self-evident and it is tempting to assign the related responsibilities to individual context-consuming applications. Unfortunately, this confuses the user, complicates application development and hinders context-aware semantic computing as a research discipline. In this article, we outline context-aware semantic computing research topics and the state-of-the-art mobile application development frameworks of special interest to us, acknowledging best practices for accessing and modeling sensor context. From the integrated point of view, context-aware semantic computing is demonstrated in terms of a software component called context engine. In order to better understand how theory is tied with practice, we also introduce a simple context engine prototype. Finally, we use the research background and the empirical setting to discuss the significant problems and relevant research directions in context-aware semantic processing.**

Index Terms—**Context Engine, Context-Aware Services, Mobile Computing, Semantic Computing**

I. Introduction

Acknowledging the user context provides a natural way to focus user attention and the use of resources in applications. For instance in mobile applications, user position, time, calendar activity, and task establish a convenient starting point for filtering, organizing, and providing access to relevant information and tools.

Pioneering research in context-aware computing research dates back to early 1990s [7]. Since then, studying and building context-aware systems have been first tackled in application-specific manner, then in terms of reusable toolkits, and finally, on infrastructure level [19]. For various reasons, however, the rate of infrastructure-level deployment and adoption in production systems is still catching up. Significant technological progress has been made, in particularly in mobile applications [34]. Simply looking at application volumes, it is fair to say that contemporary mobile sensor frameworks establish the de facto technology driver for context-aware computing.

Still, despite the research results and technological advancements, it is not self-evident how context should be realized and what is the role of sophisticated context-aware computing in the application ecosystem(s). From the perspective of context-aware computing, the current sensor APIs and frameworks provide rather low-level access to sensor information, which in practice suggests that each application deals with the context as it sees fit. This confuses users and hinders the development of more abstract context-aware computing.

A modern reincarnation of the middleware for context-aware computing is a system called a context engine. In brief, the main task of a context engine is to filter and refine the contextual clues, e.g., for recommendation applications [30].

Through this notion of the context engine, context-aware computing gets closely affiliated with a multidisciplinary research topic called semantic computing [39]. In brief, semantic computing is about computing with (machine processable) descriptions of content and (user) intentions [22]. Aligned with Semantic Web technologies [46], this provides the methodological and technological baseline for modeling, understanding, and computing with the user context (cf. e.g. [43]). Significant research problems, however, need still to be properly addressed before the promise of context-aware semantic computing can be fulfilled.

In this article, we outline context-aware semantic computing research topics and the state-of-the-art mobile application development frameworks of special interest to us, acknowledging best practices for accessing and modeling sensor context. From the integrated point of view, context-aware semantic computing is demonstrated in terms of a software component called context engine. In order to better understand how theory is tied with practice, we also introduce a simple context engine prototype. Finally, we use the research background and the empirical setting to discuss the significant open problems and relevant research directions in context-aware semantic processing.

The main contribution of this article is to review the related sensor and context modeling research in order to systematically characterize the role of context-aware semantic computing in (mobile) applications, and to use this setting to discuss the related significant research and engineering questions.

Considering (our) future research, we believe that context-aware semantic computing will have an increasingly significant impact in application development. In addition to mainstream mobile computing, perhaps two of the most prominent application areas with a large volume of industrial applications include Web of things and the Industrial Internet paradigm [48] [14].

Our current work stems from the ongoing Marie Curie ITN research project MULTI-POS, Multi-technology positioning professionals (Grant agreement no. 316528, 2012-2016) where we study context-aware semantic processing.

The rest of this article is organized as follows: in Section 2, we outline the background of our work, highlighting the current technology driver and the best practices for modeling context. In Section 3, we present context engine architecture and a simple prototype implementation. Equipped with the research background and implementation experience, we then discuss the related open research and engineering questions in Section 4. Finally, in Section 5 we conclude the article.

II. Background

Context can refer to any information that can be used to characterize the situation of an entity, where an entity can be a person, place, or physical or computational object [6].

Contextual information may include physical information such as accelerometer data, virtual information such as calendar events, recognized patterns such as observed user activities, and predictions such as weather forecasts. In the abstract sense, context can be used to reduce the computational complexity of problem solving by restricting the search space – in turn decreasing the number of irrelevant end user choices.

A. Related Research and Basic Concepts

The term context-aware (computing) appeared first time in early 1990s, with the beginning of context-aware system research [7]. In addition to solely computing with respect to time and place, context-aware systems can capture many other things as well, such as places, things, commitments, and user knowledge and preferences [30]. A typical application area is context-aware search, which includes the phases of data acquisition, context reasoning and state updates, and contextualized output [44].

The main components of a context-aware system include context providers and context-aware services, perhaps associated with service locating services or brokers [19]. In applications, the computing context, the user context, and the physical context are often differentiated [7]. Processing contextual information is carried out by a component called context interpreter, and the relevant data is stored in a context database. The basic activities include context assertion, i.e. making contextual information available, and context retrieval, i.e. exploiting the context in an application [30]. Reasoning with the context is typically based on logic programming [5][20].

In brief, we may identify three complementary approaches on how the context providers acquire contextual information [7][35][8]:

- *Direct sensor access*, where sensor information is directly read from the sensor APIs.

- *Middleware infrastructure*, which introduces a layered architecture that enhances reusability and provides concurrent sensor access. Instead of accessing directly the raw data from sensors, an intermediate software layer manages sensorial data.

- *Context server*, which in addition allows gathering information from remote data sources and distributing the costs of measurements and computations.

In any case, direct sensor access is not usually feasible since sensor access needs to be encapsulated for multitasking, concurrency etc.

In principle, context-computing tasks may be delegated to a software component called context engine [30]. For purposes of this article, we say that a *context engine* is a software component, which integrates and refines the generalized (sensor) context, the related services, and the user preferences, for the benefit of individual (user) applications. Note that the term context broker is sometimes used for a similar architecture [8].

Typical tasks of a context engine include acting as a local context provider, providing logical context interpretation, accessing external context providing services, and managing an archived sensor information database e.g. for minimizing battery consumption and user preferences. Note that these tasks typically exceed the boundaries of individual applications.

Acknowledging the close relationship between context and sensor information, the notion of a "sensor" is typically generalized. We may acknowledge at least three different types of sensors providing contextual data [4]:

- *Physical sensors* are the most frequently used sensors, capable of capturing physical data (e.g. position, orientation, and acceleration).

- *Virtual sensors* provide contextual information from applications and services. Virtual sensors may further be based on local or external data sources (e.g. user calendar vs. weather service).

- *Logical sensors* provide new contextual information by combining and computing information from physical and virtual sensors.

Considering past research, known context-aware frameworks and systems include Context Broker Architecture (CoBrA), Context-Awareness Sub-Structure (CASS), CORTEX, Gaia, Context Management Framework, and Context Toolkit, which have introduced many of the elements related to context-aware computing [4]. Besides query requests, (logical) reasoning my also be founded on event-based processing [33].

Today, vendor-specific physical sensor middleware frameworks establish the major technology driver in mainstream context-aware computing. This has a major impact both in application development and in the current strategies of modeling context.

B. Current Technology Driver: Physical Sensor Context

A nice overview of the current state-of-the-art sensor technologies can be compiled by looking at the widespread mobile platforms, Android and iOS, and considering the various Web-based cross-platform development tools.

Android developers can make use of contextual information in several ways [1]. The first approach is using the Android Sensor Framework, which includes the motion sensors (e.g. accelerometers), environmental sensors (e.g. temperature) and position sensors (e.g. orientation sensor). It is also possible to access location information with Location API and other additional location services, such as Geofence API to alert user or applications when the user is entering a certain region.

iOS developers can access similar kinds of sensor information [2], with the chief exception of using Objective-C instead of Java.

In addition to device-specific interfaces, various browser APIs are also being developed. Accepting the obvious challenges in generalizing the sensor context of different operating systems, an interesting research perspective on context providers is established by cross-platform tools. These abstract the details of the various platforms, aiming to allow implementation of an application and its user interface for several mobile platforms more efficiently [34]. Table 1 lists the most popular cross-platform development tools, pointing out what sensor information is currently available.

The need for standard access to application context has been also acknowledged by the related standardization organizations, namely the World Wide Web Consortium (W3C). In particular, the standardization of the so-called Open Web Platform includes several browser APIs that can be used in device and application independent manner for acquiring context [47].

It is interesting to observe that in most applications, developers must access and exploit sensor information directly, i.e. without the explicit notion of context engine. Further, the sensor information is mostly related to particular mobile device; any negotiation with additional context providing servers takes place in application-specific manner and is not directly supported by the (sensor) toolkits.

C. *Modeling Context*

Even if the mobile development frameworks do not yet provide integrated means for context-aware computing, various theoretical modeling approaches exist. We may identify several major strategies for modeling context [4][7], including *key-value models*, *object-oriented models*, and *ontology-based models*. Further, context can be defined in various ways [11].

Currently, there is no commonly agreed standard model or systems for sensing contextual information from various sources to enable reuse across various middle-ware systems and frameworks [4]. Ontology-based models, however, seem to offer many desirable properties such as information alignment, dealing with incomplete or partially understood information, domain-independent modeling, and formally working with context model of varying level of detail [8]. Adopting an context ontology standard might be beneficial but require global consensus on the matter.

Perhaps the most widely known sensor ontology is the W3C Semantic Sensor Network (SSN) ontology. SSN was developed based on reviewing 17 existing sensor or observation-centric ontologies [9]. In order to normalize the ontology and support its adoption with other ontologies, the SSN ontology is aligned with the general DOLCE Ultra Lite upper ontology, providing concepts such as PhysicalObject, Event, Situation, and Region.

According to the SSN ontology, sensors may have properties such as accuracy in certain conditions, or may be deployed to observe a particular feature (see Figure 1) [29]. While abstractions or extensions are applicable, the SSN ontology in practice emphasizes the aspects of physical sensor networks.

Some context ontologies, however, by design do acknowledge the generalized logical (sensor) context. A prime example is the Service-Oriented Context-Aware Middleware (SOCAM) architecture, which aims providing efficient infrastructure support for building context-aware services in pervasive computing environments [19].

In SOCAM, context modeling is carried out in OWL ontologies based on two-level information architecture: the general context concepts are captured in the common upper ontology and application-specific concepts in domain ontologies (see Figure 2). This approach suggests using upper-level context ontology, in addition to general top-level alignment ontology, for integrating various kinds of domain ontologies, suitable for explaining their role in providing context.

TABLE I.
APIs SUPPORTED BY MAIN CROSS-PLATFORM DEVELOPMENT TOOLS (ADAPTED FROM [34])

Tool / API	*Rhodes (JS)*	*Phone-Gap (JS)*	*Mo-Sync (JS)*	*Mo-Sync (C, C++)*	*Dragon-Rad*
Accelerometer		X	X		
Barcode	X	X			X
Bluetooth	X	X		X	
Calendar	X	X	X	X	X
Camera	X	X		X	
Capture		X	X	X	X
Compass		X	X		
Connection		X	X	X	
Contacts	X	X			X
Device	X	X	X	X	X
File	X	X	X	X	
Geolocation	X	X	X	X	X
Menu	X				X
NFC	X	X	X	X	X
Notification	X	X	X	X	
Screen Rot	X	X		X	
Storage	X	X	X	X	X

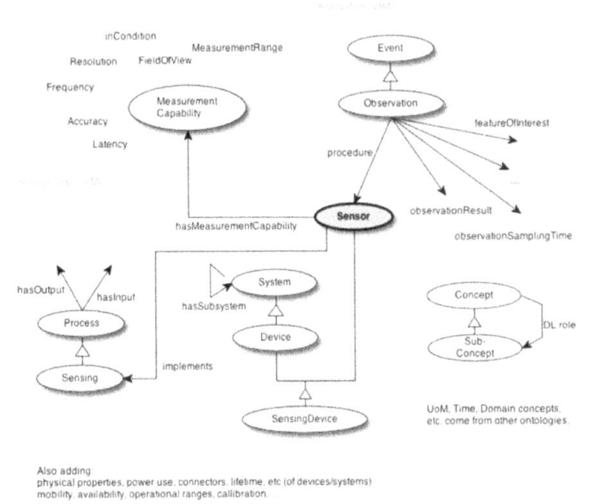

Figure 1. Overview of the SSN ontology structure prior to its modularization and alignment [29]

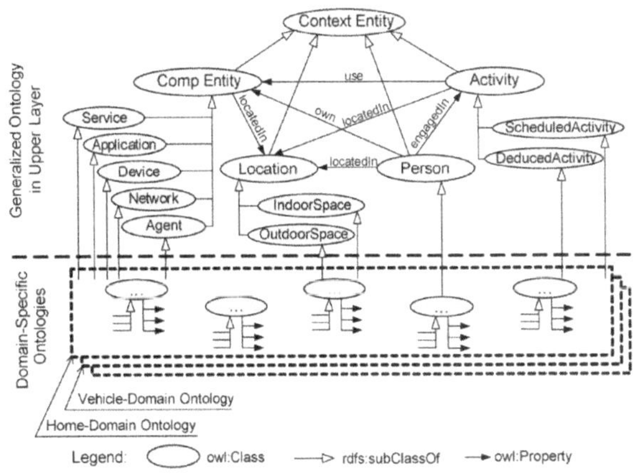

Figure 2. Class hierarchy of the upper (SOCAM) ontology [19]

It is worth observing that both of the referred ontologies above are static by design: They provide a fixed structure for observations (etc.) that is assumed to be true and which does not change overtime. Indeed, a considerable practical challenge lies in managing imprecise, uncertain, or evolving information. While the significance of this topic is widely acknowledged in the related research [41][3], related standardization is still underway [28].

III. Context Engine

It is quite difficult to study context-aware semantic computing and context engines based on very abstract definitions. To make discussion more concrete, let us next first specify a certain kind of context engine and then illustrate the chief properties of a related prototype implementation. The context engine architecture is novel but of course influenced by the aforementioned, related research.

A. *Main Properties and Abstract Architecture*

In brief, a context engine accepts the overlapping responsibilities and tasks of the local context provider and (logical) context interpretation, which typically exceed the boundaries of individual applications. The essential tasks of a context engine include providing context information to the applications via various logical queries in terms of a standard I/O interface, and managing user preferences.

The chief communication mechanism between the context engine and the applications is the context ontology. Any domain-specific knowledge is captured in terms of references to domain-specific ontology modules and user preferences.

Individual applications do not necessarily have to fully understand the knowledge base of the context engine for making simple queries or asking questions about the current context, and vice versa. For instance, a simple telephony application might only need to know whether the user's activity status is currently "working" and if the user is in a business meeting or not.

Note that domain-specific knowledge, i.e. how to actually utilize context in applications is not a responsibility of the context engine (cf. Figure 2). It does not have to understand application specific ontologies either. When needed, any extralogical computation (including heuristics, predictions, etc.) can be delegated to other services.

Simplified context engine architecture is depicted in Figure 3. In brief, the end user interacts with an application, which executes user activities and accesses contextual information through the context engine. Typical user applications include information management and communication applications, such as calendar, messaging and telephony applications, and novel software agents.

The context engine implements the context engine (service) Application Programing Interface (API). When context-aware semantic processing is needed, the user application requests context engine services. To fulfill these requests, the context engine has access to local context providers and possibly to external services. In addition to asking individual sensor values, a context interpretation query might ask the context engine to interpret and infer additional information about a given context, e.g. asking the known weather prediction (or archived value) for a given place at a given time. This might involve requests to external services.

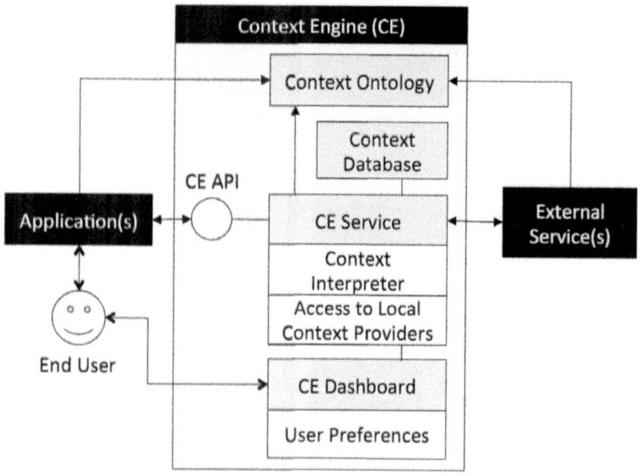

Figure 3. Simplified Context Engine Architecture

To provide internal (sensor) context archives, the context engine maintains a context database. This can be used, e.g., to analyze and optimize context engine behavior. Note that the applications may also depend on external services in their internal design.

From the perspective of the end user, the context engine also manages global user preferences that are taken into account in context-aware semantic computing. For instance, the user might prefer not accepting certain kinds of phone calls outside the office hours. For this purpose, the context engine provides the user a dashboard GUI, for defining appropriate context engine settings – or explaining how to extract the information from sensor data. The user preferences might be considered as a rule system that refers to the context ontology and user tasks.

The end user dashboard might also be used for providing extra or overriding information, e.g. to check out how context affects a particular applications, or for overriding physical sensor context (perhaps "lying").

Notably extensions to the context engine include an event listener service (e.g. notify application when specific contextual event takes place) and shortcuts for certain kinds of commonly needed queries. More complex context engines might also extend the related knowledge bases and add extralogical services to the content engine I/O interface.

B. *Experimental Context Engine Environment*

We anticipate that eventually, a context engine (of a mobile device) is a service provided by an appropriate sensor framework, including an operating system level utility similar to personal details or privacy settings. When Internet connectivity can be assumed, the main alternative is providing context engine as a webized service.

Further, considering the current mobile application ecosystem(s), it seems likely that context engines are a business for large and established Internet service and application providers, simply due user base, credibility and critical mass of applications.

In the meantime, however, it is instructional to outline a research prototype deployable to a particular device that allows us to study both the concept and implementation of context-aware semantic computing. This allows us also to learn from the developer and the user experience, and enables discussing context-aware semantic computing research questions in a concrete setting.

Figure 4 presents the main view of a Java-based context engine dashboard prototype running in an Android emulator. The user interface includes the essential functionality to start and stop the context engine service and to provide custom properties to the context ontology. Note that in this case, the context engine service has been physically deployed in a mobile device; a design stance that we will later challenge since the applications only need some access to the API and the dashboard.

Indeed, from the implementation and deployment perspective, a major design decision lies in exposing the CE API to the applications. In an Android environment, a standard approach would be to deploy the context engine as a CE background service bound to a CE dashboard activity equipped with a graphical user interface. In this architecture, any application that wishes to utilize CE services would have to either bind to the CE service, or communicate with the CE activity via the so-called intent messaging. While this approach is clearly the most powerful one within the Android environment, it would require that each application is a native Android application which somewhat complicates experimental development.

For research purposes, we have adopted an alternative implementation strategy. In our case, the context engine includes a web server that allows publishing the CE API over HTTP, based on the NanoHttpd server implementation [15]. This allows prototyping the context engine quite flexibly, and supports experimenting and analyzing the context engine in various real and in simulated environments.

In itself, a sole context engine is of course not useful. Figure 5 depicts a sample Javascript application executed within the default browser of the Android emulator. In brief, the application depicts user location using the Google Maps API [17] and shows the activity status. Note that due to the default browser's Javascript security restrictions, the application needs to be downloaded through the localhost.

Behind the scenes, the application communicates with the context engine prototype via the HTTP-based Context Engine API, which allows accessing the local, built-in (generalized) sensor information. These include a subset of the Android sensor API and the custom properties communicated via the dashboard interface.

In cases when key-value sensor data is not sufficient, contextual information can be semantically bound together via the sample context OWL ontology. Put another way, the classes of the context ontology can be populated by the individual sensor information retrieved from the environment. In principle introducing any of the sensor APIs (cf. Table 1) is also straightforward.

In a production environment, controlling the applications' access to context information would require additional management controls. Recall that when installing a native android application, similar information is asked from the user, e.g. for granting access to user location or contacts.

Since the Google Maps API is used in rendering the map view, some user data is exposed to Google by the sample application. The (user of the) context engine can either choose to accept this, or refuse using the application altogether. While several map providers exist, it seems likely that all free services include terms that allow the service providers to collect usage data in order to improve

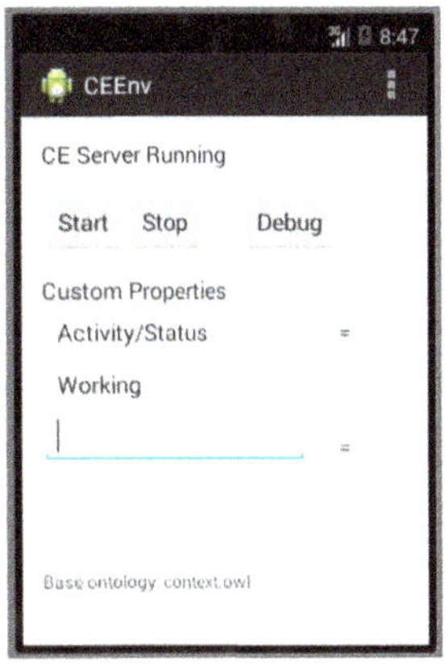

Figure 4. A context engine dashboard prototype

Figure 5. A sample browser application accessing the context engine

the user experience and to provide value-added services to their customers.

Once started, the context engine provides applications two ways to access context-aware processing services. The first approach is straightforwardly asking specific, most recent raw sensor information using a HTTP GET request. In this case, the context ontology is only used as a sort of information architecture, for application and context engine (key-value) communication. The second, more powerful approach is formulating a query in SPARQL, using an Android port of the Jena framework [13]. With the help of reasoner services, e.g. transitive or OWL reasoning, this allows logical context (ontology-based) interpretation beyond mere syntactic queries.

When compared to using the built-in Android sensor API – in addition to the interface design – a major feature of the context engine prototype is that it can provide a single entry point to all sensor information. This allows analyzing context-aware processing, refining and optimizing the use of contextual information, and considering various implementation strategies, above the level of individual applications.

In addition to accessing the explicit context providers, the context engine can also exploit the usage patterns of the applications to infer the properties of the current context. For instance, with proper training data, the current status (Working) might be statistically inferred with certain degree of belief from the user logs so that the user would not have to explicitly enter the status at all.

IV. Significant Research Problems

The research background suggests that there is a theoretical need for a context engine component that mixes the responsibilities of context-aware and semantic computing. Empirical work verifies that this is also doable in several types of common application systems, including sensor networks and mobile machine platforms.

In the general case, however, several practical and theoretical challenges still remain, including:

Deployment. Deploying a context engine requires not only providing it to a device but also exposing it to applications – and attracting application developers' interest in using it.

In principle, access to a context engine can be provided on the operating system level (such as the Android Sensor Framework [1]), cross-platform development tool level (such as PhoneGap [34]), browser platform level (such as the navigator browser API [45]), on Internet service level (such as Google API [17]), or as a "yet another" HTTP service (such as our Context Engine prototype).

The device independent approaches provide obvious flexibility of devices and platforms but might require Internet connectivity, lack device-specific features, and raise privacy concerns.

Efficiency. In a production environment, it is not self-evident how the context ontology should be populated since accessing sensor information consumes processing and energy resources. The context engine should thus somehow optimize its performance, typically in terms of a trade-off between accuracy and costs [37]. One approach is to use a context database to cache recent sensor readings and/or to choose the cheapest matching sensors for better performance [31]. Further, deploying a complete query endpoint or a reasoner into a mobile device introduces its own overhead [13][5].

A potential solution is exploiting external computing services. In cases when the role of external context providers is particularly significant or Internet access must anyway be assumed, this might in fact suggest delegating certain context engine responsibilities to an external service altogether (cf. [12]). Relying onto external providers is also in line with the business logic of the major Internet service and product providers (c.f. [18]).

Note, however, that deploying e.g. the reasoning and the database modules of the context engine as an external service does not completely remove the need for local components with local computing costs [34]. This is particularly true when accessing local sensors.

Privacy. From a very practical point of view, the main utility of the context engine lies in the fact that it provides (within device or user session) a "centralized" access point to context information. This allows it to provide individual applications far better and more abstract information about the user context than each application could possibly do on its own. Note that in addition to using the explicitly registered context providing services, the context engine may also keep logs of the regular applications usage, e.g. to statistically infer contextual information.

When detailed personal information is managed, potential privacy issues are of course raised [8]. While dependency on external services may significantly help in providing more efficient context engine services of better quality, the obvious challenge lies in controlling and managing access to this information [24].

The baseline level of privacy is established by the related technologies and policies, such as submitting and storing sensitive information anonymously and securely [23]. A significant dependency lies in managing user preferences and matching these with the end user agreements when using various kinds of applications and services. Note that involves not only the context engine but also the individual application components (cf. [17]).

Understandability. In principle, it is not technically too difficult to provide a sophisticated rule system so that users could quite flexibility assert rules for acquiring and exposing their personal information to applications and managing how context is used in applications. For instance, consider adding a complete user preference rule component to our context engine. Such system, however, would be close to full-fledged logic programming and might be quite difficult use and understand in full detail [21].

More on the end user side, the adaptation due context-aware processing is another potential issue since it can be very difficult to users to recognize which parts of the application were adapted due context-aware computing in the background. To minimize the problems, adaptation might be visualized and analyzed during design [16]. Quality of contextual information is also a potential issue and may require managing additional metadata for quality control [26].

Problems of adaptive systems are well understood in personalized search systems, where increased adaptation may e.g. guide the decision making of the inexperienced users, but be perceived as too restrictive by expert users [25]. This seems to insist a tradeoff between the level of application adaptation and user control.

Semantics. Finally, while logical queries based on sensor information using a fixed ontology suffice for many tasks, a fundamental challenge is introduced by the very notion of context itself: Some contextual properties can be derived from others and thus, context is not a fixed concept in the first place [10].

For instance, the user location (e.g. Office) can sometimes be reliably used to (statistically reason and) predict the user activity status (e.g. Working), and vice versa.

Further, sensor and other information sources evolve over time, which should also be taken into account in semantic modeling. In particular, when regulations or organizational processes undergo changes at the workflow level, so does the notion of context. This may involve introducing new terms explaining context, or worse; using the old terms with a new meaning.

Thus, the design of the context ontology should ideally reflect the fact that some contextual properties may depend on each other, and that the context ontologies evolve over time. Alternatively, ontologies can also be used to support and evaluate the quality of statistical reasoning [36]. While using an alignment or top-level ontology seems indeed necessary, it may not be sufficient unless further semantics required in the evolution (e.g. same as, broader than) and statistical reasoning (e.g. evidence for, statistically independent) are encoded as well. Indeed, semantically modeling the aforementioned challenges, include a variant of the frame problem [38] and schema-level information evolution [32], which might well be called the hard problems of semantic computing.

V. Conclusion

Access to contextual information provides computational advantage in theory and in practice. In this article, we have outlined key elements of contemporary context-aware semantic computing. To make the discussion more concrete, we have also introduced a simple context engine prototype environment.

Intuitively, the insight of context-aware semantic computing is quite clear: information about the proper context can significantly improve the user experience by enabling the design of more efficient applications and help in optimizing the related computation beneath. However, when the related theoretical and engineering dependencies are analyzed in more detail, the single objective of context-aware semantic computing gets broken down into several, evidently competing design requirements [40].

Instead of a single problem, we thus have many. To address this observation, we have acknowledged several significant research questions in the area, including deployment, efficiency, privacy, understandability, and semantics.

Looking at the specific research problems related to context engine implementation, the topics of efficiency and access, understandability, and privacy deserve special attention. In principle, a local installation of the context engine gives best control over user privacy. In practice, however, the design choices and the user agreements of individual applications may easily invalidate this assumption. Local installation also means computation and memory overhead, and of course increases the risk of a single-point failure.

When Internet connectivity can be assumed, the idea of decentralizing the context interpretation and sensor (etc.) database management tasks seems like a viable design stance. This also potentially provides the context engine the ability to coordinate e.g. pattern recognition, classification, and context ontology evolution activities among users groups and sharing and reusing sensor data, effectively providing more efficient and better user experience. It seems likely, however, that this takes place at the expense of user privacy, even if it might offer only a limited access to the local sensors.

Strictly from the semantic computing point of view, the question how to properly model the related semantics, coined by the context ontology, is also highly relevant. Even quite simple use case scenarios point out that assuming fixed context ontology is an oversimplification, and that evolution at the level of domain-specific context ontology components have to be assumed at some point. Further, when learning, classification, and prediction algorithms are taken into account, it seems rather obvious that particular sensor information may appear either in the role of "physical" or "logical" sensor, e.g. in relationship with most recent sensor data and a particular prediction algorithm. This suggests introducing also evidence-based relationships (etc.) in the context ontology.

Thus, due to the complexity of the topic, it is unrealistic to assume that a single best solution exists for context engines and hence context-aware semantic computing in general. Instead, one must be satisfied with special-purpose approaches, e.g., finding a compromise between easy deployment and privacy, and between expressivity and understandability. From the perspective of context engine standardization, this of course requires prioritizing the design objectives, and/or acknowledging several context engine profiles and modes.

We believe that the large-scale adoption of context-aware semantic computing is inevitable, and is likely to take place in terms of the mainstream Internet service and product providers. Either way, context-aware semantic computing will have profound impact in applications.

References

[1] Android Developer. Location and Sensors APIs.

[2] Apple Developer. Features – iOS Technology Overview.

[3] J.C. Augusto, J. Liu, P.J. McCullagh, H. Wang, and Y. Jian-Bo, "Management of uncertainty and spatio-temporal aspects for monitoring and diagnosis in a Smart Home," *International Journal of Computational Intelligence Systems*, 1(4) 2008, 361-378.

[4] M. Baldauf, S. Dustdar, and F. Rosenberg, "A Survey on Context Aware Systems," *Int. J. Ad Hoc Ubiquitous Computing* 2, no. 4, 2007, 263–277.

[5] K. Broda, K. Clark, R. Miller, and A. Russo, "SAGE: A Logical Agent-Based Environment Monitoring and Control System," *Proceedings of the 3rd European Conference on Ambient Intelligence*, AmI-09, 2009, pp. 112-117, Springer.

[6] J. Brown, P., Nigel Davies, Mark Smith, and Pete Steggles, "Towards a Better Understanding of Context and Context-Awareness," *Handheld and Ubiquitous Computing*, edited by Hans-W. Gellersen, pp. 304–307. Lecture Notes in Computer Science 1707. Springer Berlin Heidelberg, 1999.

[7] G, Chen and D. Kotz, "A survey of context-aware mobile computing research," Technical Report TR2000-381. Dartmouth College, 2000.

[8] H. Chen and T. Finin, "An ontology for a context aware pervasive computing environment," *IJCAI Workshop on Ontologies and Distributed Systems*, Acapulco MX 2003.

[9] M. Compton, P. Barnaghi, L. Bermudez, R. Garcia-Castro, O. Corcho, S. Cox, J. Graybeal, M. Hauswirth, C. Henson, A. Herzog, V. Huang, K. Janowicz, W.D. Kelsey, D. Le Phuoc, L. Lefort, M. Leggieri, H. Neuhaus, A. Nikolov, K. Page, A. Passant, A., Sheth, and K. Taylor, "The SSN Ontology of the W3C Semantic Sensor Network Incubator Group," *Journal of Web Semantics*, 2012.

[10] U. Christoph and J. von Stülpnagel, "Context Detection on Mobile Devices," *Second Workshop on Context-Systems Design, Evaluation and Optimisation (CoSDEO 2011)*, in conjunction with the 24th International Conference on Architecture of Computing Systems (ARCS) in Como, Italy, February 22nd - 25th, 2011.

[11] A. Dey and G. Abowd, "Towards a better understanding of context and context-awareness," *Workshop on the What, Who, Where, When and How of Context-Awareness* at CHI 2000, 2000.

[12] O. Droegehorn, "Optimizing background-communication of mobile Devices and Sensors to drive End-User Services," *IEEE 22nd International Symposium on Personal, Indoor and Mobile Radio Communication*, 2011.

[13] e-Lite. Apache Jena on Android.

[14] Forschungsunion, "Securing the future of German manufacturing industry: recommendations for implementing the strategic initiative INDUSTRIE 4.0," Final report of the Industrie 4.0 Working Group. Forschungsunion & Acatech, 2013.

[15] GitHub. NanoHttpd.

[16] J. C. Georgas, A. van der Hoek, and R. N. Taylor, "Using Architectural Models to Manage and Visualize Runtime Adaptation," *Computer*, October 2009.

[17] Google. Google Maps API.

[18] Google. Google+ API.

[19] T. Gu, P. Hung Keng, and Z. Da Qing, "A Service-Oriented Middleware for Building Context-Aware Services," *Journal of Netw. Comput. Appl.* 28, no. 1, 1-18, 2005.

[20] M. Hatala, R. Wakkary, and L. Kalantari, "Ontologies and rules in support of real-time ubiquitous application," *Journal of Web Semantics*, Special Issue on "Rules and ontologies for Semantic Web", vol. 3, no. 1, pp. 5–22, 2005.

[21] A. Hoffmann, *Paradigms of Artificial Intelligence: A Methodological & Computational Analysis.* Springer-Verlag, August 1998.

[22] *International Journal of Semantic Computing.* World Scientific Publishing. Available at

[23] P. Jagtap, A. Joshi, T. Finin, and L. Zavala, "Preserving Privacy in Context-Aware Systems," *Fifth IEEE International Conference on Semantic Computing,* 2011.

[24] X. Jiang and J. A. Landay, "Modeling Privacy Control in Context-Aware Systems," *IEEE Pervasive computing*, pp. 59-63, 2002.

[25] A. Kamis and M.J. Davern, "Personalizing to Product Category Knowledge: Exploring the Mediating Effect of Shopping Tools on Decision Confidence," *Proceedings of the 37th Hawaii International Conference on System Sciences,* IEEE, 2004.

[26] E. Kim and J. Choi, "A Context Management System for Supporting Context-Aware Applications," *IEEE/IFIP International Conference on Embedded and Ubiquitous Computing*, 2008.

[27] A. Kofod-Petersen and M. Mikalsen, "Representing and Reasoning about Context in a Mobile Environment," *Revue d'Intelligence Artificielle*, vol. 19, no. 3, pp. 479–498, 2005.

[28] K.J. Laskey, K.B. Laskey, P.C.G. Costa, M.M. Kokar, M. Trevon, and T. Lukasiewicz, " Uncertainty Reasoning for the World Wide Web," W3C Incubator Group Report 31 March 2008. Available at

[29] L. Lefort, C, Henson, and K. Taylor, "Semantic Sensor Network XG Final Report," W3C Incubator Group Report 28 June 2011.

[30] P. Mehra, "Context-Aware Computing: Beyond Search and Location-Based Services," *IEEE Internet Computing* 16, no. 2, 12 – 16, 2012.

[31] S. Nath, "ACE: exploiting correlation for energy-efficient and continuous context sensing," *In MobiSys'12*, June 25-29, UK, pp. 29-42, 2012.

[32] O. Nykänen, "Semantic Web for Evolutionary Peer-to-Peer Knowledge Space," In Birkenbihl, K., Quesada-Ruiz, E., & Priesca-Balbin, P. (Eds.) Monograph: Universal, Ubiquitous and Intelligent Web, *UPGRADE, The European Journal for the Informatics Professional*, Vol. X, Issue No. 1, February 2009, ISSN 1684-5285, CEPIS & Novática.

[33] T. Patkos, I. Chrysakis, A. Bikakis, D. Plexousakis, and G. Antoniou, "A Reasoning Framework for Ambient Intelligence," *SETN 2010*, pp. 213-222.

[34] M. Palmieri, I. Singh, and A. Cicchetti, "Comparison of Cross-Platform Mobile Development Tools," *16th International Conference on Intelligence in Next Generation Networks (ICIN)*, pp. 179–186, 2012.

[35] A. Ranganathan and R.H. Campbell, "A middleware for context-aware agents in ubiquitous computing environments," *ACM/IFIP/USENIX International Middleware Conference*, Rio de Janeiro, Brazil, June 2003.

[36] D. Riboni, "Towards the Combination of Statistical and Symbolic Techniques for Activity Recognition," *IEEE Pervasive Computing and Communications*, 2009.

[37] N. Roy, A. Misra, C. Julien, S. K. Das, and J. Biswas, "An Energy-Efficient Quality Adaptive Framework for Multi-Modal Sensor Context Recognition," *IEEE International Conference on Pervasive Computing and Communications* (PerCom), Seattle, March 21-25, 2011.

[38] S. Russel and P. Norvig, *Artificial Intelligence: A Modern Approach.* Prentice Hall, 1995.

[39] P. Sheu, H. Yu, C.V. Ramamoorthy, A.K. Joshi, and L.A. Zadeh, *Semantic Computing.* Wiley-IEEE Press, 2010.

[40] W. Sitou, and B. Spanfelner, "Towards Requirements Engineering for Context Adaptive Systems," *31st Annual International Computer Software and Applications Conference*, 2007.

[41] U. Straccia, "Foundations of Fuzzy Logic and Semantic Web Languages," Chapman & Hall, USA:CRC Press, 2014.

[42] T. Strang and C. Linnhoff-Popien, "A Context Modeling Survey," *Workshop on Advanced Context Modelling, Reasoning and Management*, In UbiComp 2004 - The Sixth International Conference on Ubiquitous Computing, Nottingham/England, 2004.

[43] A. Toninelli, R. Montanari, L. Kagal, and O. Lassila. "A Semantic Context-Aware Access Control Framework for Secure Collaborations in Pervasive Computing Environments," *Proceedings of the 5th International Semantic Web Conference*, Springer, pp. 5-9, 2006.

[44] E. Yndurain, D. Bernhardt, and C. Campo, "Augmenting Mobile Search Engines to Leverage Context Awareness," *IEEE Internet Computing* 16, no. 2, pp. 17–25, 2012.

[45] WebPlatform.org. APIs.

[46] W3C Data Web Activity – Building the Web of Data. World Wide Web Consortium (W3C).

[47] W3C Standards. World Wide Web Consortium (W3C).

[48] W3C WOT. Web of Things Community Group. World Wide Web Consortium (W3C).

AUTHORS

Dr. O. A. Nykänen works as Adjunct Professor at the Tampere University of Technology, Department of Mathematics, Tampere, Finland (e-mail: ossi.nykanen@tut.fi). His research interests include semantic computing, information modeling and scientific visualization, (computer-supported) mathematics and education, and the related applications. In addition to his research and higher education activities, Dr. Nykänen is the Manager of the World Wide Web Consortium (W3C) Finnish office.

M.Sc. A. Rivero Rodriguez works as Researcher at the Tampere University of Technology, Department of Mathematics, Tampere, Finland (e-mail: Alejandro.rivero@tut.fi), within the Marie Curie ITN research project MULTI-POS. His research interests include context-awareness, semantic modelling/computing, and e-learning.

Evaluation of Augmented Reality Frameworks for Android Development

I. Marneanu[1], M. Ebner[1] and T. Rößler[2]
[1] Graz University of Technology, Graz, Austria
[2] Evolaris Next Level GmbH, Graz, Austria

***Abstract*—Augmented Reality (AR) is the evolution of the concept of Virtual Reality (VR). Its goal is to enhance a person's perception of the surrounding world. AR is a fast growing state of the art technology and a variety of implementation tools thereof exist today. Due to the heterogeneity of available technologies, the choice of the appropriate framework for a mobile application is difficult to make. These frameworks implement different tracking techniques and have to provide support to various constraints. This publication aims to point out that the choice of the appropriate framework depends on the context of the app to be developed. As expected, it is accurate that no framework is entirely the best, but rather that each exhibits strong and weak points. Our results demonstrate that given a set of constraints, one framework can outperform others. We anticipate our research to be the starting point for testing of other frameworks, given various constraints. The frameworks evaluated here are open-source or have been purchased under Academic License.**

***Index Terms*—Augmented Reality, framework, Android, evaluation.**

I. Introduction

Augmented Reality is a very promising emerging technology, growing in popularity on mobile devices. A number of research studies were published in late 2013, forecasting the future of the Augmented Reality market [1]. For instance, Juniper Research has estimated that the number of mobile AR users worldwide will steadily grow to 200 million by 2018 [2]. The AR technology has made great progress on mobile phones, and Juniper Research further predicted in 2012 that over 2.5 billion mobile AR apps will be downloaded annually to smartphones and tablets by 2017 [3].

These premises lead to the research question – "Which is currently the best open framework for developing an Augmented Reality mobile application?" Towards this goal, we evaluated open AR frameworks for Android mobile development, due to the fact that the iOS platform has been evaluated over the years and clear results can be found in Dominik Rockenschaub's Master's Thesis [4]. Another incentive for researching the Android specific open AR frameworks is the recent job market growth registered in Android development [5].

Various AR systems are available today. Based on the hardware and technical capabilities of the testing device, three main tracking techniques are defined: marker-based tracking, markerless tracking, and GPS tracking. Our research evaluates the AR frameworks using visual tracking methods, specifically markerless tracking. Marker-based tracking is also an optical method, but has become obsolete [6]. Special fiducial marks have to be created in order to be tracked. They have to be maintained over time, which is also a costly operation. Markerless based tracking on the other hand, can target any part of the surrounding environment. Any image or object can be used as a target without being tied to a specific marker. Markerless tracking is gaining more and more ground against the marker-based tracking, as it does not require generating of invasive markers. These methods extract information and characteristics about the environment, which may be useful again later.

This publication is organized as follows: the methodology is presented in section II; section III describes the setup of the testing environment and the implementation process; the results per criterion are given in section IV while the discussion analyzing these results can be found in section V; and finally we conclude the paper in section VI.

II. Methodology

Six AR frameworks have been researched and evaluated after a careful evaluation of 35 frameworks. The 6 chosen frameworks support Android development, as well as markerless tracking. An important criterion in choosing the frameworks to be evaluated is the availability of the library. Several are available for commercial use only (for the purpose to sell and profit) or are simply not available to the general public. Others are accessible only for a trial period. Some are depreciated, like in the case of Popcode, or even present language barriers as Koozyt, where information is provided only in Japanese. Table I provides the six frameworks, the development country and start year as well as availability.

TABLE I.
Evaluated Frameworks

Framework	Development	Availability
ARLab	Spain (2006)	Demo
ARToolKit	USA (2001)	Academic License
D'Fusion	France (1999)	Watermark
Vuforia	Austria (2011)	Free
catchoom	Spain (2011)	Demo
metaio	Germany (2003)	Watermark

For evaluation purposes, various criteria have been defined. The frameworks are actively tested and the testing times are recorded. Therefore, frameworks can be compared against each other given a criterion or a set of criteria.

For a better visualization of the results, a test app has been developed. As a result, different frameworks can be tested in real time, and the results are saved in a local database. The results are then presented in different visualizations.

III. Setup And Implementation

For each criterion, we simulated a testing environment and each framework was evaluated under the same conditions. Several tests have been performed per framework for each criterion, thus the average testing time is used to compare the frameworks.

A. Criteria Categories

Two main criteria categories have been identified and actively tested: environmental criteria, and target criteria. Environmental criteria represent those constraints found in the immediate neighbourhood, which are conditioned by the environment and influence the recognition of the target image. Tested here are the influence of different light intensities or that of the frameworks' performance when dealing with dark backgrounds versus bright ones. Other considered criteria include differences in viewpoint, visible target area, the mirroring effect and various distances between the testing device and the target image, as well as the noise present in the target image due to its deterioration over time.

- The evaluation of the **light intensity** criterion has been executed in special lightning conditions. The darkness criterion has been tested inside a room, in natural light at sunrise. The second test is also performed inside a room, by overexposing the target image to the direct light of a desk lamp. The third simulated event happens by closing and opening the window shades. This experiment is meant to replicate sudden changes in light intensity. The mirroring effect is reproduced by tracking the target images on a computer screen.
- The evaluation of the **viewpoint** criterion has been performed in four different perspectives, from a 45° angle to the left, right, upwards and downwards of the target image as depicted in Fig. 1.
- **Visibility** is tested by uncovering 10% of the visible area at every step. A white sheet of paper is used to cover the target image and uncover it in incremental steps as shown in Fig. 2. When the target is recognized and the view augmented, the step increasing stops.
- The evaluation of the **noise** criterion has been performed by digitally altering the target images and adding different noise levels. The test sequence includes five noisy target images as given in Fig. 3. The noise is incrementally added starting from a 10% noisy target image in steps of 20% up to 90%.
- The evaluation of the **distance** criterion has been performed by starting testing up close, at a 10 cm distance from the target image. The distance is increased at every step by 10 cm until the framework cannot recognize the target anymore.

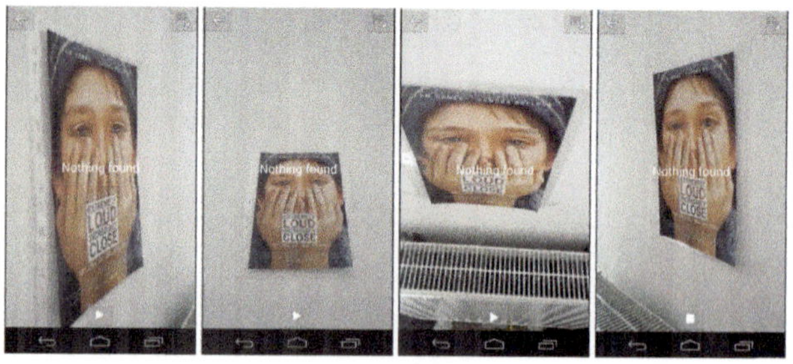

Figure 1. Viewpoint Test Sequence

Figure 2. Visibility Test Sequence

Figure 3. Noise Test Sequence

- The evaluation of the **background** criterion has been performed in two background test sequences, dark versus bright contrasts. This reproduction is meant to simulate the placement of the target image against high-contrast backgrounds.

Target criteria, on the other hand, are those target image attributes that can be configured and directly influence the detection of the target. While an image can be a target image for one framework, it might not be supported by another. For example, some frameworks are robust to aspect ratio changes or various contrast ratios. Moreover, different target images sizes and a number of special printed materials are tested. We also evaluated whether or not a considerable difference in testing time exists between detecting the original target and its grayscale.

- The evaluation of the **grayscale** criterion has been performed on a printed default size target image in grayscale. An example can be seen in Fig. 4.
- The evaluation of the **contrast ratio** is carried out on digitally modified target images as depicted in Fig. 5. Four target images are tested with a contrast value set to -50 and at each step it is incremented by 50.
- The evaluation of the **size** criterion has been executed on four different target images. Each of the six tested target images has been

reduced to 5cm, 10cm, 15cm and 20cm as shown in Fig. 6. The distance between the testing device and the target image has been constant at a default of 30cm.

- The evaluation of the **aspect ratio** requires the target images to be digitally modified. Thus, each of the six original target images is shrunk either vertically or horizontally by reducing the height, respectively the width to one third (1/3), a half (1/2) and two thirds (2/3) as given in Fig. 7.
- The evaluation of the **material** criterion has been performed in three different stages. The first case depicts the target image behind a glass window, as it would be if a poster were placed inside the side panels of a bus shelter. The second case simulates a restaurant menu scenario where the menu is laminated to protect the paper against deterioration and mishaps like spilled drinks. The third target image is printed on a glossy photo printing paper as it might be used in a scenario for displaying promotional ads or as packaging of different promotional products.

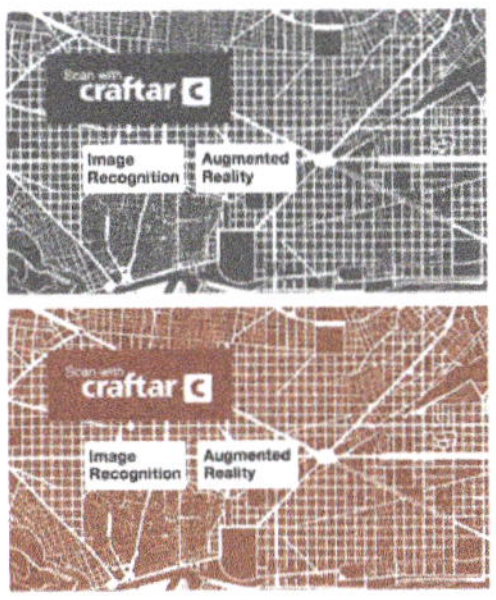

Figure 4. Grayscale Test Sequence

Figure 5. Contrast Test Sequence

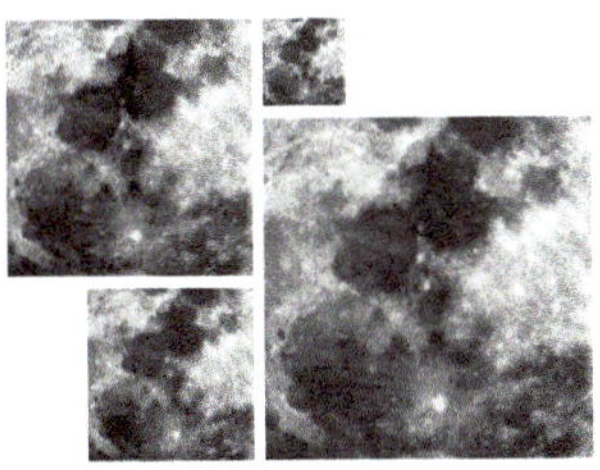

Figure 6. Size Test Sequence

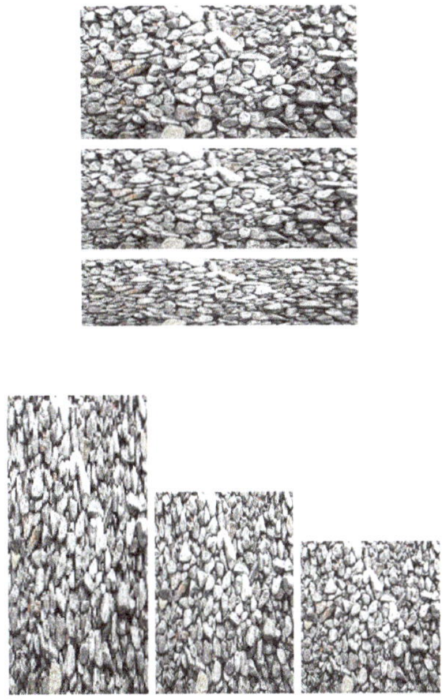

Figure 7. Aspect Ratio Test Sequence

In addition to the two mentioned categories, performance and usability criteria are defined. Each framework description makes a point to state that some basic graphical problem is resolved and the tracker is optimized. These basic issues are collected under performance criteria and it is determined whether or not they were actually solved. Here we mention the constant flicker, a visible motion blur, and the ability to deal with fast moves such that the virtual content is not lost. Registration (the accurate alignment between the real world and the virtual object) is an important factor to overcome, as well as the capability to occlude the virtual object when necessary to create the feeling that the virtual object belongs in the scene, and is part of the scene.

Furthermore, special features are implemented within each SDK for presenting a more comprehensive product. Not all features are supported by every framework. Therefore they are categorized by usability and it is determined which are endorsed. The considered features include face tracking, text detection, flash and usage of the front camera. Moreover, the ability to display the virtual content even when the target image is not in the line of sight anymore, known as extended tracking, is a useful feature in some use cases. Likewise, the possibility to track more than one target image simultaneously comes in handy.

B. *Android App*

We developed an Android app which integrates the 6 frameworks (ARLab, ARToolKit, D'Fusion, Vuforia, catchoom and metaio) and actively tests the environmental and target criteria. By actively testing, it is meant that a framework is chosen, its camera view opens and a criterion is selected for testing. When the testing environment is set, the user starts the test timer by pressing the Start button. When the virtual content is superimposed into the real world, the testing time is saved as the time needed to detect and recognize the target image.

We determined if the performance and usability criteria are supported by observing the behaviour of the frameworks in a neutral context, on the default target images (no special light conditions). Some feature support, such as face tracking and text detection are set by finding the information in the framework documentation.

Criteria and criteria categories can be added to the list of predefined criteria. Use cases can be defined by adding weighted criteria into a context.

The frameworks can be compared against each other given a constraint or a set of constraints. Furthermore, an overview for each framework is available, displaying the average testing times per framework for each criteria category.

IV. Results

In order to preserve the same testing conditions, such as sunrise light, a criterion is tested by all frameworks before moving on to the next criterion. The testing environment changes between criteria and by testing one criterion at a time for every framework, the environment is minimally altered during the testing.

A. Environmental Criteria

Light intensities are supported overall as observed in Fig. 8. On the left side are the average times for the desk lamp and screen glare light conditions. The graph on the right illustrates the testing times from the semidarkness and sudden change tests. Metaio has the lowest average testing time, followed by Vuforia with an average testing time under a second. The worst timing is registered by ARToolKit while ARLab and catchoom are steady under a second and a half.

Most of the frameworks detect 45° angles. The **viewpoint** testing times are shown in Fig. 9, the left graph provides the average testing times for the left and right perspectives, while on the right side are given the up and down viewpoints results. ARToolKit provides good testing times, nevertheless it should be considered that it recognizes only two out of the four perspectives. ARLab has difficulties detecting the target image from a 45° upward angle. The other viewpoints are easier to detect. Three out of the six frameworks detect the target image from a 45° right angle under half a second.

Detection at 10% and 20% **visibility** requires a longer processing time. Vuforia detects the target at 10% visibility somewhere between one and two seconds, being the only framework that scores this performance. For more than 20% visibility Fig. 10 illustrates that Vuforia's average time is less than a second. Catchoom detects the target image, having only 20% uncovered, in little over two seconds and faster given more visibility. ARLab and metaio need at least 40% of visible area before starting detection, while ARToolKit cannot detect anything under 80% visibility.

Vuforia and metaio have very similar times when dealing with **noise**. However, Vuforia detects a 200% level noise target image while metaio goes as high as 70%. Fig. 11 displays the average testing result for the 10 and 30% noisy targets on the left and 50, 70 and 90% noise levels on the right. The fastest times are still recorded by metaio while ARLab is the slowest detector. Four of the six frameworks have a detection time over one second. Nevertheless, Vuforia is still the most impressive while metaio outperforms D'Fusion and catchoom.

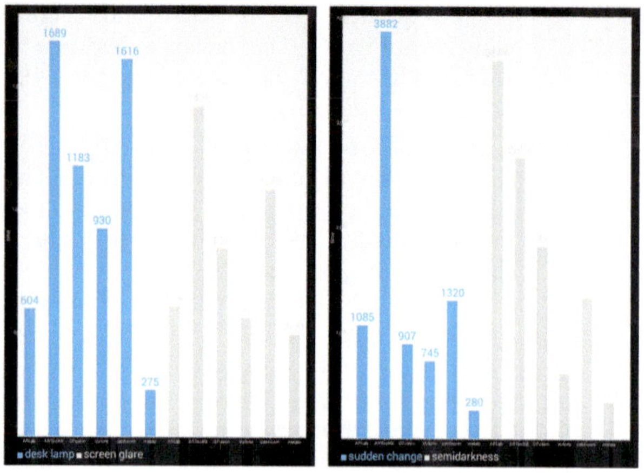

Figure 8. Light Intensities

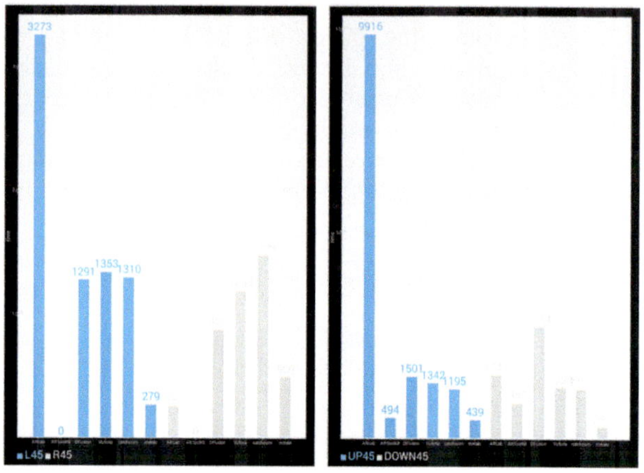

Figure 9. Viewpoint

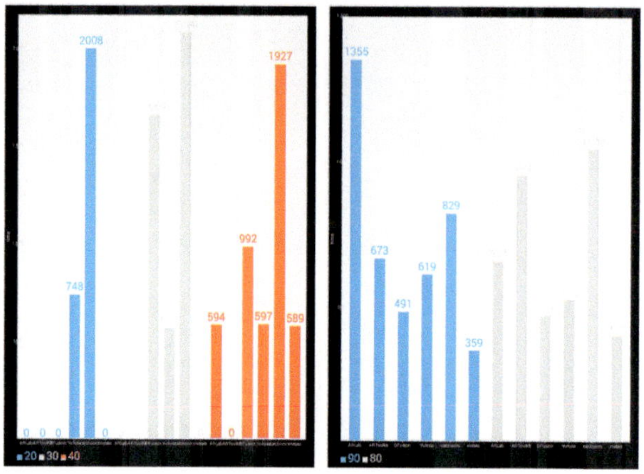

Figure 10. Visibility

Table II provides the minimum and maximum **distances** supported by each framework. Because there are too many distances to be displayed in a readable graph, Fig. 12 depicts the average over all supported distances per framework. Vuforia is portrayed as the fastest framework when it comes to the distance between the testing device and the target image. Metaio should be on the second place, if not on first place, considering that its average testing time increased because of the extra effort required to detect the target from 240cm away. The worst timing is registered by the ARLab framework.

TABLE II.
SUPPORTED DISTANCES PER FRAMEWORK

Distance	ARLab	ARToolKit	D'Fusion
10cm	yes	yes	
20cm	yes	yes	
30cm	yes	yes	yes
40cm	yes	yes	yes
50cm	yes	yes	yes
60cm	yes		yes
70cm	yes		yes
80cm	yes		yes
90cm			yes

Distance	Vuforia	catchoom	metaio
10cm			
20cm	yes	yes	yes
30cm	yes	yes	yes
40cm	yes	yes	yes
50cm	yes	yes	yes
60cm	yes	yes	yes
70cm	yes	yes	yes
80cm	yes	yes	yes
90cm	yes	yes	yes
150cm	yes	yes	yes
210cm	yes		yes
240cm			yes

Fig. 13 demonstrates the average time needed by a framework to detect the target on bright colours versus a target placed against a dark coloured **background**. The left graph illustrates the average over all testing times per framework, and the right one details the two possible cases. D'Fusion supports both bright and dark backgrounds in a constant manner. D'Fusion and Vuforia have similar times dealing with dark contrasts, while Vuforia performs slightly better in the case of brighter backgrounds. Vuforia is the only framework that has stored a better time on bright backgrounds than a darker one. As both graphs can attest to it, metaio is the leader of the background criterion.

B. *Target Criteria*

We tested both the original target image and its **grayscale** version in order to compare the testing times as shown in Fig. 14. Metaio recorded the lowest testing time, while three other frameworks tested in under a second. Catchoom and ARLab have recorded times a little over a second. As it can be seen, coloured or gray, the detection time is roughly the same. A slight difference can be observed at ARLab.

All frameworks passed the **contrast ratio** test, nevertheless not without difficulties as Fig. 15 concludes. The left graph illustrates the default target testing times (contrast level 0) and the -50 contrast level. The right side shows the average times for the 50 and 100 contrast level target images. ARLab tested poorly and metaio holds the

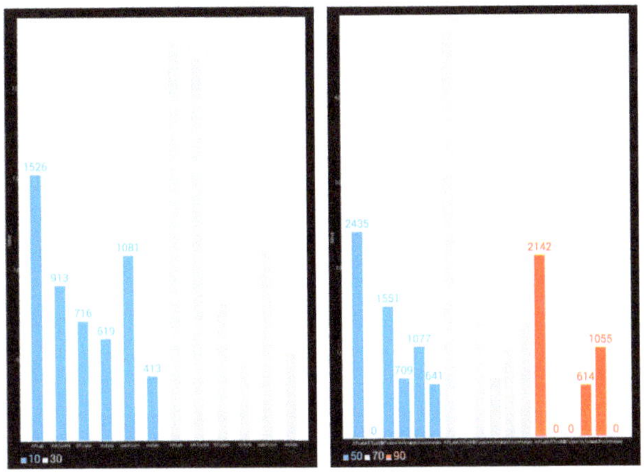

Figure 11. Noise

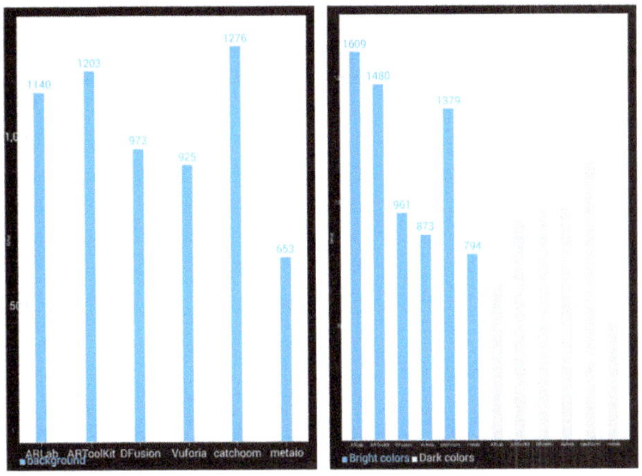

Figure 13. Background

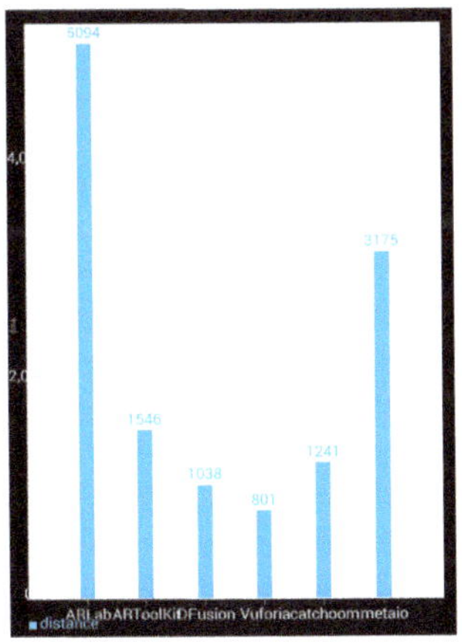

Figure 12. Distance

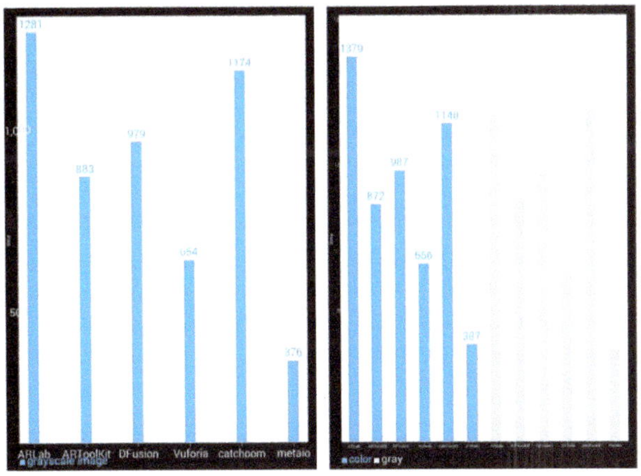

Figure 14. Grayscale

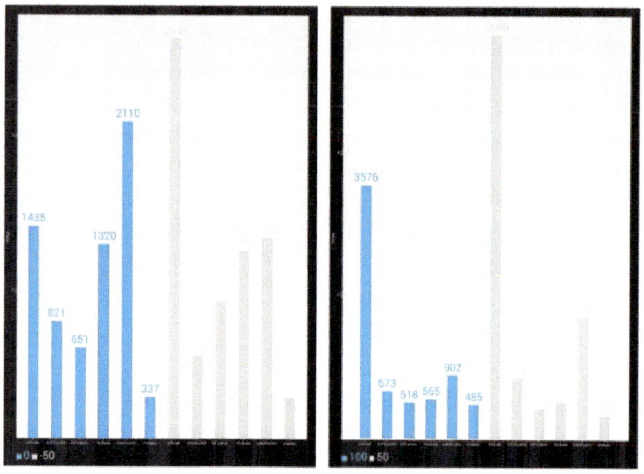

Figure 15. Contrast Ratio

fastest testing times. ARToolKit, D'Fusion and Vuforia also stored average testing times under one second. Catchoom needs on average one and a half seconds while ARLab takes double the time, close to 3 seconds and a half to detect target images of different contrast levels.

The frameworks that detect different **sizes**, keep the average testing times throughout the tests constant as proved by Fig. 16. ARLab only recognizes 10cm high target images and D'Fusion does not detect 20cm high images while ARToolKit only recognizes 20cm high targets. Metaio has the fastest time and close behind are ARLab and Vuforia, both just over half a second. Catchoom tests the highest average time and most of the various target sizes are detected under a second. Catchoom needs over a second when dealing with bigger sizes.

Metaio is the only framework that supports all six **aspect ratio** changes. ARLab, on the other hand, detects nothing. Vuforia is the closest framework to metaio. Nevertheless, metaio still has the best times and covers all constraints. Fig. 17 shows the average results for the horizontal aspect ratio on the left and the considered vertical ratios on the right.

All **materials** are supported by every framework, with the exception of ARToolKit, which was not tested on glass (technical difficulties). The left graph in Fig. 18 denotes the overall average for each framework and the right one breaks down this average by the material types: glass, glossy paper and plastic. ARLab has the best time

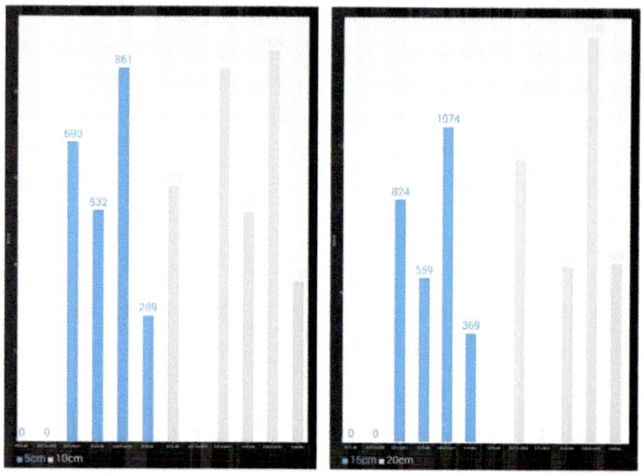

Figure 16. Size

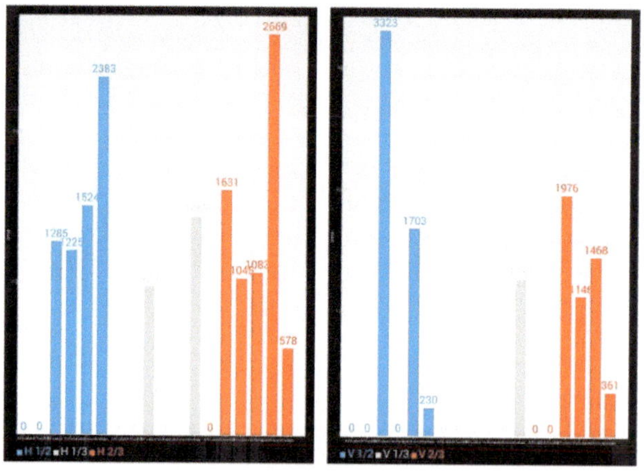

Figure 17. Aspect Ratio

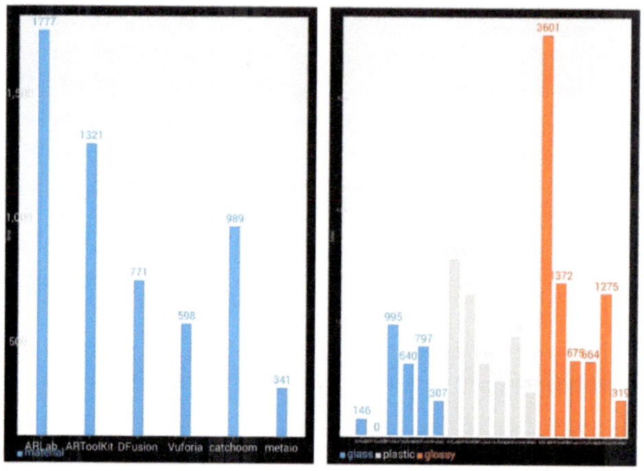

Figure 18. Material

on glass detection and the worst on plastic. Metaio has the lead on glossy and plastic while D'Fusion and Vuforia tolerate them, showing a worse performance on glass. ARToolKit keeps itself constant around the second marker.

V. Discussion

A number of different constraints have been tested on the evaluated frameworks. These constraints raise difficulties for some frameworks while others overcome them easily.

A. Scenarios

Finally, we defined a number of use cases called scenarios. A scenario is a collection of criteria considered for a specific context by acknowledging their importance through weights.

Scenario 1: Consider an indoor app for visualizing large objects like furniture, home furnishings, and appliances. A target image would be placed in the spot where the real object would be, and the virtual object is displayed on top of the target. A couple of constraints that must be considered include:

- How near or far the user can be from the target.
- Having different viewpoints for a better outlook of the virtual object.

- Occluding the virtual object by real objects if necessary.
- A correct alignment between the real scene and the virtual object.
- Using more than one target for a better placement of the virtual object, at the right place and at the right scale.
- Seeing the virtual object even when the target is not visible anymore.

For this scenario, metaio and Vuforia score the highest points. Catchoom is left behind because it does not support extended tracking.

Scenario 2: All target criteria are considered for ranking the frameworks when dealing with a magazine app scenario. An image in the magazine represents a target image for the magazine app. By hovering over with a smartphone the virtual content is revealed to the user. More constraints are considered together with:

- Gray coloured images and strong contrast images.
- Small size images.
- Images printed on different paper types.
- Recognize images given that they are not completely visible.
- Text detection to support the images.

Catchoom and ARLab have a better grasp of the environmental criteria than metaio and D'Fusion. However, metaio supports text detection and catchoom does not.

Scenario 3: The two use cases presented above are indoors scenarios. Now we also consider outdoors scenarios and suppose a company comes up with a new marketing idea and uses AR for posters creatively located on the side panels of bus shelters. A user can either be at the bus stop looking at the poster or notice it from inside a moving bus and try to decipher its message. An outdoors scenario focuses on the environmental criteria along with:

- Sudden light changes, such as the sun hiding behind the clouds.
- How far away the user can be from the target.
- Deterioration of the poster in time due to rain, wind, sun exposure and so on.
- A dynamic background such as people walking behind it or cars driving by.
- How much of the poster is visible from where the user sits.
- Fast moves as the user passes by in the bus.
- Once the poster is recognized, the virtual content is persistent even though the user is in a moving bus.

Vuforia has the lead in this scenario, followed by catchoom. Vuforia provides better support than catchoom for performance and usability constraints.

Scenario 4: Some supermarkets today use AR for promoting their products. Passing by a shelf with various products, the customer can point the camera to the boxes, select a product and initiate a game. For example bursting some bubbles, such that when a predefined number of busted bubbles is reached, the customer wins a discount for the chosen product. A couple of features must be supported and some requirements must be met, such as:

- Support for more than one target image per frame, such that the customer can choose which product he wants to play for.
- A blurred image would make it hard for the customer to play the game.
- The target image can be on a cereal box or a label on a wine bottle.
- Flash might be needed if there is not enough light in the store for the app to recognize the target.
- The label can be torn or the image on the cereal box can be ripped.

Vuforia gains the highest score, closely followed by metaio. The most noticeable difference is given by the support for fast moves; from the tests performed it has been noticed that metaio does not support fast moves.

Scenario 5: The next two use cases require special features, such as text recognition and face detection. An app exists today for translating text from English into Spanish and several other languages. Imagine you are visiting a foreign country in which you do not know the language and need help getting around. This app can be used to translate signs, window ads, and menus. An important feature that the Augmented Reality framework must support is text detection, as well as:

- The contrast between the text and the background where it is written.
- Different materials on which the target is printed, glossy paper, metal plates, glass and others.
- Handle flickering as a constant flicker of the translated word would make it difficult to read.
- How far away the user can be from the target.
- Contamination when dealing with outside signs.

Metaio and Vuforia register the best results for the relevant environmental criteria while no framework distinguishes itself when dealing with target criteria.

Scenario 6: A number of applications already exist for virtually trying on products such as glasses, hats and order them online. Take sunglasses, for example. By using such an app, the customer can browse through the catalogue, choose a pair of glasses, switch between available colours and the sunglasses are projected on the image of the face looking back from the phone. The most important feature that must be supported is face tracking, together with:

- A static background for easier face detection.
- A constant light source or the colour of the skin keeps changing.
- Fast moves lead to losing the tracking.
- The possibility of switching to the front camera.

D'Fusion is the framework that fits best the constraints imposed by such a use case, followed by metaio and Vuforia. Vuforia does not support face tracking, thus loosing points, while D'Fusion beats metaio by supporting fast moves.

Table III shows the recommended framework for each of the described use cases.

TABLE III.
RECOMMENDED FRAMEWORKS

Scenario	Framework
Interior Design App	metaio
Magazine App	metaio
Bus Shelter App	Vuforia
Supermarket Promotions App	Vuforia
Tourist Translator App	metaio
mCommerce App	D'Fusion

B. *Findings*

Testing the frameworks on the default size target from various distances brought some interesting findings. Up close, 10cm away from the target image, the target is in most cases not completely visible. However, ARToolKit and ARLab recognized the targets. This is surprising because ARToolKit needs an 80% visible area of the target, while Vuforia only requires 10%, and still could not recognize the target up close. From these findings, it is concluded that the size of the target image and the distance to it are not strongly correlated.

The light intensity tests revealed a weak ARToolKit tracker when dealing with sudden change in light conditions or a semi dark environment. Each of the considered four sequences for testing light intensities instigates environmental issues. In each case four out of the six frameworks show signs of extra work.

Another surprising result was registered by Vuforia, which can detect a target image with a 200% level of noise. Most of the frameworks can still overcome a 70%, some even 90%, but 200% is a worth mentionable achievement.

VI. CONCLUSION

This paper presented a methodic technique for evaluating open AR frameworks for Android development. A number of evaluation criteria are defined and the results published. As a prototype we developed a test app which is implemented using Eclipse with Android Development Tools (ADT). The prototype can actively test the integrated frameworks and save the testing times in a local database for further comparison.

We would like to state that no AR framework is better than another, each having its advantages and disadvantages. In some circumstances, given a set of constraints, a framework can outperform others.

In future work, we will concentrate our attention on optional features support such as multi-targets, face tracking and text detection. The app itself can be further developed to become a powerful tool of evaluation.

For a more detailed documentation of the presented methodology and results can be found in [7].

ACKNOWLEDGMENT

This work was supported by the Institute of Information Systems and Computer Media of the Graz University of Technology and financed by the Evolaris Next Level GmbH software company.

REFERENCES

[1] TechNavio - Infiniti Research Ltd. *Global augmented reality market 2014-2018*. A market research report.

[2] Juniper Research. *Mobile augmented reality: smartphones, tablets and smart glasses 2013-2018*. A market research report.

[3] Juniper Research. *Over 2.5 Billion Mobile Augmented Reality Apps to Be Installed Per Annum by 2017*. A market research report.

[4] Dominik Rockenschaub. "Entwicklung und anwendung einer systematischen vorgehensweise zur analyse marker basierter augmented reality frameworks für mobile endgeräte.", *Master's thesis*, Johannes Kepler Universität Linz, Linz, Austria, 2012.

[5] Jennifer Lynn. *iOS vs. OS: Current Job Market Is Growing Faster For Android Developers Than Apple*.

[6] Jose Dolz. *Markerless augmented reality*.

[7] Marneanu Iulia. "Evaluation of Augmented Reality Frameworks for Android Development", *Master Thesis*, Graz University of Technology, Graz, Austria, 2014.

AUTHORS

I. Marneanu is a master student at Graz University of Technology, Rechbauerstraße 12, 8010 Graz, Austria (e-mail: iulia.marneanu@ student.tugraz.at).

M. Ebner, Assoc. Prof. is with the Institute of Information Systems and Computer Media of Graz University of Technology, Inffeldgasse 16c, 8010 Graz, Austria (e-mail: martin.ebner@tugraz.at).

T. Rößler is working at Evolaris Next Level GmbH, Hugo-Wolf-Gasse 8, 8010 Graz, Austria (e-mail: Thomas.Roessler@evolaris.net).

Wavelet Entropy Algorithm to Allocate the Extreme Power Peaks in WiMax Systems

Qardi Hamarsheh, Omar Daoud, and Saleh Saraireh
Philadelphia University, Amman, Jordan

Abstract—**This work proposes a solution to overcome the effect for one of the main drawbacks of these days' wireless systems, where Multiple-Input Multiple-Output (MIMO)-Orthogonal Frequency Division Multiplexing (OFDM) combinations has been used. High peak-to-average power ratio (PAPR) arises after the OFDM stage and reduces the performance of the used nonlinear devices. Therefore, a new stage has been imposed between the MIMO and OFDM block. It is based on the entropy meaning of the wavelet transformation to trigger a proposed thresholding criterion and reconstruct the OFDM signal. As a result, the probability of high PAPR appearance will be limited and reduced; a promising result over our recently published work has been conducted; 15-25% extra reduction. This work could be denoted by MIMO-OFDM based on Entropy Wavelet Transform (MO-EWT) systems.**

The MO-EWT validity has been checked based on either numerical analysis or conducted simulation based on MATLAB; where 80% improvement of reducing the high PAPR has been achieved over the literature. These results have been reached using the same environment conditions and at additional cost and complexity of the transceivers structure.

Index Terms—**Thresholding, Wavelet Entropy, MIMO, OFDM, PAPR**

I. INTRODUCTION

The transmission capacity and spectrum efficiency are considered as vital parameters for enhancing the performance and the multimedia quality of the wireless communication systems during the past decades. Increasing both of users and market demand for the best quality was the challenge for the scholars to propose rigid systems to meet these demands. Therefore, combination between two powerful techniques has been found in the literature to offer such services with high quality and adequate spectrum allocation; Multiple-Input Multiple-Output-Orthogonal Frequency Division Multiplexing (MIMO-OFDM) systems. As a result of such combination, they making use of the multipath to enhance the system capacity instead of consider it as a deficiency. Furthermore, it is using the diversity gain to enhance the system robustness to the noise and then improving the system link reliability. Moreover, a good utilization of the frequency spectrum has been offered by using the parallel transmission of the sequential one. There are several ways to implement the MIMO systems to enhance either the spatial diversity or the diversity gain, where the OFDM is easily implemented using the fast Fourier transforms (FFT) and its inverse [1-8]. However, the wireless system performance will be affected by a main deficiency that is resulted after the inverse-FFT (IFFT) stage in the transmitter; namely, Peak-to-Average Power Ratio (PAPR). Due to the unexpected resultant large peaks, the dynamic range of the used nonlinear devices in the transceiver should be enlarged otherwise the output will be distorted and will limit the work performance of such devices. This distortion will cause some vital deficiencies that will affect the system performance, such as spectral spreading, intermodulation, effect the signal constellation. This will cause an increasing in the system cost and complexity [6,8,9].

For such problem, utilizing the PAPR will result an improvement of the overall signal to noise ratio. Various methods, techniques and algorithms have been found in the literature to ease this bottleneck problem and utilize the PAPR problem to improve the overall link budget [10-16].

As a conclusion, MIMO-OFDM technology is considered as a powerful one that is proposed to be used in the next generation of the wireless communications systems.

Imposing some other technologies to this combination has been found in the literature to improve and enhance the system characteristics, such as the wavelet transform technology. It has been used in order to improve and enhance the OFDM system under the multipath fading channel characteristics. Furthermore, the wavelet packet has been proposed to be used instead of the FFT block to enhance the both of SNR and BER, improve the spectral efficiency, the same time reduces the transmitted power, and finally can manage the frequency offset and phase noise, then reduces the inter symbol interference (ISI) and inter carrier interference (ICI) [17-23]. Furthermore, OFDM based wavelet has been found in the literature as a solution of combating the PAPR problem and then providing better transmission quality [24].

This work details a proposed work as a new usage of the wavelet transform to deal with the PAPR problem. Here, the wavelet entropies have been used to trigger a thresholding criterion to deal with the affected OFDM symbol by large PAPR values. This is in addition to give a reconstruction criterion to enhance the BER of the system. This proposition has been compared to our previously published work in [10, 24] and to some techniques in the literature such as partial transmit sequence (PTS) and the clipping algorithms.

The rest paper is organized as follows; the introduced structure of the WiMax system based on the wavelet entropy is defined in Section 2, the simulation results that validated the numerical model are presented in Section 3, while the last section summarizes the conclusion.

II. Wi-Max System Design based on Wavelet Peak Detection

As a radio resource management solution, the IEEE 802.16 standard, WiMAX, is considered these days. It offers the facility of describing the traffic profile for the users and there service need. Thus, the quality of service requirements (QoS) and the traffic characteristics could be determined to support various data transmissions scenarios [25].

The structure of the modified WiMax systems after imposing the MO-EWT is shown below in Figure 1. Here, we can divide the system block diagram into three main parts; OFDM part, the proposed algorithm and the MIMO part. For the MIMO part; Convolutional encoder is used, in addition to Quadrature Shift Keying (QPSK), and 256 IFFT points. After, the FFT block the high power peaks could be appeared due to the coherence addition at the same phase. Therefore, the proposed work has been placed at this stage to overcome the effect of the resultant high PAPR. Finally and before the transmission stage the signals will pass through Vertical-Bell Laboratories Layered Space-Time (V-BLAST) MIMO system, where it's used to enhance the system capacity/throughput expressed in terms of bits/symbol.

Not like the proposed work in the literature, the wavelet entropy has been proposed to trigger an adaptive clipping criterion in order to overcome the high PAPR effect. Furthermore, there are no overload resulted since there is no need for redundant data to be transmitted.

The proposed work is clearly explained by the depicted flowchart in Figure 2. As a start, wavelets could be defined by a small wavy function of proficiently limited duration over an average of zero. It deals with signal based on power of two and it can decompose it based on its high and low frequency bands at each level. This will be attained by using a pair of low and high pass filters to be decimated by a factor of two. For capturing different features during the temporal information detection, this process is repeated to get smaller frequency bands. Secondly, as a general, the entropy of the wavelet could be defined as:

$$P_j = \frac{E_j}{E_{total}} \tag{1}$$

E_j is the energy at each j resolution level, E_{total} is the sum of E_js where j is in the same interval; j=-N,….,-1.

The equation in (3) could be easily applied on the discrete wavelet decomposition of any signal $S(t)$ that is found in as:

$$S(t) = \sum_{j=-N}^{-1} \sum_k C_j(k)\psi_{j,k}(t) \tag{2}$$

$C_j(k)$ is the wavelet coefficient and limited to following frequency interval $2^{j-1}\omega_s \le |\omega| \le 2^j\omega_s$ [26-28].

The depicted flowchart in Figure 2 is used to ease the understanding of the process of the proposed work. It starts with scanning the resultant signal after the IFFT block, where a high peak power occasionally appears and defined mathematically as in [8]

$$\text{PAPR} = NT \times \frac{\left(\left|\frac{1}{\sqrt{N}}\sum_{n=0}^{N-1} X_n e^{j2\pi f_0 nt}\right|^2\right)}{\left(\int_0^{NT}\left|\frac{1}{\sqrt{N}}\sum_{n=0}^{N-1} X_n e^{j2\pi f_0 nt}\right|^2 dt\right)} \tag{3}$$

X_n is the data modulating the n-th is sub carrier and f_o is the nominal subcarrier frequency spacing, N is the length of FFT stage, T is the symbol duration, t is in the range of $0 \le t \le NT$.

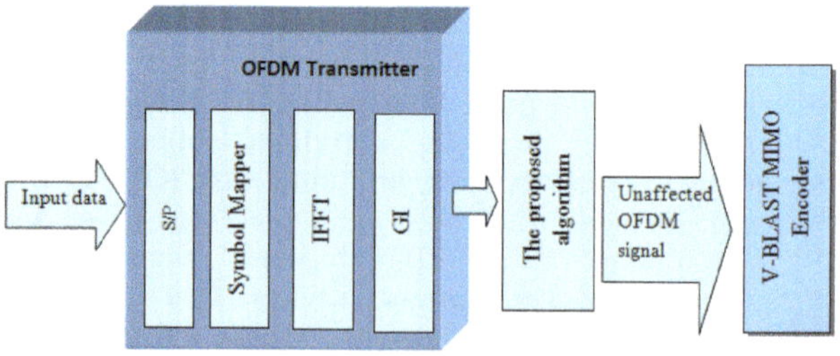

Figure 1. WiMax based MO-EWT system block diagram

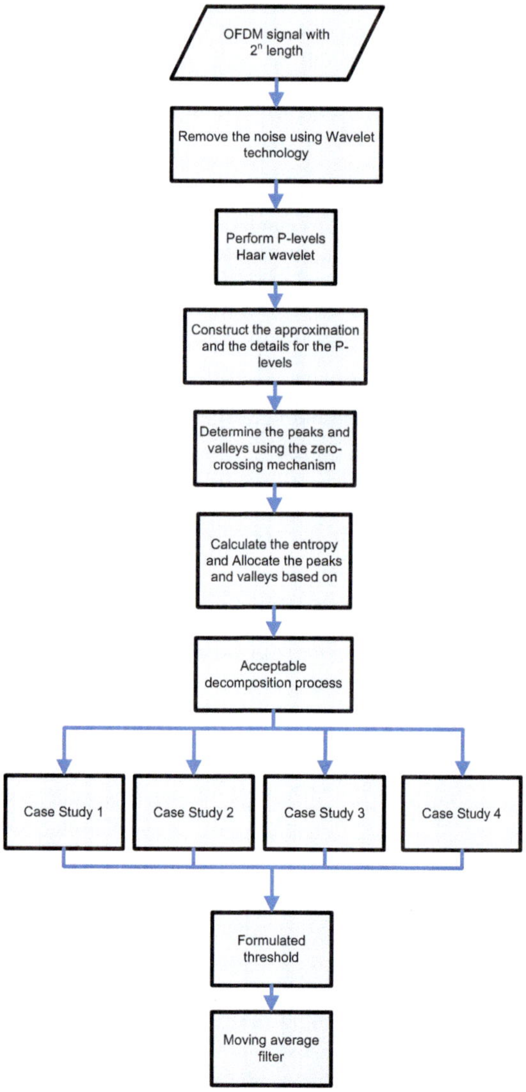

Figure 2. Flowchart of the proposed algorithm

Then the scanned signal will pass through 5 different stages as follows:

1) The preprocess stage:

In this stage, the signal will be treated from the noise using the wavelet technology, it will be decomposed into *P*-levels, and then approximates and details will be generated. This will be accomplished using Haar wavelet decomposition of a signal with *P*=8.

2) Dealing with extreme Peaks and Valleys Stage:

- In this part, the zero crossing mechanism is used to find the extreme peaks values for the founded details coefficients. This is true for local peaks and valleys detection.
- Finding the locations of such peaks in the original OFDM signal, and then store them into a different decomposition level matrix. This projection will be based on an adaptive proposed classification formula:

$$2^n * j + s \qquad (4)$$

n is the index of the decomposition level, and s value is an adaptive factor changed with changing the decomposition level.

3) Entropy calculation Stage:

Based on Shannon method, Table 1 shows the calculated entropies for both of the preprocessed OFDM signal and the decomposed P-Levels. 4 different case studies have been studied to show the proposed work powerfulness. This is in addition to check the peak detection error ratio as a factor of the comparison as shown in Figure 3.

First Case Study: This case study will take into account the details of each level to detect the True and False local extremes points.

Second Case Study: Based on achieved results in Table 1, where the decomposition acceptance shows that the decomposition is not accepted after the second level. Thus, the first two decomposition levels will be used.

Third Case Study: Using all decomposition levels that satisfy the decomposition acceptance stage.

Fourth Case Study: Based on the results found in Table 2, the peaks allocations will be based on each decomposition level separately. The error ratio plays vital role in the allocation process, as an example the false peaks in higher level with good error ratio can be excluded. For that the sharing peaks for the given packet with previous level can be detected and declared as true peaks as shown in Figure 4 for the lowest level ($P = 8$).

4) Thresholding Stage:

An adaptive thresholding formula is proposed to remove low amplitude peaks, while only the significant ones will be taken into consideration. These results are depicted clearly in Figure 5. The adaptive thresholding formula is given as:

$$\mathbf{T} = \frac{(\text{max} + \text{abs\_avg} + \text{abs\_dev})}{\text{K}} \qquad (5)$$

max defines the OFDM signal maximum value, **abs_avg** stands for the OFDM signal average value, **abs_dev** is the mean absolute deviation, and **K** is the adaptive constant that defines the type of thresholding criteria. Thus, the predetermined threshold is used as an adaptive criterion to be attuned with the different wireless systems.

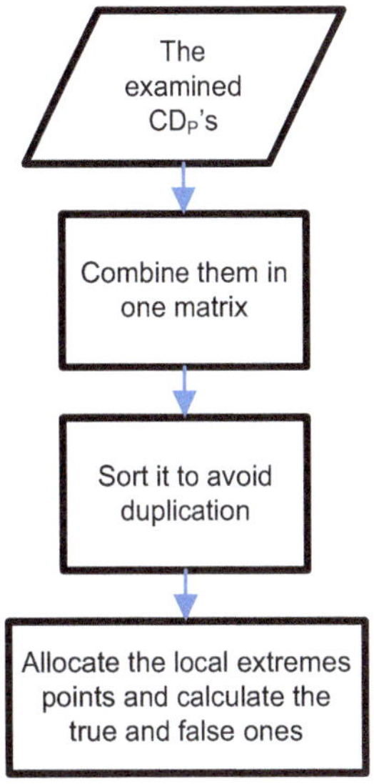

Figure 3. Flowchart of the used procedure

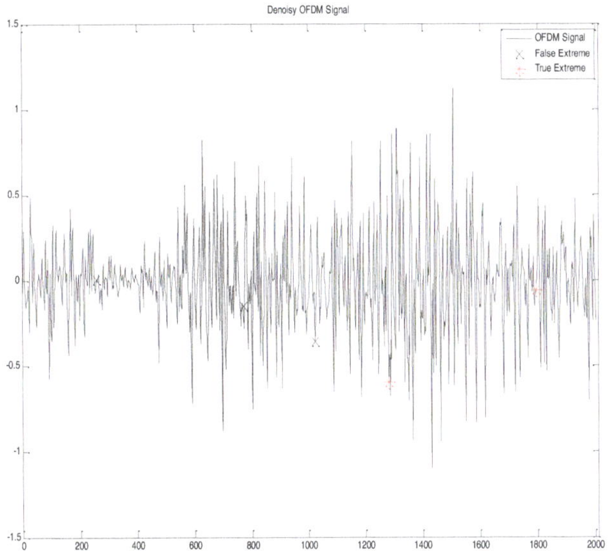

Figure 4. sharing peaks detection for decomposition level P=8

TABLE I.
PACKETS' ENTORPY VALUE

P level	Entropy				Decomposition Acceptance*
	CDp	CAp	Summation	Original Signal	
1	13.3924	60.8508	74.2432	100.4268	Accepted
2	26.5319	32.917	59.4489		Accepted
3	16.3553	20.7162	37.0715		Not Accepted
4	8.2519	9.7128	17.9647		Accepted
5	6.3765	3.2822	9.6587		Accepted
6	1.6068	0.93369	2.5405		Accepted
7	1.541	0.54864	2.0897		Not Accepted
8	0.62314	0.059486	0.68262		Not Accepted

(*)Acceptable for Decomposition when the sum of entropies of the given level (approximations and details) is less than the entropy of the level above the given.

5) Saving the original values Stage:

Figure 6 shows the results of using the moving average (MA) filter to save the original peak values to be compared with the transmitted amended ones. The suffered OFDM symbols with high peaks will be treated according to their surrounding neighbors and then replaced them to avoid the deficiencies caused by the transmission of high power peaks.

The proposed five stages algorithm shows that the wavelet entropy could be used to allocate and overcome the effect of the high peaks powers. It is clearly shown that calculated entropies in Table 1 acting as a thresholding role and gives varies results depending on the used decomposition level and on the use case study. Among the four case studies, case study number 4 gives the best results where the lowest error ratio is found and equals to 4% and 414 allocated peaks.

TABLE II.
TOTAL, TRUE AND FALSE LOCAL EXTREMES POINTS' NUMBER

Case Study Number		Number of peaks			Error Ratio %
		Total	True	False	
1		461	389	72	16
2		408	383	25	6
3		424	390	34	9
4	P-level				
	CD1	398	389	9	2
	CD2	353	347	6	2
	CD3	139	92	47	34
	CD4	53	40	13	25
	CD5	33	25	8	24
	CD6	8	7	1	13
	CD7	8	5	3	38
	CD8	5	2	3	60
Peaks Clustering		414	395	18	4

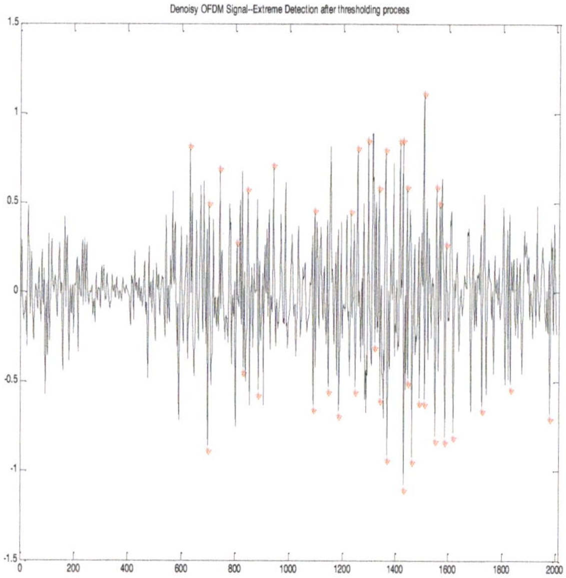

Figure 5. Extremes detection after thresholding process.

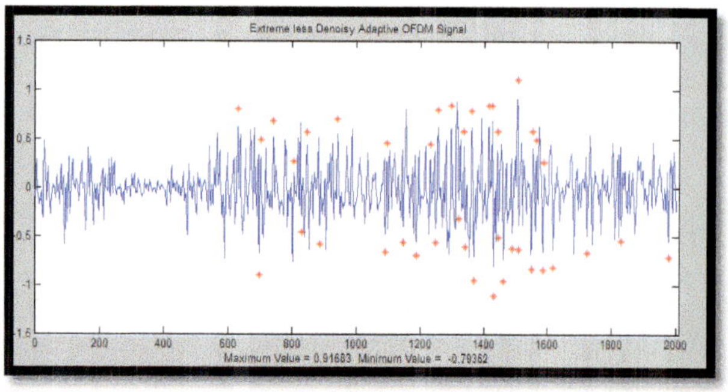

Figure 6. MA Filter result.

TABLE III.
SHARING PEAKS DETECTION AND ERROR RATIO FOR ALL DECOMPOSITION LEVELS.

P-level	Total number of local extremes points	NSEP	NRTE	$Error\ Ratio_{TE}$ % (*)
CD8	5	2	1	50
CD7	8	3	2	33
CD6	8	3	2	33
CD5	33	13	8	38
CD4	53	29	19	34
CD3	139	79	73	8
CD2	353	343	336	2
CD1	398		First level (stop sharing process)	

III. SIMULATION RESULTS AND DISCUSSION

Two main factors have been taken into consideration to check the proposed work performance; the bit error rate (BER) and the complementary cumulative distribution function (CCDF) curves. The work in defines the BER formula as

$$\text{BER}= BER_{WGN}\left(-\lambda \ln\left(\frac{1}{NQ}\sum_{q=1}^{Q}\sum_{n=1}^{N}\exp\left(\left(\frac{-SNR_q}{\lambda}\right)\right)\right)\right) \quad (6)$$

N is the total point's number of the FFT block, n is the n-th taking the values between 1 and N, Q is the amended data that passes through certain number of antennas, and λ is a system level simulation based parameter. The CCDF curves gives the system performance based on the probability of such peaks that are higher than a reference level.

A conducted MATLAB simulation limited to the following factors has been made. The results have been compared with both of the conventional techniques and our previously published work [10,22,23] under equivalent conditions.

- To be compatible with the WiMAx systems:
  - 20 MHz channel bandwidth
  - 1.152 oversampling ratio
  - 256 FFT points (200 data points)
  - QPSK modulation technique
- 2s duration real input data
- Convolutional encoder with coding rate of 1/2, and

WiMax system's performance has been checked based onto two main categories, where depicted in Figure 7 and 8; the CCDF curves and the BER curves. It is clearly shown that the proposed work has better performance than either some techniques found in the literature such as clipping technique and partial transmit sequence (PTS) or our previously published work found in [10]. This is in addition to that the MO-EWT enhances the probability that the PAPR exceeds 14dB from 2.5×10^{-2} to 1.9×10^{-1}, when the modulation technique has been change to 64QAM.

Moreover, it is observed through from Figures 8 and 10 that the use of higher order modulation techniques gives better BER than the lower order ones; the BER has been improved to be 7.5×10^{-1} while it was 1.8×10^{-1}. Thus, using MO-EWT gives extra 15% reduction of the CCDF value at 20dB threshold over our previously published work in [9].

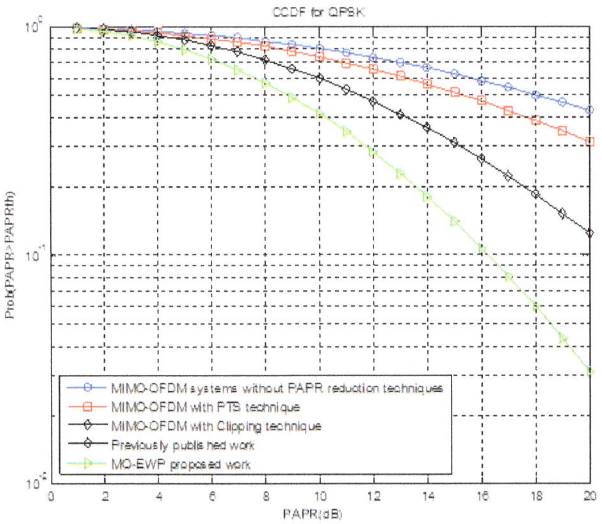

Figure 7. CCDF comparison among the MO-EWT and the work in the literature and our previously published ones using QPSK modulation Technique

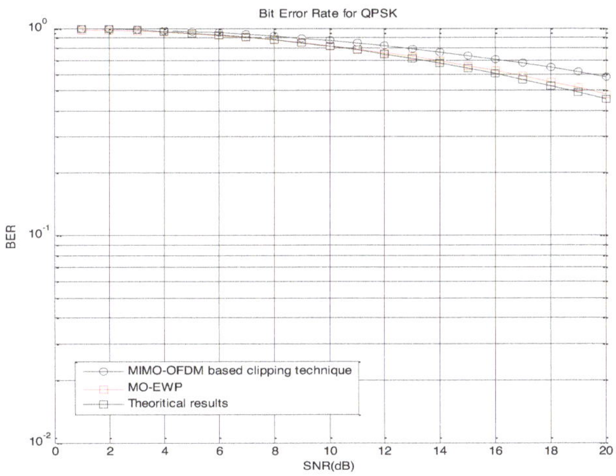

Figure 8. BER results comparing the MO-EWT performance with our previously published work using QPSK modulation Technique

IV. Conclusion

In this paper a new algorithm has been proposed not only to detect the extreme peaks but also to overcome their effects in WiMax systems, namely MO-EW**T**. This proposition making use of the wavelet entropy and the moving average filters to fulfill the procedure. In order to validate this proposition, a simulation has been conducted and checked the validity of the analytical derivation. As a result, it shows that a noticeable performance enhancement has been achieved over the work found in the literature. This enhancement has been shown based on both of BER and CCDF curves of the amended transmitted signals.

MO-EW**T** enhancement has been exposed clearly in the previous section, where the BER has been enhanced for around 60%. Based on the CCDF curves, extra 15% enhancement has been accomplished over our previously published work. This enhancement has been enlarged to reach around 80% over the work in the literature.

References

[1] Peijian Zhang, "A new scheme of space division multiplexing and its analysis", *IEEE 2nd International Conference on Power Electronics and Intelligent Transportation system*, vol. 2, pp. 407-410, 2009.

[2] Raghavendra, M.R. ; Juntti, M. ; Myllyla, M., "Co-Channel Interference Mitigation for 3G LTE MIMO-OFDMSystems", *IEEE International Conference on Communications*, pp 1-9, 2009

[3] P. Bender, *et al.*,"CDMA/HDR: a bandwidth efficient high speed wireless data service for nomadic users," *IEEE Commun. Mag.*, vol. 38, pp. 70–77, July 2000.

[4] B. Lu, X. Wang, and K. R. Narayanan, "LDPC-based space-time coded OFDM systems over correlated fading channels," *IEEE Trans. Commun.*, vol. 50, pp.74–88, Jan. 2002.

[5] Mishra, H.B. ; Mishra, M. ; Patra, S.K., Selected mapping based PAPR reduction in WiMAX without sending t[illegible] formation", *IEEE 1st International Conference on* Recent Advances in Information Technology (RAIT), pp. 182 – 184, 2012.

[6] Vitenberg, R.M., "Single Carrier Multi-Tone modulation scheme", *16th IEEE International Conference on Advanced Communication Technology (ICACT)*,pp. 568 – 574, 2014.

[7] [7]Yang, X.D. , et. al., "Performance Analysis of the OFDM Scheme in DVB-T", *Proceedings of the IEEE 6th Circuits and Systems Symposium on Emerging Technologies: Frontiers of Mobile and Wireless Communication*, vol. 2, pp. 489-492, 2004.

[8] Nee V. and Prasad R., "OFDM wireless multimedia communications", *Artech House, Boston London* 2000.

[9] *Yao Cheng, et al., " An Efficient Transmission Strategy for the MulticarrierMultiuser MIMO Downlink", IEEE Transaction on Vehicular Technology, vol. 63 , no. 2 , pp. 628–642, 2014.*

[10] Al-Akaidi M., Daoud O. and Gow J., "MIMO-OFDM-based DVB-H systems: A Hardware design for a PAPR reducing technique", *IEEE Transaction on Consumer Electronics*, vol. 52, issue 4, pp. 1201-1206, Nov 2006.

[11] Hung, Ying-Che ; Tsai, Shang-Ho Lawrence," PAPR Analysis and Mitigation Algorithms for BeamformingMIMO OFDM Systems" *IEEE Transaction on Wireless Communications*, vol. 13, no. 5, pp. 2588-2600, 2014.

[12] European Telecommunication Standards Institute (ETSI), "Digital Video Broadcasting; Second generation framing structure, channel coding and modulation systems for Broadcasting, Interactive Services, News Gathering and other broadband satellite applications", TR 102 376, V 1.1.1, 2005

[13] Digital Video Broadcasting Group, "DVB-T2 Call for Technologies", SB1644 r1, April 2007.

[14] Mansour M. M., and Shanbhag N. R., "A Novel Design Methodology for High-Performance Programmable Decoder Cores for AA-LDPC codes," in IEEE Workshop on Signal Processing Systems (SiPS), Seoul, Korea, August 2003.

[15] Hocevar D. E., "LDPC Code Construction with Flexible Hardware Implementation," in IEEE International Conference on Communications, pp. 2708 –2712, 2003.

[16] Chen Y. and Hocevar D., "An FPGA and ASIC Implementation of Rate 1/2 8088-b Irregular Low Density Parity Check Decoder", Global Telecommunications Conference, vol. 1, pp. 113-117, 2003.

[17] Han S. and Lee J., "An overview of Peak-to-Average Power Ratio reduction techniques for Multicarrier Transmission", *IEEE Wireless Communications*, pp. 56-65, Apr. 2005

[18] Tarokh V. and Jafarkhani H., "On the Computation and Reduction of the Peak-to- Average Power Ratio in Multicarrier Communications," *IEEE Transaction Communications*, vol. 48, no.1, pp.37–44, 2000.

[19] Prema, G. ; Amrutha, E., "A new MIMO-OFDM transmit preprocessing using pilot symbol assisted rateless codes to mitigate fading and wavelet based OFDM for PAPR reduction", IEEE International conference on *Signal Processing, Communication, Computing and Networking Technologies (ICSCCN)*, pp. 679 – 684, 2011

[20] Krongold S., and Jones L., "PAR Reduction in OFDM via Active Constellation Extension," *IEEE Transactions on Broadcast,* vol. 49, no. 3, pp. 258–68, Sept. 2003.

[21] Jiang T. and Zhu G., "OFDM Peak-to-Average power Ratio reduction by Complement Block Coding Scheme and Its Modified Version," Vehicular Technology Conference, vol. 1, pp. 448 – 51, 2004.

[22] Al-Akaidi M., Daoud O., and Linfoot S., "A new Turbo Coding Approach to reduce the Peak-to-Average Power Ratio of a Multi-Antenna-OFDM", International Journal of Mobile Communications, vol. 5, no.3, pp. 357-369 ,2007.

[23] Al-Akaidi M. and Daoud O., "Reducing the Peak-to-Average Power Ratio Using Turbo Coding ", IEE Proceeding Communications, vol. 153, no. 6, pp. 818-821, Dec. 2006

[24] Zhang T. and Parhi K. K., "Joint (3, k)-regular LDPC Code and Decoder/Encoder Design," *IEEE Transactions on Signal Processing,* vol. 52, no. 4, pp. 1065-1079, 2004.

[25] G. Nair, *et. al.*, "IEEE 802.16 Medium Access Control and Service Provisioning," Intel Technology Journal, vol.8, no.3, pp.213-228, August 2004.

[26] Xuyuan Zheng; Mingui Sun; Xin Tian, " Wavelet Entropy Analysis of Neural Spike Train", *IEEE Congress on Image and Signal Processing,* pp. 225-227, Tianjain, 2008.

[27] S.-Y.Lung, "Applied multi-wavelet feature to text independent speaker identification", IEICE Trans. Fundam. E87-A (4) 944–945, 2004.

[28] R. Gray, *Entrom and Infirmation theorv.* New York: Springer, 1990.

[29] 3GPP TSG-RAN-1, "TR 25.892: feasibility study of OFDM for UTRAN enhancement", March 2004.

AUTHORS

Qardi Hamarsheh is with the Dept. of Computer Engineering, Philadelphia University, Amman, Jordan.

Omar Daoud is with the Dept. of Communications and Electronics Engineering, Philadelphia University, Amman, Jordan (odaoud@philadelphia.edu.jo).

Saleh Saraireh is with the Dept. of Communications and Electronics Engineering, Philadelphia University, Amman, Jordan.

Remote Lab Experiments in Electronics for Use and Reuse

T. Zimmer, M. Billaud, M. Pic, D. Geoffroy
University of Bordeaux, Talence, France

Abstract—**This paper presents a new approach for the use of an already existing remote lab in the field of electrical engineering education. About 70 experiments representing a wide domain of electronic functions are easily accessible and can be incorporated into an existing teaching environment by an easy copy/paste action of a dedicated URL. The corresponding database structure is explained and a typical result is given to highlight the procedure to follow.**

Index Terms—**Laboratory experiments, Remote labs, Reuse of remote labs, Engineering education**

I. Introduction

In sciences education, experimental methods are crucial for understanding basic concepts students face during practical sessions. Moreover, experimental methods give students an opportunity to develop their perceptions and autonomy and, finally, to increase their interest in a given topic. In the electrical engineering (EE) curriculum, comprehension of electrical functions realized through simple circuits is a key for successful achievement of educational goals. Physical realization of the function leads to non-idealities that must be quantified by electrical measurements [1]. Furthermore, limitations of the measurement condition and the measurement resolution have to be analyzed [2]. A remote laboratory for characterizing electronic circuits has many advantages: it allows sharing high-quality instruments and full-custom circuits that cannot be duplicated in several labs for cost reasons. Net-based access allows the users to launch experiments and analyze the results any place as often as they want [3]. The remote laboratories' usefulness, acceptance and learning results have already been proven [4] and consolidated studies are in progress [5].

Setting up a remote lab demands not only knowledge in electrical engineering, but also knowledge in computer science, making the whole procedure quite complex. Many remote labs already exist and are referenced (e.g., on the lab2go platform) [6] or can be available through remote lab networks [7]. The reuse of remote labs is not straightforward. From a pedagogical point of view, the associated learning content rarely fits with what is necessary for a given lecture; in addition, the "overhead" activities such as enrollment in an existing remote lab, authentication and reservation for their employment for a given time slot very often represent very often a serious obstacle for wide use/reuse of existing remote labs.

In this paper we report a new approach for reusing remote labs. First, we present a library in which about 70 different experiments are listed and described. These experiments cover about 80% to 90% of the electronic functions that are commonly investigated during electrical engineering education up to the bachelor's level. Next, we present the organization of the database where all information about these circuits can be found. Finally, we show how each professor/teacher/instructor can access these experiments by simply copying/pasting one single hyperlink and integrating this link inside his personal teaching material. Of course, the teaching material must be available online (e.g., using OER or an LMS such as Moodle).

The paper is organized as follows: Section 2 gives an overview of the 70 experiments that can be freely used; Section 3 describes the database architecture illustrating the different experiments; Section 4 develops how a professor/teacher/instructor can integrate these remote experiments into his own learning environment by a simple copy/past operation; and finally, a conclusion is given in Section 5.

II. Available Experiments

The different electronic circuits that are usually examined during electrical engineering education can be classified into different categories:

A. Passive Electronic Circuits

In the easiest form, a passive circuit is only a dipole such as a resistor, a capacitor, or an inductor. Combining C and R components can form a passive low-pass or high-pass filter, and by adding an inductor, the three components can form a resonant circuit. All these different kinds of circuits are available for frequency domain measurement (e.g., Bode-diagram of a filter) as well as in the time domain thanks to the employment of a function generator and an oscilloscope.

B. Diodes And Transistors Based Circuits

Diodes and transistors form the foundation of any electronic circuit. Their investigations in many configurations are available:

a. Diode half wave rectifier
b. Common emitter amplifier (in different configurations)
c. Emitter follower
d. Common base amplifier
e. PNP-common emitter amplifier
f. Push-Pull output stage
g. The transistor as a switch

Most of the circuits can be investigated in both time domain and frequency domain.

C. Emitter Coupled Pair Circuits

The next class concerns the differential amplifier. For the realization of this function, different configurations are possible and can be examined:

a. Differential amplifier with long tail current source
b. Differential amplifier with active current source
c. Differential amplifier with passive load
d. Differential amplifier with active load

To get the whole picture of the differential amplifier, it can be investigated in differential mode and in common mode.

D. Op-Amp Based Circuits

Operational amplifiers are basic building blocks for many electronic functions such as

a. Comparator with Op-Amp (741), Slewrate
b. Comparator with Op-Amp (TL081), Slewrate
c. Op-Amp Amplifier in closed loop, input impedance investigation
d. Active low pass filter
e. Comparator (inverted)
f. Comparator with a threshold
g. Derivator
h. Integrator
i. Inverter, voltage gain=2
j. Inverter, voltage gain = 10
k. Voltage follower
l. Non-Inverter
m. Non-Inverter, voltage gain=10
n. Non-Inverter, voltage gain ~ 50
o. Non-Inverter, Output current limits investigation
p. Op-Amp with positive feedback
q. Schmitt-Trigger

All of these circuits can be investigated through the remote lab.

III. Architecture of the Database

For the creation of the database, we used SQLite through the SQL language. Furthermore, thanks to the PHP language and the function jQuery from Javascript we generated a table that shows information about the circuits on different levels:

The first level contains the basic information such as (see Fig.1):

a. Reference: a reference code for identifying the experiment
b. Title: a first indication about the experiment
c. Details: detailed information about the experiment
d. Component : list of the components used
e. Liens : hypertext link for access to the experiment in a different learning environment through copy/paste

The second level contains detailed information about each experiment; it can be accessed by clicking on the ⓘ-icon (see Fig.2):

Table des montages

☐	Reference	Title	Details	Component	Liens	
☐	1-1	1-1 Differential amplifier, common mode	This module is dedicated to the investigation of the differential amplifier in differential mode. A long tail resistance is used as a current source. Both, time domain and frequency domain measurement can be carried out.	RC=RE=4.7kOhm, R'=P=1kOhm, C=1µF	liens	ⓘ
☐	1-2	1-2 Differential amplifier, differential	This module is dedicated to the investigation of the common mode rejection of the differential amplifier. A long tail resistance is	RC=RE=4.7kOhm, C=1µF	liens	ⓘ

Figure 1. First level of information for the experiments

Table des montages

☐	Reference	Title	Details	Component	Liens	
☐	1-1	1-1 Differential amplifier, common mode	This module is dedicated to the investigation of the differential amplifier in differential mode. A long tail resistance is used as a current source. Both, time domain and frequency domain measurement can be carried out.	RC=RE=4.7kOhm, R'=P=1kOhm, C=1µF	liens	ⓘ

1-1 Differential amplifier, common mode - Details

Titre	max: time domain 20mV	typical: time domain	max: frequency domain: 10Hz	typical: frequency domain	DescriptionFr	Description	
Amplificateur Différentiel en mode différentiel	Vcc=10V, -Vcc=-10V, Vpeak-to-peak=10V	Vcc=10V, -Vcc=-10V, f=1kHz, Amp=1V	Vcc=10V, -Vcc=-10V, osc.Level=10mV	Vcc=10V, -Vcc=-10V, osc.Level=7mV, 20 Hz to 100kHz	Ampli-diff en mode differentiel, transistor bipolaire, charge passive	Differential amplifier, emitter coupled pair, long tail current, passive load, differential mode	

Figure 2. Second level of information for the experiments

a. Title
b. Maximum settings for time-domain measurements
c. Typical settings for time-domain measurements
d. Maximum settings for frequency-domain measurements
e. Typical settings for frequency-domain measurements
f. A description in French
g. A description in English
h. A picture-icon

The third level gives additional information for each experiment (see Fig.3):

Image 1: the circuit diagram

Image 2: a photograph of the module

Image 3: a typical measurement result in time domain

Image 4: a typical measurement result in frequency domain

Furthermore, an administrator interface has been set up that permits users to change/add/delete all the information that is in the database on the fly. Figure 4 shows the interface dedicated to adding another experiment.

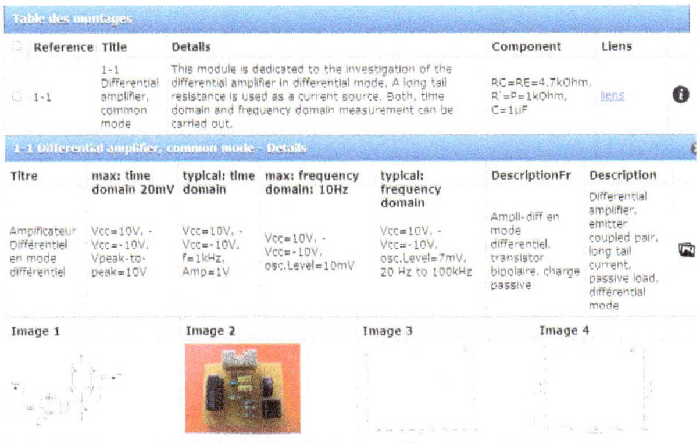

Figure 3. Third level of information for the experiments

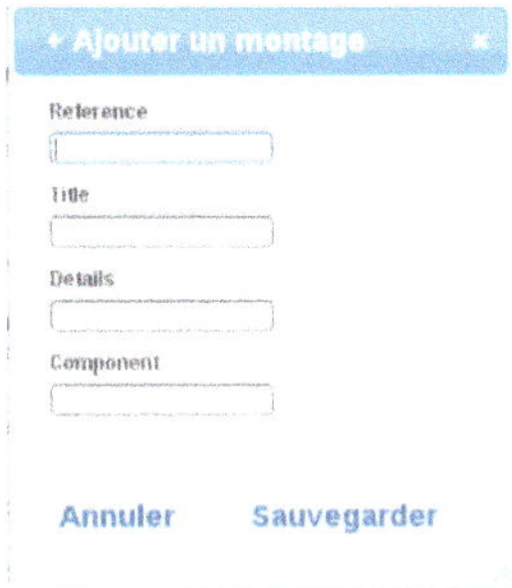

Figure 4. User interface for administration of the data base

IV. USING THE EXPERIMENTS

This section describes how to incorporate a remote experiment into a given learning environment. In the first step, we assume the professor/teacher/ instructor has made his choice from among the different experiments he wants to use. Next he clicks on the hyperlink, called "liens," seen in the first table (see Fig.1). When doing so, a window pops up indicating the link that has to be copied and pasted into the teacher's document (see Fig.5).

http://elab.creea.u-bordeaux1.fr/ELAB/docs/manip.php?experiment=XXXX

Figure 5. Typical link to copy/paste

The name "X" is different for each experiment and the user does not have to worry about it.

Now, when the student works with the teacher's learning material and clicks on the link that has been incorporated by the teacher into those learning documents, a window opens asking the students to specify the stimuli (settings for the measurement instruments, such as voltage level, to apply the frequency, the wave form and so on). An example is shown in Fig.6. These data are then transferred to our remote lab (www.real-lab.org), and the measurement is executed by applying the above stimuli.

When the measurement is completed, the results are transferred to the student's computer and are displayed. (An example is shown in Fig.7) Now the student can exploit the measured data and respond to the teacher's specific questions.

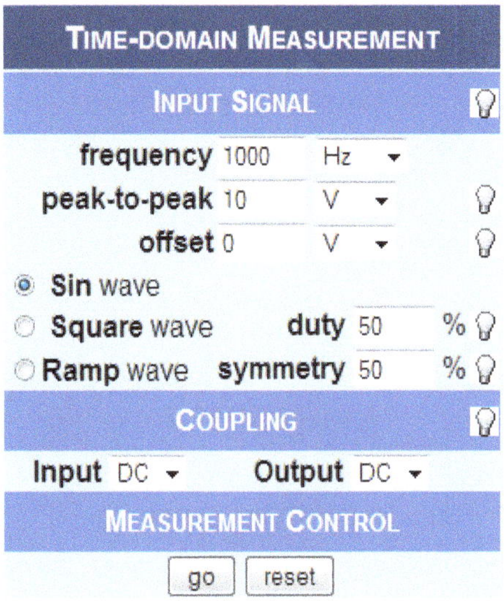

Figure 6. Measurement interface specifying the settings

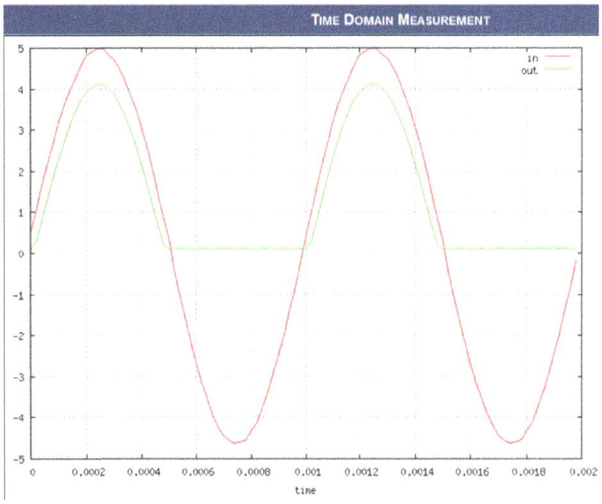

Figure 7. An example of measurement results

V. CONCLUSION

In this paper we presented a new approach about how to use an already existing remote lab in the field of electrical engineering education. Approximately 70 experiments describing a wide domain of electronic functions are easily reachable and can be incorporated in any existing remotely accessible teaching environment by a straightforward copy/paste action. An overview of the 70 experiments has been given and the corresponding database structure has been explained. In a cookbook manner, the procedure about how a professor/teacher/instructor can simply incorporate a given experiment in his private teaching documents is explained. The main advantage compared to existing solutions is that in the proposed approach, the instructor/teacher/professor has maximum freedom in constructing his course and can easily enhance the course content by adding real lab modules, so the students will benefit from the real world applications.

References

[1] M Billaud, T Zimmer, D Geoffroy, Y Danto, H Effinger, W Seifert, J Martinez, F Gomez, "Real measures, virtual instruments," Proceedings of the Fourth IEEE International Caracas Conference on Devices, Circuits and Systems, I023-1-I023-4, 2002

[2] T Zimmer, M Billaud, D Geoffroy, "Remote laboratory for electrical engineering education," International Journal of Online Engineering 2 (3), 2006

[3] N Lewis, M Billaud, D Geoffroy, P Cazenave, T Zimmer, "A distance measurement platform dedicated to electrical engineering," Learning Technologies, IEEE Transactions on 2 (4), 312-319, 2009

[4] D Lang, C Mengelkamp, RS Jäger, D Geoffroy, M Billaud, T Zimmer, "Pedagogical evaluation of remote laboratories in eMerge project," European Journal of Engineering Education 32 (1), 57-72

[5] T. Tsiatsos, S. Douka, T. Zimmer, D. Geoffroy, "Evaluation plan of a network of remote labs in the Maghrebian countries", Conference REV 2014, February 26-28, 2014, Porto, Portugal

[6] www.lab2go.net

[7] T. Zimmer, D. Geoffroy, A. Pester, R. Oros, T. Tsiatsos, S. Douka, " eSience: Setting up a network of remote labs in the Maghrebian countries" iCEER International Conference on Engineering Education and Research, Marrakesh, July 1st-5th, 2013, Morocco

Authors

T. Zimmer is with the Department of Applied Physics and Measurement Engineering at the Institute of Technology at the University of Bordeaux. He received a M.Sc. in Physics from the University of Würzburg, Germany, in 1989 and a Ph.D. in Electronics from the University Bordeaux 1, Talence, France, in 1992. From 1989 to 1990, he was with the Fraunhofer Institute, Erlangen, Germany. Since 1992, he has been with the IMS Institute, Talence, France. Since 2003, he has been a professor at the University of Bordeaux. At the IMS lab, he is the leader of the research group "Nanoelectronics." He is a cofounder of XMOD Technologies and a senior member of IEEE. He has published more than 200 peer-reviewed scientific articles, 1 book and contributed to 8 book-chapters. His education interests concern multimedia, open and distance learning and the implementation of ICT (information and communication technologies) in lab courses. In particular, he was the coordinator of 3 European projects dedicated to remote labs.

M. Billaud is with the Computer Science Department at the Institute of Technology at the University of Bordeaux, France. He received a PhD in Applied Mathematics (computer science specialty) from the University of Bordeaux 1, France, in 1985. He joined the Institut Universitaire de Technologie de l'Université Bordeaux 1, where he teaches mainly about operating systems, programming, and networks. His research interests include formal semantics, graph rewriting, and e-learning. He has been involved in various e-learning projects in Bordeaux since 2000.

M. Pic was a third year student when doing this work at the Computer Science Department at the Institute of Technology at the University of Bordeaux, France.

D. Geoffroy is with the Electrical Engineering Department at the University of Bordeaux, France. He received an MSc in Applied Physics from the University of Orsay, Paris, in 1980. He is currently an associate professor in the Department of Electrical and Electronic Engineering at the University of Bordeaux and is also the head of the Pedagogical Lab Center of this department. His special interests include multimedia, open and distance learning, and the implementation of information and communication technologies (ICT) in practical courses.

22

Exploring Smartphone Addiction: Insights from Long-Term Telemetric Behavioral Measures

Chad Tossell[1], Philip Kortum[2], Clayton Shepard[2], Ahmad Rahmati[3], Lin Zhong[2]
[1] US Air Force Research Laboratory, Wright-Patterson AFB, OH, USA
[2] Rice University, Houston Texas, USA
[3] Apple Corporation, Cupertino, CA, USA

Abstract—**This study examined smartphone user behaviors and their relation to self-reported smartphone addiction. Thirty-four users who did not own smartphones were given instrumented iPhones that logged all phone use over the course of the year-long study. At the conclusion of the study, users were asked to rate their level of addiction to the device. Sixty-two percent agreed or strongly agreed that they were addicted to their iPhones. These users showed differentiated smartphone use as compared to those users who did not indicate an addiction. Addicted users spent twice as much time on their phone and launched applications much more frequently (nearly twice as often) as compared to the non-addicted user. Mail, Messaging, Facebook and the Web drove this use. Surprisingly, Games did not show any difference between addicted and non-addicted users. Addicted users showed significantly lower time-per-interaction than did non-addicted users for Mail, Facebook and Messaging applications. One addicted user reported that his addiction was problematic, and his use data was beyond three standard deviations from the upper hinge. The implications of the relationship between the logged and self-report data are discussed.**

Index Terms—**Smartphones, Addiction, Human Behavior, Technology Social Factors.**

I. Introduction

Smartphone use is becoming ubiquitous. According to recent statistics, over one billion people worldwide own at least one of these devices [1] and they use them for a wide number of tasks, from placing a phone call to checking email, surfing the web, and listening to music [2]. These technologies have been adopted at a rate faster than any other in history [3] and the web is currently accessed more from smartphones than any other type of device [4].

The iPhone and similar smartphones have been described as addictive technologies [5]. Survey-based research has shown that many respondents would give up brushing their teeth, having sex, exercising, wearing shoes, showering, and eating chocolate instead of living without their iPhone for the same period of time [6]. There have been a number of empirical studies relevant to addiction to smartphones (see [7] for a review) and the prevalence of mobile phone addiction varies widely from study to study, with mobile phone addiction rates reported in the range of 0-38%, depending on the study and the scale used [7]. Researchers also differ in the extent to which mobile phone addiction is actually problematic. Some argue that mobile phone usage is rewarding and, like any other rewarded behavior, the addiction to mobile phones is quite prevalent, but not problematic [8]. This line of thought suggests that smartphones, like the internet, provide access to addictive content (e.g., cybersex, gaming, etc.) but are not the source of the addiction itself [9,10,11]. Other researchers suggest that mobile phone addiction is problematic and the use and abuse of the technologies can be detrimental [12,13,14,15].

The variability in the literature regarding smartphone addiction has been attributed to the vagueness of the conceptualization of technology addiction [7,16] and problems associated with confounding behaviors and consequences in clinical studies [17]. Primarily, research applied to understanding smartphone addiction has leveraged survey-based and other ethnographic methods. Indeed, over 18 different scales have been developed to assess the psychological variables underlying the addiction to mobile phones. Most of these scales have been noted as problematic to at least some degree (see [17] for an analysis). Because of the difficulty in exactly defining addiction, some researchers have adopted less controversial descriptions of these behaviors, including "problematic mobile phone usage" and "smartphone dependency". Although these descriptions do not carry some of the negative connotations associated with the clinical label of addiction, we chose to use the term precisely because it describes behaviors that are at once reinforcing *and* potentially problematic.

Self-reports about smartphone use are commonly used in these clinical studies to ascertain the behaviors and consequences associated with participants' levels of smartphone usage. Yet, there are noted problems with the accuracy of self-reports in general [18] and with recalling behaviors associated with smartphone usage in particular (e.g., compared with how they actually used their phones according to telemetric data [19]). Although the literature is replete with clinical studies of mobile phone addiction assessing individual differences in its onset and manifestation levels, few studies provide a realistic examination of behaviors associated with the addiction captured with logged (telemetric) data.

The purpose of this study was to address this shortcoming with a naturalistic and longitudinal analysis of smartphone addiction using a blended approach including both survey and telemetric data. We took an in-depth look at 34 undergraduate students and how they used their smartphone over the course of one year. All of their real usage was recorded via an unobtrusive, in-device logger to reveal patterns associated with potentially addictive behaviors associated with the technology. These data were complemented with survey responses to understand each user's (self-perceived) addiction levels and how those manifested in real-world usage patterns. Such an understanding can be helpful for assessing the severity of smartphone addiction and providing a more precise assessment of the specific aspects of smartphones that could be particularly addictive.

A. Research Questions

This exploratory study examined the existence of smartphone addiction across applications on users' devices. We were interested in two fundamental questions. First, how does self-reported smartphone addiction relate to monthly smartphone use? We hypothesized that self-reported addicts and non-addicts would have different frequency of use patterns, with addicts using their devices more frequently and for longer durations. Second, are there ways to assess addictive behaviors via logged (telemetric) data?

II. Methods

This study applied a quasi-experimental approach using naturalistic and longitudinal usage data collected over a one-year period. A full description of the broader methodology used in this study is described by [19].

A. Participants

A total of 34 students (19 male, 15 female) participated in the research. These students had diverse academic majors, socioeconomic levels, and ethnicities. Ten of the participants attended a community college and the other 24 students attended a major university in Houston, Texas. We purposefully selected students that did not previously own a smartphone to control for experience with the device. However, all participants owned a laptop and used computers frequently for collegiate studies. All students maintained grade point averages (GPAs) over 3.25.

B. Materials and Measures

iPhones that ran iOS 3.1.3 were provided to each subject free of charge over the one year study period. A custom logger [20], which operated as a background process and did not interrupt usage, was installed on each iPhone. Data were automatically captured every night with no user interaction. The data we collected included all application launches, the duration of the application launches, and when the application launches occurred (i.e., date/time stamps). More information was collected from several applications, including how many text messages were sent/received, the URLs visited over Safari, and the number of contacts in each participant's Contacts application. Most of the social data collected (contacts, text messages, email, phone calls, etc.) were obfuscated to negate privacy concerns.

A 15-question survey, the Smartphone Addiction Measurement Instrument (SAMI), modeled after the Cellular Phone Addiction Scale (CPAS; [21]) and Internet Addiction Test (IAT; [22]) was administered at the end of the year-long study (Table 1). Participants rated each item on a 5-point Likert scale ranging from never (1) to always (5). The IAT is a well-known instrument validated psychometrically across cultures [23, 24] and the CPAS has also been used in several countries [21]. It should be noted that both scales are somewhat dated; thus, the authors made substantial changes to the items. Additionally, open-ended questions and yes/no questions were also included in the survey to help interpret some of the logged data and understand the nature of any reported addiction. We refrained from introducing novel interfaces, experimenter-constructed tasks, and research-related meetings over the course of the study in order to decrease participant reactivity.

C. Procedure

After completion of an IRB approved consent form, smartphones were distributed to participants. The phones had unlimited text messaging and data services, along with 450 rollover minutes of voice service. Participants were not told how to use their device and no information on the specific purpose of the study was given, except that we were recording their usage data in an anonymized manner to understand smartphone usage. Participants were required to use the instrumented iPhones as their primary mobile phone during the entire one-year study period. At the end of the year, we administered the addiction survey. Students who completed the study were allowed to keep the iPhone as added compensation for their participation.

III. Results

A total of 21 of the 34 participants (62%) agreed or strongly agreed they were addicted to their iPhones. We grouped these users together based on their agreement to at least some level of addiction to their smartphones and labeled this group "SA" for Self-reported Addiction. Of these 21 participants, 12 were male and nine were female.

The other 13 participants (i.e., NAs for Non-Addicts) did not agree at any level that they were addicted to their iPhones (i.e., they strongly disagreed, disagreed, or neither agreed or disagreed). Seven of these users were male and the other six were female. We did not find any notable differences in the demographics within either of these groups.

One participant within the SA group considered his addiction problematic and, as we describe in more detail below, his usage data was considered outlying. We removed his data from the comparative analyses between SAs and NAs, but focus on his usage in a separate section. Every other user in the SA reported his or her addiction was not problematic.

SAs differed from NAs in responses to several items on the SAMI, primarily in their perceived ability to control the craving to check their devices and their desire to spend as much time as possible on the device (Table 1). SAs also perceived more anxiety when they could not turn on their devices or check their favorite app. Many items did not yield differences. For instance, participants generally agreed their iPhones were difficult to turn off once on and useful for withdrawal/escape.

A. Differences in recorded usage

The self-reported addicts (SAs) and non-addicts (NAs) both used their phones frequently over the year-long study period. As can be seen in Table 2, SAs spent twice as much time on their phone compared to NAs. The former also launched applications much more frequently (nearly twice as often) compared to their NA peers. This difference is not driven by the number of applications installed by users as there were not significant differences between SAs and NAs in the number of applications installed.

TABLE I.
ADDICTION ITEMS AND RESPONSES BY SELF-IDENTIFIED ADDICTED (SA) AND NON-ADDICTED (NA) INDIVIDUALS. RESPONSES ARE ON A 5 POINT SCALE, WHERE 5 IS 'STRONGLY AGREE'

	SA		NA		
	M	SD	M	SD	p*
Inability to Control Craving					
You have been told that you spend too much time on your smartphone	4.14	1.01	3.92	1.89	.90
You find yourself engaged on the smartphone for longer period of time than intended	3.31	1.27	3.49	1.02	.93
You can never spend enough time on your smartphone	3.25	1.43	2.10	1.35	**<.001**
You have attempted to spend less time on your smartphone but are unable to	3.17	1.50	2.35	1.67	**.003**
You can control when you check applications	2.71	0.36	3.92	1.81	**<.001**
Checking your smartphone satisfies a recurring urge	3.02	0.73	1.99	1.03	**<.001**
Feeling Anxious & Lost					
You experience discomfort when your phone is running out of battery life.	3.30	0.59	3.23	0.37	.88
You feel anxious if you have not checked a favorite app or switched on your smartphone for some time.	3.92	0.89	1.88	0.57	**<.001**
You find it difficult to switch off your smartphone	3.7	1.03	3.29	0.80	.08
You feel lost without your smartphone	2.19	0.96	2.01	1.19	.71
Withdrawal/Escape					
You have used your smartphone to communicate to others when you were feeling lonely	3.97	1.34	3.72	1.07	.57
Your smartphone is a consistent companion	3.61	1.11	3.24	1.20	.46
You have used your smartphone to make yourself feel better when you were feeling down	1.32	0.79	`1.21	0.76	.94
You often fill your dead time with smartphone use	3.92	1.06	3.42	0.78	.07

In order to understand these differences between SAs and NAs, some applications were grouped based on category as defined by Apple for parsimony. For example, all gaming applications were combined. Independent t-tests with Scheffe adjustments for multiple comparisons revealed a small subset of highly-used applications was used more by SAs according to our logged data (Table 2). SAs launched Mail, Facebook, Messages, and Safari much more frequently than NAs and spent more time on Mail, Facebook, Entertainment and Safari as compared to NAs. Applications within categories such as Games and Education did not discriminate user groups, though SA and NA differences in the duration of game use approached the .05 alpha significance level ($p = .06$).

In order to better understand how four of these specific applications were accessed longitudinally, Time per Interaction (TPI) rates were calculated for Mail, Facebook, Safari, and Messages applications and analyzed for each user. TPIs represent the number of seconds users spent on particular applications for each launch (TPI = Duration in Seconds / # Launches). Lower TPIs reflect application usage that is shorter in duration and more frequently launched. Conversely, longer duration usage with less frequent application launches would yield higher TPI. NAs show higher TPI for Mail (p=.015), Facebook (p<.001) and Messaging (P<.001), while no difference between NAs and SAs exist for Safari TPI (p=.63). TPIs across the duration of the study are shown in Figure 1.

TABLE II.
USAGE DATA MEAN DIFFERENCES AND STANDARD DEVIATIONS FOR SELF-REPORTED ADDICTED AND NON-ADDICTED STUDENTS. (*SCHEFFE ADJUSTED FOR MULTIPLE COMPARISONS)

	Addicts		Non-Addicts		
	Mean	SD	Mean	SD	p*
Hours/Week	8.02	4.13	4.32	4.80	**<0.001**
App Launches/Week	1059.55	287.71	592.81	249.76	**0.001**
Apps Installed	139.67	50.77	103.50	41.32	0.22
Apps Uninstalled	44.08	50.38	32.94	27.96	0.2
Education (min/wk)	2.79	1.33	2.09	3.25	0.39
Education (launches/wk)	2.19	2.46	1.86	1.07	0.65
Email (min/wk)	46.16	23.56	27.99	15.48	**0.03**
Email (launches/wk)	60.86	43.44	31.97	19.31	**0.04**
Entertainment (min/wk)	5.94	10.20	16.45	15.62	**0.04**
Entertainment (launches/wk)	6.66	10.07	1.61	1.15	0.09
Facebook (min/wk)	77.87	51.24	24.76	23.13	**.008**
Facebook (launches/wk)	45.81	29.77	20.96	22.47	**0.03**
Games (minutes/wk)	104.30	106.70	81.18	94.34	0.06
Games (launches/wk)	14.89	12.23	19.39	33.15	0.67
Text Messages (minutes/wk)	161.30	93.21	119.13	80.22	0.19
Text Messages (launches/wk)	189.17	99.88	87.18	46.77	**0.004**
Web (minutes/wk)	70.66	36.61	33.86	34.39	**0.01**
Web (launches/wk)	34.95	17.51	15.77	12.32	**0.005**

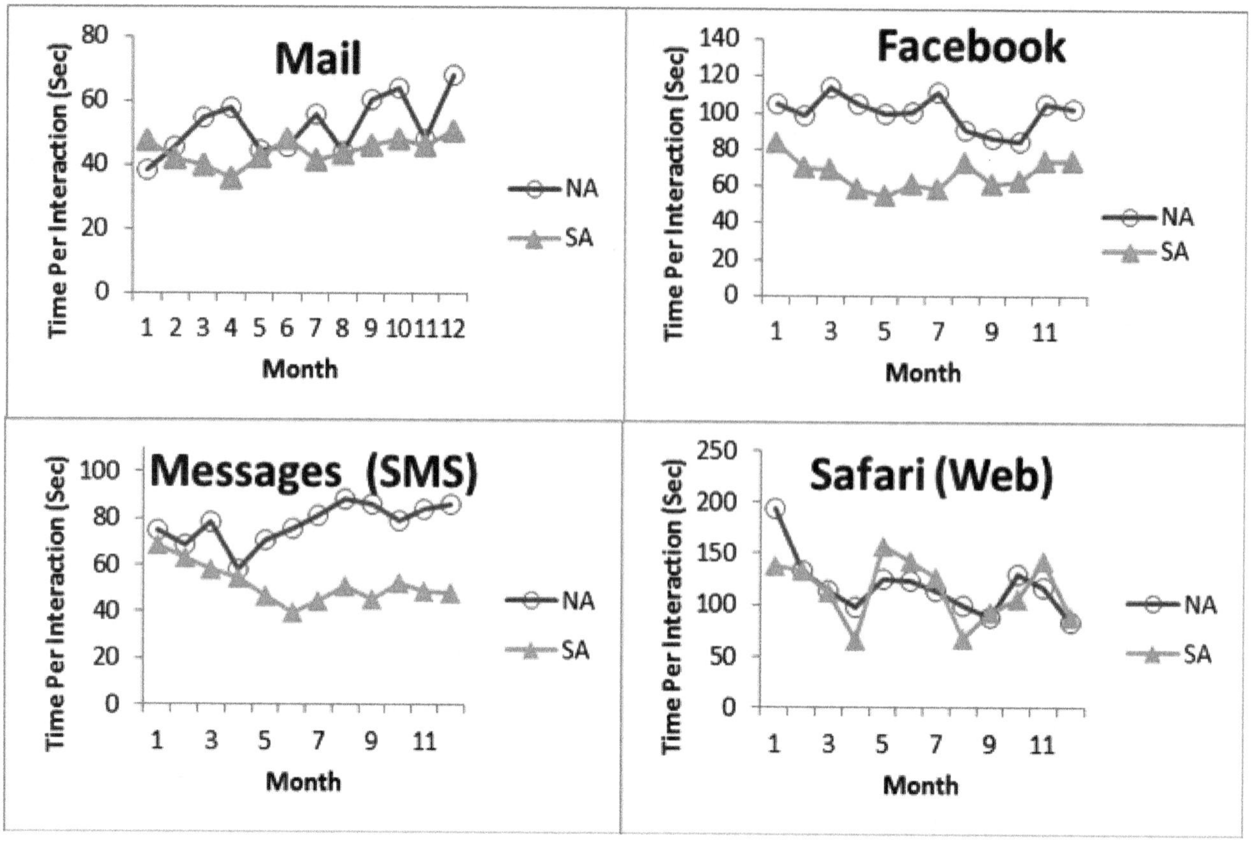

Figure 1. Time per interaction (in seconds) by month for Mail, Facebook, Messages, and Safari Applications.

B. Problematic addiction

Although there was high variability across users, the one individual who reported a problematic addiction to smartphone use was well beyond three standard deviations from the upper hinge in volume of usage. This participant reported that checking the smartphone was an uncontrollable urge and yielded an average of 122 application launches per day, equivalent to launching an application every eight minutes over the course of a normal 16 hour day. The Web was the most frequently launched application, and use was concentrated on just a few specific sites (one was adult oriented). These few sites were visited over 3,000 times throughout the course of the study (8.2 visits/day) starting almost immediately after we handed out the phones. Perhaps most surprising was this participant's rate of sending and receiving text messages. During the five busiest months, this participant averaged 220 text messages sent and received per day (one every four minutes).

IV. DISCUSSION

We found that more than half of our users self-reported an addiction to their iPhone to at least some degree. These users launched applications on their devices much more than the other users that reported they were not addicted. Additionally, their average usage was two times more than their NA peers. At the broadest level, this study suggests that users who accessed their device more and for longer periods of time are aware of their addictive behavior and, in general, did not think it was problematic. Additionally, lower TPI rates exhibited by SA users for mail, Facebook and Messaging suggest that smartphones seem to afford short and fragmented interactions perhaps leading to habitual patterns of usage that may become difficult to control over time.

Participant reports suggest this is due to habitual checking behaviors, a finding supported by other research [25]. However, an alternate explanation is that NAs could have used these applications for more productive reasons related to the function of the application than did SAs. For example, using the Mail application to view a new message or compose an email is a more utilitarian use of the application that requires more time relative to habitual checking of any new emails that have arrived. It might also be that SA participants could have saved longer tasks such as composing or reading a long message for larger format computers (e.g., a laptop).

We are unable to say for certain what is driving the differences in TPI rates in these most popular applications. However, it appears that lower TPIs are one important indication of smartphone addiction. Time-inconsistent preferences and automaticity are two theories that provide underlying support for this type of habitual and unconsciously automatic behavior [26, 27]. Web TPI did not show this trend, and this might be because the longer average session times for the web attenuate the effect.

Specific application usage discriminated users that perceived themselves as more addicted to their smartphones compared to others. To our surprise, game-playing did not discriminate addicted users from non-addicted users. Instead, the web, Facebook, SMS, and Mail were the applications that were used more by SAs compared to NAs. It could be argued that the SA users in this study were not addicted to these apps at all, but rather were instead exhibiting a (normal) desire to stay connected to their friends, and that NAs were actually low on social needs of this type. This persistent checking of the phone might be viewed by these SAs as being analogous to maintaining a conversation in a room full of friends. This might also suggest that SAs were simply more connected socially than NAs. However, this hypothesis was not reflected in the number of friends they contacted via their phones. Survey and logged data showed no differences between SAs and NAs in their number of Facebook friends ($p = .83$) or Mail contacts ($p = .79$). Of course, they could have been more connected in the type or depth of these relationships; we did not collect these kinds of data.

Telemetric data did reveal that SAs opened either Facebook or Mail applications before other applications when using their iPhones. Specifically, SAs accessed at least one of these two applications immediately after turning on their iPhones 72% of the time. NA interactions with their iPhones that started with one of these two applications only occurred 48% of total usage. This behavior pattern suggests that SA users were habitually checking these two applications, as compared to NAs who launched a wider range of applications immediately after accessing their device.

The problematic SA user reported that phone checking satisfied an uncontrollable urge, but some level of this urge was evident across all SA participants. In a sense, the smartphone itself did not seem like the source of the addiction, but rather the content to which the phone provides access. Similar to previous studies on internet addiction [24], the smartphone seemed to be a vehicle to addictive content and not the cause of the addiction itself.

We believe that this is a difference without a distinction - smartphones allow for continuous access to the Internet, and so users can utilize the technology satisfy their urges to access content in most personal and professional environments. In this sense, the fact that the smartphone is ubiquitously available could feed the addiction to an individual's content of choice. Previous addictions that could only be fed at a desktop can now be satisfied whenever and wherever one gets the urge.

Although the benefits to these devices to support humans are well established (e.g., in education [28] and medicine [29]), our results show that this continuous access could afford more addictive behaviors in a wider variety of settings. Indeed, users no longer need a sustained period of time, access to electrical outlets, or a full computer to get their email or Facebook "fix" but can simply reach into their pockets and pull out an engaging device to achieve the same end. With this in mind, smartphones as a persistent enabler should be of note to clinicians. To some degree, we believe that the device, and the related content providers, share some of the responsibility for these emerging addictive behaviors. Content is constantly updated, email is received at a steady pace and social media posts appear with high regularity, and this information push promotes these behaviors. Consider the time before this always-available platform; the newspaper arrived once a day (or twice in large cities), mail was delivered once a day and social conversations were confined to those occasions when you happened to be in the physical presence of your close acquaintances. Smartphones, and their access to constantly changing information has altered people's access to information, and this may be why these addicting behaviors have begun to emerge.

On the other side of the coin, both NA and SA participants reported their devices were helpful to fill dead-time, keep in touch with their social network when lonely, and act as their continuous companion. These positive aspects of smartphones should not be ignored. Smartphones provide access to helpful resources in a number of contexts, in addition to addictive content, that can be helpful for a wide range of problematic thoughts and behaviors. Our research agrees with [30], suggesting that smartphones can be harnessed to promote positive behaviors and, perhaps, in the near future, be used as a sensor to help to this end. We purport that qualitative research like [30], as well as combined quantitative and qualitative studies described here, will be helpful to make this future a reality.

A. Limitations

This was an exploratory study using a blended approach and there are several limitations to the results presented above. First and foremost is the small sample size. Clearly, the 34 users assessed here are quite small in number and not completely representative of the larger population of smartphone owners in terms of their demographics and context of use. Indeed, college students are a particularly vulnerable group to this type of addiction [31, 32]. However, the telemetric usage data presented here provides more depth and precision than typical survey-based research and helps to mitigate the small sample size [19]. The sample size also provided enough power to detect differences between groups.

Second, we relied on self-reports for parts of our data collection. There are well-known concerns about these kind of data [18], but believe those problems are mostly mitigated in this research because we did not ask for estimates of quantifiable behaviors (e.g. "How often do you check your message per day?") as those data were collected by the device itself. Rather, we asked more qualitative questions that require human interpretation to answer (e.g. "how difficult do you find it to switch off your smartphone?"). Finally, as pointed out in the introduction, there are particular challenges with the term "addiction". We use the term in this paper, because it is the term used in much of the literature describing excess use of technology. Indeed, even the measurement scales used to assess this overuse have the word addiction in their titles [21, 22]. We make no claims to the clinical diagnosis of smartphone addiction, as it is not currently included in the DSM-5, but instead rely on its colloquial use as a common description of behaviors that are at once reinforcing *and* potentially problematic.

V. Conclusions

In conclusion, we explored the topic of smartphone addiction using an emerging naturalistic and longitudinal telemetric approach. These data should be interpreted with caution due to the small sample size and potentially more vulnerable demographic. Still, this study provides a first glimpse into the addictive nature of smartphones as indicated by real usage data combined with survey data. This kind of behavioral logging data provides more precision regarding what addicted users actually do on their devices.

Future studies, with larger sample sizes, should employ a psychometric approach to understand the latent constructs that underlie these behaviors, the behavioral profiles of SAs, and the psychological consequences of having a pathological-level of smartphone addiction. Correlating psychometric data with specific aspects of smartphone usage seems to be an interesting way ahead.

Finally, this research raises particular questions about future directions for clinical psychology. Should technology and analytics be used to help predict and treat behavioral disorders? It could be that these kinds of analytics collected from a smartphone could be used to diagnose or predict the onset of certain disorders and provide interventions and treatments.

REFERENCES

[1] Strategy Analytics (2012). Worldwide Smartphone Population Tops 1 Billion in Q3 2012. Wireless Smartphone Strategies.

[2] Do TMT, Blom J, Gatica-Perez D. Smartphone usage in the wild: a large-scale analysis of applications and context. Proceedings of the 13th international conference on multimodal interfaces 2011:353-360.

[3] Farago P. (2012). iOS and Android Adoption Explodes Internationally.

[4] Newark-French C. (2011) Mobile apps put the web in their rear view mirror.

[5] Ofcom (2012) Communications Market Report. United Kingdom Office of Communications.

[6] Telenav (2011). Survey Finds One-Third of Americans More Willing to Give Up Sex Than Their Mobile Phones.

[7] Perez EJP, Monje MTR, De Leon JMRS. Mobile phone abuse or addiction. A review of the literature. Adicciones 2012; 24(2);139-152.

[8] Griffiths MD. Internet and video-game addiction. In Essau CA, ed. Adolescent addiction: Epidemiology, assessment and treatment. London: Academic Press. 2008; 231-267.

[9] Griffiths M. Does Internet and Computer "Addiction" Exist? Some Case Study Evidence. CyberPsychology & Behavior 2000; 3(2): 211-218.

[10] Roberts JA, Pirog SF. A preliminary investigation of materialism and impulsiveness as predictors of technological addictions among young adults. Journal of Behavioral Addictions 2013; 2(1):56–62.

[11] Yellowlees PM, Marks S. Problematic Internet use or Internet addiction? Computers in Human Behavior 2007; 23(3):1447-1453.

[12] Backer-Grøndahl A, Sagberg F. Driving and telephoning: Relative accident risk when using hand-held and hands-free mobile phones. Safety Science 2011; 49(2): 324-330.

[13] Beranuy M, Oberst U, Carbonell X, Chamarro A. Problematic Internet and mobile phone use and clinical symptoms in college students: The role of emotional intelligence. Computers in Human Behavior 2009; 25:1182-1187.

[14] Billieux J, Van der Linden M, Rochat, L. The role of impulsivity in actual and problematic use of the mobile phone. Applied Cognitive Psychology 2008; 22: 1195-1210.

[15] Chóliz M. Mobile phone addiction: a point of issue. Addiction 2010;105:373-374.

[16] Park WK. (2005) Mobile Phone Addiction. In: Ling R, Pedersen P, eds. Mobile Communications: Re-negotiation of the Social Sphere. London: Springer-Verlag, pp. 253-272.

[17] James D. (2012) Problematic use of mobile phones: Measuring the behaviour, its motivational mechanism, and negative consequences. Doctoral dissertation, Queensland University of Technology Brisbane, Australia.

[18] Podsakoff PM. MacKenzie SB, Lee JY, Podsakoff NP. Common method biases in behavioral research: a critical review of the literature and recommended remedies. Journal of applied psychology 2203; 88(5): 879-903.

[19] Tossell CC, Kortum P, Shepard CW, Rahmati A, Zhong L. Getting Real: A Naturalistic Methodology for Using Smartphones to Collect Mediated Communications. Advances in Human Computer Interaction 2012: 1-10.

[20] Shepard C, Tossell C, Rahmati A, Zhong L, Kortum P. LiveLab: Measuring Wireless Networks and Smartphone Users in the Field. ACM SIGMETRICS Performance Evaluation Review 2010; 38(3): 15-20.

[21] Koo HY. Development of a cell phone addiction scale for Korean adolescents. Journal of Korean Academy of Nursing 2009: 39(6): 818-828.

[22] Young KS. Internet addiction: The emergence of a new clinical disorder. CyberPsychology and Behavior 1998; 1(3): 237-244.

[23] Khazaal Y, Billeux J, Thorens G, Khan R, Louati Y, Scarlatti E, et al. French validation of the Internet Addiction Test. CyberPsychol Behav 2008; 11:703-6.

[24] Widyanto L, Griffiths M. Internet Addiction: A Critical Review. International Journal of Mental Health and Addiction 2006; 4(1): 31-51.

[25] Lee YK, Chang CT, Lin Y, Cheng ZH. The dark side of smartphone usage: Psychological traits, compulsive behavior and technostress. Computers in Human Behavior 2014; 31: 373-383.

[26] Hoch SJ, Loewenstein GF. Time-Inconsistent Preferences and Consumer Self-Control. Journal of Consumer Research 1991: 17(4): 492–507.

[27] Morsella E, Bargh JA. (2011). Unconscious action tendencies: Sources of 'un-integrated'action. In: Decety J, Cacioppo JT, eds. The Oxford handbook of social neuroscience, New York, NY: Oxford University Press, pp. 335-347.

[28] Guy R, Chack C, Ngesi N, Mehigan TJ, Pitt I, Simbulan IS, Schleser M, Williams PW. (2010) Mobile Learning Reviewed. In: Guy R, ed. Mobile learning: Pilot projects and initiatives. Santa Rosa, CA.: Informing Science Press, pp 59-74.

[29] Ozdalga E, Ozdalga A, Ahuja N. The smartphone in medicine: a review of current and potential use among physicians and students. Journal of Medical Internet research 2012: 14(5).

[30] Dennison L, Morrison L, Conway G, Yardley L. Opportunities and challenges for smartphone applications in supporting health behavior change: Qualitative study. Journal of Medical Internet Research 2013; 15(4): e86.

[31] Jenaro C, Flores N, Gómez-Vela M, González-Gil F, Caballo C. Problematic Internet and cell-phone use: Psychological, behavioral, and health correlates. Addiction Research & Theory 2007; 15(3): 309–320.

[32] Kandell JJ, Internet Addiction on Campus: The Vulnerability of College Students. CyberPsychology & Behavior 1998; 1(1): 11-17. G. Eason, B. Noble, and I. N. Sneddon, "On certain integrals of Lipschitz-Hankel type involving products of Bessel functions," *Phil. Trans. Roy. Soc. London*, vol. A247, pp. 529–551, April 1955.

Authors

Chad Tossell is with the US Air Force Research Laboratory, Wright-Patterson AFB, OH, USA. (e-mail: chad.tossell@us.af.mil)

Philip Kortum is with the Department of Psychology at Rice University, Houston, TX 77005 USA. (e-mail: pkortum@rice.edu).

Clayton Shepard is with the Department of Electrical and Computer Engineering at Rice University, Houston, TX, USA. (e-mail: cws@rice.edu).

Ahmad Rahmati is with the Apple Corporation, Cupertino, CA, USA. Rice University, Houston, TX, USA. (e-mail: ahmad@rahmati.com).

Lin Zhong is with the Department of Electrical and Computer Engineering at Rice University, Houston, TX, USA. (e-mail: lzhong@rice.edu).

23

Recognizing the Importance of Brand Awareness on E-commerce Sales while Shopping on Internet: Empirical Analysis of European Countries

Bader M. Al-Fawwaz, Atallah M. Al-Shatnawi and Wafa S. AlSharafat
Al al-Bayt University, Jordan

Abstract—**The reason to conduct this research is to investigate and recognize the importance of brand awareness on E-commerce sales while shopping on Internet. Empirical analysis of seven top Internet purchasing European countries is done by using the report of (Eurostat's 2011). The units of analysis are those individuals who are residing in Denmark, Germany, France, Italy, Netherland, Sweden and Norway making purchases from the Internet. Data is collected from the report of Eurostat's for the ten years from 2002-2011. To find out the relationship between Independent and Dependent Variable in the online business world, the study has done statistical analyses which are implemented by using E-views 5 software. The main motive to carry out the study is to find out the impact and importance of brand image/ awareness in the minds of online shopper, when they go for shopping on Internet. Findings of this study illustrate significant positive relationship between brand awareness an E-commerce sale. Findings are beneficial for those companies which are working on Internet and selling their products or services, so with the help of results of this study they can be able to know European online shopper's mind and can develop certain strategies in order to sell the most.**

Index Terms—**Brand awareness, E-commerce, Europe Internet shopping, E-business.**

I. INTRODUCTION

Internet is a novel way to shop products and services online. There are plenty of web stores available online through which customers can conveniently make purchases. Nielsen (2012) stated that spending on Internet is increasing in the entire European countries. Furthermore, the Internet advertisement spending is on its growth and particularly the growth in Internet advertisement is observed as 12.1% in Europe. Overall the trends in online shopping are not only increasing but people are more tend to trust on Internet e-commerce sites and make purchases. If in the case someone is looking to buy air tickets he/she can simply log on to the airline websites and purchase the E-tickets. Further, if that consumer is looking for some accommodation in any part of the world, that consumer can simply make purchase by going on the web and surfing different hotel rooms. Shopping was never as easy as this is in today's world (Harfoushi et. al 2013).

There are some other modes of communication such as telephone. However, Internet is such a way of communication which is very convenient to use and a cheap way to communicate in any part of the world. Furthermore, the Internet provides a platform to users where they can interact with the dealers which are sitting in other parts of the world and make purchases. Internet shopping is becoming same like traditional shopping, if you cannot touch the product, however you have complete details given with all the specification of the product, you can simply navigate with that specs of the product and make purchase accordingly. As Internet is an easy platform for every user to access, therefore number of whole sellers, dealers and retailors are now communicating on Internet platform, having business on Internet and make business contracts which are totally based on Internet medium.

Although the Internet market in Europe is under developing stage; most of the people avoids shopping from an E-store. Internet users predict many forms of Internet advertisement which includes the commercial advertisement and electronic advertisement (Ann, 1999). The numbers of Internet users in Europe are recorded 545,057,209 (Internet world stats, 2014). If we see the figure of Internet users we can see it is growing in European countries. Although number of shoppers in Europe is not very much, however this trend is increasing day by day. 60% of the people who use Internet in Europe have never shopped over Internet (Nielsen, 2012). The progression of the World Wide Web (WWW) and Internet have caused the formation of novel system in trade dealings- electronic transaction or shopping on web (Ling, 2010).

In European countries the number of merchants or firm creating opportunities over Internet is increasing. As use of online banking, E-ticketing, booking of rooms in restaurants and hotels, digital products and purchase of vehicles are on boom nowadays. People like to browse for different services and products through Internet in Europe. That is why it is becoming an essential strategy for any business or firm to create a homepage and promotes it over Internet. As buying a car is just on your click that is why people prefer to use this technology rather than going on a different showrooms and search for a desired car.

The primary focus to conduct this study is to identify and check brand awareness of web consumers in European countries. The primary focus of the study is to analyze the impact of brand awareness toward E-commerce sales in Europe. By studying consumer reactions or attitudes towards shopping over Internet it is easy to find out those factors which are deteriorating the rate of shopping over Internet in the European context. Although consumer attitudes or behavior varies from person to person, however by the support of literature we can find out this phenomena. This paper will be helpful for those businesses and organizations who want to market their products over Internet in the area of Europe. The merchants who want to increase their sales over Internet. As by studying the reactions of consumers over Internet, it will be helpful for merchants, organizations, sellers in Europe to find out such factors which can enhance and boost up their sales by fulfilling the need of online consumers in Europe.

II. Literature Review

The Internet has provided business of all size, with opportunities to expand their market in order to sell the products. Like traditional market, it is a market which allows seller to sell the products online. Internet has shaped an advanced method in which users connect and intermingle. Sellers have a great deal to advantage by using and attaching the control of social networking to improve their global promotion plan (Pookulangara, 2011). Progressively customers are using Internet as a pre-purchase instrument where they can interact with other users, can read the reviews and get the information about particular products and services (Adjei, 2010). Internet marketing is entirely different from the traditional marketing. In traditional marketing you have to go to the specific store, outlet in order to purchase a specific item. However, shopping over Internet is entirely different, what you have to do is just visit any E-store, Online Store, E-Shop, or Online Shop and there you can find out your product what you are looking for. E-commerce delivers a detailed description of what E-commerce is, how it's being showed and achieved, and how to measure its main chances, boundaries, matters, and jeopardies all in the computing commercial setting (Turban, 2009). These new selling techniques and changes give a new paradigm for sellers, merchants, organizations, firms to sell their products in a different field with targeted users. By creating the web pages of their respective organizations it leads them to get in touch with the targeted clients or consumers.

A. Consumer Behavior while Shopping on Internet

Number of studies has been found which have done research on this topic. Several studies (Trusov, 2009), (Dutton, 2011), (Casey, 2011) and (Michel, 2011) all have conducted empirical research about consumer behavior online. The purpose of these researches is to find out the difference between online and offline shopping and marketing concepts. Majority of the above studies is conducted on the price and brand choice of different consumers which are making online purchases, and it was found that mix of price and brand information is also involved when a customer is making purchase on the Internet. Trusov (2009) study finds that although the loyalty of a customer with brand is a strong phenomenon observed in the traditional marketing concept, however this phenomenon is less when consumer is making purchases virtually. Yet most of the customer is always looking for trustable brands when about to shop online and it ease their nerves regarding the quality of the product and they can simply trust on that brand.

Interpersonal trust and economic conditions are some important variable which affects the consumer behavior to shop from an Internet store. Internet shopping not only involves with a person to merchant trust, but it also involves the trust between person to computer system through which safe and sound transactions would be take place (Turban, 2009). A study shows empirically that "on line shopping behavior is indeed affected by interpersonal trust and the economic condition of the on-line shopper. These variables independently affect the styles or behavior of an online shopper who is shopping on World Wide Web. Checking consumer behavior while shopping on Internet has its own significance. As by checking consumer behavior while shopping on Internet, companies and sellers who are working online can increase their sales and operations. With reliable information about the consumer behavior while shopping products or use services online they would be able to design better promotions and programs which would be keyed to motivate consumers for shopping online.

B. Role of Brand Awareness

The rapid increase in E-commerce provides new challenge for marketers, as online consumer adopts new technologies; their attitude is changing (Dutton, 2011). This study will describe brand awareness in the framework of online shopping. There are certain differences while shopping in traditional market and in online market. E-commerce normally absences human friendliness and friendliness, meanwhile it is more objective, unidentified and automatic than old-style face-to-face business (Hassanein, 2007).

By brand loyalty it means that consumer is familiar with brand and they preferred that particular brand while shopping products or use services on Internet. Exactly like in traditional market brand awareness put impact on buying behavior of consumers, and consumers become loyal to that particular brand. Same in the case here, brand awareness put positive impact on buying behavior of consumers on Internet. People become loyal to that specific brand from which they are familiar with. A result of studies shows that shopper believes in a web store was a noteworthy forecaster apparent web sureness exploration meaning for product material via the online store (Hahn, 2009). Results of another research finds that the occurrence nonappearance of product's name disturbs web consumers' supposed threat, but in the contradictory way to that predictable (Huang, 2004). These all findings and studies show that brand awareness has a significant role while purchasing products and use services on Internet.

III. Problem Statement

By reviewing the literature it has been found that there are several studies which have for checking the consumer awareness only with brand awareness. There are very few studies which are conducted in the context of Europe. Furthermore, the two variables brand awareness among consumers and E-commerce sale; very few have used them as a variable in the studies. Although, some has studies these variables as independent variables which put ultimate effect of consumer behavior, but very few has studies them by taking as an independent and dependent variables. So the gap lies here that very few studies have conducted in European context.

IV. Research Question

To what extent brand awareness put impact on e-commerce sales while a consumer shops through Internet in Europe?

V. Theoretical Framework

The Theoretical Framework of "Recognizing the Importance of Brand Awareness on E-commerce Sales while Shopping on Internet: Empirical Analysis of European Countries".

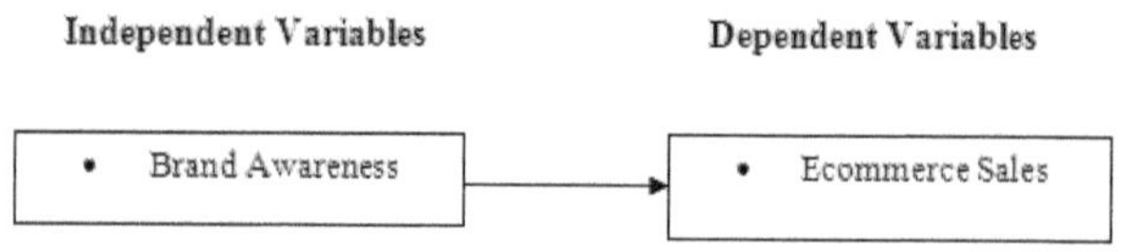

VI. Research Methodology

This section discusses the research methodology which is utilized to find out the relationship between independent variables and dependent variables. The methodology of this paper is quantitative and based on primary as well as secondary data collection. Secondary data is collected from 07 countries which are located under the territory of Europe. The names of countries are Denmark, Germany, France, Italy, Netherland, Sweden and Norway. Sample size is 77. The data is collected from those countries which are on top of spending and making online purchases on Internet. The data for independent variables which is brand awareness is collected from the consumers through online survey by sending them questionnaire. For dependent variable, the data set is collected from the reports of (Eurostats, 2011). Also Emails were sent to consumers in Europe in order to collect the data of brand awareness and to collect the right data. When contacted to Gallup it was found even they did not collected this information so this research was carried out by personal effect. Total 77 observations have been recorded of the independent variable. The research paradigm for conducting this research is deductive and approach is positivist. Questionnaire was distributed among 82 respondents out of which 77 were received back. So response rate is 94% which is enough to carry on the research.

Data was first recorded in Microsoft Excel, and then it was exported to E-views 5. After exporting to E-views 5, regression is run by using dependent variable which is E-commerce sales and independent variable is brand awareness. Different results which include descriptive statistics such as mean, median and kurtosis are received which all are in good order. Further regression results also achieved which are also in good order and results showing significant relationship. Moreover; the correlation matrix is generated which also showing strong positive relationship between the independent and dependent variable.

VII. Results & Discussions

In the following section the results are given which received from E-views. Firstly in this section descriptive statistics detail is given, which include mean, median, standard deviation, skewness and kurtosis. Secondly further the results of regression and correlation are analyzed.

In Table I, descriptive statistics is mentioned as per independent and dependent variable. The total numbers of observations are 77 which are from 07 countries of Europe, and data is collected for ten years from 2002-2011. The Sample of the paper discloses that mean of brand awareness is 170.4400; and the mean of dependent variable which is E-commerce Sales is 18.66200. Also the standard deviation (S.D) of classified advertisement is 171.1560; and the standard deviation (S.D) of dependent variable which is E-commerce Sales is 21.02158. In addition to Mean and Standard Deviation if we see the probability results than the probability of brand awareness is 0.002877 and the probability of dependent variable which is e-commerce sales is 0.002598.

Table II illustrates the Matrix of correlation in which correlation findings of every variable is displayed. As the independent variable is brand awareness, and one dependent variable i.e. E-commerce Sales, all correlations results are mentioned. Correlation is actually defined as the degree of relationship amongst two main variables. Result of correlation amongst the variable itself such as brand awareness is strongly significant or 1 and vice versa. A correlation result amongst brand awareness and E-commerce sales is 0.79 which displays strong relationship. Similarly, correlation results amongst classified and E-commerce sales are 0.79 which too displays strong relationship. Similarly correlation result for the dependent variable itself is 1, and when we see the correlation result between E-commerce sales and brand awareness it is 0.83 which depicts a very strong relationship with each other.

TABLE I.
Descriptive Statistics

	BRAND AWARENESS	ECOMSALE
Mean	170.4400	18.66200
Median	137.0000	17.85000
Maximum	834.0000	89.50000
Minimum	4.000000	0.900000
Std. Dev.	171.1560	21.02158
Skewness	0.81025	1.48963
Kurtosis	1.983698	5.125896
Jarque-Bera	8.125896	24.589652
Probability	0.002877	0.002598
Sum	13589.00	11502.00
Sum Sq. Dev.	35926.85	198523.00
Observations	77	77

TABLE II.
Correlation Matrix

	BRAND AWARENES	ECOMSALE
BRAND AWARENES	1	0.798563598756
ECOMSALE	0.836985623642698	1

TABLE III.
Ordinary Lease Square Regression Statistics

Dependent Variable: ECOMSALE				
Method: Panel Least Squares				
Date: 05/01/14 Time: 05:41				
Sample: 2002- 2011				
Cross-sections included: 7				
Total panel (balanced) observations: 77				
Variable	**Coefficient**	**Std. Error**	**t-Statistic**	**Prob.**
C	2.987469	1.897526	3.356987	0.0002
BRAND AWARENESS	0.159752	0.025986	3.125986*	0.0015
R-squared	0.996354	Mean dependent var		18.69678
Adjusted R-squared	0.907526	S.D. dependent var		17.25698
S.E. of regression	3.256896	Akaike info criterion		3.547865
Sum squared resid	869.36589	Schwarz criterion		5.002589
Log likelihood	155.68952	F-statistic		287.6982
Durbin-Watson stat	1.369365	Prob(F-statistic)		0.000000

Note: Shows significance at 1% level of significance

Table III portrays through Ordinary Least Square (OLS) regression. The overall constant value of probability is 0.0002 which shows it has very high or strong important influence of the entire model. Particularly when we see the probability value of brand awareness it is 0.0015 which is also less than the level of significance and it also shows strong positive significant impact on the dependent variable. T-stats also depict the fitness of model and tell the significance of the relationship. As per table (3); the t-stats of brand awareness is 3.12 which is more than 2 so it displays significant positive influence of brand awareness on dependent variable which is E-commerce Sales. The rate of R square is 0.996 which is more than 0.90 which displays that there is the effect of multicollinearity amongst the independent variables. As we have only one independent variable so we do not need to see the value of R-square because R-square is only seeing when there are two or more than two independent variables.

VIII. Conclusion

The primary motive to carry out this study is to investigate and recognizing the importance of brand awareness on E-commerce sales while shopping on Internet. This is basically the empirical analysis of seven top Internet purchasing European countries. From last couple of decades Internet is growing rapidly and today Internet is accessed by plenty of users sitting in their homes around the globe. This revolutionary platform not only eases the communication between the Internet users, but also the e-commerce websites are developed which are attracting consumers to make purchases. This study is targeting seven countries of Europe which are on the top of online purchases as compared to other European countries. These countries are Denmark, Germany, France, Italy, Netherlands, Sweden and Norway. Data is collected for ten years 2002-2011 from the report of (Eurostats, 2011). The independent variable of this study is brand awareness, for which data is collected through a questionnaire. Results indicates that brand awareness is positively associated with the online purchases and further with the E-commerce sales, Further regression results depicts strong significance between both explanatory and dependent variable, It displays that there is a strong positive significant relationship amongst brand awareness and dependent variable which is online shopping. Findings indicate that companies should work on Internet with proper brand names, and they should implement such strategies through which they can enhance the credibility of their brands. Also companies which are performing Internet business activity they need to create awareness of their brands in the minds of customers, so that more customers attracts towards their products or services and in this way they can have more e-commerce sales.

References

[1] Adjei, M. T., Noble, S. M., & Noble, C. H. (2010). The influence of C2C communications in online brand communities on customer purchase behavior. Journal of the Academy of Marketing Science, 38(5), 634-653.

[2] Ann E Schlosser, S. S. (1999). Survey of Internet users: Attitude towards Internet advertisement. Journal of Interactive Marketing, 13.

[3] Casey, E. (2011). Digital evidence and computer crime: forensic science, computers and the Internet. Academic press.

[4] Dutton, W. H., & Blank, G. (2011). Next generation users: The Internet in Britain.

[5] Eurostats. (2011). Online ad spend across Europe. France: http://www.ec.europa.eu/eurostat/.

[6] Hahn, K. H., & Kim, J. (2009). The effect of offline brand trust and perceived Internet confidence on online shopping intention in the integrated multi-channel context. International Journal of Retail & Distribution Management, 37(2), 126-141.

[7] Harfoushi, O; AlFawwaz, B; Obeidat, B; Obiedat, R and Faris, H. "Impact of Internet Advertisement and its Features on E-Commerce Retail Sales: Evidence from Europe", Journal of Software Engineering & Applications, 2013

[8] Hassanein, K., & Head, M. (2007). Manipulating perceived social presence through the web interface and its impact on attitude towards online shopping. International Journal of Human-Computer Studies, 65(8), 689-708

[9] Huang, W. Y., Schrank, H., & Dubinsky, A. J. (2004). Effect of brand name on consumers' risk perceptions of online shopping. Journal of Consumer Behavior, 4(1), 40-50.

[10] Internet World Stats, I. w. (2013, December 31). Internet world stats. Retrieved 06 19, 2014, from Internet world stats: http://www.Internetworldstats.com.

[11] Ling, K. C., Chai, L. T., & Piew, T. H. (2010). The effects of shopping orientations, online trust and prior online purchase experience toward customers' online purchase intention. International Business Research, 3(3), P63.

[12] Michel Lombera, I., Chuang, Y. T., Melliar-Smith, P. M., & Moser, L. E. (2011, June). Trustworthy distribution and retrieval of information over HTTP and the Internet. In INTERNET 2011, The Third International Conference on Evolving Internet (pp. 7-13).

[13] Nielsen. (2012). An Uncommon sense of the Consumer. Nielsen.

[14] Pookulangara, S., & Koesler, K. (2011). Cultural influence on consumers' usage of social networks and its' impact on online purchase intentions. Journal of Retailing and Consumer Services, 18(4), 348-354.

[15] Turban, E., Lee, J. K., King, D., Liang, T. P., & Turban, D. (2009). Electronic commerce 2010. Prentice Hall Press.

[16] Trusov, M., Bucklin, R. E., & Pauwels, K. (2009). Effects of word-of-mouth versus traditional marketing: findings from an Internet social networking site. Journal of marketing, 73(5), 90-102.

Authors

Bader M. Al-Fawwaz is an Assistant Professor at the Department of Information Systems, Prince Hussein Bin Abdullah College for Information Technology, Al-albayt University (Jordan).

Atallah M. AL-Shatnawi is an Assistant Professor at the Department of Information Systems, Prince Hussein Bin Abdullah College for Information Technology, Al-albayt University (Jordan).

Wafa S. Alsharafat is an Assistant Professor at the Department of Information Systems, Prince Hussein Bin Abdullah College for Information Technology, Al-albayt University (Jordan).

Affordances in Mobile Augmented Reality Applications

Tor Gjøsæter
University of Bergen, Norway

Abstract—**This paper explores the affordances of augmented reality content in a mobile augmented reality application. A user study was conducted by performing a multi-camera video recording of seven think aloud sessions. The think aloud sessions consisted of individual users performing tasks, exploring and experiencing a mobile augmented reality (MAR) application we developed for the iOS platform named ARad. We discuss the instrumental affordances we observed when users interacted with augmented reality content, as well as more complex affordances rising from conventions from media content, AR and the traditional WIMP paradigm. We find that remediation of traditional newspaper content through the MAR medium can provide engaging, pleasing and exciting user experiences. However, the some of the content still suffers from being shoveled onto the MAR platform without adapting it properly. Finally, we discuss what content was most successfully mediated to the user and how the content impacts the user experience.**

Index Terms—**augmented reality, mobile computing, human computer interaction**

I. Introduction

Mobile augmented reality has become increasingly viable through less expensive (and recently free) systems development kits (SDK) and ever increasing technical capabilities of mobile devices. Wagner et al [1] pioneered "…the first completely stand-alone Augmented Reality client on a PDA." (p. 2) on handheld devices in the early 2000s. Since then stand-alone AR has gradually become more common as computing power on PDAs and smartphones increased.

A mobile augmented reality (MAR) or handheld mobile augmented reality (HMAR) application is in its most basic form a mobile device with a display to show an augmentation and registration technology and software that "...have the following three characteristics: 1) Combines real and virtual 2) Interactive in real time 3) Registered in 3-D". [2]

With the smartphone revolution as a backdrop, we decided to start developing a MAR application for distributing and presenting AR content in a newspaper – Verdens Gang – with a large circulation in Norway. The software we developed in collaboration with the newspaper iterated from concept and prototype, to the stable release now available on the AppStore. Through the iterations, we saw increasing requests for reuse of traditional media content the newspaper already had in its possession. These requests resulted in a final version of the application that supports remediating of content on markers in a newspaper.

The term remediation is coined in Bolter et al [3] book "Remediation – Understanding new Media". *Remediation* can be understood as taking existing content (pictures, videos, etc) adapted to a specific media outlet – for instance a printed newspaper – and remediated through a different medium, in our case an augmented reality application.

Trough a video based user study we have uncovered how remediation of traditional newspaper content has shaped our MAR application and how different affordances of the content itself affect the user experience of the application.

II. Researching MAR

Augmented reality as a term is broadly interpreted and to some degree diluted across media and academia alike. In this paper, we use Azuma's [2] definition of AR mentioned in the previous section to position this work. We can with confidence claim the ARad application to be augmenting reality as it incorporates Azuma's three characterizing properties.

In this paper, we focus on the interplay between the augmentation (the content) and how the user experiences the remediated content in our application.

This paper contributes to further the understanding of AR as a tool in a mobile multimedia context. Additionally it provides a detailed overview of the design of our application and provides design guidelines theoretically aligned towards affordances in human computer interaction (HCI).

III. MAR as a Medium

Bolter and Grusin [3] sees augmented reality as a hyper mediated visual space and "...the insistence that everything that technology can present must be presented at one time - this is the logic of hypermediacy." This definition is fitting since augmented reality seeks to enable us to see the world how we want it, with just the right amount of information. Macintyre et al [4] argues strongly for AR in general being a new media experience mainly because of a "...fluid blend of the physical and the virtual, and the inevitable tension between them, offers rich dramatic possibilities that are impossible in any other medium."

In Interactions [5], Bolter et al argues that the field of media studies is not responsible for providing design guidelines that the HCI community. However, as content and the interaction is becoming increasingly intertwined one might look to polyaesthetics [6] to shed light on aspects of new media. Polyastechics suggests media studies should be concerned with contributing knowledge about the aesthetics of content since the interaction between content and interface creates the user experience.

We believe that *content in relation to the interface provides the complete user experience*. Poor content in an application with excellent user interface delivers a poor user experience and visa versa. This is paramount in the ARad application, as the content to some extent becomes the user interface itself.

AR technology promotes the ideal of a pure and fully transparent user experience, where the interaction with the medium is transparent to the user and just the experience of the content remains. Although some transparency is achievable on a handheld device, Rosenblum et al [7] believe "...if AR is to realize its full potential, hand-held form factors, de- spite much of the hype they are receiving now, simply are not adequate." (p. 445) We agree with Rosenblum to some degree. The handheld form factor can never create the same immersion that a totally transparent AR solution can. On the other hand, even with all its hype MAR has utility until this paradigm shifts.

Pavlik et al [8] gives an overview of the emergence of mobile AR in journalism in general. Content creators for the applications they identified (Aurasma, Layar, Wikitude World Browser, Junaio and Blippar) may benefit from the findings in this paper.

In Virtual Realities, Schmalstieg et al [9] discuss the first mobile augmented reality ad developed in 2007 by HIT Lab NZ for the Wellington Zoo. The application delivered a single static 3D model of a zoo animal augmented on a printed marker in the newspaper. They report that for mobile AR advertisement the "...most challenging aspects have been the content creation and application distribution, not the application programming." In addition, Schmalstieg et al points out, "Most of the published AR research has been on enabling technologies (tracking or displays, etc.), or on experimental prototype applications, but there has been little user evaluation of AR interfaces."[9].

In this paper we hope to contribute further to a field where there has "...to date been very little user-based experimentation in augmented reality." [10, p. 3]. Similarly, Zhou et al [11] call for more practical applications. In a more recent and extensive literature review De Sá [12] gather that "...despite the appeal and the growing number of services and applications, very few guidelines, design techniques and evaluation methods have been presented in the existing literature."

The motivation to write this paper comes from a wish to understand the nature of MAR applications, specifically the affordances in these applications. Accordingly, we aim to describe affordances and specify a set of design guidelines one may apply when *remediating* content on the mobile augmented reality platform.

IV. AFFORDANCES IN AR APPLICATIONS

Our application augments traditional media with virtual content. It enables us to create a new user experience for readers consuming traditional print media. By identifying how our MAR application *affords interaction* with this content, we seek to contribute prescriptive knowledge in the form of design guidelines intended for practitioners and academics alike in the field of interactive augmented reality mobile technologies.

To provide meaningful prescriptive knowledge about affordances we have adopted the sociocultural perspective of Kaptelinin et al [13]. This view, and its borrowing of "web of mediators" from Bødker [14]allows us to look at affordances in AR with a respect to the cultural aspect.

MAR is different from other media channels, and the content chosen for remediation needs to be adapted to the AR interaction paradigm.

Previous experience with different content types needs to be taken into account when designing user experiences for MAR. It requires the understanding of the combination of known usability conventions and conventions of aesthetics directly related to the content we are presenting. In this paper, we aim to identify how we can successfully present the content in an engaging manner on MAR.

The framework proposed by Kaptelinin et al [13] allows us to look at affordances from an individual perspective with regards to culture. Through Kaptelinin et al literature review, they find "A growing number of studies in HCI and related areas call for re-defining the notion of affordances to include social and cultural aspects of human interaction with the world." [13] Kaptelinin et al approach is named mediated action perspective and "is concerned with how humans act in their cultural environments, rather than with how animals act in their natural habitats." [13].

If we allow the discussion about affordances in HCI to include these aspects, we can study IT-artifacts with the cultural aspect in mind. This is fruitful in this study since we are creating an artefact relying on existing conventions from a wide range of interaction and content paradigms. These include the WIMP paradigm, the AR interaction technique, existing media content and smartphone conventions.

Through Bødker [14] they identify auxiliary affordances that take into account "complex relations within webs of mediation". This may be understood as there is often a need to perform indirectly or directly related actions to achieve an outcome. They identify maintenance affordances, aggregation affordances and learning affordances as examples of affordances technology may employ to achieve its intended purpose.

In this study, we will describe the *instrumental affordances* of our application. Further we look at the complex affordances that arise with bringing culture into understanding of applications.

Learning affordances: What steps do the users go through to learn how to interact with an AR application, and how we might improve these affordances.

Maintenance affordances: What needs to be taken care of to allow applications to function as effective mediators (maintenance).

Aggregation affordances: How is the application intertwined with analog and digital artifacts to achieve its outcome. Aggregation affordances illustrate the fact that some applications must be combined with other artifacts to achieve its indented purpose. AR`s character is to use the environment to provide augmentations and in this regard we aim to identify the essential aggregation affordances of a MAR application.

In addition, we propose a domain specific affordance - *Remediation affordances*: This affordance encapsulates the affordances associated with the content. Specifically, how we re-communicate the existing affordances of content through an interface. In our case, it deals with bringing the content from a known interface a to a different medium. The act of remediation creates a number of boundaries and as well as new possibilities. The aim is to identify successful remediation affordances for a given medium or interface. This is at the heart of mediated action, taking into account the culturally bound when designing.

V. Development of ARad

To address the problem formulated previously we developed several proof of concept prototypes on a variety of mobile platforms (Symbian, Windows Mobile and iOS) starting fall 2010.

An iterative process with several prototypes led to the application now freely available in the AppStore. In the following sections, we will describe the rationale behind the design choices we transferred to the interface, structure and content of ARad.

A. The structure and interface of ARad

ARad is intended to be a MAR application that enables users to seamlessly experience AR content in print media. The interface should be intuitive and enable the users to move rapidly between the different content types. ARad achieves this through a considerable amount of pre-fetching of data in load-screens. Rather than loading content for each marker, the user can download a collection of content from one content provider and experience it in a newspaper or any other printed medium by turning the page.

ARad is developed as a content browser; this means that we provide a range of content providers or campaigns represented as icons on the first screen. The application architecture is straightforward. When ARad starts up it checks for new content and content-providers. It loads new content-providers and links touchable icons to content. When the user touches an icon, it downloads the corresponding content and loads the media assets into memory while the user see a loading-screen. When the user has touched the play icon, a notification screen informs the user to direct the phone towards a marker. When the user points at a marker associated with the content, the augmentation appears. The user can leave this view by aiming the device at something untrackable and press a X icon. They then return to the screen representing the different types of content and may choose content from a different content provider (Figure 1).

This structure we believe enables the perception of ARad being a content browser for different types print media categorized as either campaigns or content-providers. Campaigns are not restricted to a template format, and commonly has some custom code implemented. Content-providers are identified as media outlets that generate new content rapidly based on templates, in contrast to campaigns that remove their custom content when its life cycle is complete.

VI. Content in ARad

The specifications for the interface and content in ARad came from newspaper in Norway (Verdens Gang). The newspaper wanted a way of using existing images, videos, illustrations and 3D material already in their possession. Marshall McLuhan [15] remarked that the content of a new medium is always the content of another medium. In keeping with this prophecy we inadvertently started the process of *remediation* old media through new media (MAR). Images, videos, illustrations and 3D material related to news or advertisement already in their possession was adapted to a format ARad could display and remediate.

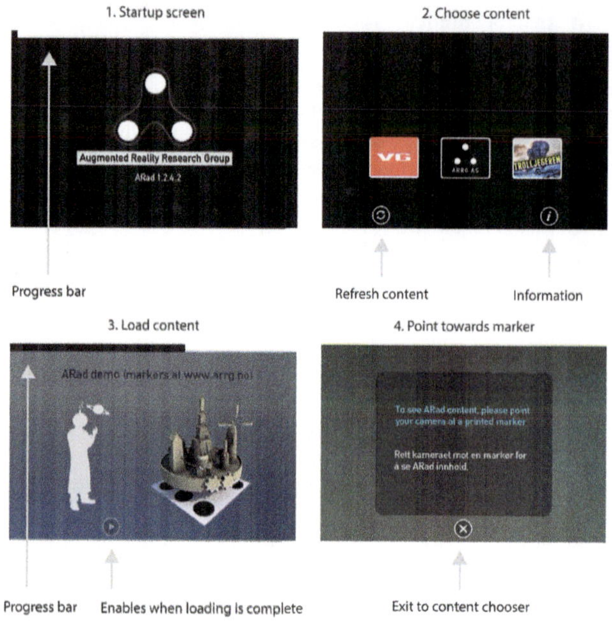

Figure 1. Different screen elements in Arad.

A. Content types

ARad can handle several different types of content using templates. The templates can be updated on the dynamically using an XML-like specification language and supports the quick turnaround in a newspaper. When new content is available we flag it in an index file and when the user reload the content new content is downloaded to the application. ARad supports the following content types:

- Image slideshows with or without audio
- Videos with audio
- Dynamic 3D content with predefined points of interaction
- Custom content

The specification relays paths to marker-, texture-, image-, audio-, video- and mesh-data for the application to load and how to organize and orient it in relation to markers. The custom content needs to be submitted to AppStore for review as it contains executable code.

1) Image slideshows with or without audio

We configure this with a list of images and corresponding single ambient track for background music. Other soundtracks can be tied to an image to provide a narrative or sound effects to amplify the user experience. The user can use arrow icons to maneuver in the image carousel.

2) Videos with audio

The video template points to a MPEG-1 file with an attached audio file. The user can play or pause the video when it is playing on a marker or in full screen. The content is a short interview with a performing artist.

3) Dynamic 3D content with predefined points of interaction

The 3D content template loads a set of 3D objects onto a marker and allows tap interaction with the 3D model. The 3D objects have a predefined idle animation with audio. Animations for interaction can be accessed using tap interaction.

All the different content types can be viewed in something we have called full-screen mode. This means that the user can hit a button to view the images, movies and 3D content without the marker in the cameras view, and keep interacting by tapping and dragging the content. This gives the user freedom similarly to the Freeze-set-go interaction method proposed by Lee at al [16].

VII. METHOD

The aim of the evaluations is to provide descriptive knowledge about - how users use the mobile AR application - and prescriptive knowledge about design - in this case a reflection on what works and how to improve what already works - about this IT-artifact in accordance with Iivari [17] types of knowledge attainable from design science. This study utilizes the framework for studying and evaluating augmented reality games described by Gjøsæter et al[18]. The framework details a step-by-step approach to conducting a study of AR games. This approach "...works particularly well with AR games since they combine virtual information with the physical environment: while video recordings in addition to screen captures allows the researcher to observe both what happens on screen as well as what happens in the physical space, Think aloud provides insights into the players' problem solving processes." (p. 78). While this study is not concerned with the gaming aspect of AR, we are concerned with what happens in virtual as well as physical space as well as insight into the users experience of the AR system.

A. Data collection methods

Think Aloud (THA) [19] is an evaluation technique that takes place during testing; users are instructed to verbalize their actions and thoughts throughout the evaluation. It is a more informal evaluation technique than other common evaluation techniques used in HCI such as cognitive walk-through and heuristic evaluation [20]. Rather than trying to uncover piecemeal design errors as these pragmatic methods excel at, this approach allows us to observe and investigate a uninterrupted user experience through the captured data.

Some previous studies of using think-aloud as an evaluation tool and observational evaluation technique for AR systems can be found. Dünser et al [21] employ think aloud to assess how young children interact with augmented reality books. Liarokapis et al [22] uses think aloud to evaluate and discuss implementation of different interaction methods to AR games.

We chose to do a concurrent Think Aloud, where the participants talked during gameplay. In addition, a retrospective Think Aloud and interview session was performed post session. This session served as a debriefing and allowed users to discuss their overall experience of the content.

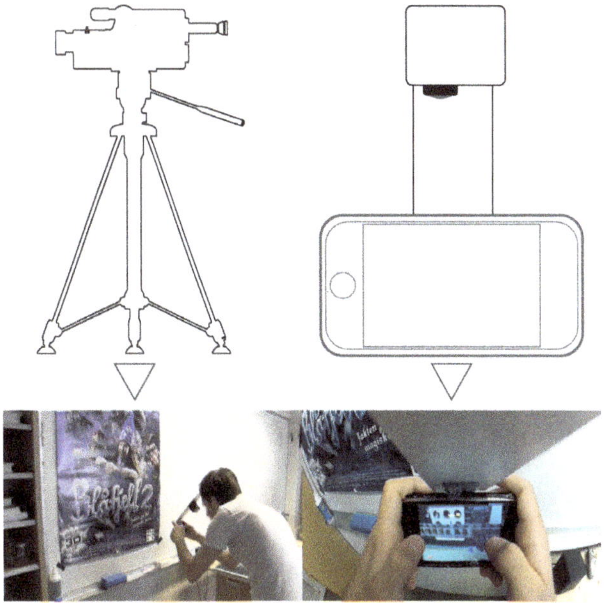

Figure 2. Tripod with custom camera rig

B. Capturing video of mobile augmented reality

Video-based qualitative research has in recent years increased in popularity within the HCI field. A key advantage to this method is that it can help capture "...aspects of social activities in real-time: talk, visible conduct, and the use of tools, technologies, objects and artifacts"[23].

We used guidelines provided by Heath et al [23] and Gjøsæter et al [18] to prepare for, and undertake the video recording. Care should be taken to ensure proper audio capture to facilitate transcription of interaction, clear video of the interaction in proper lighting and correct angles and other technical aspects of video recording. We digitized and synchronized the data in high resolution to allow us to see the participant's interaction with the context as well as the content (Figure 2). Some of the figures in this paper are formatted as cartoon strips. This format supports communication [18] of the activities, as well as the speech occurring during the sessions. The sessions were captured using a small camera attached to a rig at an angle to the handheld device screen. A portable tripod camera was used to film the user's interaction with the tangible printed markers.

We believe the combination of THA and video recordings provide a good foundation for revealing affordances and qualities with the interfaces. The video recordings provide detail about the worldly context and the user interface on the mobile device. In contrast to simply an audio recording of the Think aloud, where you would lose essential data about for instance how they are pointing and orienting the device. In some cases it may also be difficult to ascertain what user interface elements the users are referring just from audio recordings. Additionally, one can use still images, and still images in sequence from the video recordings to directly illustrate the system and underpin interesting findings.

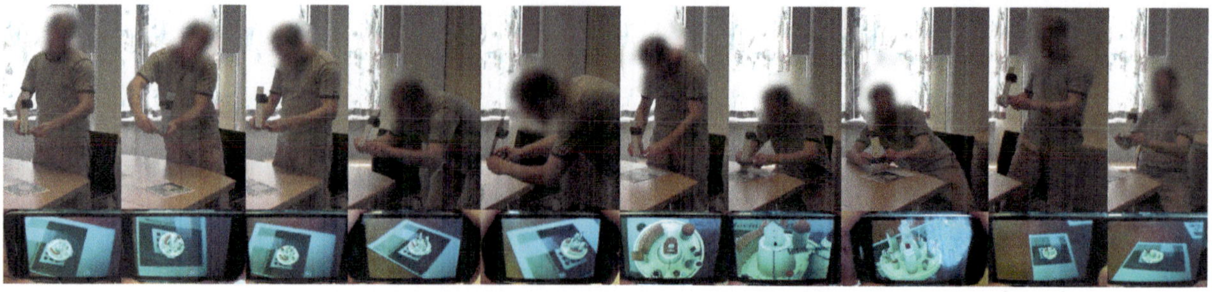

Figure 3. A user exploring 3D content

C. Structure of evaluations

A pilot study with a female primary school teacher aged 25 preceded the final evaluations. The pilot study was conducted to rehearse and uncover problems with the structure of the evaluations. We gathered data about gender, age, profession and familiarity with information technology, smartphones and tables. The final evaluations were conducted with seven participants – five male and two female. With an average age of 29 years (min: 26, max: 34), the participants were faculty members (an accountant, three PhD students, a post doc student) and others (a high-school teacher and an engineer). When asked "How would you characterize your experience with information technology?" on a Likert scale from "1 – Very little" to "8 – Very much" the participants averaged 4.8 (4 min, 6 max). In regards to "How would you characterize your experience with tablets and smartphones?" the participants averaged 5.0 (3 min, 7 max).

We label the evaluators from R1 to R7. The author's assistant is labeled A, and the author is labeled T in the transcripts. The sessions ranged from 25 minutes to 45 minutes of dual camera video. A free form retrospective think aloud followed each session, lasting between 10 and 30 minutes. The corpus for this study total 51 transcribed dual camera video clips.

During the sessions, the participants experienced six remediated content types.

- A video (Figure 4)
- An interactive image slideshow (Figure 6)
- A 3D interactive castle (Figure 3, 7)
- A 3D interactive troll (Figure 5, 7)
- A game related to the Troll (not described in this paper)
- A game related to a different campaign (not described in this paper)

VIII. Findings

ARad is remediating in nature, and we choose to look at it from this perspective in this study. When disseminating sequences of interaction that relate to the overall experience we may reveal how remediation of content can potentially provide new, better, more fun or engaging ways of interacting with print media.

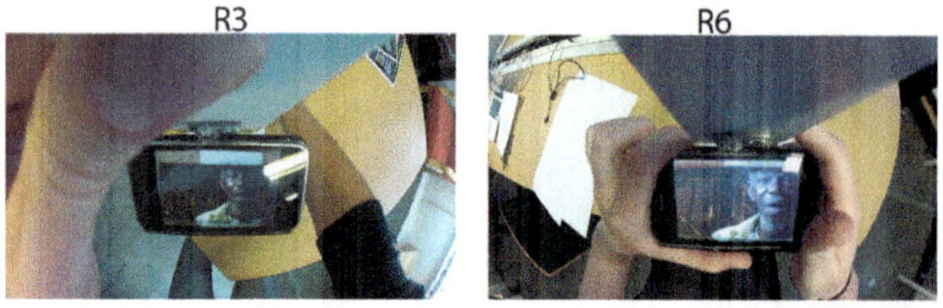

Figure 4. While interacting with 2D content, the users tried to align the content appearing on the marker to iPhone display.

A variety of other topics on the details of usability is readily available in the data. However, this study focused on the user experience and different affordances of content remediated through a handheld augmented reality application.

We use design research principles [24] in combination with established evaluation methods within the field of HCI to help us communicate and analyze the user experience of the content and interface of ARad through illustrations and quotations in context. The findings below are represented as direct quotes in italic and illustrations from the concurrent think aloud. Quotes are referenced in the text as Q with the quote number following.

A. Video content

During the interaction with the video content, we made observations in regards to how the users experienced this content.

The novelty factor of a digital video being experienced augmented on paper can not be understated (Q1, Q2). These utterances were spoken when the participants viewed the movie on the marker for the first time.

Q1 - R2: "Ah! Now it works. Cool!"

Q2 - R4: "It is attached to the paper there. That was pretty cool."

Some of the evaluators found it to be unfamiliar to watch movies in this manner and tried to align it the movie naturally to the screen (Q3).

Q3 - R6: "The only thing here is that we need to aim to show the movie in the right size."

Other participants did not utter this directly, but from the video material we can observe the participants align the video naturally after some time experimenting, and eventually bringing the movie to full screen as illustrated in Figure 4.

Some participants did not mind watching the video at odd angles. Others did not align the video to the display perfectly, but kept the movie in view for long sequences of interaction, and avoided getting the video out of view or at odd angles.

It may be beneficial to show content into alignment when we detect that the user is trying to align any flat content to the screen. Users want to see flat content in full screen mode. It may be some subtle property with AR-video we have yet to determine, but the evaluators seemed to be more excited about AR-video represented in this way than images.

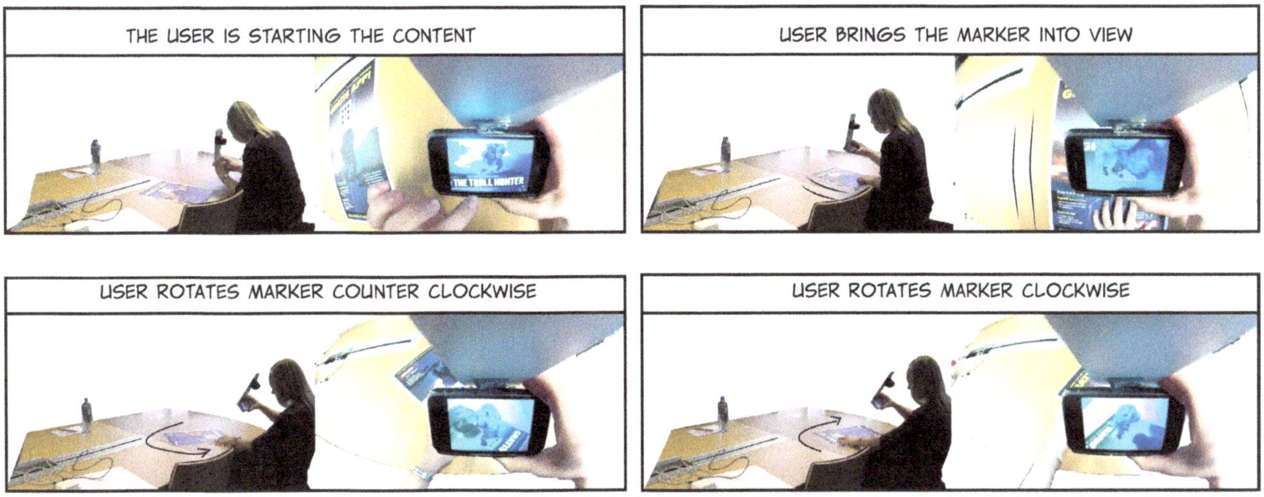

Figure 5. Observation of a user using the piece of paper with the marker on it to manipulate the 3D content.

B. Image content

We observed the participants recognize the image content as images. The concept of a slideshow we created was immediately recognized and articulated (Q5).

Q5 - R3: "*Sigh* now I've been through the entire slideshow."

Q6 - R2: "I don't know that it improves it, that the images are lying on top of the paper."

The participant does understand the concept of the application, but he questions the value of AR-mode to display images, and clearly expresses boredom through body language and sighs (Q5). Q6 shows the user not seeing the point, expressing doubts about the experience this feature adds to the image content in any way.

Q7 - R6: "There is nothing wrong with experiencing it in this manner, but you are used to it being attached to the screen."

This evaluator is decidedly neutral in discussing this way of viewing images (Q7). She walks through the motions of the task, but there is nothing suggesting fun or enjoyment when interacting with the image content. To view images in this manner does not add anything significant to experiencing of seeing image content.

We find that remediating flat still-images in this form serves little purpose as users express reservations about its merit. It serves its purpose of shoveling images to the users; they all understand the concept and diligently consume the images. We did not observe any immersive moments with the user using this content in contrast to the 3D content and video content where they conducted themselves differently.

C. Experiencing 3D content

When users interacted with augmented 3D content, we observed different reactions. We could also observe users perceiving the 3D content as significantly different than the 2D content. Technically, the image content is more or less the same, a mesh with a texture on it; particularly one user got excited about 3D content to a greater extent than 2D content (Q8, Q13, Q14, Q15).

Q8 - R7: "Wow! This is something completely different. This looks like game element perhaps? What is it?"

They spent a significantly more time exploring the 3D content than the flat image and video content. Users engaged with the model and looked at it from different angles, eager to discover features and details. This may suggest that the exploration of the 3D model itself is a captivating and fun activity (Figure 3).

Q9 -T: "Is there an angle you prefer?"

R3: "This had depth, definitely."

R3: "It's more fun than standing over, it makes it flat. The fun part is that it is 3D."

R3: "It's tempting to touch it. I want to play with the windmill."

The evaluator notes that watching the model from above gives the model less depth (Q9). This is a problem when developing AR content for print media. In this application, we achieve optimal tracking when the user has entire marker in view and a fair amount of high contrast pixels available for tracking. However, the 3D content itself achieves the most depth in conjunction with the print media at steeper angles. This is one of the main issues with freeform interaction with MAR content. It is difficult and time-consuming for a designer to develop content that is equally appealing from all angles and distances. Remediated content perhaps to a greater degree is created for other purposes, like effect shots in movies or informational content, rather than content to be viewed from above.

Q10 - R1: "It's a 3D model of a troll."

R1: "Stands still breathing."

R1: "Waiting for something, Norwegian Christian's mans blood."

The users notice subtle movement (Q10) in the content and express more emotions and attention when interacting with the Raglefant troll.

Q11 - R1: "Being that it is an animal, it is tempting to look at him, in eye height."

Q12 - R6: "I'm thinking I want to see him from a different perspective. I want to view him from the front."

Some users want a deeper interaction with this content than the 2D content; they seem to require more interaction (Q11, Q12). We suspect the fact that it is a humanoid creature; the experience may be degraded by the fact that the content is not trying to interact with the users. It appears inanimate and behaves unnaturally.

Q13 - R5: "I'm trying to press it, since that is what the theme from before."

R5: "YARRR!" (Participant mimics the noise of the troll)

R5: *Laughing*

Q14 - R7: "Oh oi! This is a Raglefant yes!"

The user exclaimed when the troll appeared in sight.

Q15 - R7: "Hello hello!"

1. Image slideshow on marker

2. Image slideshow in fullscreen

Figure 6. Image slideshow on marker (top), in full screen (bottom)

R7: "It got a bit annoyed when scratching it on the belly."

R7: "Nice!"

These users experienced fun, R5 (Q13) mimicking noises, almost forgetting the experimental setting and R7 (Q14, Q15) exclaiming and talking to the model. The 3D content, not surprisingly lends itself to greater immersion than the 2D content.

Despite a full screen icon being present throughout the interaction with the two content types, most of the users did not use it after first trying it out on the moving castle (Q16). One participant uttered this about going into full screen mode:

Q16 - R4: "Let's to full screen mode."

R4: "Now its in full screen, but that is not an improvement."

R4: "Its a lot cooler to not see it in full screen."

Those that used it found the interaction in this mode to be less compelling and to some degree pointless (Q17).

Q17 - A: "In contrast to what you mentioned before about that you want to see images filling the screen. How about this? Does this feel more natural?"

R6: "This is more like an object you should explore. You get the feeling that this object is sitting on the paper on the table, and that you are filming it in a way."

The users may feel this way because it is fun to "film" intriguing objects. If they think they are filming in 3D modus, this lends to the user experience while viewing it without its real world context lessens the immersion.

D. *Physical interaction beyond the device with 3D content*

How the users relate to the augmentations in physical space sheds light on how users perceive the content in combination with the markers.

1) *Interacting with the troll*

We observed different approach to interaction when the users engage with the troll than other content types. Two of the users tried to touch the augmentation in itself (Fig.

1. 3D castle on a paper marker

2. 3D castle in fullscreen mode

3. Raglefant on marker

Figure 7. 3D content in Arad

7), they swiped their fingers across the area where the augmentation appeared and expected some reaction. We could not observe this interaction with the 3D castle.

This leads us to believe that natural interaction for content on markers should also allow the users see some response from the interface when performing this type of interaction.

2) *Exploring 2D content*

During the evaluations of the flat content, all the users except one used their bodies and arms to align markers to the view of the mobile device. The users found the angle where the content was right side up and looked at it from that angle. None of the users wanted to see 2D content up side down (Fig. 4,6).

3) *Exploring 3D content using markers*

During the exploration of 3D content, only three participants actively engaged with the markers (Fig. 5).

- R7 rotates troll marker and castle marker.
- R5 rotates castle marker.

- R2 rotates the castle marker.

This observation may lead us to believe that many users may not engage with the marker at all, and their most natural way of interacting with 3D content will be by using the device when they are novice users of the technology. The affordance of markers and the content surrounding the markers should invite to interaction to enhance the user experience of 3D mobile AR content.

E. *Mixing paradigms*

Late in the development cycle the approach to experiencing "flat" content were changed. Instead of showing images and videos on the marker tracked in 3D first, the flat content – images and video will appear flat in full screen mode first.

This mimics the known paradigm of QR codes and came as a request by the media outlet. The intention was that users would find QR code interaction more familiar than the AR paradigm, where content appears tracked on the marker first. Users could touch a small icon, and 2D content would appear augmented on the marker. The evaluators had different views of this functionality. Through analyzing the videos we can observe how the participants chose to experience the video and image content.

The task presented to the users before experiencing the image and video data was as follows: They should point the phone at the marker and maneuver between full screen and marker mode (Fig. 6), tap through the slideshow and play/pause the video.

When we analyze one participant thinking aloud we observe that starting the movie in full screen mode is confusing (Q18, Q19).

Q18 - R4: "A film just appeared here. Did I do it correctly then do you think?"

Q19 - R4: "I have... I got it right there somehow. That was very cool."

A: "Do you know how? Can you reverse the process?

R4: "I'm not sure. I got it right. Something happened afterwards. It did not happen before when I wanted to do it. Now it happened, now its fastened to the sheet of paper. And when I press play it plays"

A: "Ok"

R4: "That was pretty neat. If I press, then I get it in full screen there. Now I understand what I am doing. I didn't understand it previously. No, I have to find the marker there yes. And then I can click play, and take it to full screen from there. And it goes elegantly to full screen. It took some effort to understand."

The user does not understand how the film appeared in full screen. The connection between the marker and content is unnoticed by the user. After some trial and error, R4 manages to point the phone towards the marker when exiting full screen mode. Since he does not need to point the phone at the marker while in full screen mode, he has lost his point of reference

We asked the participants during the retrospective THA about the full screen-first approach "What do you think about the movie and image slideshows start in full screen mode". Some users responded negatively (Q20, Q21, Q22).

Q20 - R3: "It steals from the AR experience because the first thing that shows do not show the potential it has. If you didn't expect it you maybe hadn't... Why should I do anything with it?"

Q21 - R4: "I think it should have started down on the screen, not in full screen."

Q22 - R5: "I would have started it in world-modus because ... when a movie plays, or a slideshow is running you are experiencing the content already. I don't need to make more of itself, there and then."

Some participants were indifferent to it, but acknowledge its potential to confuse (Q23).

Q23 - R6: "It may be a little confusing in the start if you have to do something to make it be on the marker."

R6: "But I did not think about it really. It is perfectly fine."

However, two of the seven participants liked that the content started in full screen mode, and disliked the idea of having to point towards a marker (Q24, Q25).

Q24 - R1: "I thought it loaded very quickly, and if you are not needed to hold the camera towards the symbol, was very all right. So I liked that very much."

Q25 - T: "If you could choose, would you like it to start in full screen or on the marker?"

R7: "Full screen then."

The idea of having content pop up in full screen mode first seems not to be favored by the evaluators. However, the option to exit from AR-mode to full screen seems to be welcomed (Q26).

Q26 - R2: "I really like the functionality of being able to go in and out of full screen."

T: "What do you like about that?"

R2: "That you are independent of the marker when you have started it."

R2: "In addition to that the content is presented in a better way."

It may be tempting to align AR close to a medium like the familiar QR language where the user point a phone on a marker and content fills the screen. However, in this case we find that it creates inconsistencies for the user, and makes them confused.

Some experienced difficulties when pointing the camera towards a marker initially. Eventually, they made the cognitive leap from content to the marker. The idea of a marker did not become obvious until the content appeared on the marker itself. So having content appear in full screen right away may provide the user little chance of making the connection between the marker and content.

F. *Preferred content*

During the retrospective think aloud session users elaborated what content they found most enjoyable. The respondents answered that they favored the 3D content as well as the video content.

The users seemed to favor the 3D content. This is in line with the observations in regards to how they acted when experiencing the different types of content. The image content most users found uninteresting. However, we find it intriguing that the video content also was represented among the respondents as the favored content.

IX. Implications

By analyzing the findings we can summarize some design implications for future attempts to remediate content in this manner.

There is a need to develop a metaphor for interacting with content beyond the touch screen, people intuitively seeks to touch an augmentation. Hurst et al [25] also points out the need for gesture interaction in AR via finger tracking. The data reveals that interaction does not need to be complex. Simple tracking of a swipe gesture in front of the camera may be enough to support greater immersion with AR content.

We can see that the users attempt to engage more directly with humanoid and animal creatures and find it awkward that they appear lifelike, but do not interact naturally with the user. This is similar to the findings of Wagner et al [26] where they found that a virtual character not relating to them made them uncomfortable and could even be perceived as offending.

When creating 3D content to be displayed on markers in print media, we find that users will try to optimize their viewing angle to the content. However, when dealing with content directly remediated from another medium care should be taken – if possible – to reorient the content. This is to ensure that it appears nice when viewed from a downward angle.

When users interact with 2D content, we find that they prefer to start viewing it on a marker. In some cases, they bring it to full screen mode to take a closer look. However, when the user is trying to align the content to the display area, some automation should take place to optimize its viewing properties so the user can experience it in its native format on screen.

As it stands now, markers have not acquired the affordance for rotating. People recognize markers, but they do not use them to their full potential yet. Adding instructions to the markers themselves to afford rotation and manipulation may improve their affordance and encourage users to interact with content beyond maneuvering the mobile device itself.

We find that image content suffers most from being remediated in this manner; no users preferred this content over video or 3D content. Most of the users preferred the 3D content, and some users genuinely enjoyed watching and interacting with the video.

It has become increasingly clear that the user experience of remediated AR content is closely related to the affordances of the content itself. Text, images and video have clear and (perhaps culturally) deeply set affordances, 3D content on the other hand does afford a true new media experience. This may be self-evident to any practitioner in AR or MAR, to anyone else it may not.

X. AFFORDANCES

Based on analysis of the findings we have identified affordances related to Kaptelinin´s instrumental affordances in addition to the learning, maintenance, aggregation, and our own niche affordance of remediation.

Firstly it is beneficial to clarify the basic instrumental handling and effecter affordances of our application. This gives an impression of the possibilities for interacting with this technology.

The application supports handling through tangible markers, GUI widgets on the display, and the by maneuvering the mobile phone itself. These handling affordances let us effect the augmentation through interaction. We may handle the device (move the device), the GUI (use buttons to move to the next image) and the markers (in the world) to handle the augmentation to affect the augmentation layer as we observe the users do eventually. The effecter affordances come through the device, GUI and marker. The users affect the augmentation primarily through maneuvering the device. The GUI allows users to play and pause videos, move between images and view animations inherent in the 3D content. We find that users rarely handle the markers to affect the augmentations.

Ideally the handling and effecter affordances should be easily and intuitively picked up, in some cases this how it transpires. Most users intuitively point the phone at the marker and starts interacting, while some users require directions to start the interaction itself. Hence we recognize the need to identify the "web of mediating" affordances that make MAR viable.

A. Maintenance affordance:

Maintenance affordance is related to tasks one must perform to operate the IT artifact. As we use CV in the tracking algorithm, both the markers and the environment the markers operates in requires maintenance. This comes in the form of maintaining good enough lighting and maintaining the integrity of the markers. This means they cannot be crumbled too much, and the lighting must be sufficient to support tracking. This is an affordance that is not particularly easy to communicate. Room lighting is not something one immediately thinks of as a problem when interacting with MAR. Human eyes adapt easily to changing lighting conditions, in sharp contrast to CV tracking algorithms.

B. Aggregation affordances:

Markers have yet to become a convention for point of interaction. Even though the participants interacted with printed surfaces with markers on them, not everyone understood that markers reveal AR content. Markers in themselves poorly afford the desired interaction, and the participants seldom read on-screen instructions. This is an obstacle for AR content in print media at this stage. Newspapers, and sheets of paper do not afford engaging manipulation.

C. Learning affordances:

Even with onscreen instructions telling the users to point the camera at the marker, this simple instruction would not be absorbed. This is similar to a finding in a study of AR games [18]. The study makes a point of users having trouble understanding what object in the world the AR application is referring. It must be made exceedingly clear what real world objects the application interacts upon.

When the content appeared on the marker, affordances related to translating and rotating the objects were immediately recognizable using the phone to rotate the view. Very few of the participants wanted to rotate the printed media to experience the content from different angles.

Seldom we find that other AR applications try to shed light on how they work. According to Bødker[14], some applications need to mediate how they work, in contrast to invisible computers. AR relies heavily on computer vision (CV), for a user to be able to correct errors in CV it is useful to understand the limitations of the CV algorithm. To enjoy an AR application we may need a representation of its internal status more clearly. We do provide users with instructions on where to point the marker initially, but we decided against giving the users information about how well they are tracking the markers. This was a conscious decision. We felt that users might focus on achieve optimal tracking, rather than consuming content if we provided them with visualization of the current performance of the tracking.

D. Remediation affordances:

The findings emphasize the need for users to bring the content into a known interaction paradigm. Users often tried to align 2D content to the display insofar that it resembled a familiar way of consuming this type of content. Users liked the idea of being independent of the marker, by enabling them to enter full screen mode to move the device freely while still seeing the content. In the trials we had the content appear in full screen directly, some users found this to be unintuitive as they expected the content to be first in 3D.

E. Design guidelines for content remediated through handheld AR:

It is necessary to note that these guidelines are intended for handheld augmented reality. They may not have much merit in other AR applications. However, we word them to be usable in general AR applications.

1) Learning affordances

- Clearly afford what objects the application is referring to in the physical world to support learning.

This can be achieved by use visual clues that make it clear how the software understands the world.

- Represent the internal state insofar that users may learn how the CV algorithm performs satisfactory.

Signal the inner workings of the tracking algorithm.

2) Maintenance affordances

- Afford actions needed to adequately maintain an environment suitable for augmenting.

We believe this may be achieved by introducing functions to the hardware: external sensors for sensing light and providing information about the environment to enable users to take steps to improve the performance of the application.

3) Aggregation affordances

- Afford the relationship the device has to the trackable.

Visual clues can be used to make the users direct the device at the marker. Bear in mind that textual clues may not be perceived.

4) Remediation affordances

- Afford easy transition between the known interaction and the augmented interaction of remediated content.

Users will try to view 2D-content as they are used to in other media, we support this by allowing users to enter a mode where they are not dependent on the marker.

- 3D content should afford their value from natural angles.

As 2D content can easily be viewed from a bird's eye view, this is not the case for 3D content. 3D content needs to be designed explicitly to support viewing from a bird's eye view.

- Afford the same entry point for interaction even though the content is different

We experienced that our effort to align the interaction closer to a QR paradigm would confuse the users rather than improve the user experience.

- Humanoid character that afford interaction should allow interaction with the user

Users may become uncomfortable or provoked if humanoid characters that afford interaction are unresponsive.

XI. Conclusions

This study describes an application using AR to augment print media with remediated content. We find that remediation of video content in an AR context seems promising. Users find it at times awkward to view flat (2D) content in a marker-context and like the functionality allowing them to bring flat content into a full screen mode. Image slideshows in the AR context has little merit amongst the users and contributes little to an overall positive user experience of the application. 3D content provides users with the greatest level of immersion and users engaged more directly with this type of content.

A consequence of MAR being a new medium it may be tempting to leverage existing maladapted content to help saturate this new platform with content. Case in point being the idea of using a QR-code metaphor, and 2D images because it is familiar and easy. We believe this degrades the user experience, firstly because it makes the MAR medium to some degree pointless and it adds little to the user experience.

Affordances related to aggregation, learning, maintenance and remediation is described in this paper. These affordances and can be used in combination with design guidelines to assist design of remediating MAR applications. These guidelines can be utilized to create a better user experience of content for the MAR platform.

Overall we conclude that the ARad application gives a fun user experience, and we argue that the true potential in remediating content to MAR lies in the user experience of 3D content.

XII. Future Research

In this study we have looked at remediated content from a newspaper to the MAR domain. Future research would be into the affordances of MAR games. The maintenance and learning affordances we expect to remain similar. However, in regards to aggregation and the remediation of game content and interaction we expect some differences. Game developers adapting games for the MAR domain may seek to remediate concepts and content from game development to the MAR application. It would be interesting to investigate if this is the case, and if so what shape and form the content and conventions take, and to identify design guidelines for game interfaces in MAR applications.

References

[1] D. Wagner and D. Schmalstieg, "First steps towards handheld augmented reality," presented at the ISWC '03 Proceedings of the 7th IEEE International Symposium on Wearable Computers, 2003, p. 127.

[2] R. Azuma, "A survey of augmented reality," *Presence-Teleoperators and Virtual Environments*, vol. 6, pp. 355–385, 1997.

[3] J. D. Bolter and R. Grusin, *Remediation - Understanding New Media*, Paperback edidtion. The MIT Press, 2000, pp. 1–290.

[4] B. Macintyre, J. D. Bolter, E. Moreno, and B. Hannigan, "Augmented reality as a new media experience," presented at the

Augmented Reality, 2001. Proceedings. IEEE and ACM International Symposium on, 2001, pp. 197–206.

[5] J. D. Bolter, M. Engberg, and B. Macintyre, "Media Studies, Mobile Augmented Reality, and Interaction Design," *interactions*, ACM, Jan-2013.

[6] M. Engberg, "Writing on the world: augmented reading environments," *Sprache und Literatur*, pp. 67–78, Feb. 2012.

[7] L. J. Rosenblum, S. K. Feiner, S. J. Julier, and J. E. Swan, "The Development of Mobile Augmented Reality," in *Expanding the Frontiers of Visual Analytics and Visualization*, no. 24, London: Springer London, 2012, pp. 431–448. http://dx.doi.org/10.1007/978-1-4471-2804-5_24

[8] J. V. Pavlik and F. Bridges, "The Emergence of Augmented Reality (AR) as a Storytelling Medium in Journalism," *Journalism & Communication Monographs*, 2013.

[9] D. Schmalstieg, T. Langlotz, and M. Billinghurst, "Augmented Reality 2.0," in *VIRTUAL REALITIES*, no. 2, Wien: Virtual Realities, 2011.

[10] J. Swan and J. Gabbard, "Survey of user-based experimentation in augmented reality," presented at the Proceedings of 1st International Conference on Virtual Reality, 2005.

[11] F. Zhou, H. B.-L. Duh, and M. Billinghurst, "Trends in augmented reality tracking, interaction and display: A review of ten years of ISMAR," presented at the ISMAR '08: Proceedings of the 7th IEEE/ACM International Symposium on Mixed and Augmented Reality, 2008, pp. 193–202.

[12] M. De Sà and E. F. Churchill, "Mobile Augmented Reality: A Design Perspective," in *Human Factors in Augmented Reality Environments*, no. 6, W. Huang, L. Alem, and M. A. Livingston, Eds. New York: Springer, 2013, pp. 139–164.

[13] V. Kaptelinin and B. Nardi, "Affordances in HCI: toward a mediated action perspective," presented at the CHI '12 Proceedings of the SIGCHI Conference on Human Factors in Computing Systems, 2012, pp. 967–976.

[14] S. Bødker and P. B. Andersen, "Complex mediation," *Human-Computer Interaction*, vol. 20, no. 4, Dec. 2005. http://dx.doi.org/10.1207/s15327051hci2004_1

[15] M. McLuhan, *Understanding Media - The extensions of man.* McGraw-Hill, 1964.

[16] G. A. Lee, U. Yang, Y. Kim, D. Jo, K.-H. Kim, J. H. Kim, and J. S. Choi, "Freeze-Set-Go interaction method for handheld mobile augmented reality environments," presented at the VRST '09: Proceedings of the 16th ACM Symposium on Virtual Reality Software and Technology, 2009.

[17] J. Iivari, "A paradigmatic analysis of information systems as a design science," *Scandinavian Journal of Information Systems*, vol. 19, no. 2, 2007.

[18] T. Gjøsæter and K. Jørgensen, "Combining Think Aloud and Comic Strip Illustration in the Study of Augmented Reality Games," presented at the NOKOBIT 2012, 2012, pp. 1–21.

[19] K. A. Ericsson and H. A. Simon, "Verbal reports as data.," *Psychological review*, vol. 87, no. 3, pp. 215–251, 1980.

[20] A. Dix, J. E. Finlay, G. D. Abowd, and R. Beale, *Human-Computer Interaction (3rd Edition)*, 3rd ed. Prentice Hall, 2003.

[21] A. Dünser and E. Hornecker, "An Observational Study of Children Interacting with an Augmented Story Book," presented at the Edutainment'07 Proceedings of the 2nd international conference on Technologies for e-learning and digital entertainment, Berlin, 2007, pp. 305–315.

[22] F. Liarokapis, L. Macan, G. Malone, G. Rebolledo-Mendez, and S. de Freitas, "A Pervasive Augmented Reality Serious Game," presented at the Games and Virtual Worlds for Serious Applications, 2009. VS-GAMES '09. Conference in, 2009, pp. 148–155.

[23] "Video in Qualitative Research," 2010.

[24] A. R. Hevner, S. T. March, J. Park, and S. Ram, "Design science in information systems research," *Mis Quarterly*, vol. 28, no. 1, pp. 75–105, 2004.

[25] W. Hürst and C. Wezel, "Gesture-based interaction via finger tracking for mobile augmented reality," *Multimedia Tools and Applications*, pp. 1–26, Jan. 2012.

[26] D. Wagner, M. Billinghurst, and D. Schmalstieg, "How real should virtual characters be?," presented at the ACE '06 Proceedings of the 2006 ACM SIGCHI international conference on Advances in computer entertainment technology, 2006.

AUTHOR

Tor Gjøsæter is with University of Bergen, Norway.

25

Online Temperature Control System

A. Ikhlef, M. Kihel, B. Boukhezzar, A. Guerroudj and N. Mansouri
Université Constantine 1, Constantine, Algérie

Abstract—**In this paper, a remote temperature control system is proposed. The physical system is controlled in real time through an Internet network. For educational purposes, the students use only a web browser to tune and test a PID controller via a shared user interface. The PID parameters are calculated using basic experimental Ziegler-Nichols tuning rules. After the hardware and software experiment description, the remote online experiment is presented and the results are given.**

Index Terms—**Electrical engineering, laboratories, online learning, PID controller, temperature system.**

I. Introduction

The great success of the Internet has put new life into distance learning. Recently, many institutions have proposed internet-based courses and e-learning educational programs. e-Learning and internet-based tools and methodologies are active research fields.

Experiments are an important part of educational programs, particularly in engineering and experimental sciences such as physics and chemistry. However, the high cost and/or complexity of the experiments prevent the students from working in optimal conditions. In these cases, remote laboratories seem to be the best solution. The main aim of these laboratories is to share the same experimental equipment between a number of users spread throughout a given geographical area. Moreover, they provide remote access to hardware experiments without the necessity of people being displaced. Furthermore, other benefits are obtained by saving time and reducing the number of days of lab inactivity.

Remote labs capitalize on the large diffusion of the Internet around the world. Essentially, they exploit web-based platforms to give an interactive interface to the users. This interface interacts directly with the remote hardware via requests-answers sent on the network. Many architectures and platforms are used to achieve these objectives. From the user side, only a simple web browser (e.g., Internet Explorer, Mozilla, Firefox) with an appropriate plug-in is necessary (e.g., JRE for applets-based interfaces, runtime engine for a LabVIEW based remote experiments).

In the domain of electrical engineering, many remote experiments and online laboratories have been designed [1-4]. Some of them deal with fundamental electronics experiments.

The aim of this paper is to present the design details and test process of an online remote temperature system control experiment. Beyond its classical aspect, this choice is motivated by the fact that the physical system is simple but relevant. Both identification and online PID controller design are included in the remote experiment.

This paper is organized as follows: Section II presents the hardware and the software environment. In Section III, the objectives, technical content and procedures for the remote experiment are presented. This is done either from the experiment designers' or administrators' side or from the remote-student side. The remote experiment test is presented in Section IV. Finally, conclusions and some perspectives are given.

II. Experiment Description

The work described in this paper was performed within the framework of the Tempus e-Sience project (Euro-Maghribeen project financed by the European community) in which the university, Constantine1, is a partner. The main objective of this project was the use of new technologies to develop e-Learning in the electrical engineering field, especially the implementation of remote labs for experimental work.

A. Hardware Part

In this part, the design of an interactive network structure is represented. This structure allows students to supervise and control the temperature in an experimental system using a PID controller. This architecture allows the students to change the desired temperature value and the PID controller parameters during the experiment from any terminal connected to the Intranet or Internet network. Different kinds of network architectures can be used for an iLab [5].

The proposed network architecture is represented in Fig. 1, and the system used for this experiment is shown in Fig. 2. It is composed of a temperature measurement system, an acquisition card and a server. All these parts are connected according to the control loop given in Fig. 3. The block diagram shows a closed loop control system. The measured temperature is compared with a reference value. The difference between these two values is processed by a PID controller, which initiates actions to drive the difference signal toward zero. Supervision and control of the system output are performed through a graphical user interface (GUI) developed in the Lab server.

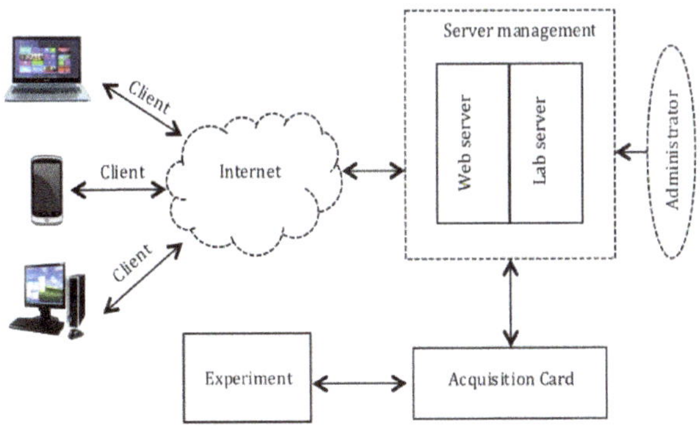

Figure 1. Online lab architecture

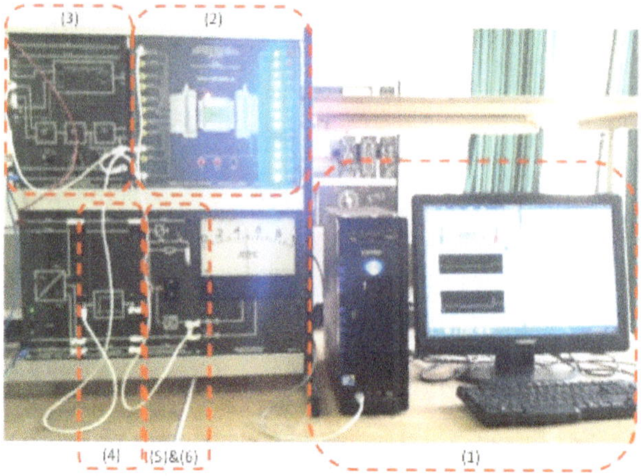

Figure 2. Online lab architecture

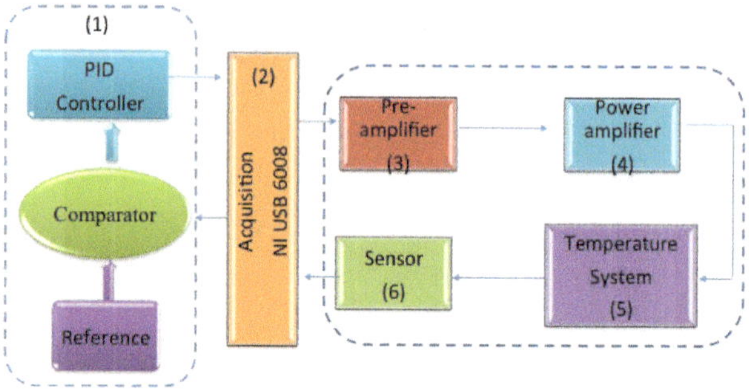

Figure 3. Experiment block diagram

B. *Software Part*

The acquisition card used in this work is a NI-6008 card, which is compatible with the LabVIEW development environment [6]. Therefore, this environment is used for data-acquisition, remote monitoring, and control of different experiment parameters via a GUI.

We noticed that the LabVIEW environment can control experiments connected to the Intranet or the Internet network in real time using LabVIEW webserver. It also offers the possibility of controlling experiments remotely from a web browser. In this case, the only required software for the remote clients is the LabVIEW Runtime Engine, which can be downloaded freely from the National Instruments website.

As exhibited in Fig. 1, this environment has a lab server that manages the experiment and a web server that handles client access. Real time remote monitoring and control are effectuated by using the LabVIEW web-publishing tool. In this case, the developed VI (Virtual instrument) is accessible on the local or the Internet network via web browsers. To access the VI remotely, the IP address of the server and the name of the shared VI should be known by the remote client. At the same time, the published VI must always be running on the Lab server. The remote user regains control of the experiment only after requesting and receiving acceptance from the server. During the experiment, the students can start and stop the program execution whenever they want. They can change the desired temperature value and the PID controller parameters at any time. Moreover, it is also possible to store all the experiment results data from the graph in "spreadsheet" format.

III. Temperature Control Remote Experiment

The temperature subsystem is involved in many everyday life systems such as air conditioning units, irons, electric ovens and many other industrial systems [7]. The proposed remote experiment concerns a temperature system control. The heat is generated by thermal resistance.

A. *Remote Experiment Objectives*

The aim of the experiment can be summarized in three objectives, which constitute the student work steps:

- Temperature system identification from experimental data.
- Open-loop Ziegler-Nichols PID tuning and test of the temperature system.
- Closed-loop Ziegler-Nichols PID tuning and test of the temperature system.

B. *Preparing the Practical Work*

Before the physical remote experiment, the students are provided with experimental data obtained from a step response of the temperature system. The students are asked to plot the system time answer and to confirm its S-shape. The expected curve is shown in fig. 4.

Once the curve is obtained and plotted, the students should identify the parameters of a first-order delayed system: static gain K, time constant τ and delay τ_d. The transfer function of the model is the following form [8]:

$$G(s) = \frac{Ke^{-s\tau_d}}{1+\tau s} \tag{1}$$

C. *Remote Experiment Content*

The experiment content can be divided into two parts: Open-loop and closed-loop PID tuning and test.

1) Open-loopZiegler-Nichols PID tuning and test:

Once the experimental model is obtained during the student preparation, a PID controller is directly obtained using Ziegler-Nichols rule for an open-loop identified system [8]. The parameters K_p, T_i and T_d are a function of the answer curve slope *a* and the model delayτ_d.

From the users' side, the remote students have an interface to directly introduce the PID parameters and to make a real remote test on the physical system via a web browser. This interface allows the students to follow the regulated physical system response on the web browser in real time. The student has to do an analysis to adjust the true parameters if a wrong result is obtained as saturation or overshoot.

The student interface is shown in Fig 5. For both open and closed loop cases, the implemented controller on the measurement server using LabVIEW has the following transfer function

$$C(s) = K_p\left(1 + \frac{1}{T_i s} + T_d s\right) \tag{2}$$

2) Closed losed-loop Ziegler-Nichols PID tuning and test:

For the closed loop case, the process to be controlled is put in a closed loop with a variable gain K_0. As shown in Fig 6, the gain is tuned until the system starts to oscillate. The value of K_0 causing the start of oscillation and the oscillations period T_0 are then used to tune the three PID parameters [8].

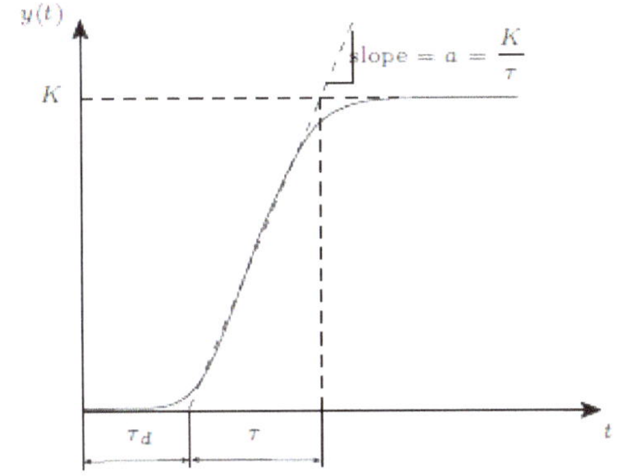

Figure 4. Open-loop answer of a delayed first order system

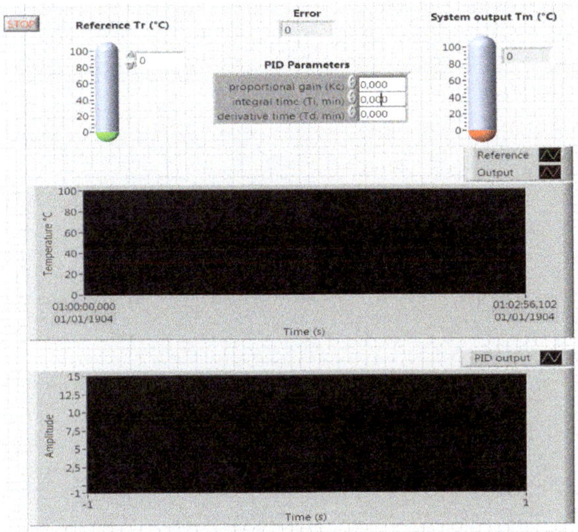

Figure 5. Student interface to tune the PID controller.

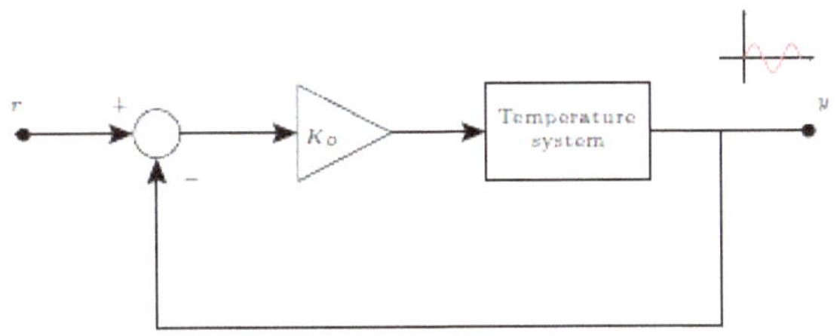

Figure 6. Oscillating closed-loop response with proportional action.

IV. Experiment Results

To illustrate the performance of the experimental lab described in this paper, the effect results of different PID parameters on the system response have been investigated. Fig. 7and Fig. 8 represent the temperature system responses when the PID parameters are tuned from the Zeigler-Nichols open loop (ZNOL) and closed loop (ZNCL) methods, respectively. The interface that controls the system via a web browser is shown in Fig. 9.

V. Conclusion

An online remote temperature control system is proposed for educational purposes. A simple web browser is used by the students to access the physical experiment. Experimental Ziegler-Nichols tuning rules for a PID controller were tested in real time. The control system was successfully tested online. As a perspective, an iLab shared architecture will be used to share the experiment on the Internet network.

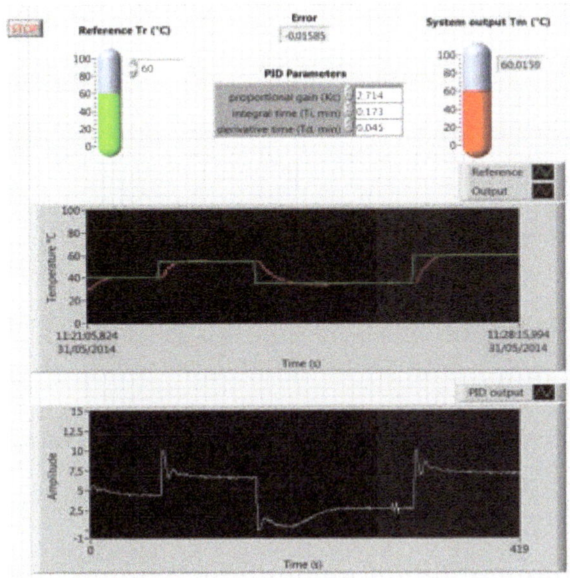

Figure 7. Temperature system response with PID parameters tuned from ZNOL.

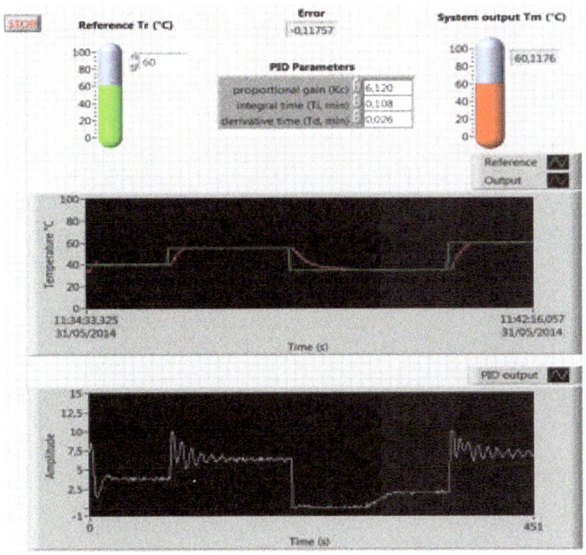

Figure 8. Temperature system response with PID parameters tuned from ZNCL

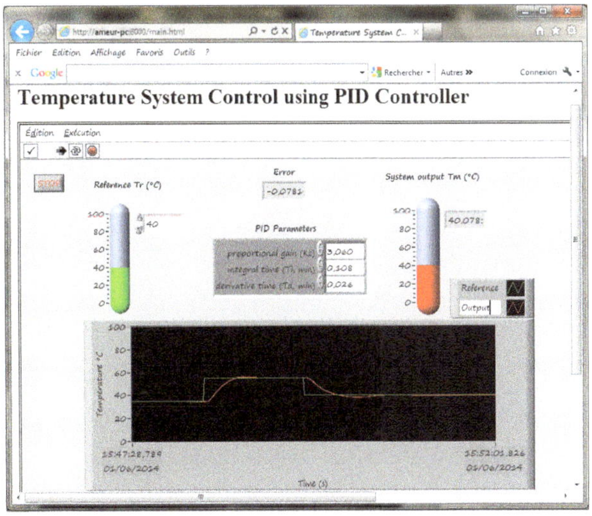

Figure 9. Graphical user interface as displayed on remote client web browser.

References

[1] T. A. Fjeldly and M. S. Shur, LAB ON THE WEB Running Real Electronics Experiments via the Internet. Wiley-interscience, 2003.

[2] A. Leva and F. Donida, "A remote laboratory on PID autotuning," Proc. 17th IFAC World Congress. Korea, pp. 8147-8152, July 2008.

[3] B. Aktan, C.A. Bohus, L.A. Crowl, and M.H. Shor,"Distance learning applied to control engineering laboratories," IEEE Transactions on Education, Vol. 39, pp. 320–326, 1996.

[4] P. H. Wu and C. H. Kho, "The design and implementation of a remote automatic control laboratory: using PID control as an example," Tamkang Journal of Sience and Engineering, Vol. 11, pp. 219-228, 2008.

[5] V. Judson et al, "The ilab shared architecture: a web services infrastructure to build communities of internet accessible laboratories," Proceeding of IEEE, Vol. 96, No. 6, pp. 930-951, June 2008.

[6] LabVIEW software platform, URL: http://www.ni.com/labview.

[7] P. Zhang, Advanced Industrial Control Technology. 2010.

[8] G. F. Franklin, D. J. Powell, and A. Emami-Naeini, Feedback Control of Dynamic Systems. Prentice Hall, 2002

Authors

A. Ikhlef is with Universités de Constantine, Laboratoire d'Automatique et de Robotique, Faculté des sciences de la technologie, Université Constantine 1, Algérie (e-mail: ameikhlef@yahoo.fr).

M. Kihel is with Universités de Constantine, Laboratoire Micro-systèmes et Instrumentation, Faculté des sciences de la technologie Université Constantine 1, Algérie (e-mail: kihel_mouloud@hotmail.com).

B. Boukhezzar is with Universités de Constantine, Laboratoire d'Automatique et de Robotique, Faculté des sciences de la technologie, Université Constantine 1, Algérie (e-mail: b_boukhezzar@hotmail.com).

A. Guerroudj is with Universités de Constantine, Laboratoire d'Automatique et de Robotique, Faculté des sciences de la technologie, Université Constantine 1, Algérie (e-mail: abdou.guerroudj@yahoo.fr).

N. Mansouri is with Universités de Constantine, Laboratoire d'Automatique et de Robotique, Faculté des sciences de la technologie, Université Constantine 1, Algérie (e-mail: nor_mansouri@yahoo.fr).

Virtual ATM: A Low Cost Secured Alternative to Conventional Mobile Banking

Shabnam Shahreen Sifat and Ali Shihab Sabbir
Independent University, Bangladesh, Dhaka, Bangladesh

***Abstract*—Mobile banking has become immensely popular among customers as a suitable method for money transaction. Banks are assertively adopting this mode. It is playing a vital role in availing banking services in remote areas where placing branch or ATM booth is not economically feasible. Mobile banking is so far the easiest way of expanding banking coverage. But there is huge possibility of fraud in case of mobile banking as authentication & all type of transaction information is provided via mobile phone including pin. So mobile banking is not intended to be used for big transactions where ATM transactions are suitable because of its two layer authentication system. In this paper, a new system is introduced that provides ATM service without traditional booths but two layer authentications with a tiny OS independent device has been introduced named VATM. This paper discusses how this system works using a low cost device made of micro-controller & CDMA module for communicating with bank for authentication which is used as an alternate of Automated teller machine for providing two layer authentications.**

***Index Terms*—Agent assisted Banking, Electronic Banking, Mobile Banking, Secured Banking.**

I. Introduction

A sound & efficient banking system drives economic growth of a country. Especially in developing countries taking banking services to the doorsteps of multitude is a challenge. For these reasons banks are adopting technology-driven electronic banking services for expanding coverage. However World Bank research shows 2.5 billion of world's adult population is still unbanked, which is about half of the total population [22]. According to the International Telecommunication Union worldwide mobile subscriptions reached nearly 7 billion which is equivalent to 95.5 percent of the world population by May 2014 [21]. So mobile banking has huge potential of being means of providing financial services to this vast unbanked population.

Juniper Research estimated 590million mobile banking users worldwide in 2013, where almost 60% of this total number belongs to Chinese banks [2]. According to Forrester Research (May 2014), there were 51 million mobile banking customers in Europe in 2013 [5]. And Javelin Research shows 95 million mobile bankers in US in 2013 and it is also projected that it will reach 149 million by the end of 2018 [6].

Kenya is one of the world leaders in mobile banking section. According to Communications Commission of Kenya (CCK), April 2014, there are 26 million mobile banking subscribers in Kenya which is 83% of its total

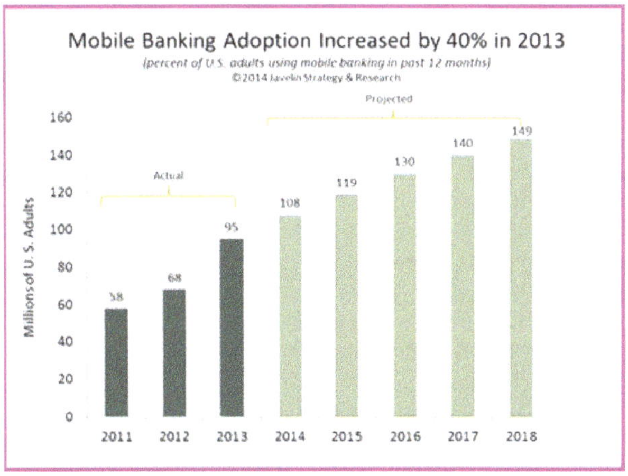

Figure 1. Growth of mobile banking in U.S. [6]

mobile phone subscribers. Safaricom's M-PESA is the biggest mobile banking provider in Kenya and Tanzania. According to Mobile Transaction, one quarter of the 44 billion dollar economy in Kenya runs via M-PESA. M-PESA is now expanded to Afghanistan, South Africa, India & Eastern Europe [4].

In Bangladesh, almost 87 percent of the total population of 165 million is still outside the network of conventional banking[11]. Majority of these unbanked population lives in rural areas. According to Bangladesh Telecommunication Regulatory Commission (BTRC), total number of mobile phone subscribers in Bangladesh reached 114.8 million by the end of January 2014 which is almost 70 percent of total population [3]. So mobile phone can become an effective and convenient vehicle of expanding banking services in remote areas through mobile banking. In Bangladesh among 28 MFS (Mobile Financial Service) licensed banks, 20 are operational.

BKash is one of the leading MFS in Bangladesh. It is a joint venture between Brac Bank Ltd. and Money in Motion LLC, USA. Bkash was launched in mid 2011 and grew to 11 million registered accounts with 80,000 agents by the end of 2013. It provides Cash in/out, Payments, International Remittances, P2P payments, Bill Pay [1].

Dutch Bangla Bank Ltd -DBBL mobile banking is the mobile financial service of Dutch Bangla Bank Ltd. which was also launched in 2011. Cash in/ Out, International Remittances, Salary Disbursement, P2P payments, Mobile top-off, Bill Pay services are provided by DBBL mobile banking. Islamic Bank Ltd's MCash, United Commercial Bank Ltd's UCash, Trust Bank Ltd's mobile money, ONE Bank Ltd's OK mobile banking are also pioneers in MFS in Bangladesh. According to Bangladesh bank, number of mobile banking subscriber reached at 16.7 million by the end of June 2014 [7].

Though mobile banking is attracting customers, researches show that, it is yet to earn full confidence from the multitude [21]. The lack of confidence is attributable to security gaps of the system. Various fraud incidents, which got wide media coverage, are making them skeptical about adopting mobile banking.

II. Literature Review

Research shows that low-income & low-literate people of developing countries can adopt mobile banking services [10]. Research ICT Africa found that, mobile banking can become one of the best ways of reaching unbanked population of Africa where main barrier is the distance to banking services or ATMs [8]. In Nigeria mobile banking is customer's second choice after ATMs for money transaction [9]. But now mobile banking services like Bkash, UCash are allowing customers to deposit & withdraw instantly from their agents [11]. This model is getting more popular now-a- days as it's bringing the bank next to the customer's door. Mobile money transfer like Bkash has a good prospect in developing country like Bangladesh and can contribute to socio-economic development in near future [12].

Shaik S. Ahamad et al. proposed a m-banking solution which ensures secured end to end communication channel and end to end application security from the UICC to the Remote Bank Server via Mobile Equipment by generating all digital signatures in a tamper proof hardware to ensure end to end security at the communication layer and at the application layer [13].

Though mobile banking has huge prospect, adoption rate of mobile banking is still very low [19]. One of the major reasons behind is security risks like theft of mobile device & risk of being hacked [20].

Security aspects of mobile banking have been addressed in the following papers:

Aditya K. Tiwari et al. introduced biometric authentication system that enables face authentication in which the user sends his/her encrypted image to the bank server for verification. After banks approval the user's face becomes one of the keys to access the account [14].

H. Elbehiery et al. developed an android application for mobile banking with the help of face recognition and ciphering algorithms for increasing security [15].

Bindu Rama Rao invented a system for agent assisted mobile banking. This system facilitates mobile fund transfer, deposit & withdraws. These operations are conducted by one or more agents needed. In this system the user connects to the bank server with a fund transfer request using his/her handset and receives a one-time token. Then the user approaches to the respective agents to complete the transaction [16].

Manoj & Satish propossed a cost effective symmetric cryptography based SMS banking system where message is encrypted using AES & MD5 hash algroithm. Messages from user mobile are¬ sent in encrypted format which are decrypted by bank server. Bank will also send encrypted secured messages to the customers which are decrypted in his handset [17].

Karun presented a GPRS protocol to establish secure connection between bank servers & mobile devices with two major components. First is basic client server handshake and second is transfer of SPG message using created secure tunnel & exchanged cipher suites as well [18].

III. Our System- VATM

A. *Overview:*

People having access to ATMs are more comfortable using them due to their secured authentication process. An automated teller machine identifies its customer not only with ATM card but also with pin number provided by the customer. So for transaction it requires two layer of authentication. One is customer must insert his/her valid ATM card in ATM machine and the other is the pin number, which is to be provided through the machine. Thus the security is attributed from two independent sources.

In case of mobile banking customers are identified with mobile number & pin. For this all the information provided to the bank use customer's mobile phone which questions the security of the transaction. That's why customers are facing different frauds. Frauds are cloning customers SIM card or stealing mobile phones and accessing their accounts. They can track customer's pin number by installing different sniffing tools.

But avoiding mobile banking is not a solution to this problem because rapid expansion of banking coverage is easily possible through mobile banking because placing an ATM booth depends on the usage & traffic because it requires huge investment as ATM machines are expensive, it requires air conditioned infrastructure, installation of network and maintenance, regular cash replenishment & security measures. So placing ATM in rural areas is not a good option.

To acquire ATM service a customer needs to have an ATM card of the bank he/she deals with. Then he/she needs to find an ATM booth which supports that bank. Inserting ATM card & providing PIN in ATM machine he/she get access to his/her account. Then transaction request is made. Machine provides the requested amount to the customer & also returns the ATM card. VATM will provide the same service but in different procedure. System components are given below:

Mobile phone: In VATM process ATM card is replaced by mobile phone which is used to send primary withdraw request from customer's mobile to bank.

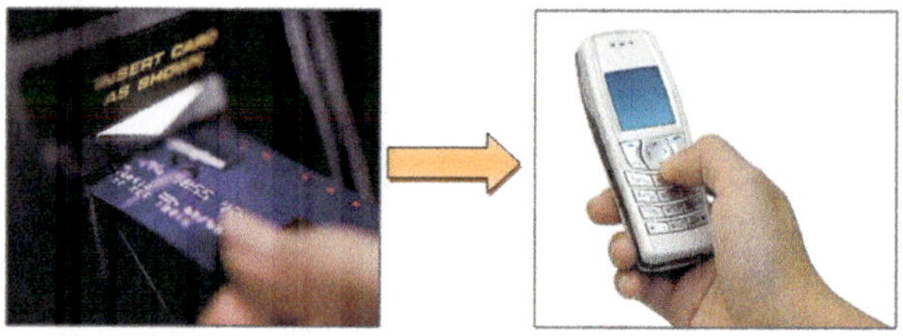

Figure 2. ATM card replaced by mobile phone

VATM device: ATM machine is replaced by VATM device. VATM device is used for sending PIN to bank's server for authentication.

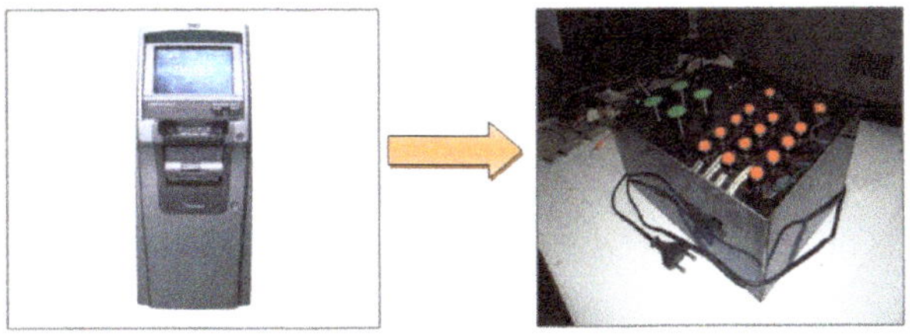

Figure 3. ATM replaced by VATM device

Bank's agent: Unlike available ATM service money will be provided to customer by bank's agent.

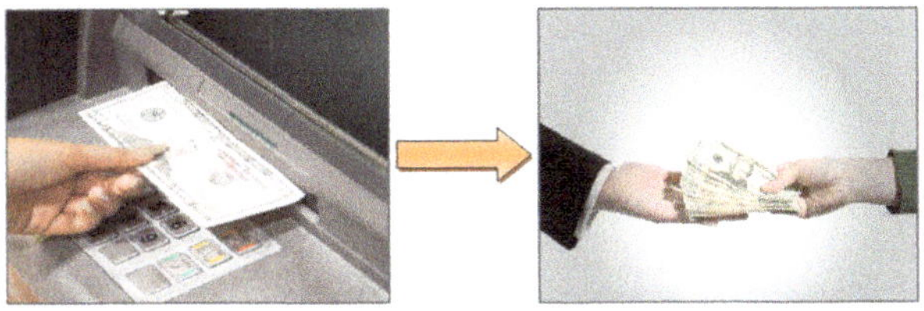

Figure 4. Automated teller replaced by agent

B. *Withdraw process:*

VATM withdraw process has four major steps.

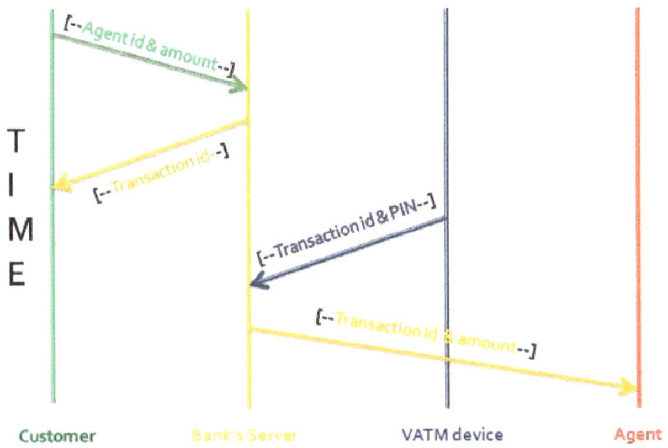

Figure 5. Schematic diagram of message transaction

Description of each process is given below:

Step 1: First customer needs to find nearby authorized agent of the bank who enables VATM service. Every bank agent will have a unique ID. Customer will have to type agent's ID and amount he/she wants withdraw in the message option of his/her mobile, then send it to the bank's server. Bank will recognize that customer and the agent by his/her mobile number which is used to send request for withdrawal. Authentication will be done in bank's server. If the customer have sufficient balance for withdrawal server will generate a nonce transaction ID for that request.

Step 2: In this step customer receives a SMS from bank containing transaction ID of the request he made. But in case of invalid request customer will receive a SMS containing the reason. Like insufficient balance, wrong agent id etc.

Step 3: After getting transaction ID customer will have to enter that transaction ID & his/her PIN in VATM device. Through VATM device transaction ID & PIN goes to the bank's server in form of a SMS for authentication.

Step 4: After authenticating the entered PIN & transaction ID bank server sends a SMS to the agent's mobile containing transaction ID & amount to pay. Agent verifies the customer with the transaction ID and pays the amount to the customer. In case of wrong PIN customer will be notified through a SMS.

C. *Flow Chart:*

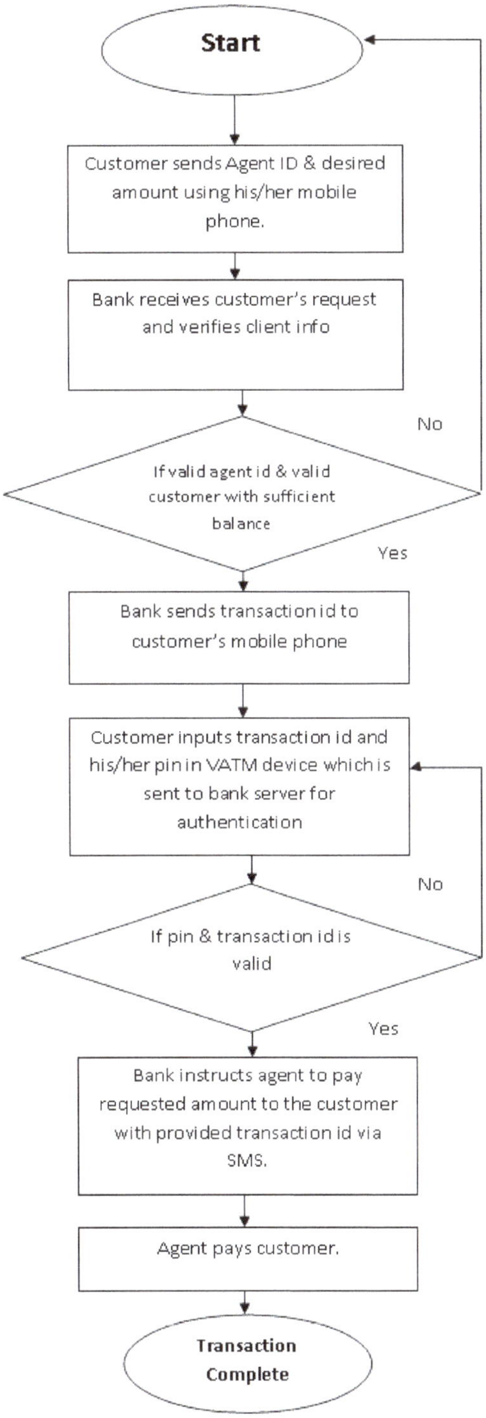

Figure 6. System flowchart

D. System Architecture

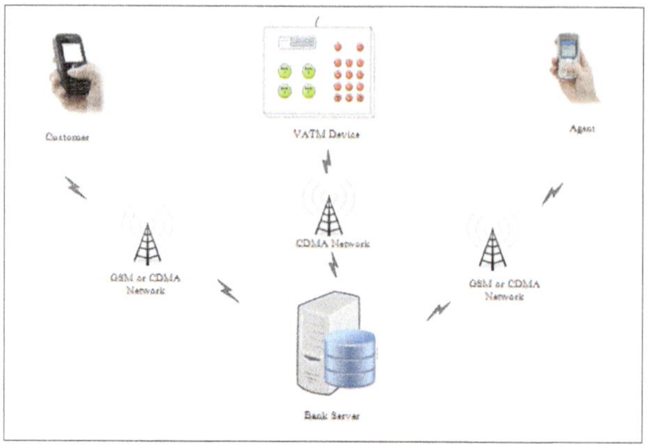

Figure 7. System Architecture

E. Security:

Proposed system tries to address the security flaws of existing mobile banking services. This system follows the two layer authentication process of automated teller machine. In first layer withdraw request is sent from customer's mobile phone. Bank identifies valid customer with his/her phone number and replies back with an one time use transaction ID which will be valid for a certain period of time. Then in second layer customer inserts the transaction id & pin number in an OS independent device called VATM which is placed in agent's shop. Than bank again verifies customer with his/her pin & identifies the request by the received transaction ID. If everything matches bank ask the agent to pay the customer who holds the transaction ID. In this process customer's mobile phone is only used for sending withdraw request & the VATM device is responsible for transmitting pin. This process protects the customer's secret pin from being sniffed as it is never sent via any device which can be programmed or have operating system.

This system also ensures data confidentiality over network as the device transmits the pin using CDMA network which is more secured than GSM network. So there is no chance of being intercepted by intruders.

F. Device Description:

The VATM device includes the following units:

1. On board interface to communicate with customer
2. A small LCD display
3. A microcontroller
4. CDMA Module

Description of each unit given below:

1. On board interface to communicate with customer: A user friendly interface has been created so that user can easily understand how this machine can be used. There are total 18 buttons in this machine. Ten of them are labeled as 0, 1, 2, 3, 4, 5, 6, 7, 8, and 9. These are basically the digit buttons of this machine. Label ← is used for backspace button & ↓ is used for enter button. Left button from the top is reset button & right one is on/off button. All these buttons red colored. There are four green colored buttons. These buttons stands for each bank. Main function of these buttons is sending data provided by the user to the bank which is labeled on button.

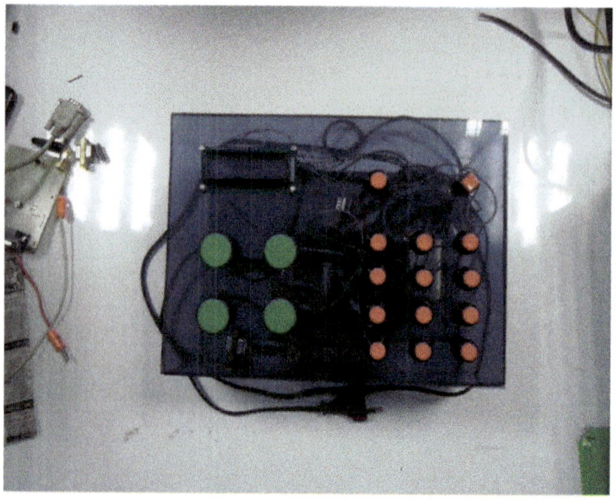

Figure 8. VATM device Top view

Figure 9. VATM device side View

2. A small LCD display: There is a LCD display in the device. Prompt for transaction id & pin is shown in the screen. User can see what he/she is entering for those fields. First field is for transaction id & second for PIN. After entering transaction id user needs to press enter button to enter pin. But "*" sign is shown for each digit entered in the PIN field. And after pressing send button a message "Done" is shown on the LCD to ensure successful completion. After that the machine again gets ready to process another task and prompts for transaction id & pin.

3. A microcontroller: In this device we have used an 8-bit micro-controller, PIC16F877A which is connected with all buttons, LCD display & CDMA module.

4. CDMA Module: CDMA module is used in this system for communicating with bank's server via sending SMS from this device.

IV. ADVANTAGES OVER TRADITIONAL ATMS & M-BANKING SERVICES

- Cost of VATM device is significantly low.
- No need to think about traffic & usage for placing VATM point so banking service can be expanded radically and even in rural areas.

- Not need to create extra account like Bkash because this service can be availed with existing bank accounts.
- As the function of VATM is to send a SMS to the bank server so multiple banks can be served with one machine.
- No risk of sniffers as pin is never transmitted via mobile phone. This VATM device makes the process more secured by providing two layered authentication like ATM.
- No matter if the mobile phone is lost.
- Installation of network & maintenance is not required as VATM device will use available CDMA network.
- Placing VATM device does not require any air conditioned room. It is a tiny device which can be placed anywhere.
- No need to worry for regular money transfer as money will be provided by bank's agent.
- No need to invest for providing security (e.g. security guard, closed circuit camera).
- Eco friendly as it consumes negligible amount of electricity.

These are the reasons why VATM process could be a better choice for providing banking services. Every step is unambiguous & can easily be understood by a general customer. VATM process can easily be operated compared to ATM machine. So this process will attract general people more. VATM process can work as a tool to expand banking service in remote areas.

V. CONCLUSION

The proposed VATM model provides mobile banking service following the two layer customer authentication process of ATM to make the transaction more secured. VATM service can be provided easily with available CDMA network and simple and compact OS independent device. If we think in bank's prospect VATM would be the most cost effective, transparent & efficient way of providing banking service. In customer's point of view VATM would be easier, affordable & convenient way of availing banking service.

REFERENCES

[1] Ahmed Jamal et al, Financial Stability Report 2013, Financial Stability Department, Bangladesh Bank, Dhaka, Bangladesh, Issue 4, June 2014

Windsor Holden, Press Release: Mobile Banking Users to Exceed 1 billion in 2017, Representing 15% of Global Mobile Subscribers, [Online].

[2] BTRC, Mobile Phone Subscribers in Bangladesh January 2014, [Online].

20143. BTRC, Mobile Phone Subscribers in Bangladesh January 2014, [Online].

www.btrc.gov.bd/content/mobile-phone-subscribers-bangladesh-january-2014 , last access 18 October 2014

Lucy Harris, 7 April 2014, M-PESA: How Kenya took the lead in mobile money, [Online]. Available:

http://www.mobiletransaction.org/m-pesa-kenya-the-lead-in-mobile-money/ , last access 18 October 2014

[3] Forrester, 15 May 2014, Forrester: Tablet Banking In Europe To Overtake Mobile Banking By 2016, [Online].

[4] Javelin Strategy & Research, 19 Mar 2014, Mobile Device Boom Led to 74,000 New Mobile Bankers per Day in 2013, [Online]. Available.

[5] GSMA: Mobile for Development Impact, Country overview: Bangladesh, GSMA Intelligence, August 2014

[6] Alex Comninos, Steve Esselaar, Ali Ndiwalana & Christoph Stork, Towards Evidence-based ICT Policy and Regulation M-banking the Unbanked, Research ICT Africa, Volume ONE 2008 Policy Paper 4

[7] Dr. Edwin M. Agwu & Dr. Adele-Louise Carter, Mobile Phone Banking In Nigeria: Benefits, Problems and Prospects, in Int. Journal of Business and Commerce, Vol. 3, No.6: Feb 2014[50-70]

[8] Indrani Medhi et al, Mobile-Banking Adoption and Usage by Low Literate, Low-Income Users in the Developing World, in Third Int. Conf., IDGD 2009, Held as Part of HCI International 2009, San Diego, CA, USA, July 19-24, 2009, pp 485-494. doi: 10.1007/978-3-642-02767-3_54

[9] hypertag, Banking to Unbaked, [Online].

[10] Md. Mehedi Hasan & Md. Rayhanul Islam, Assessing Service Quality Of Mobile Money Transfer In Bangladesh:A Case Study On bKash, Int. Journal of Innovative research & development, Vol 2, Issue 7, July 2013

[11] S.S. Ahamad, A Secure and Reliable Mobile Banking Framework, in Proceedings of the 48th Annual Convention of Computer Society of India- Vol II, 2014, pp 741-748

Tiwari et al, Biometric Authentication for Mobile Banking Security (May 17, 2014). Available at SSRN:

[12] Hussam Elbehiery, M. Saied Abdelwahab, Ghada Abdelhady. Secure Mobile Banking. in J Am Sci2014;10(1):135-142](ISSN: 1545-1003)

[13] Bindu Rama Rao, System for agent assisted mobile funds transfer and mobile banking US 8639602 B2, Jan 28, 2014

[14] Manoj V. Bramhe & Dr. Satish R.Devane, Cost effective and secured mobile banking solution using symmetric cryptography, in Int Jr of Advanced Computer Engineering and Architecture, Vol. 1, No. 1, June 2011

[15] Karun Madan, Secure GPRS solution for mobile banking: an efficient security protocol, in Int. J. of Electrical, Electronics and Mechanical Controls, ISSN (Online): 2219-7501, Vol 2 Issue 1, Jan 2013

[16] Chian-Son Yu, Factors affecting individuals to adopt mobile banking: empirical evidence from the UTAUT model, Journal of Electronic Commerce Research, VOL 13, NO 2, 2012

[17] Mohammad Rokibul Kabir, Factors Influencing the Usage of Mobile Banking: Incident from a Developing Country, World Review of Business Research, Vol. 3. No. 3. July 2013 Issue. pp 96–114

[18] International Telecommunication Union, The World in 2014: ICT Facts & Figures, [Onilne].

[19] Asli Demirguc-Kunt & Leora Klapper, Measuring Financial Inclusion, The World Bank, Development Research Group, Finance and Private Sector Development Team, April 2012, 17, 2009. Published as resubmitted by the authors on May, 16, 2009.

[20] Gimun Kim et al., Understanding dynamics between initial trust and usage intentions of mobile banking, Information Systems Journal, Volume 19, Issue 3, pages 283–311, 18 DEC 2007.

AUTHORS

Shabnam Shahreen Sifat is with the Dept. of Computer Science & Engineering, Independent University, Bangladesh, Dhaka, Bangladesh.(e-mail: sifu_05@yahoo.com).

Dr. Ali Shihab Sabbir, is with Dept. of Computer Science & Engineering, Independent University, Bangladesh, Dhaka, Bangladesh. (e-mail: alisabbir@gmail.com).

Investigating Students' Perceptions on Mobile Learning Services

Mohammed Amin Almaiah, Masita @ Masila Abdul Jalil
University Malaysia Terengganu, Terengganu, Malaysia

Abstract—**M-learning is a form of learning, which has similarities and differences with the electronic learning (e-learning). It is mainly based on the use of the mobile wireless technologies that allow for learners to easily access learning materials anytime he desires and anywhere, whether on campus or off campus. Therefore, this creates a new flexible learning environment in the context of different learning settings. Students' perception of such technology is one of the most important factors for successful adoption of m-learning in the higher education environment. This study is conducted to investigate the perceptions of students in University Malaysia Terengganu (UMT) to move towards applying m-learning in their studies by using their mobile devices and to explore their expectations on mobile learning services. A total number of 91 undergraduate students majoring in computer science participated in the study. The findings show that the students have positive perception towards mobile learning and would like to use their mobile devices for both learning and administrative services.**

Index Terms—**Mobile learning (M-learning), Electronic learning (E-learning), mobile wireless, students' perceptions.**

I. Introduction

The uses of mobile devices and related technologies have increased significantly in the recent years because of their capabilities to be used in many fields. The use of devices such as mobile phones, smart phones and PDA have found their way into the learning and educational sector as well. In addition, students and teachers in the universities and education institutions around the world are increasingly using the mobile technology to access learning materials, internet access and facilitate learning in new and innovative ways [19]. M-learning is the next form of e-learning that support learning anywhere and anytime which depends on communication technologies such as GSM, WAP, GPRS and Bluetooth. Over the past few years the number of studies has increased for the adoption of m-learning in many countries such as USA, Asia, Britain, Scandinavia, and Australia [3]. University of Florida supports the idea of using m-learning through providing access to fast wireless network using mobile devices on campus [6]. Thus, the utilization of m-Learning technology offers benefits to both students and educational institutions involved [10]. In addition, Pointed out that more focus should be directed towards the importance of m-learning and its uses among universities students because it provides many of services at any time, any place. Therefore, the higher education institutions must plan in future to provide a more flexible learning environment to meet the needs of new generations of students who are increasingly attracted towards mobile technology.

Despite there are many studies have been conducted to investigate the use of m-learning in higher education institutions, but it is still limited. However, the rate of adoption of the higher education institutions for the use of m-learning is still slow; this means that even though it has many advantages, there are many obstacles that limit their acceptance within the higher education environment [1]. One of the main reasons that are attributed to the poor usage of m-learning technology is the lack of sufficient studies that explore the factors that influence m-learning acceptance in the higher education institutions [22]. Therefore, this study is conducted to investigate the students' perceptions and expectations on m-learning services in the higher education institutions.

II. Problem Statement

The popularity of mobile devices is increasing significantly day by day, as many learners are using mobile technology in their learning environment [16]. M-learning has many features such as flexibility of learning anytime and anywhere which have brought new changes in learning and education environment. Therefore, this feature enables students to take advantage of their free time while they are outside the classroom to complete their study and homework [24]. In addition, students who are waiting for their flight at the airport, they can use wireless mobile devices such as smart phone, PDA and Tablet PC to access lecture materials or download an assignment or interact with their instructors or friends.

Despite m-learning has been developed fast, there is need to investigate the elements that have influence m-learning acceptance among students in higher education institutions [14, 21]. Without considering the importance of m-learning acceptance among students and explore students' readiness levels to use mobile learning can cause ineffective use of m- learning devices. Therefore, study of the perceptions of students for using m-learning may help the success of the adoption of m-learning in higher education environment. Chen et al. [7] pointed out that a better understanding of the students' requirements will help the decision maker to adopt m-learning successfully.

Few empirical research studies on the use of m- learning in higher education institutions have been reported [39]. Some of these studies have suggested some of the factors that influence m-learning acceptance in higher education institutions. However, there is still gap in the field of m-learning adoption; where many researchers have called for further investigation and research in the field of adoption of m-learning in higher education institutions [21, 11, 14, 15]. Further research is required to investigate the students' perceptions of the application of mobile devices and m-learning services in higher education institutions. Hence, the study here, aims to explore the students' perceptions of the application of m-learning services in higher education environments.

III. Related Works

Several of studies have conducted the adoption of m-learning in many countries such as USA, Britain, Scandinavia, and Australia [1]. Andrew and others [3] discussed groups of directions in an effort to investigate the use of the m-learning in the higher education including: fostering interactivity on-campus using mobile devices, investigation the needs of students and strategies for low-cost m-learning use.

Pointed out that more focus should be directed towards the importance of m-learning and its usage among students in universities because it provides many of the services at any time and any place. Korucu and Alkan studied the relation and differences between m-learning and e-learning and the use of m-learning in the higher education in turkey. The results showed that the use of m-learning technologies has become of the most important technologies that support learning because it offers learning any time, any place. In addition, m-learning will contribute greatly to the improvement of teaching and learning in the future. Another study conducted by Haag [9], focused on m-learning delivery courses to provide the students with an alternative platform to continue their learning after leaving the classroom. The findings showed that seventy percent of the participants preferred m-learning courses version more than e-learning courses.

In addition, yusup discussed the status of iPad used as a device for learning among primary school teachers in Malaysia. A survey was conducted in which 93 teachers participated. The teachers were provided with iPads to help them in the program of online distance-learning. The results indicated that the majority of teachers' prefered to use iPad as compared to laptops during the teaching and learning process and trust that iPad is a very important learning device to share information among colleagues.

IV. M-learning Vs. E-learning

Despite E-learning and M-learning have similar elements, but there are some differences in terms of technology, learner access and mode of communication. TABLE (1) compares the aspects of E-learning and M-learning which were derived from a literature review of M-learning [4, 13, 17].

TABLE I.
"Comparison between E-learning and M-learning"

Features	E-learning	M-learning
protocol	Web-Based	WAP-Based
Network	Wired	Wireless
Accessibility	Anywhere	Anywhere and Any-time
Connectivity	Intranet or Internet	Mobile Networks, GSM, GPRS, UMTS and CDMA.
Devices	PC or Laptop	Mobile Phone, Smart Phone and PDA
Instructor to Student communication	Asynchronous Delayed Communication	Synchronous Instant Communication

V. M-learning Services for Higher Education Institutions:

M-learning is considered to be the next step of e-learning by the use of wireless mobile devices and other communication technologies to enable instructors and learners to conduct learning process anytime and anyplace. There are two main services that m-learning offered which are: (1) learning materials services such as content materials (2) administrative services such as registration and schedule students as shown Figure 1.

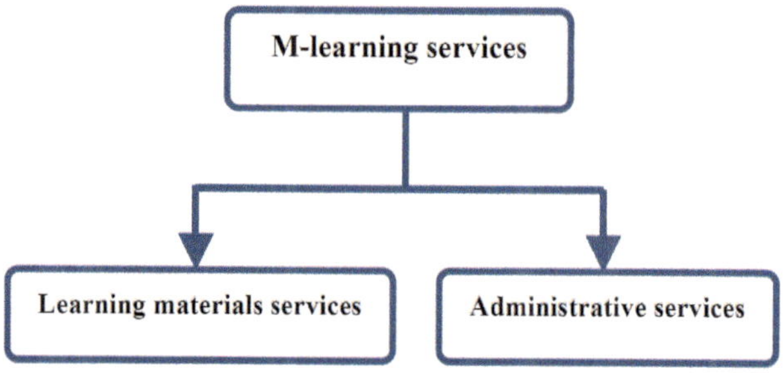

Figure 1. "M-learning Services Types"

Several M-learning studies in higher education institutions examined the manner in which wireless mobile devices are implemented in teaching and learning methods of various universities. Alzaza and Yaakub [2] conducted a study to investigate the importance of the use of m-learning services among Malaysian higher education students. The results showed that the courses registration and result exams are the highest rank and followed by library services, schedule services, and admission services.

Botzer and Yerushalmy [5] conducted a project to investigate the relevance of cellular application called "math4mobile". Researchers used this project "math4mobile" to understand the opportunity of adopting personal and mobile technologies for learning functions, particularly the benefit of mobile phones for facilitating mathematical learning. The project contained group effort activities such as making use of cellular video camera to capture certain images, making use of MMS (multimedia message) for sharing video clips among the participants, making use of mathematical applets to create graphs which provide representation of information and making use of SMS (short message system) to send and exchange verbal messages and mathematical objects. The study found that the positive engagement of the students in the project and these devices able to enhance the learning process.

VI. Methodology

A questionnaire was designed to investigate the perceptions of the students towards applying m-learning. The instrument was adapted from Trifonova, Georieva and Ronchetti which contains four sections. The first section contains general information to collect data about users' demographic such as gender, age and year of study. The second section contains multiple choice questions to find out about the availability of mobile devices, accessing the internet by mobile devices and students' knowledge and experience with mobile technology media. In the third section, there are developed 11 statements with a five-point Likert scale ranged from 1-Strongly Disagree to 5-Strongly Agree designed to explore the students' perceptions towards applying m-learning. Finally, the fourth section based on the scales 1 to 3 where 1-Useful, 2-Neutral and 3-Not Useful to explore the students' expectations on m-learning services.

The questionnaire was distributed to the first, second and third year Computer Science students, School of Informatics and Applied Mathematics, University Malaysia Terengganu (UMT). A total number of 100 responses were obtained. Nine questionnaires were discarded due to being incomplete answers.

VII. Data Analysis and Results

A. *Respondents profile*

As shown in Table II, 50.5% of the respondents were male and 49.5 were female. The majority of respondents 81.3% were aged between 18-22 years while 22-26 years were 18.7%. In terms of year of study, most of the respondents 38.5% were the second year followed by first year 35.2% and then third year 26.4%. This indicates that the results represent perceptions and expectations of different levels of students.

B. *Availability of Mobile Devices*

As shown in Figure 2 the results indicate that the majority of respondents 68.1% have smartphone, 18.7% of respondents have mobile phone, while 11.0% have tablet Pc and 1.1% have PDA and other devices.

C. *Students' Knowledge and experience of Mobile Technology and M-learning*

The study found that the respondents were familiar with mobile technology. The questions 1-3 asked respondents if they access internet using their mobile device. As shown in Figure 3 the results showed that 89% of the respondents accessed the internet using their mobile device.

In addition, 81% of respondents used their mobile device to access the internet on a daily basis as shown in Figure 4.

Furthermore, the results showed that 68.1% of the respondents were paying money to access the Internet using their mobile devices; this indicates normal price of accessing the internet using mobile devices. Question 4 asked respondents if they used mobile device to learn computer science. The results showed 52.7% of respondents used their mobile devices to learn computer science and 41.8% did not.

The last two questions 5-6 asked respondents if they have heard about m-learning and their opinions of m-learning. The results indicated that 58.2% of respondents knew about mobile learning and 41.8 did not. The majority of respondents 82.4% answered that m-learning is good idea and would like to use it, while 9.9% answered good idea but I would not like to use it, whereas 1.1% don't think it is good idea, and 6.6% answered others as shown in Figure **5.**

TABLE II.
"Respondents Profile"

	Classification	Frequency	percent	cumulative percent
Gender	Male	46	50.5	50.5
	Female	45	49.5	100.0
Age	18-22	47	81.3	81.3
	22-26	17	18.7	100.0
Year of study	1st	32	35.2	35.2
	2nd	35	38.5	73.6
	3rd	24	26.4	100.0

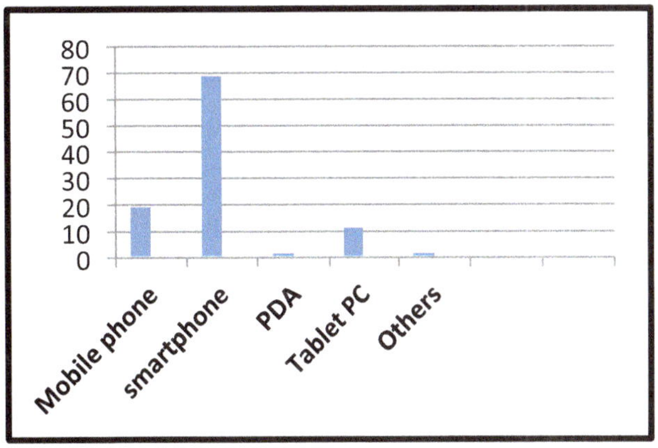

Figure 2. "Availability of Mobile Device"

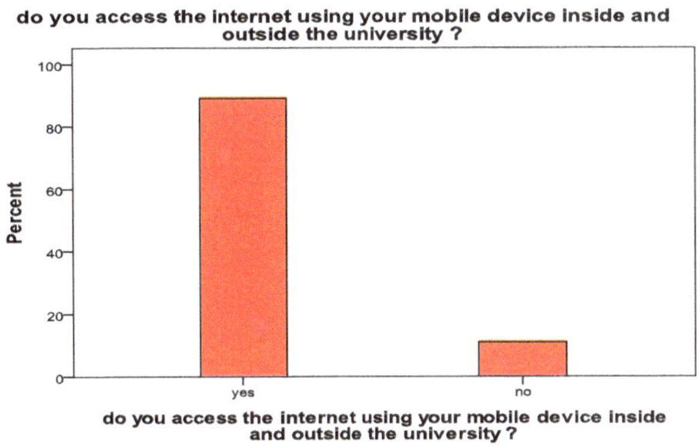

Figure 3. "Internet accessibility using mobile device"

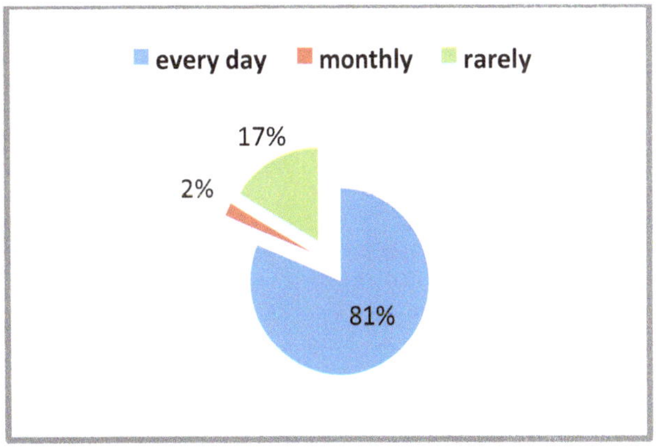

Figure 4. "Times of internet access by mobile device"

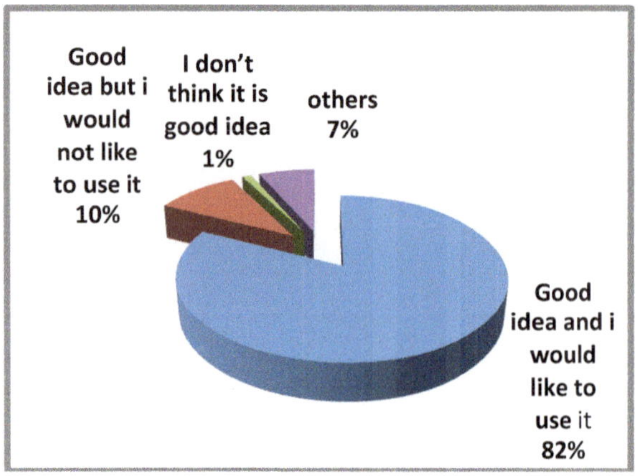

Figure 5. "Students opinions about m-learning"

D. Students' perceptions towards applying M-learning

Table 3 contains 7 statements to explore the students' perceptions for applying m-learning services.

The first statement asked respondents "if the mobile device useful to accesses lectures online", the responses mean was (M= 4.0220), this indicates that the students "agree". In addition, 47.3% of respondents agreed that the mobile device useful to access the lectures online, 31.9% strongly agreed,15.4% neutral,2.2% disagreed and 3.3% strongly disagreed as shown in Figure 6.

The mean in the second statement was (M= 4.0330) this indicates that the students "agree" when they used mobile device will increase the flexibility to learn inside and outside classroom. Figure 7 shows 47.3% of the respondents agreed, 34.1 strongly agreed, 9.9% neutral, while 5.5% disagreed and 3.3% strongly disagreed.

The results of the third statement (M= 3.9011) indicates that the students "agree" with using mobile device in teaching and learning will make the learning process easier and enjoyable. As shown in Figure 8, 51.6% agreed, 23.1% strongly agreed, 19.8% neutral, whereas 3.3% disagreed, 2.2% strongly disagreed.

TABLE III.
"STUDENTS' PERCEPTIONS ON M-LEARNING"

Statement	N	Minimum	Maximum	Mean
1. I think it is useful to access lectures online using my mobile device.	91	1.00	5.00	4.0220
2. I believe that using a mobile device will increase the flexibility to learn inside and outside the classroom.	91	1.00	5.00	4.0330
3. I think implementing and using M-learning as a part of teaching and learning process will make the educational process easier and more enjoyable.	91	1.00	5.00	3.9011
4. I think that using M-learning will help me to get good grades.	91	1.00	5.00	3.4835
5. Implementing M-learning will enable me to have independent learning.	91	2.00	5.00	3.7692
6. I think M-learning will add value to e-learning.	91	2.00	5.00	4.0110
7. I would be interested to learn using this method in the future.	91	1.00	5.00	3.9780

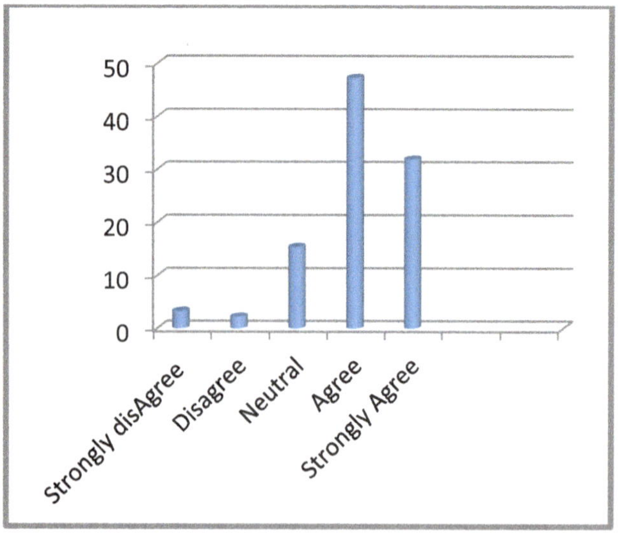

Figure 6. "Useful to access lectures online using mobile device using mobile device".

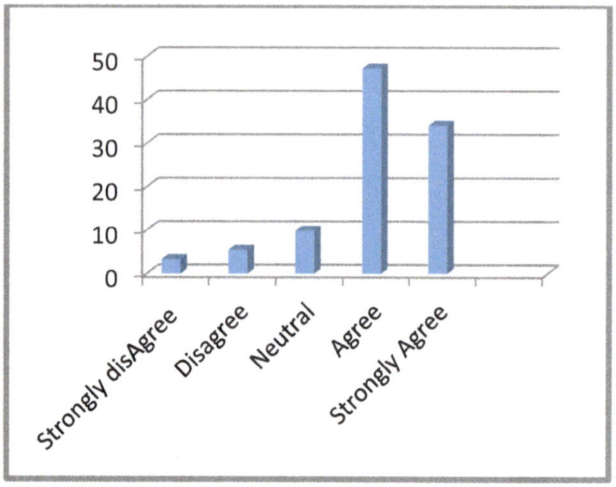

Figure 7. "Flexibility to learn using mobile device"

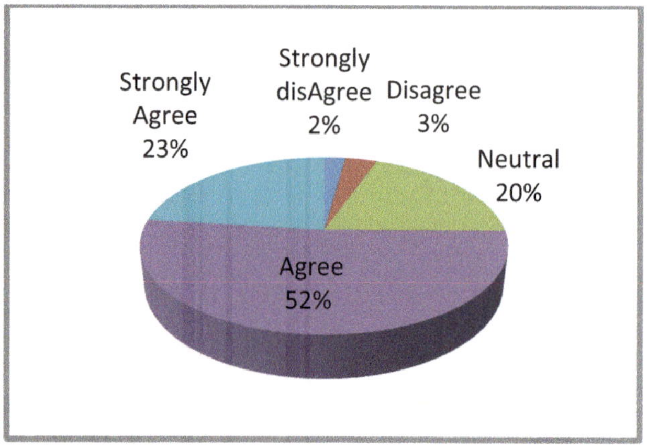

Figure 8. "More easier and enjoyable using mobile device"

In the fourth statement the mean was (M= 3.4835) this indicates the students between "agree" and "neutral" for using m-learning will help them to improve grades. 42.9% neural, 37.4% agreed, 11.0% strongly agreed, 6.6% disagreed and 2.2% strongly disagreed as shown in Figure 9.

The next statement asked "If m-learning enable students' independent learning", the responses mean was (M= 3.7692) this indicates between "agree" and "neutral".

As shown in Figure 10 that 51.6% of respondents agreed, 16.5% strongly agreed, 24.2% neutral and 7.7% disagreed.

In the statement six the responses mean was (M= 4.0110) this indicates that the students "agree" that m-learning add value to e-learning. Majority of respondents 75.8% agreed, 15.4% strongly agreed, 3.3% neutral and 5.5% disagreed as shown in Figure 11.

The last statement asked "If the students interested to learn using m-learning in the future", the mean was (M= 3.9780) this indicates the students agree to use m-learning in the future as learning process. As shown in Figure 12 that 82.4% of the respondents agreed to use m-learning in the future, 11.0% strongly agreed and 2.2% neutral, disagreed and strongly disagreed.

E. Students' expectations of M-learning services.

The study aims to explore students' expectations on mobile learning services that they would like use through their mobile devices. The study found as shown in Figure 13 that the majority of respondents 83.5% preferred access learning contents online using their mobile device and 81.3% for exam results were the highest ranked, followed admission status 75.8%, library service 71.4%, course registration 70.3%, schedule services 67.0%, to collaborate with other students and financial balance 58.2% and the lowest rank was to collaborate with lecturers 54.9%.

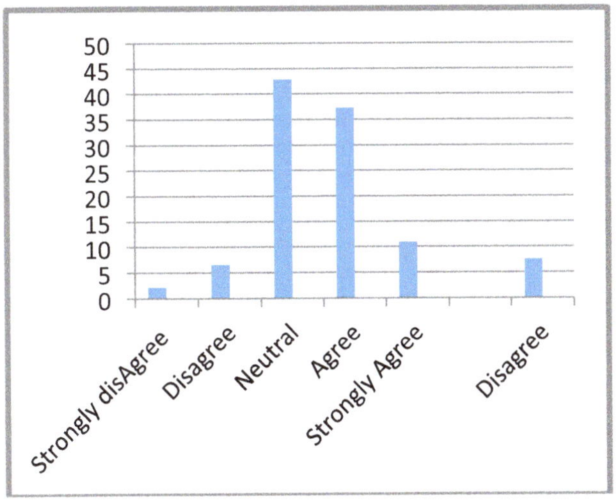

Figure 9. "Improve students' grades using M-learning"

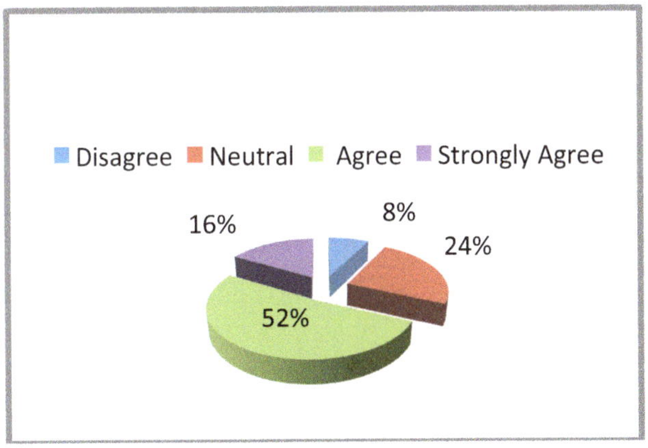

Figure 10. "Learning independent using m-learning"

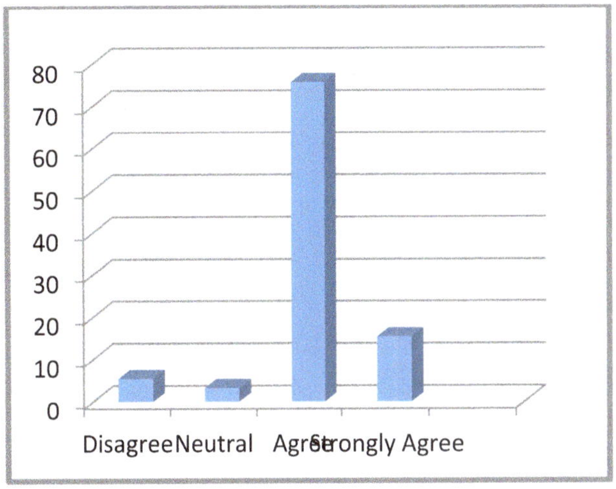

Figure 11. "M-learning add value to e-learning"

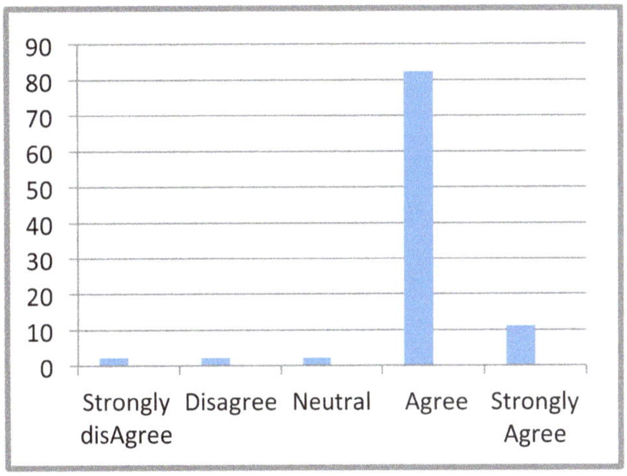

Figure 12. "Use m-learning in the future"

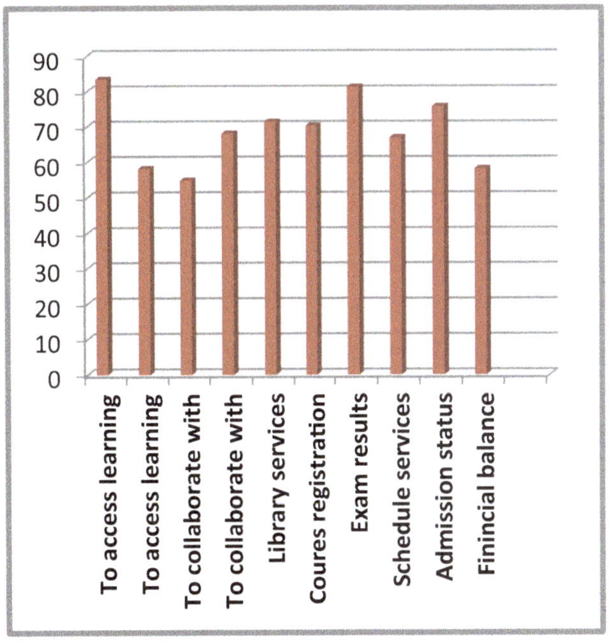

Figure 13. "M-learning services"

VIII. CONCLUSION

This study sought to explore students' perceptions of the application of m-learning and to investigate the students' expectations of m-learning services through their studies. Based on the results, the majority of students have positive perceptions on the use of m-learning. The results of this study indicate that the students may access to lectures online by using mobile devices more easily and useful. The results also indicated that mobile learning will help to increase flexibility of learning inside and outside classroom because the students' can access to learning materials anywhere and anytime. As in this study, most of the students expect that the use of m-learning will help them to improve academic achievement and enable them to independent learning.

The results also indicated that students perceive that m-learning will play an important role of in learning and will add value to e-learning in the future. The results showed that the majority of students would like to use this technology in the future. Finally, the study found highly ranked m-learning services the expected to be used by students include online access to learning materials and administrative services such as retrieving exam results and library services etc.

However, the use of mobile wireless technologies will become the choice for the students in higher education environment in the future; there is still need to further investigate the critical success factors that influence m-learning acceptance among students.

REFERENCES

[1] Ally, M., & Prieto-Blázquez, J. (2014). What is the future of mobile learning in education?. *RUSC. Universities and Knowledge Society Journal*, *11*(1), 142-151.

[2] Alzaza, N. S., & Yaakub, A. R. (2011). Students' awareness and requirements of mobile learning services in the higher education environment. *American Journal of Economics and Business Administration*, *3*(1), 95.

[3] Andrew Litchfield, Laurel Evelyn Dyson and Elaine Lawrence, Agnieszka Zmijewska "Directions for M-Learning research to enhance active leraning," in Ascilite Singapore 2007.

[4] Attewell, J. (2005). Mobile Technologies and Learning: A Technology Update and m-Learning Project Summary. *Technology Enhanced Learning Research Centre*. London, UK: The Learning and Skills Development Agency.

[5] Botzer, G. and Yerushalmy, M. (2007). Mobile Application for Mobile Learning. In *Proceedings of the International Conference on Cognition and Exploratory Learning in the Digital Age* (CELDA), September 7-9, Algarve, Portugal.

[6] Cisco. (2012). *University Embraces Bring-Your-Own-Device with Wireless Network* [Online].

[7] Chen, C. M. , & Chung, C. J. (2008). Personalized mobile English vocabulary learning system based on item response theory and learning memory cycle. Computers & Education, 51(2), 624-645. http://dx. doi. org/10. 1016/j. compedu. 2007. 06. 011

[8] Fernandez, V., Simo, P., & Sallan, J. M. (2009). Podcasting: A new technological tool to facilitate good practice in higher education. *Computers & Education*, *53*(2), 385-392.

[9] Haag, J. (2011, January). From elearning to mlearning: the effectiveness of mobile course delivery. In *The Interservice/Industry Training, Simulation & Education Conference (I/ITSEC)* (Vol. 2011, No. 1). National Training Systems Association.

[10] J. Hoffmann and N. Miner, *Tailored Learning: Designing the Bled that Fits*, Massachusetts: American Society for Training and Development, 2009, pp. 182.

[11] Hussin, S. , Manap, M. , Amir, Z. , & Krish, P. (2012). Mobile Learning Readiness among Malaysian Students at Higher Learning Institutes. Asian Social Science, 8(12), 276-283.

[12] Korucu, A. T., & Alkan, A. (2011). Differences between m-learning (mobile learning) and e-learning, basic terminology and usage of m-learning in education. *Procedia-Social and Behavioral Sciences*, *15*, 1925-1930.

[13] Laouris, Y. and Eteokleous, N. (2005). We need an Educationally Relevant Definition of Mobile Learning. In *Proceedings of the 4th World Conference on mLearning*. October 25-28, Cape Town, South Africa.

[14] Liu, Y., Li, H. , & Carlsson, C. (2010). Factors driving the adoption of m-learning: an empirical study. Computers & Education, 55(3), 1211–1219.

[15] Mcconatha, D., Praul, M., & Lynch, M. J. (2008). Mobile Learning in Higher Education: An Empirical Assessment of a New Educational Tool. *Turkish Online Journal of Educational Technology*, *7*(3).

[16] Park, Y. (2011). A Pedagogical Framework for Mobile Learning: Categorizing Educational Applications of Mobile Technologies into Four Types. International Review of Research in Open & Distance Learning, 12(2).

[17] Traxler, J. (2007). Defining, Discussing and Evaluating Mobile Learning: The moving finger writes and having writ.... *The International Review of Research in Open and Distance Learning*, 8(2).

[18] Trifonova, A. N. N. A., Georgieva, E. V. G. E. N. I. Y. A., & Ronchetti, M. A. R. C. O. (2006). Determining students' readiness for mobile learning. In*Proceedings of the 5th WSEAS International Conference on E-ACTIVITIES*(pp. 20-22).

[19] UNESCO, *Policy Guidelines for Mobile Learning*, 2011.

[20] Virvou, M. , & Alepis, E. (2005). Mobile educational features in authoring tools for personalised tutoring. Computers & Education, 44(1), 53-68.

[21] Wang, Y. S. ,Wu, M. C. , &Wang, H. Y. (2009). Investigating the determinants and age and gender differences in the acceptance of mobile learning. British Journal of Educational Technology, 40(1), 92–118.

[22] J. Wishart and D. Green. (2010, June 29). *Identifying Emerging Issues in Mobile Learning in Higher and Further Education* [Online].

[23] Yin, C., David, B., & Chalon, R. (2009, August). Use your mobile computing devices to learn-Contextual mobile learning system design and case studies. In*Computer Science and Information Technology, 2009. ICCSIT 2009. 2nd IEEE International Conference on* (pp. 440-444). IEEE.

[24] Yusup, Y. (2014). Preliminary study on teachers' use of the iPad in bachelor of education program at a private university in Malaysia. *TechTrends*, *58*(2), 14-19.

AUTHORS

Mohammed Amin Almaiah and **Masita @ Masila Abdul Jalil** are with the School of Informatics and Applied Mathematics, University Malaysia Terengganu, Terengganu, Malaysia (mohammad7ups@yahoo.com masita@umt.edu.my).

mLUX: Usability and User Experience Development Framework for M-Learning

A. Dirin[1] and M. Nieminen[2]
[1] Haaga-Helia University of Applied Science, Helsinki, Finland
[2] Aalto University, ESPOO, Finland

***Abstract*—in this paper, we propose the mLUX framework, a model based on the user-centered design (UCD) framework, which is specifically for the development of m-learning applications. We present the results of four case studies conducted to develop m-learning applications in which the proposed mLUX framework was applied. The main goal of the proposed mLUX framework is to ensure that the stakeholders, especially students, recognize that m-learning applications are learning media that fulfill their essential educational requirements. We begin by reviewing the literature on the contributions to mobile learning usability made in various conferences and journals from 2002 to 2010. This review helped identify and recognize the methodology used to develop mobile learning applications during this period. Based on the literature review, as the empirical case studies, four m-learning applications are examined to demonstrate the performance of the proposed framework for the development of m-learning applications. We apply three distinct measurement criteria to assess the performances of the mLUX framework. This paper also argues that emotional factors, such as the user's enjoyment, adjustability, and reliability, are significant design issues in m-learning.**

***Index Terms*—mobile learning usability, user experience, user-centered design**

I. INTRODUCTION

Contemporary smartphones are complex devices that use a wide range of advanced technologies. In addition to their communication capabilities, these devices are able to execute many other functions and capabilities, even multitasking. M-learning applications are examples of multitasking applications that enable learners to learn and access learning resources regardless of time and place both during classroom training and in off-site learning.

The functionality of m-learning applications is expanding to wide range of new possibilities in addition to improved video, audio, animation, and synchronization capabilities in internal or external applications, such as calendar and email. As Traxler [1] stated, mobile learning using handheld computers is an obviously immature pedagogical technology; nevertheless, it is developing rapidly. However, mere technology is not enough. There is an obvious need to design these applications so that they are user-centered way and have high usability. However, is usability sufficient to motivate users to use the mobile learning application regularly? In the era of thousands of mobile apps available in platform-specific application stores, m-learning applications must engage students in using them.

This paper reviews the extant literature to determine how the expanding possibilities for m-learning development have been addressed in previous studies. Additionally, this paper proposes enhancements to the development framework of user-centered m-learning applications. Changes in students' media consumption present challenges to learning activities. Therefore, it is essential that students be emotionally engaged with the learning application, which is achieved by making the application enjoyable to use. Students then will be motived to use the application. The m-learning application should be able to compete with other entertainment applications on the smart phone [2], such as games, for the students' attention.

II. OBJECTIVE AND SCOPE

The objective of this paper is to present the mLUX framework and demonstrate how it has been used to produce acceptable, usable, and error-free m-learning applications. A qualitative approach [3] was used to create the concept and assess of the m-applications usability and performance.

In order to increase our understanding of the development process of m-learning and activities, we use the case study approach [4][5][6]. Regarding technology, from the broad range of different information appliances that can be used in m-learning (e.g., e-readers, tablet computers, and laptops), this study is limited to smartphones.

III. METHODS AND DATA

This paper is based on review of the literature on mobile learning usability. The review collected data mainly from papers published in m- and e-learning journals and conferences from 2002 to 2010.

This paper also refers to a study [2] on the contemporary expectations of students regarding learning environments that are facilitated by contemporary smartphones. The study was conducted as a comparative usability evaluation [7], and it used the "thinking-aloud" method. Moreover, this study uses data gathered from four successful m-learning development projects that were conducted according to the mLUX framework process. These four projects are the case studies examined in this paper.

The framework focuses on usability and the user's experience throughout the development process [8]. In the process of designing the mLUX framework, the designers collected the user data, analyzed it, and proposed a design based on processed data. In all phases of the design and development, the lead users [9] were directly involved. In developing the mLUX framework, we utilized appropriate methods in each development phase, the details of which are described in the following sections.

IV. Related Research

In addition to the ISO 9241-11's definition of usability, Nielsen[10] defined it as a quality attribute that assesses how easy user interfaces can be used. The definition of usability is based on how efficiently users are able to learn to use the system and how easy it is to memorize the essential interaction steps, perform tasks without errors, demonstrate error-free activity, and experience satisfaction with the use the application. In the field of human-computer interaction (HCI), researchers have developed sets of usability guidelines to help designers produce usable systems [11]. The development of mobile learning usability has evolved over time through different paths, which are addressed briefly in the following paragraphs.

The usability of mobile learning applications often requires the assessment of multi-tier usability. Vavoula [12] considered mobile learning applications on three different levels: the micro level (concerned with the evaluation pf usability), the meso level (concerned with the evaluation of learning), and the macro level (focused on the evaluation of integration with the existing educational contexts).

The importance of usability for mobile applications was noted as soon as mobile phones became capable of executing third party applications. In the early 2000s, mobile devices had small screens, low processing power, and low memory capacity; moreover, their data exchange rates were expensive. However, solutions to these issues have increased the use and development of m-learning applications, which have fostered the importance of usability of mobile learning applications has increased [13][14][15][16][17].

The analysis of published papers on m-learning from 2002 to 2010 revealed that seminal work on usability assessments was produced during this period. Initially, the concept of m-learning was a research topic confined to research at universities . However, this trend rapidly changed as phones became smarter, and the cost of mobile data communication dropped significantly. This trend is also reflected in studies on m-learning usability from 2002 to 2005. Kukulska-Hulme [18][19] inspected a dozen mobile learning applications from the perspecitive of usability, which led to the identification of failures in the general usability of mobile learning applications. During this period (2002–2005), methods used to assess general usability, such as heuristic evaluation and usability testing, were applied.

This trend, however, changed in 2004. Usability assessments are no longer sufficient. An increasing number of methods and guidelines for developing m-learning applications have been proposed. These include Syvänen's [20] recommendations on pedagogical usability as well as Keinonen's [21] argument for the changing role of mobile devices. As they become more personal and portable, it becomes increasingly difficult to predict what users might do. Kukulska-Hulme [18] recommended that education practitioners utilize new technologies in educational settings. Soloway [22] focused on learner-centered design, and Mostakhdemin-Hosseini [23] investigated a framework for a scenario-based mobile learning application. Pehkonen and Malliou [24][25] advocated a user-centered design in m-learning application development. Adaptive mobile learning application development was also proposed during this period by Vainio [26], Malliou [25], and Jäppinen [27]. This development was furthered by Seong's guidelines for mobile learning usability, specifically m-learning [28]. Koole's FMRE framework [29] was based on the context of information. In 2007, Magal-Royo [30] and Nikkoi [26] pointed out that there were no specific or suitable criteria to evaluate m-learning platforms. They recommended that traditional usability methods should be combined with in-situation methods in order to assess m-learning applications.

Many attempts have been made to improve the quality of m-learning applications. Uden [31] utilized activity theory [32][33][34][35] to overcome the complexity of the relationships involved in designing a mobile learning environment. Uden utilized the activity theory by clarifying the purpose of the activity (i.e., exploration of the motives and goals of the activity system) and used it as the unit of analysis. The activity system is the unit of analysis, which gives a context and meaning to seemingly random individual events.

Despite the transition in the focus of the research on usability of m-learning applications, there was little indication of specific issues related to the factors in user experience that deal with the emotional side of application usage and the engagement of users. Nielsen and Norman [36] defined user experience as something that encompasses all aspects of the end-user's interaction with the company, including its services and its products. Hassenzahl [37] defined user experience as a consequence of the user's internal state (e.g., predispositions, expectations, needs, motivation, mood, etc.), the characteristics of the designed system (e.g., complexity, purpose, usability, functionality, etc.), and the context (or the environment) within which the interaction occurs (e.g. organizational/social setting, meaningfulness of the activity, voluntariness of use, etc.). The case studies presented in this paper are based on Brusilovsky's classification of educational components [38]. Brusilovsky classified the courses of virtual universities into four main components: presentation, activities, communication, and administration. Presentation refers to functions that are related to the delivery of new materials. Activities refer to activities that students need to perform. Communications refers to interactions between teacher and students or between students. Finally, administration refers to activities related to registration, record keeping, and so on.

V. User-Centered Design Principles And Processes

User-centered design (UCD) and the development of interactive systems and devices have gained increasing importance in service and product development [39]. Gould [40][41] argued that in order to design a usable system, users should be continuously involved in the development process and the design should be modified based on their feedback. User-centered design reduces costs and improves usability because it continually focuses on the customer's needs as early as possible in the development process. User requirements are the focus of all stages of development cycle. ISO 9241-210 [42] defined the human-centered design processes for interactive system design as follows: 1) cooperative design: designers and user involved in all stages; 2) participatory design: users' occasionally participate in the design process; 3) contextual design: design based on the actual context. User-centered system design (UCSD) focuses on usability not only in the development process but also throughout the system's life cycle [8]. Human-centered design [43] and interactive interaction design [44] are aspects of the same issue, but in practice, they are used synonymously for UCD [45]. Figure 1 presents Wallach and Scholtz User-Cenetered design activities [46].

A. Scope

In the scoping phase, the goals and the constraints of the projects are defined. The goal of the design should be addressed in advance, such as designing a new functionality for the existing application or redesigning the existing functionality. Moreover, in this phase, the roles of the players are analyzed and a common ground between existing product concept and research results is created.

Scope in m-learning:

Regarding the scoping activity in m-learning, there is a need to define the aim and role of the developed system. In addition to defining the roles of the players in the m-learning system, the analysis addresses the extent to which the m-learning system covers the learning activities by asking the following questions: 1) Is there a need to design a standalone application? 2) Should the m-learning application enhance the existing e-learning platform? 3) Is the aim is to support the traditional classroom-based education offering? Similarly, the context of use of m-learning needs to be defined. This includes, for example, the analysis of where the application is used the most and the need for an internet connection.

B. Analyse

In the analysis phase, the main aim is to reveal the attributes of the user, such as the tasks that they currently or expect to handle in their future applications. This information is often collected through usability inspections or by various user-study methods.

1) Analyse in m-learning

The students and teachers are the main users of the m-learning application. In the analysis, we learned the current ways that students and teachers use to deal with educational tasks and their expectations of the forthcoming application. The results of the case studies revealed that the user-study methods, such as web questionnaires and semi-structured interviews, were adequate to collect information from the participants. The web questionnaires enabled us to collect information about the ways in which students and teachers deal with their educational tasks. This method was preferable because the users could provide the information at their convenience. The semi-structured interviews were used to learn about the participants' current and future expectations.

Moreover, to analyze the collected data, we used transcript coding and users and task analysis methods. Transcript coding helped identify details about the actions and the environment in which these actions were often took place.

C. Design

The main goal of the design is to transform the findings into concepts and tangible prototypes.

Design in m-learning:

The concept of the potential application is presented as a learning activity scenario. A learning scenario describes the learning activities that the students and teachers carry out in specific real-life learning environments. For early feedback from the users, we asked students and teachers to review the scenario and share their remarks about the concept with designers. This yielded information about users' further demands for the proposed concept of a mobile learning application. Finally, based on a revised scenario, a low and high fidelity prototype was designed and developed accordingly.

D. Validate

Validation concerns applying various usability methods to ensure that the design follows the initial requirements and supports the required functionalities.

Validate in m-learning

The specific validation of m-learning is carried out with students and teachers. They were asked to assess the low and high fidelity learning prototypes in a usability lab. Controlled validation enabled the explicit delivery of feedback from students and teachers to the developers. M-learning specific tasks included the following: teachers upload course and lecture material for students' use; students are asked to view, download, and use the learning material; and students are asked to submit their assignments through the m-learning application. Nielsen's heuristic evaluation guidelines were used to assess the consistency of the m-learning application. Heuristic evaluation was used to evaluate all applications that students and teachers used for learning purposes. This enabled the analysis and evaluation of the performance, usability, and user experience in the use of the combined m-learning application.

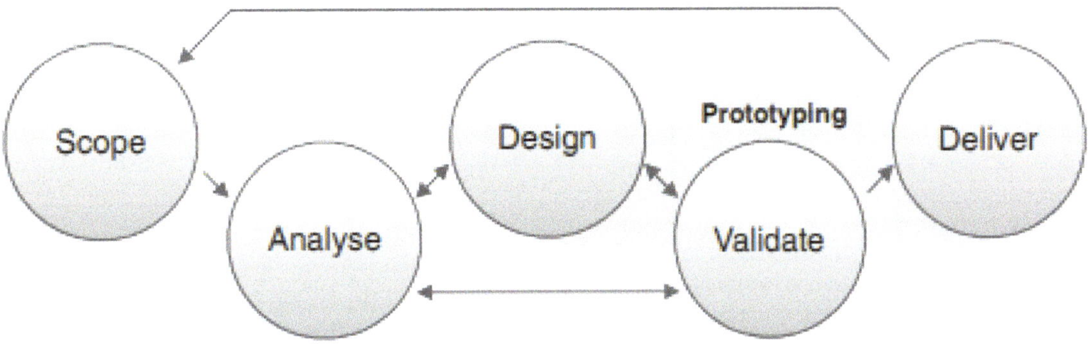

Figure 1. User-centered design activities according Wallach and Scholz [46]. These activities are the basis of the mLUX model.

E. Deliver

In the delivery activity, the implementation team handles the outcome of the design and the related documentation.

1) Deliver in m-learning

The design and development documentation of m-learning application consists of the following: 1) results and documents of user studies (i.e., students and teachers); 2) detailed list of user requirements (students and teachers); 3) the learning scenarios and the proposed learning prototype concepts; 4) the usability assessment test report.

VI. The MLUX Model: Constructing A User-Centered Design Module For M-Learning

In this section, we explore the mLUX, which is the framework used to develop the m-learning application. In addition to the model presented by Wallach and Scholz [46], the ISO 9241 standard (parts 11, 1998 and 210, 2010) is used to define the important characteristics of this framework, including the stakeholders involved in m-learning application development and the definition and analysis of the context of use.

The proposed mLUX framework consists of three conceptual parts: 1) the role-players in the m-learning application; 2) the context of use of the m-learning application, and 3) the m-learning application development process. Figure 2 illustrates the mLUX framework.

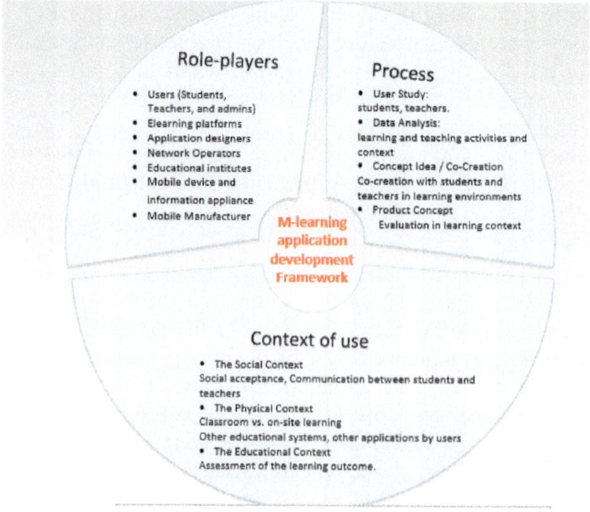

Figure 2. The proposed framework for developing m-learning applications (mLUX)

A. Actors and roles in M-learning applications

Our previous studies [47] [ICAICTE, 2014] found that numerous role-players (e.g., students, teachers, administrators, educational institutes, etc.) are involved in the design, development, and usage of m-learning applications. It is essential to identify the needs and expectations of the role-players because they directly or indirectly influence the use of the application. The mLUX framework includes the direct involvement of the role-players in the application design and development phases. Subsequently, the role-players are directly or indirectly involved in the design and development of the m-learning application.

The students, teachers, and administrators are the key users of m-learning applications. The needs, expectations, and the motivations of these key players affect the features of the m-learning application For example, a teacher provides course materials through uploading assignments. Similarly, students download the material, or they may submit their answers to the assignments through m-learning applications. Moreover, the administrative personnel support students and teachers in their educational activities in various ways, such as guiding them throughout course registration activities and providing course certificates through the m-learning application.

In addition, many invisible role-players, individuals, organizations, systems, and technical systems are involved in the design and development of m-learning applications. These invisible role-players are not the actual users of the application, but they have a significant influence on the design and usage of the m-learning application.

The mobile learning application enhances the existing e-learning platforms, which are a significant role-player in the design and development of m-learning applications. Therefore, the designers have to determine the features that must be implemented in the m-learning application.

The designers and developers of m-learning applications are invisible role-players. They create the concept and select appropriate development technologies. Moreover, the design and development of m-learning applications often incorporates various experts, such as user study experts, concept designers, and software developers, whose design activity affects the overall m-learning application.

The network operator's role in the design and development of m-learning applications is that of an intermediary between the content provider (teachers) and the content users (students). Therefore, the network operator's constraints and regulations, such as the format of the exchange data, are compulsory in m-learning application development. The network operator is an invisible role-player and has a direct influence on the usage of the m-learning application. Therefore, network operators' support in m-learning content deliveries is vital, especially when the number of m-learning application subscriptions increases.

Educational institutes are the primary agents in the practice and usage of m-learning applications, which making them a strong role-player in the design and development of m-learning applications. The educational institutes' demands and expectations in m-learning applications need to be addressed properly. An important example is the influence of the cost of the m-learning utilization on the course. However, the institution's decision to use m-learning is based on its overall institutional aims, its access to the infrastructure required by the m-learning application, and its appreciation of the pedagogical trends that support m-learning and [18]. In addition to providing financial support for m-learning applications and content development, educational institutes have the important role of promoting and encouraging their staff to use m-learning applications in their courses.

Mobile device/terminal manufacturer. This invisible role-player is indirectly involved in formulating the usage and development m-learning application. M-learning applications target the mobile devices of students and teachers, but the latter affect the usage of the application. Despite some convergence, the technological features of the mobile devices still vary significantly. For example, some devices fully support HTML5 specifications, whereas others only partially support them. Therefore, the designers and developers have to consider the inconsistencies in technological and operating systems in the mobile devices used by the students and teachers.

B. The context of the use of m-learning

The main role-players of m-learning applications such as teachers and students carry out their educational tasks in different times and environments. Therefore, the mobile learning is not just about learning with a portable device, but across contexts[48]. These include the following contexts.

1. *The social context* in this study refers to social acceptance of the m-learning application as a learning medium by students and teachers. In addition, knowledge transformation is a social process, which occurs when there is interaction among learners. M-learning provides social presence and awareness as the learner is situated in different learning contexts such as formal location e.g. classrooms, semi-formal locations e.g. libraries, and informal locations e.g. residences[49]. M-learning enables and supports communication between students and teachers regardless of the physical location.

2. *The physical context* refers to a non-human aspect of the context, the accessibility and the performance related to the physical aspects of the environment. Mobile learning applications aim to overcome the location and time constraints, which occur in traditional classroom and e-learning platforms. In a traditional classroom, knowledge is shared and transformed inside the classroom: Students and teachers share and exchange the information inside the classroom to create knowledge. However, students may be physically located in different places where teachers provide mobile learning content and services. Therefore, m-learning applications overcome the location constraints; Students and teachers have access to educational resources regardless of the location through their mobile devices.

3. *The educational context* is about the surroundings where the learning happens. These surroundings are composed of the learning outcomes, the subject matters of the learning topics, learning assignments and material, as well as the pedagogical approaches.

C. m-learning application development process

In the mLUX framework, the process used to develop the application is divided into four main phases: 1) The elicitation phase, in which user studies are conducted with students, teachers, and administrators; 2) The data analysis phase, in which the collected data from students and teachers are analyzed. In this phase, the most important needs and features of the students, teachers, and administrators are identified. Various methods are used to analyze the data, such as transcript coding, User and environment analysis is conducted to classify the needs and features; 3) The idea creation or the concept designs phase, in which the potential m-learning application features are presented as scenarios, followed by the design of the prototype of the potential application; 4) the product concept design phase, in which a usability test is conducted on the proposed m-learning application prototype after which the prototype is revised based on the latest feedback, after which is the product is ready to implement.

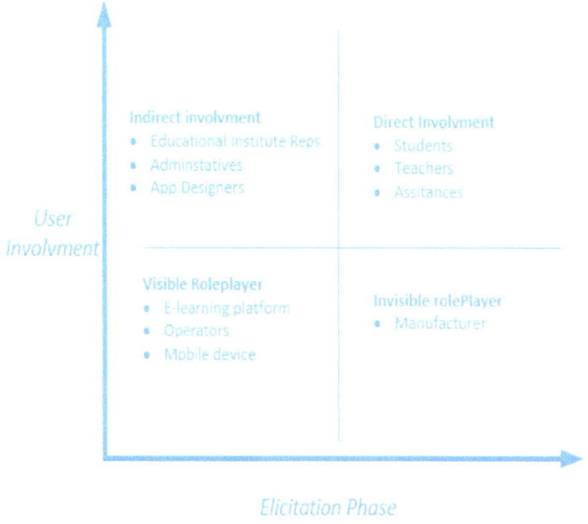

Figure 3. Role-players involvement in m-learning application design and development

1) User Study

In the user study phase, the primary goals of the m-learning application and the characteristics of each stakeholder are identified. Moreover, in this phase, the user study is conducted with the role-players who are directly or indirectly involved with the design and usage of the application. Furthermore, the main goal of the user study is to determine the needs and expectations of the students and teachers regarding the m-learning application. User representatives, such as students, teachers, and administrative personnel, are selected based on the context of use of the application. Therefore, it is essential to determine whether the application should be a new "standalone" or whether it should enhance existing e-learning offerings.

Eason [50] classified users as primary, secondary, and tertiary. Primary users use the system constantly, secondary users use the system occasionally, and tertiary users make decisions about the system. In m-learning application design and development, the role-players are classified as:

- Direct involvement in the elicitation phase:
  - Primary users, such as students, instructors, and teachers
  - The secondary users, such as administrative personnel
- Indirect involvement (resourcing, supporting, constraining)
  - Tertiary users, such as representatives of the application's owners and commissioners
- Application designers

Figure 3 presents the summary of the role-players involved in m-learning application development at various phases.

In this phase, the designers learn how students and teachers carry out their educational tasks and the context in which these tasks are performed. The following are examples of the questions that aimed to reveal the basic needs and expectations:

- How is the target course currently implemented?
- Is the lecture offered as an online course or face-to-face?

- How do students and teachers currently carry out educational tasks, such as the reading material, the assignments mode, and type of assignments?
- What smart devices do student and teachers have experience with?
- What is the overall knowledge that students and teachers have regarding smart devices?
- How often do students and teachers commute between school and home?
- Where do students and teachers often need to have access to course content?
- How would users like to change the current implementation of the course?
- What are the preferred ways to learn and handle educational tasks and activities?
- How can they use their mobile devices in their educational tasks?

Beyer [51] recommends that six to twenty users should participate in user studies, which this study defines as the number of users. In this study, we often asked six to eight students and teachers to participate in the user study phase. Six to eight users are recommended because of the diversity of the mobile devices used by stakeholders and the complexity of the application content, which is used by learners.

2) Data Analysis

In this phase of the mLUX framework process, the designers interpret the data collected in the previous phase. The designers apply various data analysis methods, such as interview transcript coding. The user and task analysis reveals the ways in which students and teachers currently handle educational tasks. Moreover, the designers determine the environments in which these tasks are carried out. This information is vital for developing a robust m-learning application. Figure 4 presents the analysis of the tasks and environment.

Interview transcript coding helped to identify the actions and activities of students and teachers in current learning and teaching formats. Moreover, the data analysis phase revealed users' expectations of the potential m-learning application. The results of the analysis were used to classify the tasks and the environments in which the students and teachers performed the educational tasks. Kujala [52] also promoted the application of a user and task matrix when different stakeholders used the same application.

3) Idea/Co-Creation Design

In the concept design phase, the data categorized in the previous phase are expressed as learning and teaching scenarios. The scenarios contain the functionality and features of the potential m-learning application. Kyng [53] recommended that the design be easily understood by all potential stakeholders in the elicitation phase. Therefore, in a possible concept design method, scenarios are selected because students and teachers are able to understand the functionalities and deliverable features of the targeted application more clearly. The scenarios are written so that the proposed user experience factors are expressed in different educational components. The content of the scenarios is divided into four educational components:

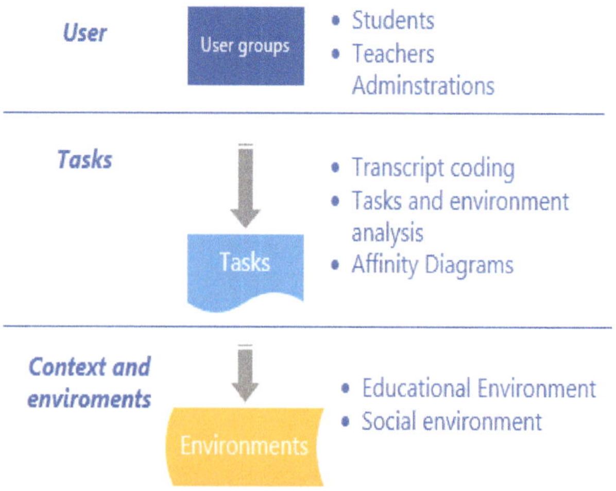

Figure 4. Analysis of users, tasks, and environments in m-learning application development

presentation, activities, communications, and administration. In each component, the requirements of the students and teachers are expressed as a short story.

The following are examples of the features that are written in the scenarios:

- The kind of courses that the m-learning application supports
- The preferred material formats in the m-learning application
- Lab/home assignments and their formats
- The m-learning application is specific to the type of communication, such as chat, email, and so on.
- Locations where the application is used

To validate the learning application, three to five students and teachers share the prepared scenarios. After the students and teachers review the scenarios, the designer conducts a semi-structured interview to collect their feedback on the proposed concept. This feedback is analyzed, and the designer revises the scenarios according to the results. After validating the m-learning application concept, low-fidelity or high-fidelity application prototypes are designed based on the revised scenarios.

4) Product Concept

The application prototype is assessed by the students and teachers. The assessments are often carried out in the usability lab. The goal is to assess the functionalities and performance of the application from the perspective of learning. We apply Nielsen's Heuristic evaluation guidelines, post-interviews, and think-aloud methods in the assessment. Students and teachers go through sets of predefined tasks during the sessions. For example, the teachers may be asked to upload the lecture presentation, and the students may be asked to upload the presentation. The designers analyzed the collected feedback and modified the prototype accordingly. The revised prototype of the m-learning application was then ready for development. Figure 5 presents the proposed mLUX framework process that applied and evaluated in m-learning application development.

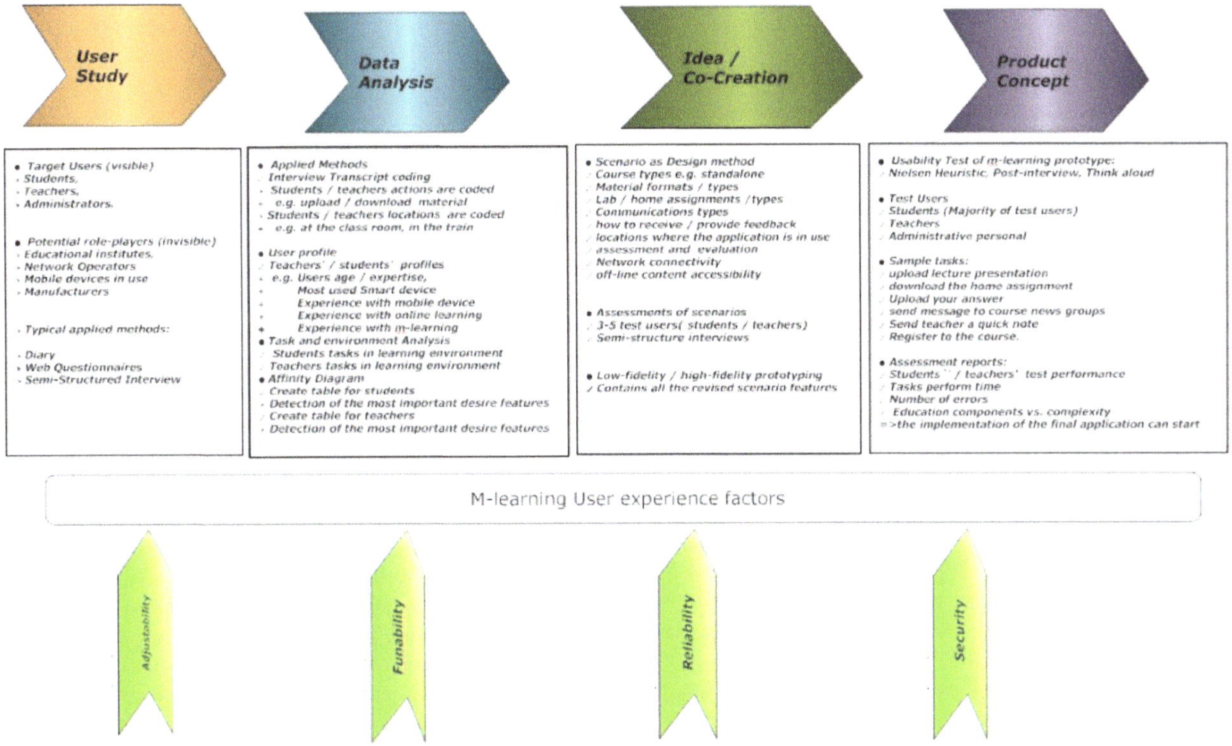

Figure 5. The mLUX framework applied in the case studies

The proposed mLUX framework process was designed and developed so that the m-learning application would be usable, efficient, and error-free. Moreover, the framework aims at ensuring that the students, teachers, and administrators would have a positive experience because the application is enjoyable, adjustable, reliable, and secure. These factors of user experience ensure that the users of the m-learning application, such as students and teachers, are emotionally engaged with the application so that, for instance, students feel that the learning is enjoyable. Moreover, by ensuring that the user interface is adjustable and customizable, the students feel that they are engaged in the application's design. Therefore, students consider the application a familiar learning environment. Additionally, because the application is secure and reliable, students and teachers trust it, which reduces concerns about losing important data, such as submitted assignments, and the teachers trust that the submitted feedback reaches the intended students.

VII. Framework Evaluation Criteria

The proposed mLUX framework has been assessed by conducting four case study projects. The criteria for the mLUX framework assessments [54] are as follows:

1. The m-learning applications' usability and user experience are assessed from the perspectives of learners and teachers. The assessment is based on Brusilovskys' educational components. Brusilovsky divided the course delivery in a virtual university into four main categories: presentation, activities, communications, and administration. In each category, the adjustability, delightfulness, reliability, and satisfaction of the features were assessed. Table 1 presents the framework measurement criteria based on educational components and factors of usability and user experience.
2. The assessment of usability of the m-learning application by students and teachers was based on traditional usability assessment methods. The evaluation of m-learning usability was based on ISO 9241 part 11 and part 304, which recommend that the measures of usability should include effectiveness, efficiency, and satisfaction. A general metric was used to measure the usability as a quality metric, such as the success rates. Students' and teachers' performance was also measured in completing the tasks. The measurements are based on the time required to perform the tasks or the number of errors. In addition, the students' and teachers' subjective satisfaction with the application performance and functionality was measured.
3. Semi-structured interviews were conducted with the developers, designers, and the owners of the application. The aim of this assessment was to collect further information about the performance of the mLUX frameworks from people who had applied the framework as a method of application development. Examples of questions are the following:
   - Did you find the mLUX framework helpful in the design and development of the m-learning application? How easy was it to create the concept by applying the mLUX framework?
   - How many errors were identified in the scenario phase of your m-learning application design?
   - What are your overall impressions of the mLUX framework as method for application development?

TABLE I.
CRITERIA OF FRAMEWORK MEASUREMENT

Educational Components	Usability and User Experience Factors	Criteria
Presentation	Adjustability	Supports learning medium / personalization
	Delightfulness	Facilities and acceptance of the tool
	Reliability	Supports communication, and technology adequacy
	Satisfaction	Performs the required task
Activities	Adjustability	Supports the learner to perform the basic learning activities
	Delightfulness	Supports the learner to perform tasks without difficulties
	Reliability	Supports the learner to perform his task securely
	Satisfaction	Successfully performs the educational activities
Communications	Adjustability	Supports the performance of basic communication related tasks
	Delightfulness	Facilities and acceptance of the tool
	Reliability	Supports communication and technology securely
	Satisfaction	Successfully performs the task securely
Administration	Adjustability	Supports the learners requirements
	Delightfulness	Supports communication and technology securely
	Reliability	Successfully performs tasks securely
	Satisfaction	Successfully performs tasks

For example, to assess the presentation component, we measured the degree to which users were able to personalize and customize the m-learning application user interface and content based on personal preferences. Therefore, the assessment indicates whether the application supports the adjustability element in the presentation component. In this phase, we evaluated the user experience and usability by asking users to perform predefined tasks using the application. The case studies used different contexts, such as Java applications, adaptive learning, context-aware nursing applications, and so on. Our evaluation assessed only educational components that were relevant to the application's context. For example in the educational game application, the testing was mainly focused on the presentation and communication components, whereas in driving the m-learning application we assessed and evaluated all the layers.

VIII. FRAMEWORK EVALUATION: CASE STUDIES

The mLUX framework was applied in four different case studies. The aim was to assess the performance of the proposed mLUX. The mLUX assessments composed of three distinic evaluations 1.The application usability assessment 2. The case study application assessment from educational components perspective, and 3. The designers and developers feedback.

A. Case 1: M-learning application for introduction to Java programming course

In this case study project, we applied the mLUX framework to design and develop a mobile learning application prototype for a Java programming course at the HAMK University of Applied Science.

The case study was initiated by conducting a user study with students and teachers to learn their needs and requirements. Several methods were applied, such as a diary, questionnaire, and semi-structured interviews, and six students (n = 6) and two teachers (n = 2) were used. The user study begin with collecting background information about the users, such as the type of mobile phones, how often they used their mobile phone, and where and where did they use their mobile the most often. We then created a user profile based on the collected background information. We the invited the students and teachers to participate in short semi-structured interviews. The discussions in the interview sessions were recorded for further analysis. We applied transcript coding, user and task methods to explore and classify the users' needs and requirements In the next step, two scenarios were created—one for the students and another for the teachers. The scenarios contained the requirements and the features that we identified in the data analysis phase. The scenario was then shared with the participant students and teachers for their review and assessment. In this round, we asked three students and one teacher to review the scenario. Finally, the scenarios were revised based on the reviewer's feedback. The potential high-fidelity application was designed and implemented based on the revised scenarios. The prototype was assessed by conducting a usability test with the potential students and teachers at the media lab at the HAMK Applied Science University.

In the usability test, the users were given several predefined tasks to carry out one by one. A semi-structured interview was conducted after the user performed all the predefined tasks. The objective of the semi-structure interview was to learn about the experiences of the students and teachers in evaluating the prototype of the mobile learning application. Figure 6 presents the application's home screen.

Figure 6. M-learning application for the Java course

The application assessment process and the assessment results are discusses as follows:

1) Usability and user experience assessment

We conducted different usability test sessions for the students and teachers in the media lab at the Hamk University of Applied Science. Six students (n = 6) were asked to participate in the first test session, and two teachers (n = 2) were asked to participate in the second test session. The users were given distinct sets of tasks to carry out within a given time. Examples of the given tasks are as follows:

- Login to the application with the given credentials.
- Upload and download today's lecture notes.
- Upload the assignments, submit your answers, and provide feedback for students assignments.

The assessment results indicated that the users were able to carry all the given tasks successfully. These results were obtained although this test was the users' first experience in using a third party mobile application in their devices. Moreover, the test results demonstrated that the students and teachers were excited about have the application in all courses at the university even though the data exchange cost was relatively high for students. Interestingly, the users' initial mental model of the m-learning application was that they had to carry out all the educational activities via short messages (SMS). However, this perception changed as soon as they experienced the applications on the test devices. Additionally, the users expressed appreciation of the simplicity of the user interface on the mobile learning application: "*I liked the application, it was easy to find the needed feature, you just need to go the categories*." Similar responses were made by several students and teachers during the post-interview sessions. The users, however, complained that the input medium, that is, the keypad was not convenient for writing a long text.

2) Educational activities assessment

In this evaluation process, the students and the teachers were asked to perform a predefined task to assess the educational components of the application. Table 1 presents the assessment criteria and the process. The results of detailed assessments of each educational component are beyond the scope of this paper. The following are examples of the assessment process and the results. For example, in the *presentation component*, the students were asked to download lecture materials, such as lecture notes. The students demonstrated that they were able to download the lecture notes successfully, although the technology did not support many proposed functions, such as videos. *The teachers uploaded assignments in the activity component, and the students were asked to upload the assignment and submit their answers through the m-learning application. The teachers provided feedback on the students' assignments, such as accepted or failed.*

The assessment analysis revealed that the students did not enjoy using the functionality in the activity component because of the keypad was inconvenient. On the other hand, the teachers enjoyed providing instant feedback on the student's assignments even when they were walking. *In the communication component, the students and teachers were asked to use the following features (chat, SMS, email) to communicate with students or other classmates.* The students and the teachers found the communication functionalities interesting, especially the capability of direct and secure communication with their peers and course staff. Furthermore, the students and the teachers found the email functionality in communication component particularly efficient because the course participants' emails were accessible on the m-learning application. Therefore, they believed that sending an email to a classmate or teacher was simple and fast because it involved on a few clicks. Similarly, in regarding the administration component, *the students and the teachers were asked to perform predefined tasks related to this component, such as registering to the next available course*. The students and the teachers responded favorably to the functionality of administration component.

3) mLUX: The developers experiment

The design and development of this case study project was carried out by two groups of students at the HAMK University of Applied Science. The first group consisted of four students (n = 4) who had previously studied usability engineering related courses. In the project, they had the role of designers. This group conducted the user study with the students and the staff. They carried out the data analysis and designed the application concept, which resulted in the scenarios. The second group consisted of two students (n = 2) who had previous experience in application development. They had the developers' role in this case study. A workshop was conducted to discuss the mLUX development process with both groups. The team especially appreciated the systematic approach of the framework used to achieve the goal. Additionally, the participants recognized that the divisions were an asset in this framework. Moreover, the participants believed that having the scenario of a design methodology helped the students, teachers, designers, and developers to conceptualize the overall application clearly. They showed their wiliness to utilize the methodology in future projects in responses such as "*I liked it, clear and helpful to come up with the prototype,*" which were repeated by the developers and the designers.

B. Case 2: Adaptive m-learning Application for driving school

The second case study was designed to develop an m-learning application for driving school candidates. In this project, we applied the proposed mLUX framework process to designing and developing an adaptive m-learning application. The project was conducted at the Haaga-Helia University of Applied Science during the autumn of 2012, and it continued during 2015. The application was originally commissioned by the Haaga Driving School in Helsinki. The early version of the concept and implementation of the application were published in CSEDU 2014 [55]. This mobile learning application enabled the driving license candidates to fulfil the driving schools' mandatory theory requirements. Additionally, the instructors at the driving school were able to obtain instantly full reports on the students' performances in driving and theory. Furthermore, the application notified the instructors if the students made critical mistakes during the driving sessions. In Case Study 1, the application was considered an enhanced tool for existing face-to-face lectures. In Case Study 2, however, the application was a standalone m-learning application. The students and instructors first participated in various user study methods, such as web questionnaire and semi-structured interviews. The students in this case study were mainly driving school applicants who aimed to get a driving license. We conducted the user study with seven users (n = 7) four of which (n = 4) were male and three (n = 3) were female. In the next phase of the project, we analyzed the collected data by applying methods, such as transcript coding, to explore the real needs and expectations of the students and instructors. Figure 7 presents the role-players who were involved in the case study 2.

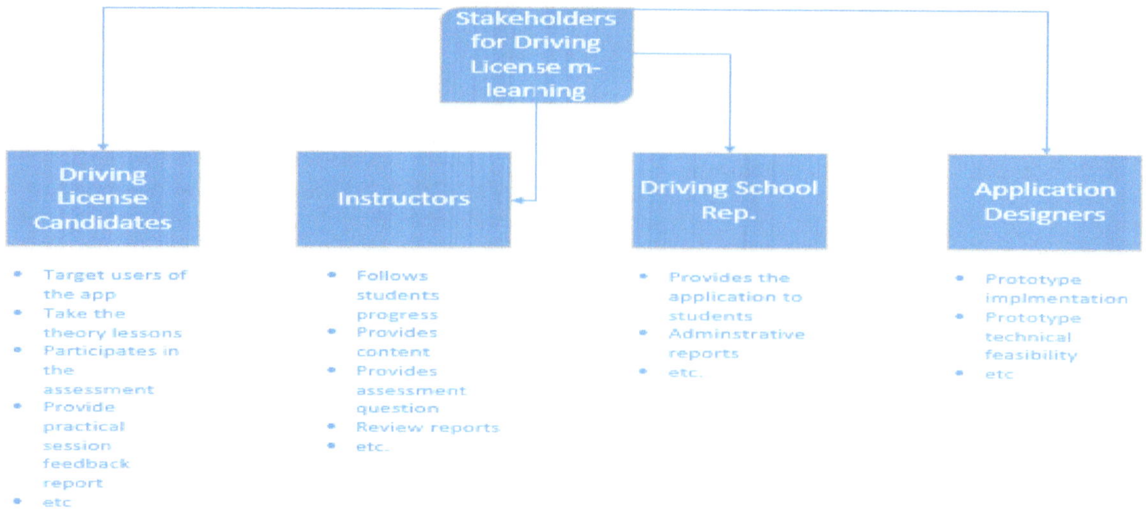

Figure 7. Users' involvement in different phases

Two different scenarios were written for the next phase. The first scenario reflected the implementation of the application based on the driving license candidates' requirements, and the second scenario contained the features required by the instructors. The scenarios were then shared with six students and instructors (n = 6) of whom three users were new and had not been consulted in the elicitation phase.

The users were asked to read the scenario and then to participate in a short semi-structured interview. After all the users reviewed the scenarios, the designer analyzed the collect feedback and revised the scenario accordingly.

We proceeded with this case study by designing a high fidelity prototype based on the revised scenario. Finally, we conducted a usability test with potential users. Figure 8 presents a screenshot of candidates application user interface.

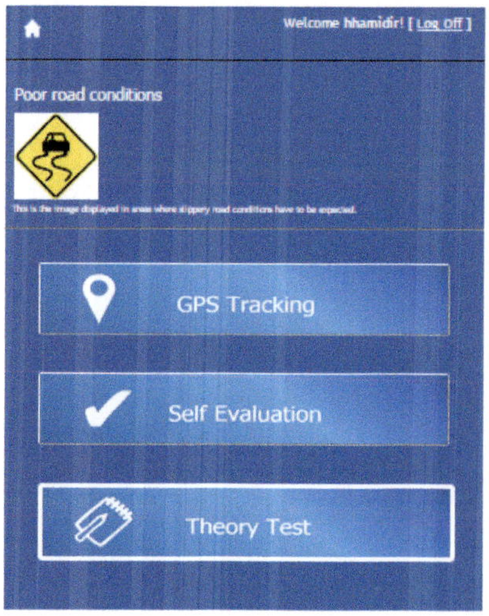

Figure 8. The m-learning application adapted for driving license candidates

1) Usability and user experience assessment

In this case study, we conducted a usability test with five potential test users (n = 5) and two instructors (n=2) in the media lab at the Haaga-Helia University of Applied Science. The users were asked to perform ten (n = 10) predefined tasks using the application. After the test sessions were completed, short semi-structured interviews were conducted to collect data on the users' experiences with the application. We also recorded all the actions that the users performed in the usability test session for further analysis. Unlike in the previous case study, the users had a mental model of the mobile learning application because they were already familiar and had experience with third-party mobile applications in their smart devices. The analysis of the test sessions indicated that the users were able to perform all predefined tasks. The users found the application easy to use, and believed that the application contain the essential features. Therefore, the application concept matched the user's mental model of a mobile application in their smart device. The users specifically enjoyed having the content about their driving license on their mobile phones. Responses such as "great, this application is like a permanent reference book in my pocket all the time" were repeatedly given [46].

2) Educational activities assessments

The criteria for the educational components test are shown in Table 1. As in the usability test, the students and instructors were given set of predefined tasks. The tasks were prepared exclusively for each component. *The presentation component*, the students, and the instructors were able to carry out the tasks properly although some components were not implemented during the test period. Nevertheless, the application content and the User Interface (UI) were customizable based on student's performance, such as in the theory lessons and practical driving session.

Moreover, the test showed that the application supported multi-formatted content based on the students' needs and application requirements. However, because of the technological inadequacy and lack of development time, the presentation component was not fully implemented.

In the *activity component*, the test users showed their satisfaction with the clarity of the features: "the application helped me what to do next." However, the reliability factor for this component could not be tested because the prototype was not fully implemented. *The communication component* was not explicitly implemented, in which the students and teachers were instructed to use SMS or any other existing methods. In the administration component, the test users were able to perform the required task, such as printing out the student's driving activities during the last ten days. The results of the analysis revealed that the application supported the administration component features and fulfilled the criteria defined in Table 1.

3) Methodological assessments

The interviews conducted with designers and developers indicated that they were satisfied with the resulting application. They were especially satisfied that the mLUX framework reduced the number of the design errors before implementing the high fidelity prototype: "*We managed to find most of the errors already in scenario review phase.*" The designers were specifically happy with the involvement of potential users in the design phases, which helped identify the new feature. In the elicitation phase, not all the designers found all the required features: "*We identified additional features both in the scenario and in low-fidelity prototype testing.*" The other role-players, such as those who conducted the user studies, were also consulted about their motivation for using the mLUX framework: *"In the beginning using the framework work was difficult, did not know what to do, but after the transcript coding everything changed.*" The main negative feedback received from developers was that this methodology required extensive consultation with users, which would consume a great deal of time. For example, one stakeholder, who had commissioned the application, expressed overall satisfaction with the m-learning application: "*Very surprised everything works with any errors; we tested the application with students and instructors."*

C. Case 3: Business guide mobile game application development

In the third case study, a customer briefing m-learning application was developed based the on mLUX framework process. The application owner provided us with the potential requirements of the application. Hence, in this case study, no user study was conducted to define the potential features of the application. Therefore, the concept design began with three different scenarios based on the given list of requirements. These scenarios represented different pedagogical approaches to developing target application: 1) m-learning application as an educational game; m-learning application as a choice question; and m-learning applications where theoretical content is presented to users.

The application owners initially assessed the scenarios by confirming that we evaluated the proposed scenario by testing it on potential users of the applications. We shared the scenarios with five (n = 5) students and teachers. The students and teachers read the scenarios and then participated in short semi-structured interviews. In the interview session, we consulted the students and teachers about the reviewed scenarios, asking questions, such as "*What do you think about the scenarios?*" and "*Which scenario would you prefer as an application in your phone?*" We recorded the interview sessions with students and teachers for further analysis of the responses. The results revealed that the majority of the students and teachers recommended the educational game scenario as appropriate for the potential m-learning application. Hence, the educational game scenario transformed into a low-fidelity paper prototype with the help of a wire-framing tool. The paper prototype was shared once again with the students and teachers to collect further design-related feedback. The prototype evaluation was carried out with five (n = 5) students and teachers, three of which had already been consulted in the scenario review sessions. Two users (n = 2) were new test users.

The feedback collected from the evaluations resulted in modifications to the design concept. The analysis of the collected feedback indicated that the users were dissatisfied with the application's starting style, the non-responsiveness of the soft keys (buttons) in the user interface (ui) and the lack of essential functionality. The functional and "proof-of-concept" prototype was then implemented by using technologies, such as PhoneGap, HTML5, CSS3, and various JavaScript libraries. The server-side of the application was implemented using Ruby to build the API and Redis for persistent data storage.

When the application was ready and functional, the usability tests were conducted by applying the Nielsen heuristic evaluation guidelines [56]. Unlike in the usability assessment of the paper prototype, where the users carried out predefined sets of tasks, in this assessment, the users had the freedom to play the game as they wished. In order to gather feedback, the users were asked to pause the game only when they reached a specific stage.

The main goal of this assessment was to measure the consistency and aesthetics of the proposed educational game. A secondary goal was to evaluate the usability of the product. The users were asked to participate in a short interview after the test session.

Figure 9 presents screenshots of the user interface in the business game application.

Figure 9. Screenshots of the business game application

1) Educational activities assessments

In this prototype, the presentation component was not fully implemented although the users were able to select the game character. Moreover, the adjustability factors, such as setting the color and components of the game, were not appropriate for the proposed game. Nevertheless, the users were able to create a profile and select a role image. In this case study, the activity component required students to play and explore the game's features, which all users did properly. Furthermore, the users demonstrated their enjoyment in responses such as *"Wow, this is fun."* In this application prototype, we also did not implement the communication component. Therefore, we ignored this feature in the test session. The personnel administrators showed their satisfaction with the implemented features, such as create and delete, and with the ability to print users' profiles and scores. The users also showed satisfaction with the administrator component in responses: *"I can see who played and what points they gained."*

2) Methodological assessments

Because the owner of the application was in a different city, Skype meetings were conducted after each development phase, such as scenario design, low-fidelity prototype design, and high-fidelity evaluations. Therefore, the owner was considered a stakeholder in the application. The designers considered the mLUX framework an appropriate development model to achieve their goal. The designers particularly appreciated the scenario design in comments, such as *"Scenario saved many development times. I had initially planned to design an application and then check with users, now I know that users want game and not any other mobile application types"* and *"Scenario helped me see the kind of game users preferred and wanted me to implement."* In the starting phase of the project, the designers considered the scenario design an unnecessary step. However, during the project, the designers considered the framework a robust approach to development. The designers, developers, and customers expressed satisfaction with the concept of the application concept and the product. The negative feedback from the designers and developers focused on the extensive user consultancy required, which consumed a great deal of time during the design and development of the application.

D. Case 4: Context-aware nursing application (CANA)

The objective of this case study was to test the proposed mLUX framework process. Several mHealth applications exist, but none combines the essential professional tools for nursing. We tackled this problem in a Finnish elderly care facility. We applied the mLUX framework in a user study. The profiles of 12 nurses and their expectations of work-related mHealth application functionalities were identified. The results were utilized in the conceptual design of the context-aware nurse assistant (CANA) application, which combines the identified functionalities and provides context-sensitive services to consolidate nurses' work activities. The results of this case study are published in the ICUM 2015 conference proceedings [57]. Figure 10 presents sample of application user interface.

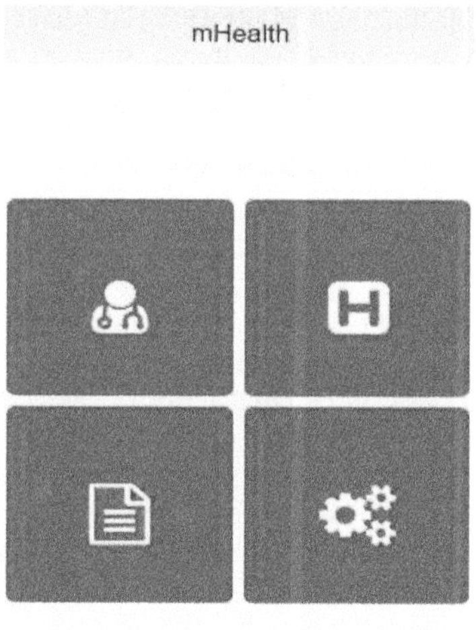

Figure 10. Screenshot of the context-aware user interface

The framework assessment conducted in this case study is briefly discussed in the following sections.

1) Usability and user experience assessment

The prototype was evaluated by four nurses (n = 4) and a doctor (n = 1) ranging age between 21 to 60 years (four females and a male), who also participated in the initial user study phase. We conducted the evaluations in one-to-one sessions with five users at elderly care facilities. The evaluation process was as follows: We gave a set of predefined tasks to the users to perform by using the application individually. We asked the users to think aloud when they carried out the tasks. After the users completed the tasks, we conducted semi-structured interviews with them to collect data on their evaluation experiences.

The objective of this usability evaluation was to measure the prototype's overall design and to assess the concept. The results of the analysis showed that all test users successfully performed the predefined tasks within the given time. The test users considered the application unique and excellent, which would ease their work significantly. Several responses expressed that "I never ever thought that I could perform my work related tasks with my mobile." The test users were surprised at the simplicity of the application. No significant usability or functional failures occurred in this usability assessment.

2) Educational activities assessments

The application in this case study was not designed for educational purposes. Hence, the assessment of the educational components features is not applicable in this project. Nevertheless, we assessed the communication and administration components, which are similar to the m-learning application communication and administration components. The users were delighted to see that through the application, they could access other coworkers in the same department: *"Nice feature; I don't have to go to the office and search for the other nurse in the department"* and *"I can see who I can contact if I need help, great."* Similarly, the results of the assessment demonstrated that the users were able to carry their activities properly and without encountering difficulties. The further assessment of the administration component showed that that users were able to view, search, and print, in addition to adding and removing users.

3) Methodological assessments

A meeting with designers and developers was organized to discuss the process of applying the ppmLUX framework process. The designers considered the process of developing the application straightforward: "It is an excellent method for designing and developing an application without benchmarking." Moreover, the application developers considered the methodology very effective although it involved many role-players. However, the designers complained about the mLUX framework's lack of design guidelines: *"I wish we had the design guidelines to help us get the better user experience for the design."* Furthermore, the designers found that the continuous feedback from users was a motivating factor: *"The scenario evaluation feedback showed that I am doing the right thing."* Furthermore, the designers found the scenario design to be a helpful method for users to conceptualize the final application easily. The designers promoted this scenario as a strength of the proposed mLUX framework process. The discussion notes indicated that all stakeholders of the application were satisfied with the outcome: *"Good model to design a mobile application; I'll apply the UCD framework in my next project."*

IX. Conclusion And Discussion

A. *How to Design M-Learning with Good UX*

The designs and development of application used in smart gadgets, such as smartphones, require special design considerations. Application designs require many iterative phases in order to make the application appealing yet hide the complexity of the application's functionalities. M-learning applications are complex because they deal with learning and learners, who have diverse needs and requirements. Therefore, the design and development of the m-learning applications studied here required the participation of the role-players in design and in development. This ensured that the m-learning application would be efficient and easy to use by both students and teachers.

Nevertheless, the literature review showed that traditional usability assessments were not sufficient for a robust mobile learning application. M-learning applications should be able to compete for students' time and dedication among the entertainment applications in the phone. The results of the four case studies in this project demonstrated that mLUX framework was acceptable for use in m-learning applications. The performance of the mLUX frame was evaluated by role-players, such as students, teachers, and the administrative personnel, designers, and developers. Moreover, the applications tested in the four case studies were assessed based on the educational components of user experience criteria from educational perspective, this assessment helps to evaluate the m-learning application as an educational medium. Furthermore, the results of the case study evaluations revealed that the students and teachers directly influenced the design and functionality of the m-learning application.

This study demonstrated that the successful structure of the mobile learning application did not depend solely on the requirements of the user and the usability of the application. A successful mobile learning application, fulfil the educational components, which include the usability and user experience. Based on the results, we believe that the development of acceptable mobile learning application should use the following three main components, which are shown in Figure 11:

1. The application's features, functionalities, and deliverables enabled the users to perform their essential educational tasks through their mobile learning application.
2. The development technology facilitates usage with a wide variety of the latest wireless and application development technologies across several platforms.

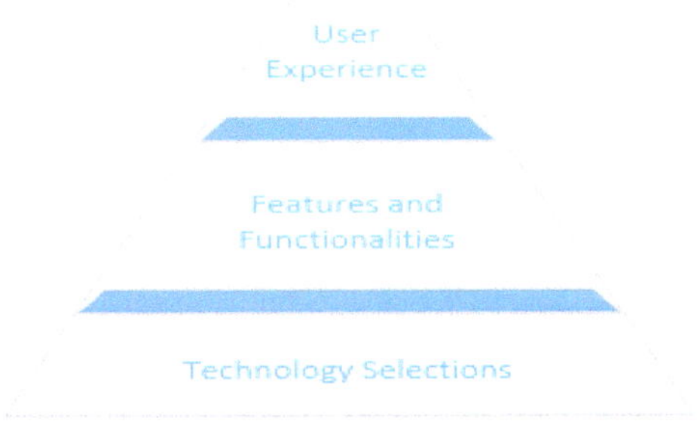

Figure 11. Components in the development of m-learning applications

The results of this study showed that the most prevalent factors in the usability of m-learning applications were adjustability, Delightfulness, reliability, and satisfaction [50]. Smartphones are supported by diverse development technologies and divergent application features. Hence, the design and development of m-learning applications must include their support across various smartphones platforms. Therefore, designers should ensure the compatibility of the m-learning application with the platforms of different smart devices [2]. Hence, the compliance of m-learning applications with smart device-supported technology [2] would result in positive user experiences with the application [40].

The mobile learning application is often used for long periods. In addition, students often need to pay extra attention to learning how to use the application as well as the assigned topic. Therefore, m-learning applications must provide positive experiences for both students and teachers. The experiences that students and teachers gain from using the mobile learning application will ensure the application's sustainability and its long-term utilization in the educational process. Furthermore, entertainment applications, such as games, compete for the time that students spend using their smartphones. These factors increase the importance of ensuring that the m-learning application user has a positive experience. In this study, the design of the m-learning application was such that the students and teachers were entertained by using it. However, the design and function of the application must avoid complexity and help users focus on the content. Factors of user experience, such as adjustability, helped the students and teachers to customize the application based on their own design preferences. The case study applications' evaluations revealed that the adjustability and customization factors are important to gain positive user experience. The adjustability helps students and teachers to consider the m-learning application as their own design. Other attributes of the application, such as delightfulness, indicated a significant effect on the user's retention and sustainability in using the application. In the context of mobile learning, a delightful and enjoyable experience means that the students and teachers are not just satisfied but even entertained by using the m-learning application.

The result indicates that the delightfulness as an important factor in m-learning. Delightfulness is achieved when the target application users, such as teachers and students, create the content of the learning application. Furthermore, when the user interface in the m-learning application is designed to be entertaining, it helps students to avoid complexity in the learning environment and yield an easy comprehension of the learning topics.

In the future, we aim at evaluating the mLUX framework from the design perspective based on the responses to questions, such as "How to create an enjoyable experience?" and "What are the m-learning application sustainable use factors?" The answers to these questions will pave the way towards more adaptive learning, personally adjusted learning experiences, and broader learning contexts.

References

[1] J. Traxler, "Defining mobile learning," in IADIS International Conference Mobile Learning, 2005, pp. 261–266.

[2] A. Dirin, M. Nieminen, and M. Kettunen, "Student capabilities to utilize m-learning service in new smart devices," in Proceedings

of the 2013 International Conference on Advanced ICT, 2013.

[3] N. Golafshani, "Understanding Reliability and Validity in Qualitative Research," Qual. Rep., vol. 8, no. 4, pp. 597–606, 2003.

[4] R. K. Yin, Case Study Research: Design and Methods, vol. 5, no. 5. 2009.

[5] K. M. Eisenhardt, "Building Theories from Case Study Research.," Academy of Management Review, vol. 14, no. 4. pp. 532–550, 1989.

[6] I. Benbasat, D. K. Goldstein, and M. Mead, "The Case Research Strategy in Studies of Information Systems," Manag. Inf. Syst. Q., vol. 11, no. 3, pp. 369–386, 1987. http://dx.doi.org/10.2307/248684

[7] R. Molich, M. R. Ede, K. Kaasgaard, and B. Karyukin, "Comparative usability evaluation," Behaviour & Information Technology, vol. 23, no. 1. pp. 65–74, 2004. http://dx.doi.org/10.1080/0144929032000173951

[8] J. Gulliksen, B. Göransson, I. Boivie, S. Blomkvist, J. Persson, and Å. Cajander, "Key principles for user-centred systems design," Behaviour & Information Technology, vol. 22. pp. 397–409, 2003.

[9] G. He and J. Yu, "Identify Lead Users by Customer Competence," in IEEM2010 - IEEE International Conference on Industrial Engineering and Engineering Management, 2010, pp. 1305–1308.

[10] J. Nielsen, "Usability 101 : Introduction to Usability," All Usability, vol. 9, pp. 1–10, 2003.

[11] J. L. Mcgrenere, "Design: Educational electronic multi-player games - A literature review," Science (80-.)., vol. 25, no. June, p. 2003, 1996.

[12] G. Vavoula and M. Sharples, "Meeting the challenges in evaluating mobile learning: a 3-level evaluation framework," Int. J. Mob. Bended Learn. 1(2), pp54-57., 2009. http://dx.doi.org/10.4018/jmbl.2009040104

[13] M. Ziefle, "The influence of user expertise and phone complexity on performance, ease of use and learnability of different mobile phones," Behaviour & Information Technology, vol. 21, no. 5. pp. 303–311, 2002.

[14] J. Gong and P. Tarasewich, "Guidelines for handheld mobile device interface design," Proc. DSI 2004 Annu. Meet., pp. 3751–3756, 2004.

[15] D. Zhang and B. Adipat, "Challenges, Methodologies, and Issues in the Usability Testing of Mobile Applications," International Journal of Human-Computer Interaction, vol. 18, no. 3. pp. 293–308, 2005.

[16] D. Corlett and M. Sharples, "Tablet technology for informal collaboration in higher education," in Mobilelearning anytimeeverywhere, 2004, vol. Bracciano, pp. 59–61.

[17] A. Hussain and E. Ferneley, "Usability metric for mobile application: a goal question metric (GQM) approach," in Proceedings of the 10th International Conference on Information Integration and Web-based Applications and Services (iiWAS '08), 2008, pp. 567–570.

[18] A. Kukulska-Hulme, "Mobile learning: a handbook for educators and trainers." 2005.

[19] A. Kukulska-Hulme, "Mobile usability in educational contexts: What have we learnt?," Int. Rev. Res. Open Distance Learn., vol. 8, 2007.

[20] P. Syvänen, Antti., & Nokelainen, "Digitaalisen oppimateriaalin mobiilikäytettävyyden arvioinnin kriteerit, 2004, Saarinen Jorma (toim). 5/2004 Evaluator. Hämeenlinna: Hämeen AMK. 2004," 2004.

[21] C. Lindholm and T. Keinonen, Mobile Usability: How Nokia Changed the Face of the Cellular phone. McGraw-Hill, Inc. New York, NY, USA ©2003, 2003.

[22] E. Soloway, M. Guzdial, and K. E. Hay, "Learner-centered design: the challenge for HCI in the 21st century," interactions, vol. 1. pp. 36–48, 1994.

[23] A. Mostakhdemin-hosseini, "Mobile learning framework," in Mobile Learning Framework. Proceedings IADIS International Conference Mobile Learning 2005, 2005, vol. 3, pp. 203–207.

[24] M. Pehkonen and H. Turunen, "Preliminary guidelines for the design of the mobile learning activities and materials.," in Europrix Scholars Conference, 2003.

[25] E. Malliou, F. Maounis, A. Miliarakis, S. Savvas, S. Sotiriou, and M. Stratakis, "The motfal project - Mobile technologies for Adhoc learning," in Proceedings - IEEE International Conference on Advanced Learning Technologies, ICALT 2004, 2004, pp. 910–911.

[26] V. Teija and A. Mikko, "A critical approach to an adaptive user interface design. Learning with mobile devices – research and development - a book of papers. Learning Skills Development Agency: (mlearn 2003)," in mlearn 2003, 2003.

[27] A. Jäppinen, M. Ahonen, T. Vainio, and E. Tanhua-Piiroinen, "Adpative Mobile Learning System: The essential issues from the design perspective. In J.Attewell & C.Savill-Smit (Eds). M-learning: an educational perspective," in MLEARN 2004 Mobilelearning anytimeeverywhere, 2004, pp. 73–75.

[28] D. S. K. Seong, "Usability guidelines for designing mobile learning portals," Proc. 3rd Int. Conf. Mob. Technol. Appl. Syst. - Mobil. '06, p. 25, 2006.

[29] M. Koole and M. Ally, "Framework for the Rational Analysis of Mobile Education (FRAME) model: Revising the ABCs of educational practices," in Proceedings of the International Conference on Networking, International Conference on Systems and International Conference on Mobile Communications and Learning Technologies,ICN/ICONS/MCL'06, 2006, vol. 2006.

[30] T. Magal-Royo, G. Peris-Fajarnes, I. Tortajada Montañana, and B. Defez Garcia, "Evaluation Methods on Usability of m-Learning Environment: Interactive Mobile Technologies," October, vol. 1, no. 1, pp. 32–39, 2007.

[31] L. Uden, "Activity theory for designing mobile learning," International Journal of Mobile Learning and Organisation, vol. 1. p. 81, 2007.

[32] L. S. Vygotsky, Mind in society: The development of higher psychological processes, vol. Mind in So. 1978.

[33] B. Nardi, "Activity theory and human-computer interaction," in Context and consciousness: Activity theory and human-computer interaction, 1996, pp. 7–16.

[34] K. Kari, "Activity Theory as a potential framework for human-computer interaction research," MIT Press, 1995. [Online]. Available: file:///C:/Users/hayes/Downloads/Activity Theory and HCI.pdf.

[35] Y. Engeström, "Activity theory and individual and social transformation," in Perspectives on activity theory, 1999, pp. 19–38.

[36] J. Nielsen and D. A. Norman, "The Definition of User Experience." .

[37] M. Hassenzahl and N. Tractinsky, "User experience - a research agenda," Behaviour & Information Technology, vol. 25, no. 2. pp. 91–97, 2006.

[38] P. Brusilovsky and M. Philip, "Course Delivery System for Virtual University," 2001.

[39] M. Nieminen, "Information Support for User-Oriented Development Organisation. Considerations based on the Construction and Evaluation of Knowledge Storage," Tesis Dr., p. 228, 2004.

[40] J. D. Gould and C. Lewis, "Designing for usability: key principles and what designers think," Communications of the ACM, vol. 28. pp. 300–311, 1985.

[41] J. D. Gould, S. J. Boies, and J. Ukelson, "How to Design Usable Systems," in Handbook of Human Computer Interaction, M. G. Helander, T. K. Landauer, and P. V. Prabhu, Eds. Elsevier Science, 1997, pp. 231–254.

[42] ISO, "Ergonomics of human-system interaction - Part 210: Human-centred design for interactive systems (ISO 9241-210:2010(E))," ISO. pp. 1–32, 2010.

[43] ISO, "ISO 9241 Part 210: Human-centred design for interactive systems," in ISO 9241 Ergonomics of humansystem interaction, 2010.

[44] J. Preece, Y. Rogers, and H. Sharp, Interaction Design: Beyond Human-Computer Interaction, vol. 18. 2002.

[45] P. Friedrich, Web-based co-design. 2013.

[46] D. Wallach and S. C. Scholz, "User-Centered Design: Why and How to Put Users First in Software Development," in Software for People, 2012, pp. 11–39.

[47] A. Dirin and M. Nieminen, "Managing m-learning application development : Roles and Responsibilities," in International Conference on Advanced ICT (ICAICTE-2014)., 2014.

[48] M. Sharples, "Big issues in mobile learning," 2006.

[49] R. M. Kekwaletswe and D. Ng'ambi, "Ubiquitous social presence: Context-awareness in a mobile learning environment," in Proceedings - IEEE International Conference on Sensor Networks, Ubiquitous, and Trustworthy Computing, 2006, vol. 2006 I, pp. 90–95.

[50] K. Eason, Information technology and organizational change. London: Taylor and Francis, 1987.

[51] H. Beyer and K. Holtzblatt, Contextual design: Defining customer-centered systems, vol. 32, no. 2. 1998.

[52] S. Kujala, M. Kauppinen, L. Lehtola, and T. Kojo, "The role of user involvement in requirements quality and project success," 13th IEEE Int. Conf. Requir. Eng., 2005. http://dx.doi.org/10.1109/re.2005.72

[53] M. Kyng, E. Nielsen, and M. Kristensen, "Challenges in designing interactive systems for emergency response," Proc. 6th Conf. Des. Interact., 2006. http://dx.doi.org/10.1145/1142405.1142450

[54] Dirin Amir and N. Marko, "Assessment of User-Centered Design Framework for m-learning Application Development," in 17th International Conference on Human-Computer Interaction, 2015.

[55] A. Dirin and M. Casarini, "Adaptive M-Learning Application for Driving Liceness Candidates Based on UCD Framework for M-Learning Application Development," in 6th International Conference on Computer Supported Education, 2014.

[56] J. Nielsen, "How to Conduct a Heuristic Evaluation," Useitcom. pp. 1–11, 2002.

[57] M. Dirin, A. Dirin, and T. H. Laine, "User-Centered Design of a Context-Aware Nurse Assistant (CANA) at Finnish Elderly Houses," in The 9th International Conference on Ubiquitous Information Management and Communication, 2015.

AUTHORS

A. Dirin is with Haaga-Helia University of Applied Science /BIT/, Helsinki, Finland.

M. Nieminen is with Aalto University, Department of Computer Science, ESPOO, Finland.

A Review of Mobile Learning Applications for Mathematics

Athanasios S. Drigas, and Marios A. Pappas
NCSR DEMOKRITOS, Institute of Informatics and Telecommunications, Net Media Lab, Athens, Greece

***Abstract*—Mobile and online learning applications become more known year after year and are used today from millions of students and educators in all over the world. Wireless mobile devices like smartphones, PDAs and tablets, could be used to benefit students' learning in or out of the classroom. In front of the idea of inclusion of mobile learning in educational process, we represent in this paper some important case studies which examine the consequence of using mobile tools and apps, as well as online applications in mathematics teaching, at all educational levels.**

***Index Terms*—Mathematics education, mobile applications, mobile learning, online applications.**

I. Introduction

The last decades, mobile technologies, such as smartphones, tablets and laptops, as well as online applications and tools, became an integral part of the lives of most teachers and students in all over the world. These devices have transformed the way that people communicate, search for information and work. The challenge for the educators and researchers was to explore how mobile technologies might be used to support learning [10].

With the term mobile learning is meant the delivery of learning to students through the use of wireless internet and mobile devices, including mobile phones, personal digital assistants (PDAs), smartphones and tablet PCs [34].

Mobile learning is an area that develops very quickly and has been considered as the future of learning [24]. Mobile devices enhance anytime and anywhere learning, providing access to learning resources, even outside the school. This flexibility makes it possible for adult learners to minimize their unproductive time, which may enhance their work-education balance [25]. Technological progress can contribute significantly to the improvement and spreading of the use of mobile learning, as handheld devices become lighter, cheaper, with better screen analysis, longer battery life and faster network speed.

Naismith et al (2004) suggested a pedagogy-model-based classification of mobile learning with six categories: 1) behaviorist, 2) constructivist, 3) situated, 4) collaborative, 5) informal and lifelong learning and 6) support for learning and teaching [23].

In recent years, researchers developed online and mobile applications to support teaching in Algebra, Geometry, Mathematical Analysis, Statistics and other areas of mathematics. Mobile math applications allows users to explore functions, providing graphical capabilities and offer many kinds of specific calculators. There are apps designed to handle measurement tasks and educational apps for practicing on numerical and mathematical skills. Technologies that provide support for mathematics on the web have also been increasing over the last decade. Online and mobile educational tools for mathematics can assist students' problem solving, enhance comprehension of mathematical concepts, provide dynamically representations of ideas and encourage general metacognitive abilities [29]. The frequent use of mobile technologies in the course of mathematics, would help students to improve their skills on the one hand, and on the other would encourage the improvement of mobile learning applications.

II. Smartphones

A. Graphs and functions

Botzer et al (2007) presented a pilot case study, where participated four female mathematics students, studying for a teaching certificate. The project was based on Math4Mobile, a cellular application for mathematics learning. This mobile learning environment includes Sketch2Go, an application which allows users to sketch graphs, increasing and decreasing functions and make visual exploration of phenomena, and Graph2Go, a graphing calculator for dynamic transformation of functions. The project included also the use of cellular video camera to record occurrences, MMS messages to exchange videos between participants and SMS to exchange verbal messages. As the researchers claimed, the contribution of the mobile environment enables the use of mathematical applications anytime and anywhere, encourage the performance of mathematical operations and enhance experiential learning [4].

Daher et al (2009) made an experiment for learning mathematics in an authentic mobile environment, which took place in an Arab middle school in Umelfahn, Israel. In this experiment participated 32 8th grade students who volunteered and owned cellular phones. Learning carried out in outdoor activities, where students could study mathematics concepts through exploration and investigation with their mobile phones. Students used algebraic midlets, from the site of Institute for Alternatives in Education (www.math4mobile.com), to see the graphs of several templates of linear functions. As the results showed, the environment of mathematics learning using mobile phones enables independent and collaborative learning in authentic real life situations, engages students in various mathematical actions and makes learning mathematics easier and faster [2].

In the same year, Daher accomplished another study to find how middle school students respond to learning mathematics with cellular phones and web applets. In this study participated 15 9th grade students. The mostly used applets by the students were Graph2Go, which enables graphs and functions drawing and Fit2Go, for linear and quadratic functions. Each one of the participants was interviewed about his or her perceptions about learning mathematics using cellular phones or applets, the differences between using cellular phones and web applets, as well all his/her attitudes about using these learning applications in the future. The participants highlighted that learning mathematics using cellular phones and applets enhances solving mathematical problems in short time, without much effort and precisely, ensures the correction of their solutions, and makes them more active in their learning [3,6].

B. The laws of arithmetic

Zhao and Okamoto (2009) from University of Electro-Communications in Japan, introduced a Mobile Mathematics Tutoring (MoMT) system for primary school students, based on individual learner's abilities. The system analyses the user's learning profile, in order to provide personalized mathematics tutoring and exercises. The system also allows discussion between students, so they can exchange ideas, experiences and questions, via email, text messaging, photos, audio recordings and videos. MoMT lays out its content using XHTML Basic, cHTML and XHTML MP. This mobile tutoring system can improve arithmetic skills and student interest in learning mathematics concepts [40].

In 2010 Diah, Ehsan and Ismail developed a promising mobile educational game for primary mathematics education." MathRush" was designed to support mathematics' learning outside of the classroom. Its framework is consisted of four parts: (1)learning theories, (2)mobile learning approach, (3)games development approach and (4)learning and education medium. The game criteria include goals, rules, competition, challenge, fantasy and entertainment. It is a fun way for children to practice basic mathematical facts such as addition, subtraction, multiplication and division. The functional areas of the prototype, such as stability, correctness and integrity of the game assets have been successfully tested on Java Wireless Toolkit Emulator platform and on Sony Ericsson K750i. MathRush is a simple but very useful game for students who just started to study mathematics [7].

Zaranis et al (2013) developed sixteen different activities for mobile teaching realistic mathematics in kindergarten education. The project was based on the principles of Realistic Mathematics Education (RME) for preschool education, which was developed in the Netherlands, targeting fundamental mathematical concepts for kindergarten students. The three levels developed during kindergarten education, aimed at counting and calculating simple additional and subtraction, calculating addition and subtraction problems, and calculating addition and subtraction using a missing variable. The applications they created used the App Inventor application development software and were tested on the Android operating system. The produced software should be user friendly to kindergarten children, without requiring reading and writing knowledge, combining animation and sound [39].

C. Algebra

Roberts et al (2011) presented the Nokia Mobile Learning for Mathematics project, in South Africa. In the 24 project, from January to June 2010, participated about 3000 10th grade mathematics learners from 30 schools across three provinces in South Africa. With the voluntary participation in this project, students and teachers had access to interactive mathematics learning materials through a mobile platform with a social media application support. Learners could work through theory sections and answer questions from a database of 10000 questions (multiple choice, true or false, spot the error, open-ended questions), categorized by topic and difficulty. Results showed that it is possible to use networking platform for teenagers' mathematics homework. However, the establishment of this kind of teaching presupposes the equal access of the students to mobile devices, in all over the country [30].

One year later, Kalloo and Mohan presented "MobileMath", a mobile learning application designed to help secondary school students improve their performance in algebra. The application, which is available on mobile phones with internet access, offers lessons, examples, tutorials, quizzes and games that support users to practice certain mathematical skills. Three evaluation studies with 57 students were carried out to examine the effectiveness of the learning application. Questionnaires, pre-tests, post-tests and interviews were used for data collection. According to the results of the analysis, most of the students enjoyed learning activities, especially games and they thought that the application helped them improve their performance in algebra [15].

D. Geometry

Wijers et al (2008) created an interesting mobile gaming and learning environment, based on geographical reality, maps and location technologies, to support 12-14 year old students. MobileMath is played on a mobile phone with a GPS receiver.The basic goal of the experiment was to help students experience mathematical concepts in the physical world. Users, playing in teams, gain points by creating virtually constructed mathematical shapes (squares, rectangles or parallelograms). The construction process was done by physically walking and clicking on the location for each vertex. To answer the research questions, they made a pilot study with 60 students in three different schools. After the game, 54 of the players completed a questionnaire. The students understood easily the goal and the rules of MobileMath. The collaboration within teams went well and there was no problem with phone using [36,37].

Kaufmann et al (2003) introduced Construct 3D, a three dimensional geometry construction tool, designed for mathematics and geometry education at high school and university. It is based on the mobile collaborative augmented reality system Studierstube, described by Schmalstieg et al in 2002. Construct 3D uses augmented reality (AR) to allow users to share a virtual space. AR allows users to see their own body and hand, as well as the results of their movements while they work. This tool offers a set of functions for construction of geometric shapes such as lines, spheres, cylinders and cones. Teachers, students, colleagues and friends who tried the system, gave feedback with comments and observations to help designers develop the application [17,18].

Tangney et al (2010), working on MobiMath, an approach to utilizing smartphones in teaching mathematics, created and presented two learning applications. "Angle

Tool" uses smartphone's accelometer to represent the rwadout of the angle at which the phone is being held. Teaching objectives of the software are based on the geometry and trigonometry curriculum. "Cuisennaire Roots" designed to explore the addition, subtraction and equivalence of fractions. The rods consist rectangles of different colour and length which represent different fraction values. The tool enhances collaborative learning experiences, organizing groups, in which students must communicate to find the correct solutions [31].

III. Tablet PCs And Ipods

A. Algebra

In 2006 in the University of Southern Queensland, Australia, Galligan et al examined how the tablet PC could be used in the lectures in Foundation Mathematics. Teachers in Algebra and Calculus, used tablet PCs for their lectures instead of writing on the blackboard or using a projector. As the teacher was solving a problem on the tablet PC, students could see the solution in their own tablet, send immediate answers or questions to the teacher and of course they could keep a record of the lecture. The experiment was very helpful for the students, as they could understand the lecturer easier and use the recorded lectures for revision. They also used tablet PCs for distance learning in higher education. Teachers trailed web conferences and online chat sessions via tablets for tutoring in the subjects Algebra, Calculus and Data Analysis. According to the students' feedback, the online sessions motivated them and enhanced communication with other students [11].

Franklin et al (2008) developed a case study in a school of Southeastern Ohio, to help middle school students to learn algebraic equations, the concept of scope, absolute value and elimination. They created math videos that could be used by the students to study anywhere and anytime, using mobile devices. In this four weeks study participated 39 8th grade mathematics students, arranged in two groups: MsClass1 group contained 22 students with multiple academic levels and MsClass2 group, with 17 students included special education students and low achievers. The materials and the software used, were iMac computers, iPod touch, Microsoft Office 2003, iMovie on the Macintosh, Adobe Photoshop and iTunes 7,6.The results showed that students in both groups loved to study with iPods in group or alone and showed significant improvement in understanding algebraic concepts [9].

In 2009 O'Connell et al proposed a pen based tablet application. Algosketch is a system for fluid pen based entry and editing for mathematics with support for interactive computation. The system supports entry of basic math symbols (∞, R, e.t.c.), math relations (≤, ≥, ≠, ≈, ∩, e.t.c.), integrals, summations, roots, trigonometric, functions and matrices. Using Algosketch enables specification and computation of mathematical expressions, specification of matrices and recognition of correct input. In order to evaluate the effectiveness of the application, six subjects interacted with Algosketch via Compaq tc4400 tablet PCs which were connected to a local area connection. The evaluation of the Algosketch prototype showed that it is a promising and useful tool, which offers pen entry that is as easy as writing on paper [27].

Kiger et al (2012) examined the effect of Mobile Learning Intervention (MLI) in mathematics education of third grade students. In their nine week study participated two experiment groups at Midwestern's Park elementary school, during the third quarter of the 2010-11 schoolyear. All school classrooms were equipped with computer, internet, mobile smartboard, iPads and iPods. According to the curricula, the learning goals of the intervention were multiplication and division facts. The comparison group, consisting of 46 students and 2 teachers, trained using usual educational techniques and the MLI group, with 41 students and 2 teachers trained using iPod touch devices, equipped with math applications such as Multiplication Genius, Mad Math, Pop Math, Flash to Pass, Math Drills, Math Magic, Flowmath and Multiplication Flashcards to Go. Due to the study's results, on average MLI students answered more questions correctly on the post-intervention test than the comparison students [20].

Zanchi et al (2013) described in their paper the Next generation Preschool Math project (NGPM). Each unit of the project compines digital games for both individual and collaborative play and non-digital activities to support students' learning. NGPM focus on two mathematics topics, subtizing and equipartitioning, providing rich mathematics learning opportunities for young children to understand concepts like number, quantity and ratio. The system comprises of digital and non-digital assets and a teacher's guide. Digital Assets' set consist eight tablet applications: Sara Skates, Birthday café, Jungle Gym, BubbleFun, Breakfast Time, Lemonade Stand, Photo Friends and ParkPlay [38].

B. Geometry

At the University of Salzburg in 2002, Markus Hohenwarter et al presented GeoGebra, a project designed to combine features of interactive geometry software and computer algebra systems. It might be used in teaching for demonstration and visualization, as a construction tool, for preparing teaching materials and a helpful tool for discovering mathematics. The system used for years a built-in a computer algebra system (CAS) for symbolic computation. They developed the symbolic features of GeoGebra 4,2 to allow students to work on fractions, equations and formulas that include undefined variables. The summer of 2013 they developed the last version, GeoGebra 4,4, which runs on iPad, Adroid tablets and Chromebooks and has been translated to more than 50 languages. In parallel they designed GeoGebraTube, a website which supports direct uploading of constructions and allows users to rate tag and comment the materials. The upcoming version of GeoGebra 5, will include a fully dynamic 3D for three dimensional geometry and graphics [13,14].

Barry Kissane (2004) presented an interactive hand held educational tablet device, the Casio ClassPad 300 [21,22]. The device includes:

- A constraint-based geometry capability (HREF1)
- A stylus for dragging and dropping expressions
- Dual windows, which allows users to select a variety of potentially windows to explore links
- A measurement bar which provides measurements of various aspects of figures that have been drawn.
- A Geometry link inside an eActivity, designed to structure and direct a user's activity

C. *Arithmetic Skills and Numerosity*

Barendregt et al (2012) developed and evaluated a mathematics iPad game using multi touch interaction. The Fingu Game was designed to develop fundamental arithmetic skills, for 4 to8 year old children, focusing on conceptual subtizing, numerosity and finger gnosis. In Fingu, one or two small sets of objects are shown and the player has to count how many objects are shown and place the same number of fingers on the screen, before time runs out. There are 7 levels of difficulty in the game and if the player answers incorrectly, he loses a heart. In the three weeks pilot testing participated 11 children, 5-6 year old, from a preschool in a small city in Sweden. The more the students played the game, the more they increased the percentage of correct answers [1].

In 2003, Bull et al proposed MoreMaths, an interaction that takes place on both a desktop PC and a handheld computer. The main tutoring session of MoreMaths is on the desktop PC, where students are provided with the appropriate learning materials, which are automatically synchronized to the handheld device, that can be used anywhere and anytime [5].

Petty et al (2007) made a 6 weeks study at Glendale elementary school in Flinton, Pensylvania, to evaluate the intervention of a tablet PC mathematics software in elementary education. Three 4th grade classes with 60 students participate in the study. They used six HP 1100 tablet PCs, equipped with the Tablet Math System. This system, which was developed at Carnegie Mellon University, provides teachers with the ability to view how students attempt problems, as well as general statistics on class' performance. Tablet Math System is consisted of two complementing applications, a thin client application installed on the tablet PCs and a web application, used by teachers to assign exercises and view students' results. Teachers had difficulty integrating the system directly into their curriculum, although they claimed that the system enhanced their teaching [28].

IV. Personal Digital Assistants (PDAs)

Ketamo et al (2002) presented a two stages study for kindergarten's geometry teaching via mobile learning. In the first stage of the study, there were three groups of 6-year-old children: the experiment group 1, consisting 21 participants who trained with a geometry game for PC using dynamic illustrations, the experiment group 2, consisting 20 participants who trained with a geometry PC game using static illustrations and the control group with 30 children. All groups were pre and post tested for their geometry skills. On the second stage of the study, two years later, there was only one experiment group with 17 pupils, who trained on an adaptive geometry game on Compaq iPaq (PDA), connected to a server with a wireless network. According to the study's results, low skills students benefits more from the mobile learning than the averagely skilled. The PDA version of the geometry game worked extremely well, although some development has to be made [19].

Wachira et al (2009) developed a study, in which participated 20 middle school teachers, enrolled in a Midwestern university's mathematics teacher preparation course. The purpose of the study was to train the teachers on inquiry based mathematics content integrating PDAs' (Personal Digital Assistants) to teach middle school mathematics students. The content of the training was based on the National Council of Teachers of Mathematics (NCTM 2000) including data analysis, ratios, geometry and measurement. At the end of the sessions, students were given a questionnaire to express their perceptions of this new project. Most of the teachers agreed that PDA was a valuable educational technology tool and would like to learn more on how to use PDAs for instructional activities [33].

V. Online Apps And Tools

In recent years there have been designed and presented several online learning applications and tools for mathematics. These learning applications could be used from the students anytime and anywhere, via mobile devices using wireless connection.

Wang et al (2003) designed an open Web-based Mathematics Education (WME) system, using standard internet technologies. The WME framework allows easy and systematic development of an unlimited set of mathematics education contents and mathematics education support capabilities. The framework includes the Mathematical Education Markup Language (MeML), which provides markup elements to represent mathematics education pages, web servers to deliver pages, WME page processors to receive MeML pages and WME services including mathematics knowledge, computation, mathematics education and pedagogy services [35].

A. *Algebra and problem solving*

AGILMAT is a web application, designed by Tomas, Leal and Domingues in 2007 to help students learn mathematics and especially high-school algebra. It is a constraint logic programming application, which allows users to generate mathematics drills automatically, depending on the grade, the topic and the difficulty chosen and provides a one line solution for each exercise. Users can also refine exercise generation by setting different values for particular parameters [32].

In 2005 Nguyen and Kulm used web-based practice to enhance mathematics learning and achievement for middle school students in Southeast Texas. In their study participated 95 students from 6 math classes, randomly assigned to one of two treatment groups. Students in the web-based assisted learning and practice (WALA) group, did their H/W (fraction operations set and converting demicals to fractions set) in the computer labs, using Javascript, Perl and HTML files, during the 3 weeks of the experiment. They could check immediately if their answers are correct and receive adapted feedback for each answer. Participants of the traditional assisted learning and practice (TALA) group, used paper and pencil to practice on the same H/W. The web-based approach of intervention brought better results and motivated students to do their homework [26].

Edwards et al (2010) presented 4MALITY, a web-based mathematics tutoring system, designed to teach mathematical problem-solving skills and test-taking strategies for 4th grade students [8]. 4MALITY adapts to the students' levels of knowledge, by using an Artificial Intelligence decision system. The system's mathematics content is based on the Massachusetts Curriculum Framework. There are four categories of learning styles provided:

- Explain questions in terms of language used.
- Mathematical computational operation.
- Test-taking and problem solving strategies.
- Visual approaches to computation.

B. Dynamic Geometry Software

Bülent Güven and Temel Kosa examined the effect of dynamic geometry software (DSG) Cabri 3D on student mathematics teachers. This three dimensional geometry tool, launched in 2004, provides an environment in which students can explore geometric relationships and construct geometrical objects. Forty student mathematics students of Karadeniz Technical University participated in the 8 week period study. Purdue Spatial Vizualization (PSV) test, which consists of 36 questions in three sections (Development, Rotations and Views), was used as pretest and posttest for data collection. The results of the tests, before and after intervention, showed that students' spatial skills improved significantly [12].

C. Mathematical Programming

Karagiannis et al (2006) designed an online software application to facilitate the education of mathematical programming. WebNetPro is an online application for students, mathematicians and operations research scientists. It is a powerful educational tool for Linear Network Programming, which can assist the teaching process of courses such as Graph Theory, Network Optimization Algorithms and Data Structures. The tool helps users to verify their calculated solutions or convert one network representation to another. WebNetPro is implemented in the Matlab computing environment uses a graphical user interface (C.U.I.) for solving problems [16].

VI. CONCLUSIONS

The purpose of this study was to examine the most representative studies of recent years, involving online and mobile applications and tools for mathematics as well as their effect in the educational process. We presented some innovate applications which are addressed by kindergarten children to university students. These apps could be used to improve arithmetic skills, numerosity, for graphs' representation, geometrical objects constraction, algebra problem solving and mathematical programming. The results of the studies revealed that online and mobile learning applications motivated the students, making mathematics course more enjoyable and interactive than the ordinary teaching practices. The work of the developers gives promises for more learning tools in the near future, creating a new educational model.

REFERENCES

[1] Barendregt, W., Lindström, B., Rietz-Leppänen, E., Holgersson, I., & Ottosson, T.: Development and evaluation of Fingu: a mathematics iPad game using multi-touch interaction. In *Proceedings of the 11th International Conference on Interaction Design and Children*, pp 204-207. ACM.(2012)

[2] Baya'a, N. F., & Daher, W. M.: Learning Mathematics in an Authentic Mobile Environment: the Perceptions of Students. *iJIM*, *3*(S1),pp. 6-14.(2009)

[3] Baya'a, N., & Daher, W. : Students' perceptions of Mathematics learning using mobile phones. In *International Conference on Mobile and Computer Aided Learning* (Vol. 4).(2009)

[4] Botzer, G., & Yerushalmy: M. Mobile application for mobile learning. In *Proceedings of the International Conference on Cognition and Exploratory Learning in Digital Age* ,pp. 7-9.(2007)

[5] Bull, S., Cui, Y., McEvoy, A. T., Reid, E., & Yang, W. (2004, March). Roles for mobile learner models. In Wireless and Mobile Technologies in Education, IEEE International Workshop on (pp. 124-124). IEEE Computer Society.

[6] Daher, W. : Students' perceptions of learning mathematics with cellular phones and applets. *International Journal of Emerging Technologies in Learning (iJET)*, *4*(1).(2009)

[7] Diah, N. M., Ehsan, K. M., & Ismail, M. (2010). Discover mathematics on mobile devices using gaming approach. Procedia-Social and Behavioral Sciences, 8, pp.670-677.

[8] Edwards, Robert W. Maloy Sharon A., and Gordon Anderson:Teaching math problem solving using a web-based tutoring system, learning games, and students' writing. Journal of STEM Education, 11(1,2), pp.82-90.(2010)

[9] Franklin, T., & Peng, L. W.: Mobile math: Math educators and students engage in mobile learning. *Journal of Computing in Higher Education*, *20*(2), pp.69-80. (2008)

[10] Frohberg, D., Göth, C., & Schwabe, G.: Mobile learning projects–a critical analysis of the state of the art. *Journal of Computer Assisted Learning*,*25*(4),pp. 307-331. (2009)

[11] Galligan, L., Loch, B., McDonald, C., & Taylor, J. A. (2010). The use of tablet and related technologies in mathematics teaching. Australian Senior Mathematics Journal, 24(1), pp.38-51.

[12] Guven, B., & Kosa, T.: The Effect of Dynamic Geometry Software on Student Mathematics Teachers' Spatial Visualization Skills TOJET, *7*(4). (2008).

[13] Hohenwarter, M.:GeoGebra 4.4–From desktops to tablets. *Indagatio Didactica*, *5*(1).(2013)

[14] Hohenwarter, M., & Fuchs, K.: Combination of dynamic geometry, algebra and calculus in the software system GeoGebra. In *Computer Algebra Systems and Dynamic Geometry Systems in Mathematics Teaching Conference. Pecs, Hungary*.(2004)

[15] Kalloo, V., & Mohan, P. (2012). MobileMath: An innovative solution to the problem of poor Mathematics performance in the Caribbean. The Caribbean Teaching Scholar, 2(1), pp.5-18.

[16] Karagiannis, P., Markelis, I., Paparrizos, K., Samaras, N., & Sifaleras, A.:E-learning technologies: employing Matlab web server to facilitate the education of mathematical programming. *International Journal of Mathematical Education in Science and Technology*, *37*(7), pp.765-782.(2006)

[17] Kaufmann, H., & Schmalstieg, D.: Mathematics and geometry education with collaborative augmented reality. *Computers & Graphics*, *27*(3),pp.339-345.(2003)

[18] Kaufmann, H., Schmalstieg, D., & Wagner, M. Construct3D: a virtual reality application for mathematics and geometry education. *Education and information technologies*, *5*(4),pp.263-276.(2000)

[19] Ketamo, H.: mLearning for kindergarten's mathematics teaching. In*Wireless and Mobile Technologies in Education, 2002. Proceedings. IEEE International Workshop on* ,pp.167-168.(2002)

[20] Kiger, D., Herro, D., & Prunty, D.:Examining the influence of a mobile learning intervention on third grade math achievement. *Journal of Research on Technology in Education*, *45*(1), pp.61-82.(2012)

[21] Kissane, B.: Interactive mathematics on the ClassPad 300. In: MAV Annual Conference 2004, 2 - 3 December 2004, Monash University, Clayton Campus, VIC, pp.259-269.(2004)

[22] Kissane B.: *Mathematics education and the iPod Touch.* In: AAMT - MERGA Conference 2011, 3 - 7 July 2011, Alice Springs, N.T. pp. 932-940 (2011)

[23] Laura Naismith, Mike Sharples, Giasemi Vavoula, Peter Lonsdale. Literature Review in Mobile Technologies and Learning. A NESTA Futurelab Series - report 11.(2004).

[24] Trifonova, A. : Mobile learning-review of the literature.(2003)

[25] Motiwalla, L. F. Mobile learning: A framework and evaluation.*Computers & Education*, *49*(3), pp.581-596. (2007)

[26] Nguyen, D. M., & Kulm, G.:Using web-based practice to enhance mathematics learning and achievement. *Journal of Interactive Online Learning*,*3*(3), pp.1-16.(2005)

[27] O'Connell, T., Li, C., Miller, T. S., Zeleznik, R. C., & LaViola Jr, J. J.: A usability evaluation of AlgoSketch: a pen-based application for mathematics. In *Proceedings of the 6th Eurographics Symposium on Sketch-Based Interfaces and Modeling*, pp.149-157. ACM.(2009)

[28] Petty, D. D.: Integration and Perception of Tablet PC Mathematics Software in Elementary Mathematics Education.Dietrich College of Humanities and Social Sciences at Research Showcase @ CMU(2007)

[29] Pierce, R., Stacey, K., & Barkatsas, A. : A scale for monitoring students' attitudes to learning mathematics with technology. *Computers & Education*, *48*(2),pp. 285-300.(2007)

[30] Roberts, N., & Vänskä, R. :Challenging assumptions: mobile learning for mathematics project in South Africa. *Distance Education*, *32*(2),pp.243-259.(2011)

[31] Tangney, B., Weber, S., Knowles, D., Munnelly, J., Watson, R., Salkham, A. A., & Jennings, K.: MobiMaths: An approach to utilising smartphones in teaching mathematics MLearn.Malta.(2010)

[32] Tomás, A. P., Leal, J. P., & Domingues, M. A.: A web application for mathematics education. In *Advances in Web Based Learning–ICWL 2007* , Springer Berlin Heidelberg, pp. 380-391. (2008)

[33] Wachira, P., Keengwe, J., & Onchwari, G.:Personal Digital Assistants (PDAs) in Mathematics Teacher Education. *AACE Journal*, *17*(2), pp.113-125.(2009)

[34] Wang, Y. S., Wu, M. C., & Wang, H. Y. (2009). Investigating the determinants and age and gender differences in the acceptance of mobile learning. British Journal of Educational Technology, 40(1), pp.92-118.

[35] Wang, P. S., Kajler, N., Zhou, Y., & Zou, X.: WME: towards a web for mathematics education. In *Proceedings of the 2003 international symposium on Symbolic and algebraic computation* ,ACM, pp. 258-265.(2003)

[36] Wijers, M., Jonker, V., & Kerstens, K.: MobileMath: the Phone, the Game and the Math. In *Proceedings of the European Conference on Game Based Learning, Barcelona,* pp.507-516.(2008)

[37] Wijers, M., Jonker, V., & Drijvers, P.: MobileMath: exploring mathematics outside the classroom. *ZDM*, *42*(7), pp.789-799.(2010)

[38] Zanchi, C., Presser, A. L., & Vahey, P.: Next generation preschool math demo: tablet games for preschool classrooms. In *Proceedings of the 12th International Conference on Interaction Design and Children*, pp. 527-530. ACM.(2013)

[39] Zaranis, N., Kalogiannakis, M., & Papadakis, S.: Using Mobile Devices for Teaching Realistic Mathematics in Kindergarten Education. *Creative Education*, *4*(07), 1.pp.1-10.(2013)

[40] Zhao, X., & Okamoto, T. (2009). A Personalized Mobile Mathematics Tutoring System for Primary Education. Journal of the Research Center for Educational Technology, 4(1), pp.61-67.

AUTHORS

Athanasios Drigas is a Director of Research at N.C.S.R. Demokritos. He is the Coordinator of Telecoms Lab and founder of Net Media Lab since 1996. From 1985 to 1999 he was the Operational manager of the Greek Academic network. He has been the Coordinator of Several International Projects, in the fields of ICTs, and e-services (e-learning, e-psychology, e-government, e-inclusion, e-culture etc). He has published more than 260 articles, 7 books, 25 educational CD-ROMs and several patents. He has been a member of several International committees for the design and coordination of Network and ICT activities and of international conferences and journals. Last but not least, he has received several distinction for his scientific work (articles, patents, projects) (e-mail: dr@iit.demokritos.gr).

Marios Pappas (MA in Inclusive Education student) is a secondary school Mathematics teacher. He is also a research associate at N.C.S.R. Demokritos, Institute of Informatics and Telecommunications, Net Media Lab, Athens, Greece. (e-mail: marios214@hotmail.com)

Online Laboratory in Digital Electronics Using NI ELVIS II+

Ahmed Naddami[1], Ahmed Fahli[1], Mourad Gourmaj[1], Mohammed Moussetad[2]
[1]Université Hassan 1, Settat, Maroc
[2]Université Hassan 2 Mohammedia, Casablanca, Maroc

Abstract—**This work is part of the e-Sience project, which is realized in the framework of the Tempus project coordinated by Bordeau University and financed by the European Union. The ''e-Sience'' project aims to create a network of online labs in the Maghrebian countries.**

The present paper presents our recent work to extend our online laboratory, the Khouribga OnlineLab at Hassan the 1st university in Morocco. It will focus on the development of a digital online laboratory supporting experiments in the fields of combinational logic circuits, sequential electronics, and digital multiplexer. The developed system combines NI ELVIS II+ and the NI digital Electronics FPGA board.

By using the interactive version of iLab Shared Architecture, students will have complete access to manage and control the system in real time.

Index Terms—**e-Sience; iLab; OnlineLab; Measurement; real time; online labs; FPGA; ELVIS.**

I. Introduction

Due to growth in the number of students at Hassan the 1st university and the expensive nature of experiments, the existing hands-on laboratories are not enough. Furthermore, since modern instruments are not being acquired, practical work is limited to conventional laboratories, so a large number of students cannot get the experiences they need. [1]

However, implementing online labs should help faculty overcome the lack of experiments by giving students access to the online labs at any time and from anywhere. For this reason, we are engaged in the ''e-Sience'' project to implement online labs with other partners, especially from France, Austria, Romania, Greece, Tunisia and Algeria.

In this paper, we present the implementation of an online laboratory in digital electronics, which is part of the second year courses in our engineering programs. This laboratory represents an ideal and integral component in engineering education. It helps students understand theoretical concepts, develop their social skills, and maintain their professional to prepare them for professional life. Our online lab will give students the chance to access courses and remote practical work via the internet.

Remote sessions are based on the development of internet technologies and science information technology that help both students and teachers access real experiments on real hardware through the internet. It presents high quality features and is easy to use. Recently, remote laboratories have widely been adopted by several institutes, especially in electrical and electronic engineering as well as digital and sequential electronics. These disciplines are the target of this work [3].

It is almost impossible to find a part of society that has not been touched by digital electronics [4]. They have become the most significant development in the history of society [5]. Digital electronics is found in all areas of air/sea/land defense, like semiconductor components, ASIC design, FPGA and CPLD programming, IP address blocking design, design of electronic cards.

The present work focuses on the development and implementation of an online digital laboratory to enhance education in the electrical and electronic engineering fields based on the "heart" of remote laboratories: NI ELVIS II+' Educational Laboratory Virtual Instrumentation Suite' [6].

II. Khouribga OnlineLab

The Khouribga OnlineLab in Hassan 1st university in Morocco is one of several laboratories developed in Tunisia and Algeria within the e-Sience project (Fig.1). e-Sience was inaugurated on the 28th of June 2013 in Khouribga and is considered to be the first laboratory developed in the e-Sience project group based on the Interactive Shared Architecture (ISA) available via http://www.onlinelab-uhp.ma/.

e-Sience is seeking to create a network of remote labs in the countries already mentioned to modernize the educational processes in science and technology [7 and 10]. It consists of

- Implementing a network of online labs in Maghrebian countries.
- Modernizing the educational processes in Maghrebian countries.
- Adapting content to the developments in science and technology.
- Taking into account the expectations of the professional world.
- Implementing practical courses for distance education.
- Creating teaching units in an e-Learning format
- Setting up and using a Learning Management System (LMS).
- Evaluating the newly created educational resources
- Using the pedagogical resources.

Since its inception, the Khouribga, OnlineLab (Fig.2) has been cooperating with other remote laboratories. Recently, it has been cooperating with WebLab-Deusto, Bilbao, which is an open source RLMS (Remote Laboratory Management System) developed by the University of Deusto in Spain in October 2004[8]. The cooperation started through experiments with the VISIR (Virtual Instrument in the Reality) electronic platform of WebLab-Deusto from our own Laboratory by using the federation of remote laboratories (Fig.3), which is aiming to enable two or more different systems. These systems, located in different entities (universities, secondary schools, research centers), can share laboratories [9].

III. Related Work

A. Architecture System

1) Hardware Implementation

The hardware part of this online digital laboratory contains mainly the NI ELVIS II+ and the NI Digital Electronics FPGA Board.

a) NI ELVIS II+

NI ELVIS is low cost and is the most used hardware platform for developing remote laboratories in several fields: concepts in circuits, microcontrollers, controls, telecommunication, measurements and digital electronics [6]. NI ELVIS is a hands-on design and prototyping platform that integrates the 12 most commonly used instruments, including oscilloscope, digital multimeter, function generator, and bode analyzer. This work is based on the implementation of digital electronics. So, NI ELVIS has been specifically defined to use Boolean logic, digital electronics, and FPGA concepts at college and university laboratories [11].

NI ELVIS uses a USB cable to connect the workstation to the computer. It can also be programmatically accessed and controlled within NI LabVIEW, a visual programming language based on a graphical block diagram user interface.

b) NI Digital Electronics FPGA Board

The new NI Digital Electronics FPGA "Field-programmable gate array Board" is ideal for teaching digital electronics and digital design concepts. It is a circuit development platform based on the XC3S500E Xilinx Spartan-3E FPGA that was designed to help educators teach digital electronics with access to real world signals and instrumentation. It also includes the necessary I/O to teach and use the basics of advanced digital electronics. The most important characteristics of the NI Digital Electronics FPGA Board are fully integrated and very useful with the NI ELVIS platform. This integration gives the ability to build comprehensive test bed benches and mixed-signal circuits that can be designed and tested in one platform (Fig.4) [12].

2) Software Implementation

To make the digital online laboratory remotely accessible via internet, we are using the interactive version of ISA [13], based on a web service that provides a management system for sharing laboratories among different universities. It offers a complete tool to users, such as user account management, scheduling and data storage.

The ISA is a robust, scalable, open-source infrastructure that provides a framework for the development and deployment of remote laboratories. ISA is divided into different kinds of laboratories: ***Batched laboratories*** and ***Interactive Laboratories*** [14]-[15]. In our laboratory, we are using the second category, which supports real time

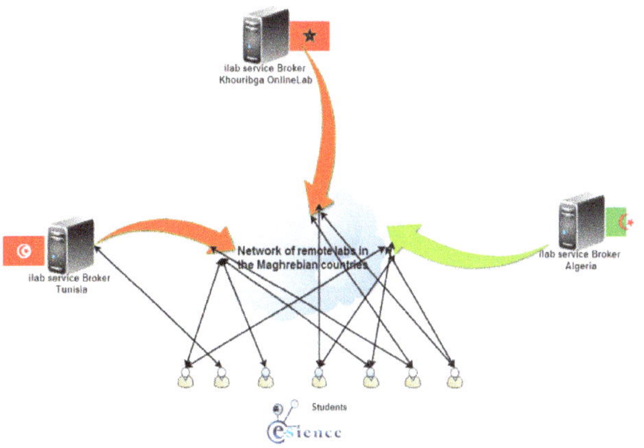

Figure 1. Network of online labs in maghrebian countries

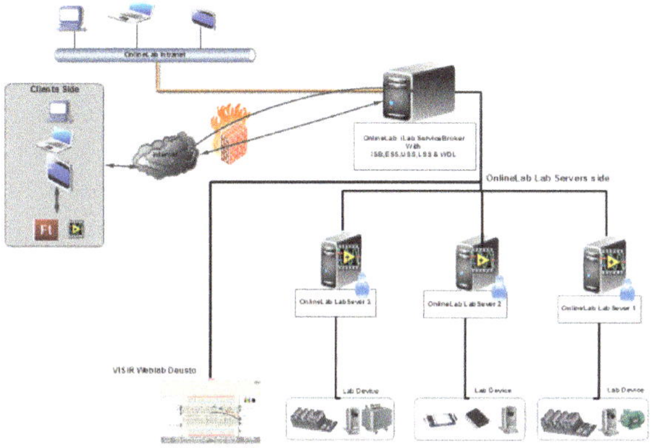

Figure 2. Khouribga OnlineLab Architecture

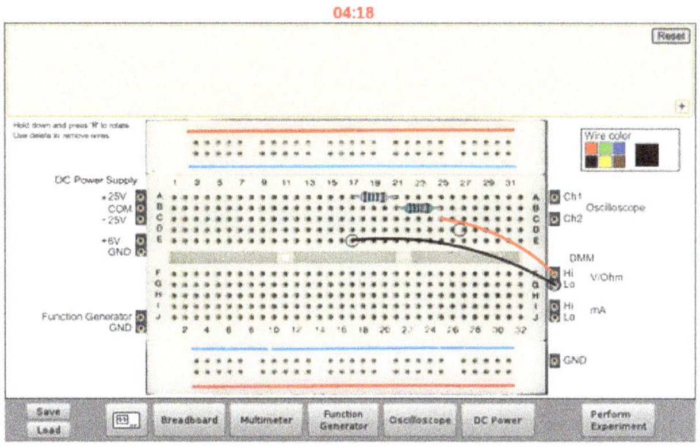

Figure 3. VISIR breadboard

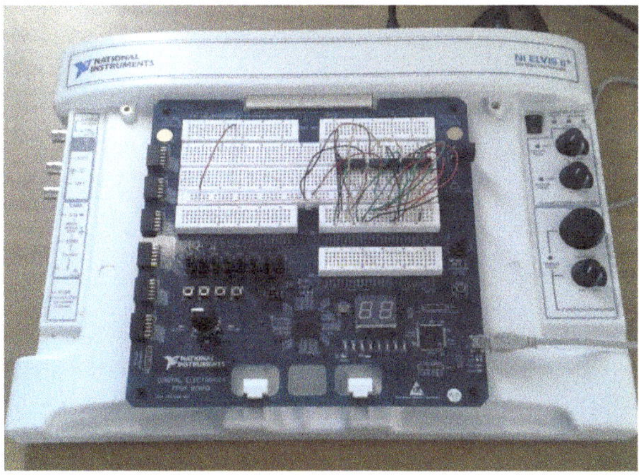

Figure 4. NI Digital Electronics FPGA Plugged into NI ELVIS II

experiments. This category is also divided into three-tiered architecture consisting of: a lab client, a Service Broker and a lab server.

The Lab Client is a user interface. In this case, it is the front panel of the LabVIEW Virtual Instrument that allows users to configure and run an experiment. The ***Service Broker*** is a web service and the core of the ISA, which provides functionality and management tools, authorizes access to experiments and data storage, and facilitates communication between the Lab Server and the Lab Client as a middleware. The Lab Server is the machine connected to the NI ELVIS. It is responsible for receiving experiment submissions, translating them into instrument and specific commands, executing those commands, and passing the results back through the Service Broker to the user. In addition to this, the interactive ISA introduces three services: the ***Experiment Storage Service*** (ESS), which allows the Service Broker**,** Lab server and Lab Client to store experiment data. The ***Lab-Side Scheduling Service*** (LSS), in conjunction with ***User Side Scheduling Service*** (USS), helps its users manage reservation of experiments in advance (Fig.5).

3) LabVIEW graphical programming

The great variety of LabVIEW's graphical system design environment's Boolean and numeric controls/indicators, together with the wealth of programming structures and functions, make LabVIEW an excellent tool to visualize and demonstrate many of the fundamental concepts of digital electronics [16]. Also, we can create interactive front panels in LabVIEW that interface with the FPGA, change the parameters, and show the results of the designs reflected immediately on the front panel (Fig.6).

IV. Experiment Setup

The online digital laboratory supports several experiments based on the FPGA. These experiments are the digital logic gates, sequential logic, seven segment display and full Adder. These experiments are remotely controlled by using the interactive version of ISA via the internet.

A. Running a remote Experiment

As we have already described, our experiment is deployed in the Interactive Shared Architecture; however, the student launches and performs the Service Broker available via http://www.onlinelab-uhp.ma/. If the user is authorized to use the experiment, he /she selects the digital logic experiment and then chooses the available time to launch it (Fig.7).

B. Digital Logic Gates Remote Experiment

Gates are elementary building blocks of digital logic circuits as well as building blocks of more complex logic circuits. These devices function by "opening" or "closing" to admit or reject the passage of a logical signal; every terminal is in one of the two binary conditions: low (0) or high (1).

The aim of this experiment is to study the basic logic gates (AND, XOR and OR) by verifying the performance of those gates using a Truth table. Students participating in this experiment observe the pulse response of the logic gates.

The basic AND gate consists of two inputs and an output. If the two inputs are A and B, the output (often called Q) is "on" only if both A and B are also "on."

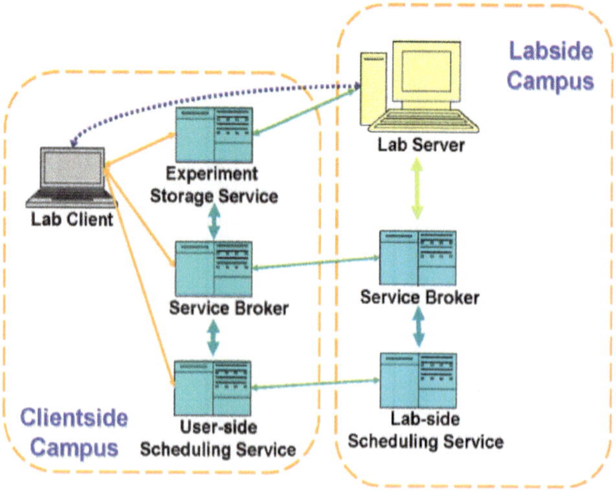

Figure 5. Interactiveilab Shared Architecture

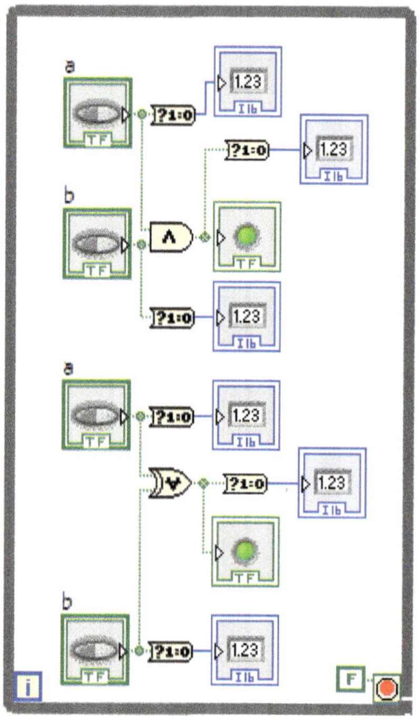

Figure 6. Block diagram of Gates

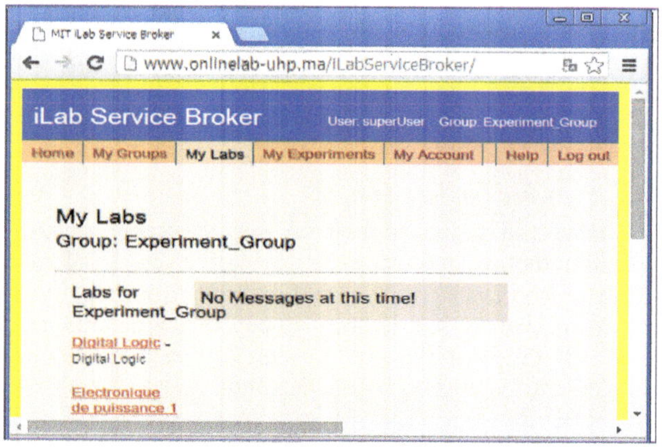

Figure 7. Available Experiments

The student manipulates and makes changes remotely by changing the values of A and B and obtains the result in the Truth table in real time (Fig.8).

C. Full Adder Remote Experiment

An interesting application to understand the logic gates AND, OR and XOR and their interconnections is the implementation of a full Adder. Full Adder is the heart of an arithmetic logic unit that presents an essential part of any computer for all mathematic operations; Adder is to connect two half-Adders and an OR gate. In this experiment, the user can demonstrate the binary addition and learn to perform and use binary arithmetic circuits (Fig.9).

D. Flip-Flops Remote Experiment

JK flip-flop is a mostly clocked logic device implemented as the master slave, which is able to store one bit of information. The objective of this experiment is to enable students to determine the logic of JK flip-flops, to observe their state transition, and then to analyze the timing diagram (Fig.10).

E. 4 bits Binary Counter Remote Experiment

A counter is a sequential circuit that goes through a sequence of states upon the application of input pulses. It is considered as the most useful and versatile subsystem of a digital branch. The objective of this experiment is to help students demonstrate the operations and recognize the characteristics of a synchronous 4 bit binary counter. There are 7 display segments that show the corresponding decimal value for the 4 states of counter (Fig.11).

V. CONCLUSION & FUTURE WORKS

This work focuses on the implementation of an online digital laboratory that provides several experiences in digital sequential circuits. It is based on the FPGA, the NI ELVIS platform, and the interactive ISA. The present implementation is contributing to the increase of remote labs in the network of online labs to help more than 1,000 students from different engineering curricula remotely perform and execute their experiments in digital electronics.

REFERENCES

[1] Ahmed Naddami, Ahmed Fahli, MouradGourmaj, Mohammed Moussetad, " Determination of the synchronous machine's parameters using the iLab Shared Architecture " ,IJOE, Vol 10, No 3 (2014).

[2] T. Zimmer, D. Geoffroy, A. Pester, R. Oros, T. Tsiatsos and S. Douka - eSience: Setting up a network of remote labs in the magrebian countries, ICEEL 2013, Marrakesh, Morocco, 1 –5 July 2013

[3] Tawfik, M. , Pesquera, A. ; Gil, R. ; Martin, S. ; Diaz, G. ; Peire, J. ; Castro, M. ; Pastor, R. ; Ros, S. ; Hernandez, R. "Shareable educational architectures for remote laboratories"TAEEconference,Vigo, 13-15 June 2012

[4] James Feher,"Introduction to Digital Logic with Laboratory Exercises" Copyright © 2009 James Feher

[5] http://www.ehow.com/about_4564073_digital-electro

[6] http://www.ni.com/

[7] http://www.esience.org/

[8] http://www.weblab.deusto.es/

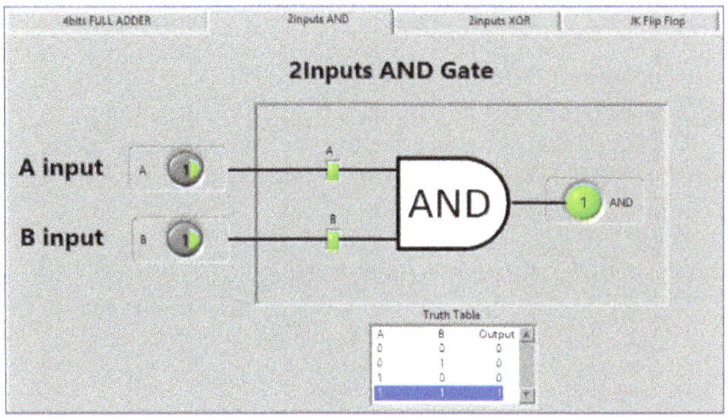

Figure 8. 2 Inputs AND User Interface

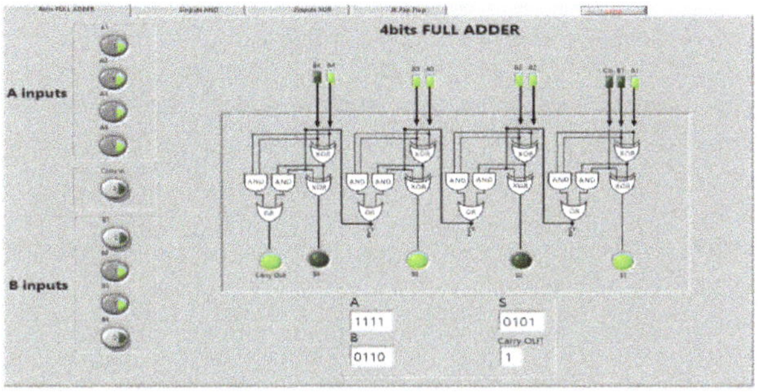

Figure 9. Full Adder User Interface

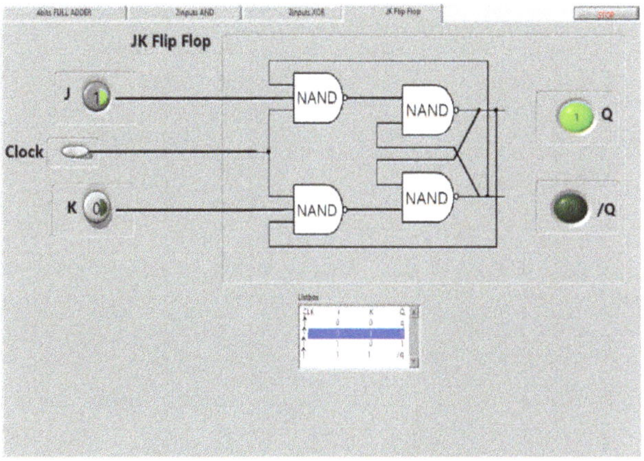

Figure 10. JK Flip-Flop User Interface

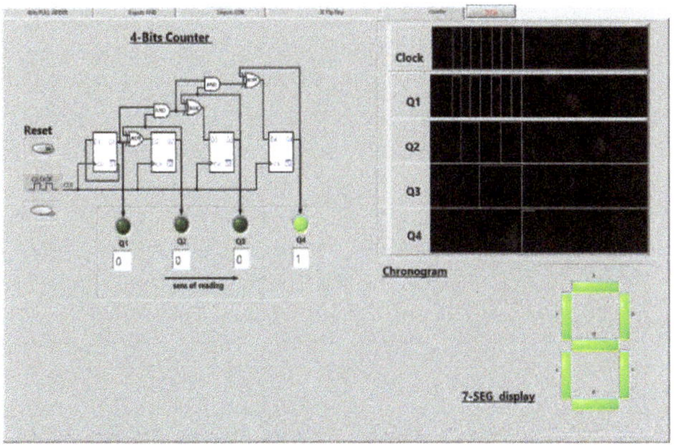

Figure 11. 4 bits Binary Counter User Interface

[9] Orduña, P. (2013). Transitive and scalable federation model for remote laboratories. PhD Thesis, Universidad de Deusto. http://www.weblab.deusto.es/pub/dissertation_pablo.pdf

[10] Ahmed Naddami, Ahmed Fahli, MouradGourmaj, Andreas Pester, Ramona Oros " Importance of a network of online labs in magrebian countries" Proceedings of the IEEE , REV14 Conference, pp.77-78, Porto 26-28 February, 2014

[11] http://www.ni.com/ni-elvis/applications/

[12] http://www.ni.com/white-paper/8596/en

[13] Harward, V.J., et al, "The iLab Shared Architecture: A Web Services Infrastructure to Build Communities of Internet Accessible Laboratories," Proceedings of the IEEE , vol.96, no.6, pp.931-950, June 2008

[14] Kimberly DeLong, V. JudsonHarward, Philip Bailey, James Hardison, Gordon Kohse, YakovOstrocsky"Three Online Neutron Beam Experiments Based on the iLabShared Architecture", Education Engineering (EDUCON), Madrid,pp .145 – 150,14-16 April 2010.

[15] Hardison, J., DeLong, K., Bailey, P., Harward, V.J., "Deploying Interactive Remote Labs Using the iLab Shared Architecture," Frontiers in Education (FIE) Conference, Saratoga Springs, New York, October 22-25, 2008

[16] Barry Paton Dalhousie ,"Fundamentals of Digital Electronics" Copyright © 1998 by National Instruments Corporation

AUTHORS

Ahmed Naddami is with the Laboratoire MATIC, Department of Electrical Engineering, The Polydisciplinary Faculty in Khouribga, Hassan 1st University-Settat, Morocco (e-mail: ahmed.naddami@gmail.com).

Ahmed Fahli is with the Laboratoire MATIC, Faculty of Sciences Ben M'Sik in Casablanca, Hassan II University Mohammedia, Morocco (e-mail:fahli@uhp.ac.ma).

MouradGourmaj is with the Laboratoire MATIC, Department of Electrical Engineering, The Polydisciplinary Faculty in Khouribga, Hassan 1st University-Settat, Morocco (e-mail: mourad.gourmaj@gmail.com).

Mohammed Moussetad is with the Laboratoire LIMAT, Faculty of Sciences Ben M'Sik in Casablanca, Hassan II University Mohammedia, Morocco (e-mail: m.moussetad@gmail.com).

A Mobile Science Index for Development

Enrique Canessa, Marco Zennaro
International Centre for Theoretical Physics (ICTP), Trieste, Italy

***Abstract*—A simple Mobile Science Potentiality Index (mSPI) is introduced as a barometer reflecting a new signal of world development through the capacity for a interconnected community to participate collectively in, and carry out, scientific studies using mobile devices. The mSP Index could be a new signal for world development.**

***Index Terms*—mobile devices, science, information technology, development**

I. Introduction

There are billions of mobile phone subscribers today around the world out pacing fixed-line Internet users with more than thousands new users added every minute. About 60% of them live in developing countries, making mobile phones the first telecommunications technology in history to have more users there than in the developed world. For example, cell phone usage in Africa is growing almost twice as fast as any other region and jumped from 63 million users two years ago to 152 million last year [2]. Most of these countries have reported impressive adoption levels of mobile phones, a phenomenon that is now creating a paradigm shift, namely, computing and rich-media communications moving from traditional PC's to mobile devices (smart-phones, tablets, netbooks and others).

Originally, Internet and mobiles were separated technologies but now they are merging together. Mobile technology is becoming a regular part of everyone's daily lives both in the developed and developing world. Advancements in the smart-phone technology have produced such powerful gadgets and applications that can compete with PC's in this 21st century. With this kind of new computer power of mobiles, analysts and programmers can now develop scientific applications to address numerous challenges. These include helping to prevent disease outbreaks, building a census or tracking agricultural stock levels. Data gathering with mobile devices can help to save time and money for organizations while also improving information accuracy including those needed from/for developing countries. The world-wide adoption of mobile devices is opening new ways for development including, directly and for the first time, the poorest regions of the world.

Data forecasting from the International Telecommunications Union [1] indicates that in only a few years half of the planet's population will have access to the Internet through a mobile device. The possibility of connecting with mobile devices to the Internet via broadband UMTS networks at reasonable subscription prices is growing fast everywhere. This means that a majority of the world's population will soon access to the Internet via their mobile devices. While this is a proxy for actual phone and Internet usage, it is nonetheless also an indirect indicator that interest on mobiles is increasing in education, academia and scientific sectors too. Motivated by this revolution on mobile access to Internet, we discuss here a simple Mobile Science (m-Science) Potentiality Index which could reflect a new signal of world development through the capacity for a interconnected community to participate collectively in, and carry out, scientific studies using mobile devices.

II. 2. m-Science and mSP Index

Mobile Science (or "m-Science" in short) is becoming an *state-of-the-art* discipline because of unprecedented technological developments being done in the field of information and communications. m-Science as such comprises three main subjects of interest with a great impact on the society: *Sensing, Computing* and *Dissemination* of scientific knowledge by the use of mobile devices. These, in turn, incorporate:

- mobile data gathering by the use of sensors and mobiles,
- the analysis and process of mobile data (by mobile devices too) and
- the access to on-line services and applications directed to nurture scientists and scholars (such as mobile access to e-Journals, podcasts, web lectures and webinars, virtual conferences, mobile collaboration tools, m-Learning, *etc*).

The release, under a Creative Commons license, of a premier open book on m-Science by the Science Dissemination Unit (SDU) of the ICTP in trieste, Italy [5], is an effort to engage the scientific community, engineers and scholars worldwide in the design, development and deployment of the newest mobile applications. The goal is to create awareness on the huge possibilities of Mobile Science as well as to motivate a new generation of learners, scholars and scientists. The book gives a balanced mix of technical detail, general overview, societal impact and a sense of the possible.

We devise here an approximated Mobile Science Potentiality Index (mSPI) to be used as a barometer reflecting the potential and capacity for a community (for example a group of researchers) in a nation or region to develop and participate in m-Science. The mSPI we introduce is based on a simple relation:

"The ratio of the mobile broadband subscriptions available by country and the amount of researchers in labor force (working full time in different fields of science) in a country". For illustrative purposes, an example of how the mSP Index may behave on a nation wide scale follows in Table I by using recent published data based on the OECD's definition of wireless broadband data (2010). The broadband subscribers per 100 inhabitants (2009) includes satellite, terrestrial fixed wireless, dedicated and standard mobiles, which certainly gives a crude estimation for mobile access to Internet in the sense given within m-Science, but it can be a useful ratio to obtain a reasonable indication of the current potentiality of m-Science in some places of the industrialized world where discrete data is only available so far. Hopefully, data for developing countries will be made available with the increasing usage of mobile devices and the participation of these countries in the internalization of science.

III. Remarks

From Table I it can be deduced that when communities have greater access to broadband facilities, then researchers (even if fewer than other work forces in a country) and broadband subscribers can participate in m-Science experiments leading to a greater mSP Index. In a thought experiment, this mSPI index may in principle score and monitor over time, a community (consisting of "not only" researchers) on a scale from 1 (fully mobile and connected community) to 0 (highly localized and isolated community). As m-Science through mobile sensing, computing and dissemination continues to become less and less expensive, more integrated and more ubiquitous, the hope is that projects in this new field will increasingly be within reach of scientists in the less industrialized world too. A world-wide mobile accessibility will help bootstrap participatory and truly societal scale networks for science research. An increasing mSPI measure could well be a useful, reliable instrument to attempt quantifying the world-wide expansion of, and interest on, the new growing m-Science discipline in years to come depicting a rapidly changing landscape.

The mobility of the phone provides some other important scientific and economical opportunities. Data collection with a mobile phone has the potential to dramatically improve services that relies on accurate and up-to-date information. The more time-critical the information, and the more remote the location, the more scientists have to gain from a mobile phone based solution. Thus, m-Science offers tremendous benefits to academic research and to society as a whole and the mSP Index could be a new signal for world development. It is surprising to note that the maximum reported mSPI value of 74,20% is for the Netherlands and the minimum is 25.57% for Japan, contrary to what it would it be expected.

Based on information extracted from the Web of Science (Thomson Reuters) database of scientific publications spanning from 1980 to 2009, the world's scientific production has grown from about 400,000 to 1,200,000 publications in the last three decades. This increment of interest in science, together with the recent technological developments in mobile technologies, is helping to make of the new m-Science a completely new research field. The potential for m-Science and science-related mobile apps, is unlimited. Since the majority of new generation devices can access the web in some way, mobile technological tools can be used to collect basic information as well as to exchange information and to access scientific computing among other services and carrying lots of benefits that permeates into society.

TABLE I.
Mobile Science Potentiality Index (mSPI) [1]

Country	Researchers in labor force per 10000 inhabitants (1999)	Broadband subscribers per 100 inhabitants (2009)	mSPI (%)
Netherlands	50	37.1	**74.20**
Austria	34	22.1	**65.00**
Switzerland	55	35.6	**64.73**
Italy	33	20.5	**62.12**
Denmark	61	37.1	**60.82**
Belgium	54	29.0	**53.70**
United Kingdom	55	29.5	**53.63**
New Zealand	44	23.2	**52.73**
Canada	58	29.6	**51.03**
Germany	60	30.3	**50.50**
France	61	30.4	**49.84**
Norway	78	33.9	**43.46**
Ireland	51	19.5	**38.24**
Sweden	91	32.4	**35.60**
USA	81	26.4	**32.59**
Australia	67	23.3	**34.78**
Finland	99	26.7	**26.97**
Japan	97	24.8	**25.57**

1 Data from OECD: http://www.oecd.org/

Mobile technology can be used in powerful new ways to carry out scientific research, analyze and share results to many. Mobile phones are, for instance, being used to gather scientific data from remote and isolated places that would be impossible to retrieve by other means. Scientists, acting alone or in groups, are starting to use mobile devices and web-based mobile applications to systematically explore interesting scientific aspects of their surroundings ranging from climate change to earthquake monitoring world-wide. Accurate and timely field data gathering is possible with mobile phones. By attaching sensors to GPS-enabled cell phones, it is possible to gather the raw data necessary to begin to understand how, for example, urban air pollution impacts both individuals and communities. While integrating a sensor into a phone and transmitting the data that it gathers to a database is posssible, doing so at low cost, on a societal scale, with millions of phones providing data from hundreds of networks spread throughout the world makes this possibility a challenging one. This is a unique m-Science occasion that deserves further attention and promotion world-wide (Canessa *et al.* 2009). At the expense of sampling a given location continuously, a sensor in a user's smart-phone can provide significant geographic coverage. Also, mobile sensors will be heavily biased towards locations in which people congregate, so for human-centric applications, sensing in mobile devices will often provide coverage exactly where it is needed most. In over-sampled locations, the precision of the sensing system can be increased by carefully averaging the readings from several nearby sensors.

All these advantages point to build the largest scientific instrument ever built, consisting of millions or billions of sensors (attached in some cases to mobiles), aggregating data on an unprecedented scale. This instrument could be truly societal scale, reaching across economic, social, geographic and political boundaries, and illuminating the corners of human activity, how our environment affects us, and how we affect our environment. Besides, supercomputing on a cellphone using this huge data will become also a reality.

The development of scientific mobile applications need certainly still to be simplified without the needs for having high computer literacy. There is a useful and simple blocks language for mobile phones as for example "App Inventor for Android" [6]. This is a visual programming environment for building mobile apps. by putting together blocks that represent the phone's functionality so, even those with no programming experience can use App Inventor to create mobile apps. On the other hand, mobile application development with the alternative Python (better known as "Python on Symbian or PyS60") is also a powerful simple code to build/prototype mobile applications rapidly with interface for telephony, messaging, multimedia, camera, sensors and location tracking.

The wide availability of cellular telephones equipped with cameras opens many other opportunities for inexpensive, portable photometric measurements suitable for educational purposes. Spectrophotometry can make more sense to students around the world when they can see light change intensity when passed through a sample than when just seen equations, sketches or the output of a computer screen or meter. Mobile social networks are rapidly being organized by young people for the dissemination and credibility of information. Besides, m-Science technology through Internet access opens up also amazing educational opportunities. With m-Learning, students can also take advantage as learning is now freed from the restrictions of location. Mobile learning programs can specially foster Physics and Mathematics m-Learning.

All of this novel m-Science activity aims to engage the scientific community, engineers and scholars world-wide in the design, development, deployment and usage of the newest mobile applications which benefits many and giving a substantial contribution to the world development. The proposed Mobile Science Potentiality Index can be used to best assess a country's development in science and mobile technology innovation and as a barometer reflecting a new signal of world development through the capacity for a interconnected community to participate collectively in, and carry out, scientific studies using mobile devices.

REFERENCES

[1] ITU - *International Telecommunications Union*, (2010)

[2] EPROM - *Entrepreneurial Programming and Research on Mobiles* of the MIT- Massachusetts Institute of Technology, (2009)

[3] OECD - *Organisation for Economic Co-operation and Development*, (2010)

[4] E. Canessa, C. Fonda and M. Zennaro, *Supporting Science in Developing Countries using Open Technologies* (2009) Eur. J. Phys. **30**, 651–660.

[5] E. Canessa and M. Zennaro (Eds.), *m-Science: Sensing, Computing and Dissemination,* ISBN 92-95003-43-8 (2010)

[6] *App Inventor for Android, (*2010)

AUTHORS

Enrique Canessa and **Marco Zennaro** are with Science Dissemination Unit (SDU), International Centre for Theoretical Physics (ICTP), Trieste, Italy.

Connecting Mobile Users Through Mobile Social Networks

Faisal Alkhateeb, Ahmed M. Manasrah, Iyad Abu Doush and Eslam Al Maghayreh
Yarmouk University, Irbid, Jordan

***Abstract*—Nowadays, social networks become popular with the emerging of web-based social networking services. Recently, several mobile services are developed to connect users to their favourite social networks such as Facebook, Twitter, Flickr, etc. However, these services depend upon the existing web-based social networks. In this paper, we present a mobile service for joining groups across communities. The originality of the work is that the framework of the service allows creating and joining social networks that are self-contained for mobile company servers. The service consists of several sub-services such as users invitation, group finding and others. Users, regardless of their disability, can use the service and its sub-services without the need to create their own accounts on social web sites and thus their own groups. We also propose a privacy control policy for mobile social networks.**

***Index Terms*—Social networks, mobile social networks, RDF.**

I. Introduction

A social network is a social structure that may consists of individuals (or organizations) called "nodes", which are connected by one or more specific types of relationships, such as friendship, kinship, common interest, financial exchange, affiliation, etc [29]. The "social network" concept has been used several decades ago [25]. But, nowadays, social networks and social networking services become more popular with the contribution of the current facilities such as the web and mobile communications.

A social networking service is usually an online service (could be a platform or a site) [5] that aim at constructing social networks or social relations among people.

A social network service essentially consists of a representation of each user (often a profile), his/her social links, and a variety of additional sub-services [Site 1]. Most social network services are web based and provide means for their users to interact among themselves over the Internet, such as e-mail and instant messaging. Online community services are sometimes considered as a social network service, though in a broader sense, social network service usually means an individual-centered service whereas online community services are group-centered. Social networking sites allow users to share ideas, activities, events, and interests within their individual networks [14].

Recently, the number of social networks services has been increased and many of them are web-based such as Facebook [Site 2], Twitter [Site 3], LinkedIn [Site 4], MySpace [Site 5], Flickr [Site 6], Hi5 [Site 7], and others.

As the number of mobile phones users is on the rise, researchers and industry directed towards mobile social networking[1].

Recently, several mobile social networking services have been developed such as Meet Now Live, sniff, Trapster,

Whrrl, Xumii, etc (refer to the related work section for more details). Nonetheless, these services depend upon the existing web-based social networking services. That is, the users can use these services to connect to social networks and keep in touch with their friends on Flickr, Facebook, MySpace, and others. Moreover, the users of mobile phones form a natural social network and, to the best of our knowledge, none of the existing mobile networking services benefit from this fact and thus building this social network.

In other words, most of the social networking are online services that emphasize on building and reflecting social relations among people, whereby users can share data/activities with others whom supposed to be their friends or friends of a friend. Most social network services are web-based and provide means for users to interact over the Internet, such as e-mail and instant messaging. However, computers are not always accessible, where mobile phones are. Therefore, a trend to turn the mobile phone into a social network is starting to arise. Especially, mobile phones provide services, such as phone calls and text messaging that are widely used. These services were designed for interpersonal communication and suitable for small groups of people, but they do not fit well for the coordination of larger groups [18], [2], [20]. Conversely, the advanced capabilities of mobile devices extend the scope of social interactions, beyond the small social group where people can now share their current location, phonebook contacts, calendar events and personal notes, and personal media (photos, video clips, etc) among others within the social group. However, such feature or flexibility imposes and raises higher awareness on privacy associated with shared information as well as the boundaries of the comfort level on what type of information can be disclosed and to whom and when. On the other hand, social websites provide various methods to users in order to easily publish their personal information on the Web. Yet, once published, users have no control on how their data should be accessed (i.e. who should access them and when/how?). In other words, most Social Networks provide a coarser-grained privacy settings restricting access to private data to those who are in the user's friends lists or user's groups such as Facebook's privacy preferences

[1] 800 million users by 2012 of mobile social networking and $60 billion in revenues [Site 8].

and Google+[Site 15] circles. Users however, need to restrict their profile access to their friends list only after they are invited. Otherwise, open/public profile (i.e unrestricted) can be accessed by anyone even though requesters are not in the user friend list yet. As a result, users still demanding more complex privacy settings as discussed in [6].

In our side, we have presented a mobile social networking service that will be used for building a social network for mobile phone users and their connections [31]. The proposed service consists of several sub-services that will be discussed in details in the paper. Our work is motivated by the fact that the number of mobile users is more than the number of Web users and users can use the service and its subservices without the need to create their own accounts on social web sites and thus their own groups. In this paper, we provide more details about the service and show how users regardless of their disabilities, for example deaf or blind persons, can benefit from the proposed service. In addition, we extend our work [31] by proposing fine-grain privacy policy for restricting or granting access to specific data on mobile social networks.

Paper Outline. The remainder of the paper is organized as follows. In Section II, we introduce RDF and the SPARQL language. Section III is dedicated to the presentation of the proposed service. We propose in Section IV a privacy control policy for the mobile social networks. We discuss other related work in Section V. Finally, we conclude our work in Section VI.

II. PRELIMINARIES

This section provides an overview of the elements that are necessary for presenting the proposed approach namely: RDF and SPARQL. RDF is basically a graph-based language, which makes it very suitable for representing social networks (as they have a graph nature) while SPARQL will be used for searching RDF data.

A. RDF

RDF (Resource Description Framework) [21] is a knowledge representation language dedicated to the annotation of resources within the Semantic Web. Currently, many documents are annotated via RDF due to its simple data model and its formal semantics. In its abstract syntax, an RDF document is a set of triples of the form `<subject, predicate, object>`.

Example 1: The assertion of the following RDF triples:

```
{<ex:person1 foaf:name "Faisal Alkhateeb">,
<ex:person1 foaf:knows ex:person2>,
<ex:person2 foaf:name "Iyad AbuDoush">,
<ex:person2 foaf:knows ex:person3>,
<ex:person3 foaf:name "Ahmad Manasrah">
<ex:person3 foaf:knows ex:person1>,
}
```

means that there exists a person named "Faisal Alkhateeb" who knows another person named "Iyad AbuDoush" who in turn knows a third person named "Ahmad Manasrah". Note that foaf [Site 9] and vcard [Site 10] (which is used in this paper), are two RDF vocabularies.

An RDF document can be represented by a directed labeled graph, as shown in Figure 1, where the set of nodes is the set of terms appearing as a subject or object in a triple and the set of arcs is the set of triples (i.e., if <s, p, o> is a triple, then s → o is an arc labeled with p).

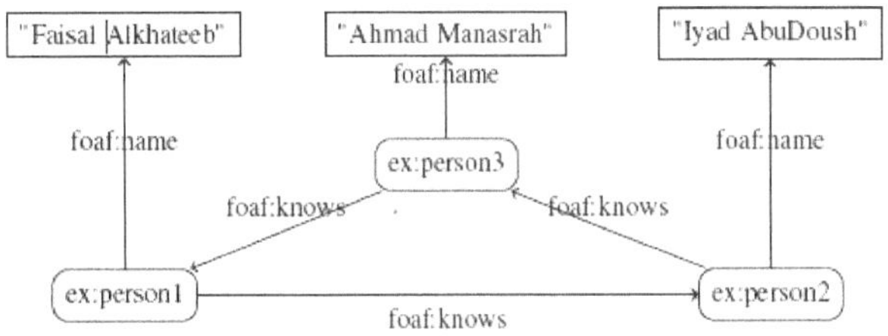

Figure 1. An RDF graph.

B. SPARQL

SPARQL is a W3C recommendation language developed in order to query RDF knowledge bases, e.g. retrieving nodes from RDF graphs [24]. A simple SPARQL query is expressed using a form resembling the SQL SELECT query:

```
SELECT B FROM u WHERE P
```

where u is the URL of an RDF graph G to be queried, P is a SPARQL graph pattern (i.e., a pattern constructed over RDF graphs with variables) and B is a tuple of variables appearing in P. Intuitively, an answer to a SPARQL query is an instantiation of the variables of B by the terms of the RDF graph G such that the substitution of the values to the variables of P yields to a subset of the graph G[2].

Example 2: The following SPARQL query:

```
SELECT ?p1 ?name
FROM <Figure1>
WHERE {
        ?p1 foaf:knows ?p2 .
        ?p2 foaf:name "Faisal Alkhateeb" .
        ?p1 foaf:name ?name .
}
```

could be used to retrieve the name of persons who know a person named "Faisal Alkhateeb". When evaluated against the RDF graph of Figure 1, the following answers will be returned:

#	?p1	?name
1	ex : person1	"Ahmad Manasrah"

III. THE PROPOSED SYSTEM

A. Design

One of the goals of the proposed service is to build a social network for mobile users. The social network will be created incrementally based upon the mobile users profiles as well as their call and message connections. To this extent, we use the RDF language for representing users' profiles and so building the social network, i.e.,

[2] When using RDFS semantics [8], this intuitive definition is irrelevant and one could apply RDFS reasoning rules to calculate answers over RDFS documents.

each user profile will be represented as an RDF knowledge base (see Figure 2). In this RDF knowledge base, the username is used for identifying (and thus searching) user profiles since it is a unique key. The obtained RDF knowledge base will be maintained in the server of the mobile company.

```
username foaf:name "Faisal Alkhateeb" .
username vcard:key <id> .
username foaf:mbox "email here" .
username foaf:address "Irbid... " .
username vcard:mobileTel "00962... " .
username vcard:car <car-number> .
username foaf:knows mobile-num2 .
username foaf:knows mobile-num3 .
  ...
```

Figure 2. An RDF graph representing a part of a user profile.

Building the social network of mobile users in the server of the mobile company has several benefits. These benefits are described next:

1) Profile activation

Any user that wants use the service and its sub-services should activate his/her profile. This is achieved through an SMS sent by the mobile company to the user in order to fill his/her information (such as his name, international identification number[3], address, car number if any, etc). The company will then provide the user with his/her username that could be a combination of his/her name and/or id.

2) Profile updating

After profile activation, the user can use his usernmae for updating his profile. In this case, the mobile company will notify his/her group members with any changes to his/her mobile number.

3) Users invitation

Any user x can send an SMS message to another user y to invite him/her to his/her group. Additionally, users' invitation service could be automatically invoked based upon some heuristics such as the number of call connections between both users and intersection between their profiles. If y accepts the invitation then an RDF triple is added to their both profiles (to the RDF knowledge base) using the `foaf:knows` property and their identification number[4]. That is, if `mobile-numX` (respectively, `mobile-numY`) is the identification URL of user x (respectively, user y), then the RDF triple `<mobile-numX foaf:knows mobile-numY>` (respectively, `<mobile-numY foaf:knows mobile-numX>`) is added to the profile of x (respectively, profile of y).

4) Instant messaging

This service can be used by a given user to send a message to the group members with whom he has connection.

5) Group finding

A mobile social network system enables people to be engaged into friendship using the mobile phone [17]. cities are expanding and changing continuously, and the people are on the move most of the time [11]. Location based services (LBS) can provide the people with services with regard to their location. The geographical position of the person is identified by matching the mobile terminal with the mobile network. The LBS can be used for several purposes: healthcare, friends making, or finding the nearest store. Hewett et al. [1] indicates that blind and visually impaired people use social networking sites mostly to communicate through chat functions and to stay connected with friends. The mobile-based social network can provide several services to people with disabilities. With the addition of the services identified in the previous section it can provide the following services:

- Identify the location of a person or a place by using their profile and locating their geographical position using LBS. For example, a blind person wants to meet his friend who is in another building in a university. The LBS will identify the position of the blind person and the position of his friend using LBS. Then these positions will be passed to the maps in the phone to help the blind person in the navigation.
- Looking for people with specific characteristics or interests. For example, a deaf person can look for all deaf people in the surrounding area who identify their profiles as public. Another example is a blind person who is interested to see if any of his friends is going to the festival tonight. The user can use this sub-service to locate the group members based on the constructed social network with the following two options: WI-FI or zone limit:
  - If the user selects the WI-FI limit option, then the service provides the user with the group members that are located around him/her via WI-FI distance limit upon request. The user can then selects one or more of the located group members and can send to them files, perform instant chatting or can call of one of them for urgent purpose. It is possible to use bluetooth limit instead of the WI-FI limit. In this case, an application should be deployed in the user mobile. The application interacts with the bluetooth to retrieve mobile names of group members located around him/her via WI-FI distance limit upon user request. Then it interacts with the service so that the mobiles names retrieved by the bluetooth are refined and are restricted to only mobiles names of the group members. Also, this option requires that the mobile name of each group member is the same as the member identification name (i.e., A unique username is provided by the company to the mobile user and it is a combination of his/her name and the mobile number).
  - If the user selects zone limit option, then the service will provide the user with the group members located in the user zone. Note that the zones are defined by the mobile phone company. For example, Yarmouk University zone is the zone reserved for Yarmouk University. This service invokes the tracking location-based service [27], [26], [4], [10], [28] to locate each group member of a given user. The name and the mobile number is provided to the user if the group member is located in the same zone of the user invoking this service. Moreover, a group member should provide the

[3] The international identification number (iin) is mandatory. For those countries that do not use iin, the international mobile phone number could be used.

[4] We can use PSPARQL [3], our extension to SPARQL, to find friend of friend.

permission to the service to know his/her location by any of other group members.

6) *Profile finding*

This service can be used when a given user has a problem with a given mobile or a car. In this case, he/she can use this service to retrieve the knowledge (profile) about the owner of that car or that mobile number as well as the group knowing him/her. This is achieved based upon the car number or the mobile number as a key for searching. The user can then call him/her or one of the retrieved group members. This is important since it often happened that one car hits another or one person cannot exit his/her car because another car closed the road. In more details, if the user selects this service then he/she has several search options:

- Searching based on a mobile number: the user can enter the mobile number (say mobile-X) of the person to search for. In this case, the service will transform the user request to the following SPARQL:

```
SELECT *
FROM <mobile-social-network>
WHERE{
      ?s vcard:mobileTel "mobile-X" .
      ?s ?p ?o .
}
```

This query will be used to retrieve the profile of the person based on his/her mobile number. It should be noticed that the retrieved person profile contains the members he knows as they belongs to his/her profile. So that, the user can communicate with one of them. Moreover, if the mobile number belongs to a mobile company y while the mobile number of the user requesting this service belongs to a mobile company x, then a request is sent from x to y.

- Searching based on a car number: the user can enter the car number (say car-X) of the person to search about. In this case, the service will transform the user request to the following SPARQL:

```
SELECT *
FROM <mobile-social-network>
WHERE {
?s vcard:car "car-X" .
?s ?p ?o .
}
```

- Searching based on a person id number: similarly the user can retrieve the profile of a given person based on his/her id number. In this case, the service will transform the user request to the following SPARQL:

```
SELECT *
FROM <mobile-social-network>
WHERE {
?s vcard:key "id-X" .
?s ?p ?o .
}
```

IV. Privacy Control For The Proposed Service

In the recent few years, some papers have investigated mobile social network related technologies, such as Dodgeball [17], [30]. However, to our knowledge, there is no paper having made an access right policy for friends and friends of a friend. In fact, most of them still depend on inherited access rights or customizable rights. One of the proposed methods in addressing such issue is to consider the mobile sharing as privacy-trust decision problem as suggested by [15], [19]. The approach takes into consideration the properties of a social relationship based on related data stored in a mobile device. The author binds the relation or strength of the social with the sensitivity of the resource to be shared at the initialization of the user profile. However, a privacy trust decision suggests that a friend can inherit his inviter access rights without any restrictions, which again raises the awareness of privacy concerns as well as the inherited access right dilemma.

Other solutions proposed for addressing privacy issues include distortion of information [7], context sensitive system adaptation [16], feedback loops [16], and mechanisms for access control [22]. However, most of these solutions typically involve a single aspect of privacy and does not decide what and how one might choose to disclose certain private information and to whom and to what extent. In addition to that, the issue of inherited access rights is still exists even though it can be adjusted later (i.e after the requester is added to the user friend list or to one of the trusted group) as illustrated in Figure 3.

Mobile social networking systems must give users sufficient control of their personal data, especially their personal information [13]. For instance, Twitter and Jaiku [Site 16] allow users to set their information private or public. However, all the user's posts will be either public or private. It would be better if there can be options that would let the user make some posts public and others private as well as deciding what to share and with whom. The traditional method in adding users/groups is portrayed in Figure 3. We can easily notice that the user initially can get access to the inviter profile unless he is added to one of the customized groups that has access rights to certain information. However, a better mechanism is needed that decides what users are allowed to view from the moment he is invited and how the whole user social network can be managed and organized. We urge that a person (in mobile social networking) can decide whom to invite via SMS. Prior to the invitation, the user can decide on what level of information he is willing to share with the new potential friend. For instance, the inviter may consider his location as sensitive for a friend who is not very closed to his social/relation, thus, the inviter can decide whether to disclose this info with the new friend or not yet.

Similarly, for his photo albums or address book. Moreover, the inviter can decide on whether the new friend can inherit the inviter rights to new friends, also he can decide if the new friend (friend of a friend) is allowed to publish the inviter's location to the invitee friends and friends of friends, etc. Figure 4 illustrates the logical sequence steps involved with the proposed method.

The proposed policy has the following structure (Figure 5) and vocabulary (Figure 6 and 7).

Where access indicates whether the user can access his inviter shared folders. Share is a feature that allows users to share the inviter info with their friends and friend of friends and invite is another feature that restricts users from inviting other users to the inviter social network. For instance, the following policy says that ahmad who has the uid=112 invites the user who has uid=111 with the following access rights. He is granted a limited access to the folder amman_gallary which indicates that the user 111 can only view the inviter's amman_gallary folder and not his friends. Moreover, he can only invited the inviter's (uid=112) friends to his social network.

```
<contents>
 <entry>
  <name>ahmad</name>
  <uid>112</uid>
 </entry>
</contents>
<policy>
 <xacl>
   <object name="access" aid="/amman_gallary"/>
   <rule>
   <acl>
    <subject>
     <uid>111</uid>
    </subject>
    <action name="limited" permission="grant"/>
    <action name="read" permission="deny"/>
    <action name="write" permission="deny"/>
   </acl>
  </rule>
 </xacl>
 <xacl>
  <object name="invite" aid="*"/>
  <rule>
   <acl>
     <subject>
      <uid>111</uid>
     </subject>
     <action name="limited" permission="grant"/>
     <action name="read" permission="deny"/>
     <action name="write" permission="deny"/>
  </acl>
  </rule>
 </xacl>
</policy>
```

V. RELATED WORK

Several mobile social networking services have been proposed in the literature. However, these services are either based on social web sites or can be used to retrieve location of persons whose profiles match some attributes. In contrast to these services, our proposed service will create a social network for mobile users. The created social network, which is resided on the mobile phone company server, will be used as a base for other sub-servies.

Chang et. al. [9] proposed an interactive location-based system that supports social services. The system can be used to locate persons whose profiles match some attributes requested by the service.

A wireless virtual social network called PeopleNet is proposed in [23]. Using PeopleNet, a user can exchange information with other users via mobile phones.

Meet Now Live [Site 11] is a mobile social networking service that can be used by barflies and club-hoppers to transmit their real-time location in and around New York City.

Sniff [Site 12] (Social Network Integrated Friend Finder) is a mobile application that enables users to locate contacts in real-time via SMS or on Facebook, providing users with a detailed map view of their friends' current locations. However, users should provide the permission to the service to allow locating them by their friends.

Trapster [Site 13] is a free service that uses mobile phones and other types of devices to allow users to share the location of police speed traps.

Whrrl [Site 14] is a mobile social discovery service designed for BlackBerry device platform. It provides users with features like local discovery, mapping and microblogging via the addition of business components like events and calendar features. "Whrrl catalogs each user's history of places visited and events attended, mapping their behavior with color-coded icons that illustrate whether the experience was positive, neutral or negative. Users may share ratings and reviews as well as instantly update their location or status within their circle of contacts, be they business or casual".

Xumii (Site 8) is a service that can be used to call and message friends across social networks, create and join public and private groups across communities, update the user's status on all his/her identities, and upload photos and snapshots to Facebook, MySpace and Flickr. Users can connect, using any standard mobile browser, to their favourite social networks. Also, they can simultaneously stay in touch with their friends on Facebook, MySpace, Twitter and Flickr, with no need to move from one website to another or one application to another.

VI. CONCLUSION

Most of the social networking services are online services that emphasize on building and reflecting social relations among people. We have presented in this paper a mobile service for building a social network for linking mobile users. The constructed social network will be maintained and updated regularly in the server of the mobile company.

Though the main objective of the proposed service is to construct a social network, it consists of several sub-services such as: group finding (a user can search the group members located in a given zone), instant message (a user can send message to his/her group members), and other services. We used the RDF language for representing mobile user profiles and the SPARQL query language for searching the RDF data representing user profiles. It should be noticed that the service is not a replacement of the existing services that can be used to connect to web-based social networks but a complementary to these services. That is, users can use the service to connect to their social groups anywhere and anytime without the need to create an account on web-based social network services. Users regardless of their disabilities, for example deaf or blind persons, can benefit from the proposed service. Additionally, to our knowledge, there is no paper having made an access right policy for friends and friends of a friend. To this end, we proposed a fine-grain privacy policy for restricting or granting access to specific data on mobile social networks.

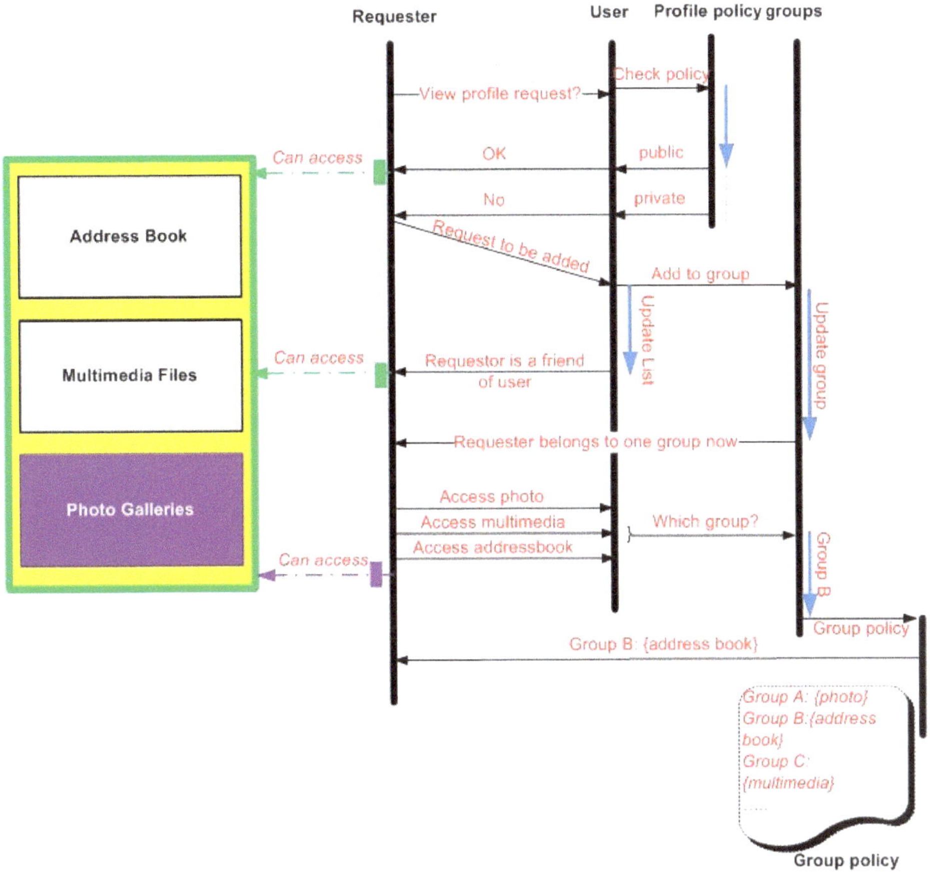

Figure 3. Traditional request (invite)/add access rights policy process.

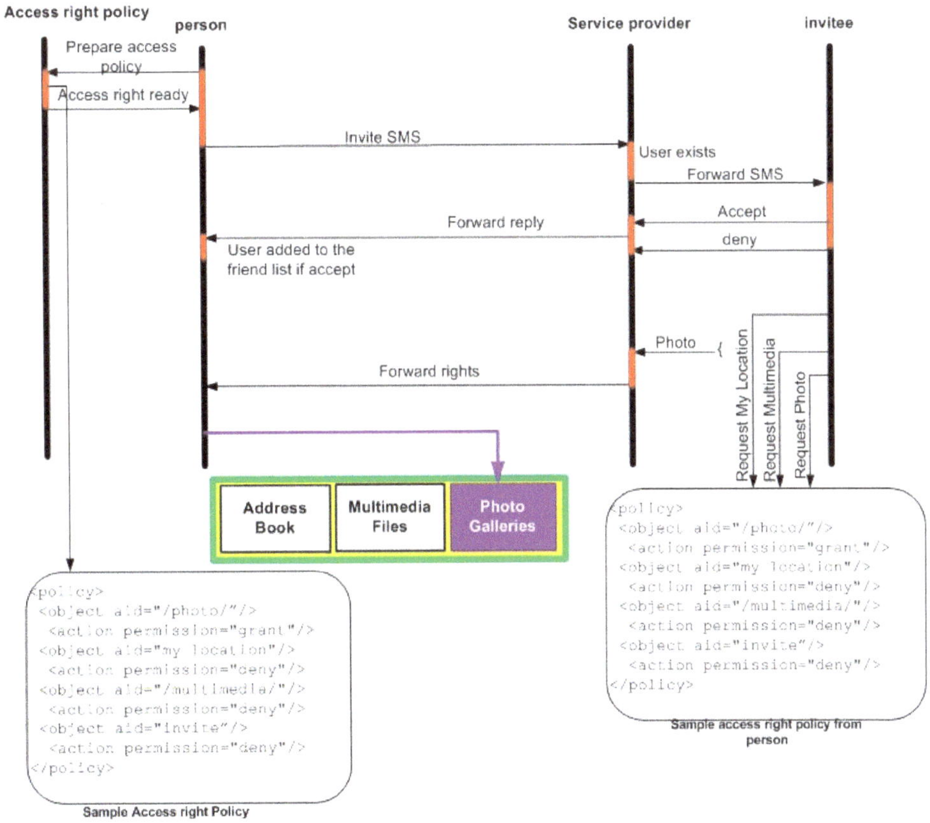

Figure 4. The proposed access rights policy process.

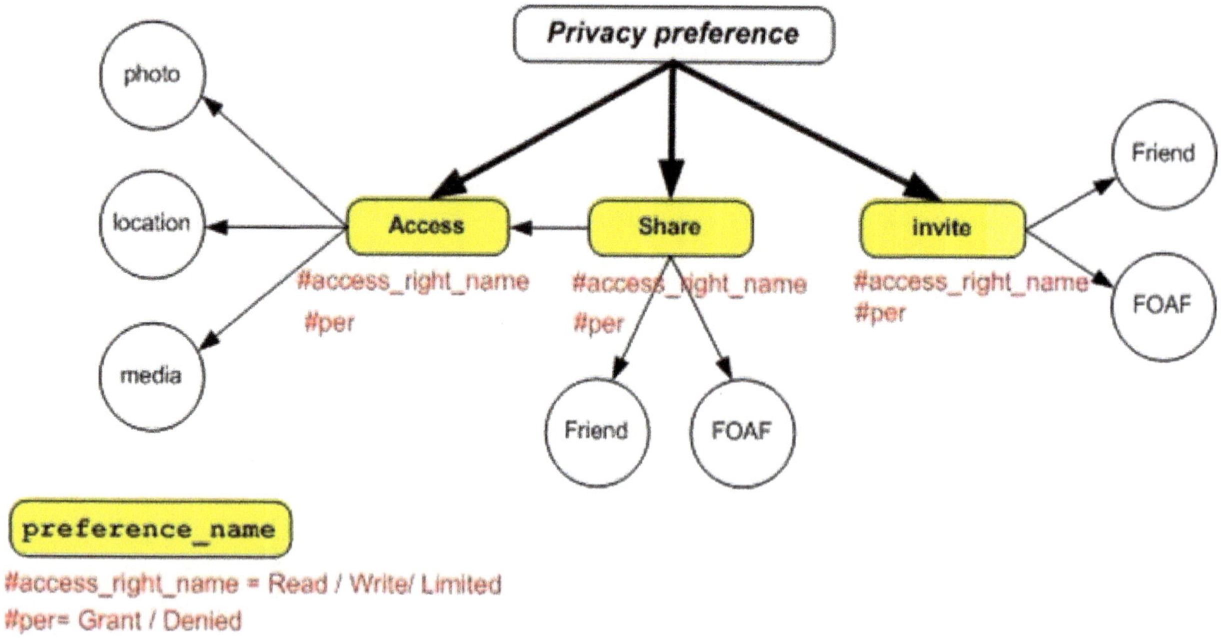

Figure 5. Policy Structure.

```
<contents>
<entry>
<name>#inviter_name</name>
<uid>#inviter_user_id</uid>
</entry>
</contents>
/* this is an access policy for the user #inviter_name who has the
#inviter_user_id for a friend who has #friend_user_id
*/

<policy>
  <xacl> // xml access control language [Site 17]
         <object name="#preference_name" aid="/#resource_name"/>
         // "/" indicate a folder name, without "/" it is a service
                <rule>
                       <acl> // access control list
                         <subject>
                              <uid>#friend_user_id </uid>
                              // this user has the following access right on
                              the #resource_name
                          </subject>
                      <action name="#access_right_name" permission="#per"/>
                      <action name="#access_right_name" permission="#per"/>
                       </acl>
                </rule>
         </xacl>
</policy>
```

Figure 6. Policy Vocabulary.Conclusion

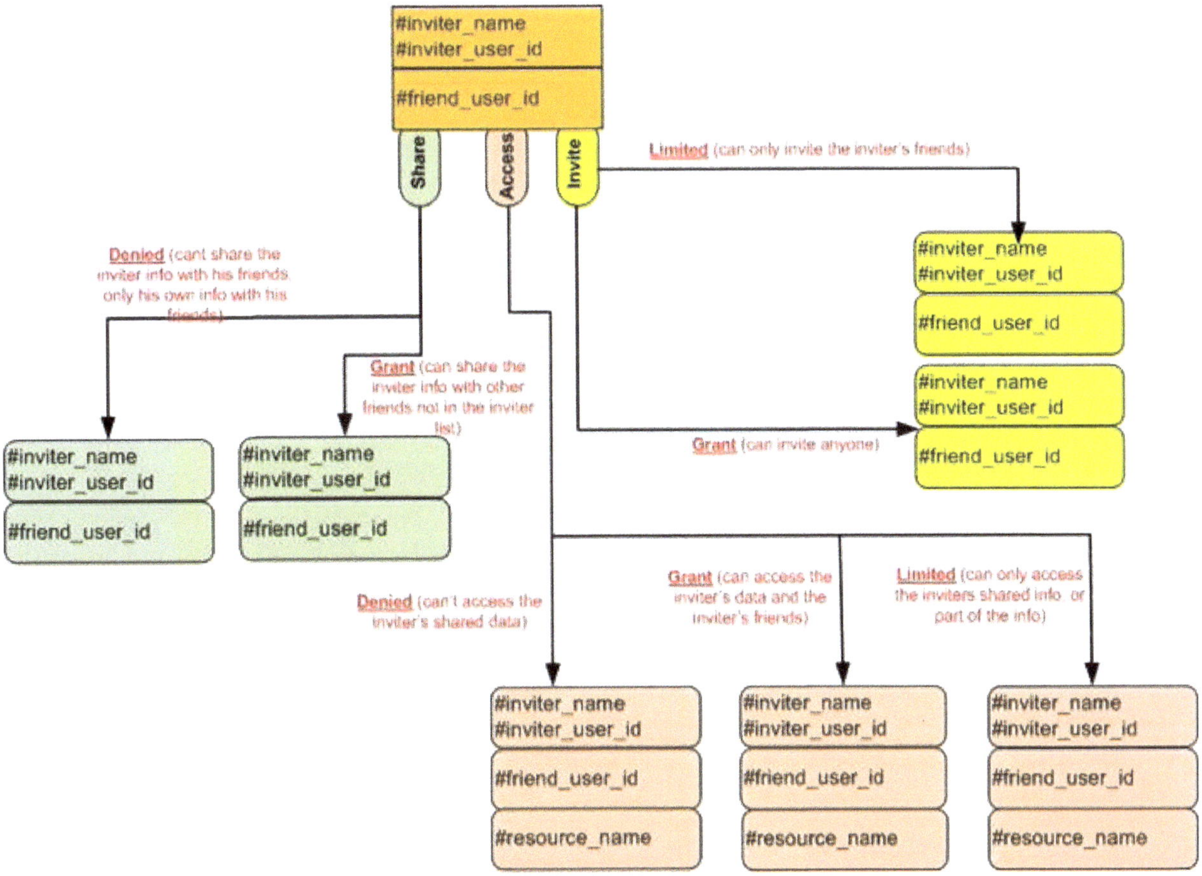

Figure 7. The use of the policy vocabulary.

REFERENCES

[1] Post-14 transitions - a survey of the social activity and social networking of blind and partially sighted young people. Technical report.

[2] Giuseppe Lugano . Understanding mobile relationships . In Human Centred Technology Workshop HCT , Pori, Finland, 2006.

[3] Faisal Alkhateeb, Jean-François Baget, and Jérôme Euzenat. Extending SPARQL with regular expression patterns (for querying RDF). Journal of web semantics, 7(2):57–73, 2009.

[4] Paolo Bellavista, Axel Küpper, and Sumi Helal. Locationbased services: Back to the future. IEEE Pervasive Computing, 7(2):85–89, 2008.

[5] Danah Boyd and Nicole Ellison. Social network sites: Definition, history, and scholarship. Journal of Computer- Mediated Communication, 13(1), 2007.

[6] Danah Boyd and Eszter Hargittai. Facebook privacy settings: Who cares? First Monday, 15(8), 2010.

[7] Michael Boyle, Christopher Edwards, and Saul Greenberg. The effects of filtered video on awareness and privacy. In CSCW 2000, pages 1–10. ACM Press, 2000.

[8] Dan Brickley and R.V. Guha. RDF vocabulary description language 1.0: RDF schema. Recommendation, W3C, 2004. http://www.w3.org/TR/rdf-schema/.

[9] Yao-Jen Chang, Hung-Huan Liu, Li-Der Chou, Yen-Wen Chen, and Haw-Yun Shin. A general architecture of mobile social network services. In International Conference on Convergence Information Technology, 2007, pages 151 –156, nov. 2007.

[10] Ying Chen, Xiaoyan Chen, Xiaocheng Ding, Fangyan Rao, and Dong Liu. Bluelocator: Enabling enterprise locationbased services. Mobile Data Management, IEEE International Conference on, 0:167, 2002.

[11] Rui Cheng, Zhuo Yang, and Feng Xia. izone: A location based mobile social networking system. In Third International Symposium on Parallel Architectures, Algorithms and Programming (PAAP 2010), pages 33–38, December 2010.

[12] Rui Cheng, Zhuo Yang, and Feng Xia. A mobile-based service for building social networks for mobile users. In Third International Conference on Intelligent Systems, Modelling and Simulation (ISMS'2012), pages 545–549, February 2012.

[13] Nathan Eagle and Alex Pentland. Social serendipity: mobilizing social software. IEEE Pervasive Computing, 4(2):28–34, 2005.

[14] Nicole B. Ellison, Charles Steinfield, and Cliff Lampe. The benefits of facebook "friends": Exploring the relationship between college students' use of online social networks and social capital. Journal of Computer-Mediated Communication, 12(4), 2007.

[15] Lugano Giuseppe. Designing for societal development: Enhancing mobile social software through social capital. In International Social capital and networks of trust congress, 2007.

[16] Scott E. Hudson and Ian E. Smith. Techniques for addressing fundamental privacy and disruption tradeoffs in awareness support systems. In CSCW, pages 248–257, 1996.

[17] Ls Humphreys. Mobile social networks and social practice: A case study of dodgebal. Journal of Computer-Mediated Communication, 13(1), 2007.

[18] Rich Ling. The Mobile Connection : The Cell Phone's Impact on Society (The Morgan Kaufmann Series in Interactive Technologies). Morgan Kaufmann, May 2004.

[19] Giuseppe Lugano. Mobile social software: Definition, scope and applications. In eChallenges 2007 Conference, pages 1434–1441, Netherlands, 2007.

[20] Giuseppe Lugano and Pertti Saariluoma. To share or not to share: Supporting the user decision in mobile social software applications. In User Modeling, pages 440–444, 2007.

[21] Frank Manola and Eric Miller. RDF primer. Recommendation, W3C, 2004. http://www.w3.org/TR/rdf-primer/.

[22] Gloria Mark, Ludwin Fuchs, and Markus Sohlenkamp. Supporting groupware conventions through contextual awareness. In ECSCW, pages 253–, 1997.

[23] Mehul Motani and Vikram Srinivasan. Peoplenet: engineering a wireless virtual social network. In in Proceedings of ACM MobiCom, pages 243–257, 2005.

[24] Eric Prud'hommeaux and Andy Seaborne. SPARQL query language for RDF. Recommendation, W3C, January 2008. http://www.w3.org/TR/rdf-sparql-query/.

[25] A.R. Radcliffe-Brown. On social structure. Journal of the Royal Anthropological Institute, 70:1–12, 1940.

[26] Bharat Rao and Louis Minakakis. Evolution of mobile location-based services. Commun. ACM, 46(12):61–65, December 2003.

[27] Jochen H. Schiller and Agnès Voisard, editors. Location-Based Services. Morgan Kaufmann, 2004.

[28] Stefan Steiniger, Moritz Neun, and Alistair Edwardes. Foundations of location based services. Cartouche - lecture notes on lbs, v. 1.0, 2006.

[29] Barry Wellman and S.D. Berkowitz. Social structures: A network approach. Cambridge: Cambridge University Press, 1988.

[30] Nina D. Ziv and Bala Mulloth. An exploration on mobile social networking: Dodgeball as a case in point. In ICMB, page 21, 2006.

[31] Faisal Alkhateeb. "A Mobile-Based Service for Building Social Networks for Mobile users". In Proceeding of the 3rd IEEE International Conference on Intelligent Systems, Modelling and Simulation (ISMS'12), ISBN: 978-0-7695-4668-1, pp.545-549, Kota Kinabalu, Malaysia, 8 – 10 February 2012.

WEBSITES LIST

Site 1: Social networking service, last visited October 20, 2011. (Wikipedia): http://en.wikipedia.org/wiki/Social_networking_service

Site 2: Facebook: http://www.facebook.com/

Site 3:Twitter: http://twitter.com/

Site 4:LinkedIn: www.linkedin.com/

Site 5: MySpace: www.myspace.com/

Site 6: Flickr: www.flickr.com/

Site 7: Hi5 : http://hi5.com/

Site 8: Xumii : http://www.myriadgroup.com/Mobile-Operators/Xumii\%20Social\%20Networking.aspx and http://xumii.com/v2/index.jsp

Site 9: Foaf vocabulary: xmlns.com/foaf/0.1/

Site 10: Vcard vocabulary :www.w3.org/TR/vcard-rdf/

Site 11: Meet-now-live http://www.meetnowlive.com/

Site 12: sniff : https://www.sniffu.com/

Site 13: Trapster : trapster.com/

Site 14: Whrrl : http://www.whrrl.com/

Site 15: Google+ : https://plus.google.com/

Site 16: Jaiku : http://jaiku.com/

AUTHORS

Faisal Alkhateeb is an assistant professor in the department of computer science at Yarmouk University. He holds a B.Sc. from Yarmouk University, 1999; a M.Sc. from Yarmouk University, 2003; a M.Sc. from Grenoble 1, 2004; and Ph.D. from Grenoble 1, 2008. He is interested in knowledge-based systems, knowledge representation and reasoning, constraint satisfaction and optimization problems, intelligent systems, Semantic Web, and Artificial Intelligence.

Ahmed M. Manasrah is a senior lecturer at the Faculty of Information Technology and Computer Sciences, Yarmouk University, Irbid, Jordan since 2010. Dr. Manasra served as the Deputy Director (Research and Innovation) and the Head of iNetmon project at the National Advanced IPv6 Centre (NAv6) in Universiti Sains Malaysia. Since 2009. Dr. Manasra started his career as a web developer at Telaterra LLC from 2003 until 2004. Between 2005 until 2008, Dr. Ahmed worked as senior research officer in NRG, Universiti Sains Malaysia. He is interested in network monitoring and security.

Iyad Abu Doush is an Assistant Professor in the Department of Computer Science at Yarmouk University, Jordan. Dr. Abu Doush has received his BSc. in Computer Science from Yarmouk University, Jordan, 2001, and his M.S. from Yarmouk University, Jordan, 2003. He earned his Ph.D. from the Computer Science Department at New Mexico State University, USA, 2009. Since then, he has been a professor of computer science, Yarmouk University, Jordan. He has published more than 20 articles in international journals and conferences. His research interests include assistive technology, intelligent interfaces, and multimodal interfaces. Other research interests include human computer interaction, computational intelligence, and collaborative virtual environments.

Eslam Al Maghayreh is an Assistant Professor in the Department of Computer Science at Yarmouk University (Jordan) since June 2008. He has received his PhD degree in Computer Science from Concordia University (Canada) in 2008, his M.Sc. degree in Computer Science from Yarmouk University in 2003, and his B.Sc. degree in Computer Science from Yarmouk University in 2001. His research interests include: Distributed systems, multi-agent systems, and runtime verification of distributed programs.

Hands-On Open Access Broadband Wireless Technology Lab

Mapping Course Outcomes to Lab Experiments

Yazan A. Alqudah[1], Todor Cooklev[2]
[1]Princess Sumaya University for Technology Amman, Jordan
[2]Indiana University Purdue University, Fort Wayne, Indiana

Abstract—**The unprecedented growth in wireless communication is offering opportunities and challenges for educators. Thanks to technology advances and job opportunities, more and more students are interested in wireless communications courses. However, bridging the gap between classroom and real-world experience remains a challenge. Advanced undergraduate communications courses typically focus more on theory. Some courses are given online, and lack hands-on experiments. Driven by feedback from industry and students, we propose practical laboratory experiments that attempt to bridge the gap between classroom and real world. The laboratory exercises take advantage of the infrastructure of deployed wireless networks and allow students to measure, and analyze data, as well as to interact. The proposed labs can be used even in online courses. This paper describes the experiments proposed, the procedures and typical results. The experiments are tied to course objective.**

Index Terms—**component; Open Access Lab, Field Measurements Experiments, Broadband Wireless, WiMAX, Path Loss, Propagation Models, and Throughput.**

I. Introduction

Mobile wireless communications technology has witnessed unprecedented popularity and penetration in the last decade. There are an estimated 6.5 billion mobile users around the world. With the popularity of wireless access comes the demand for higher data rate and ubiquitous access. Equipment manufacturers and service providers have been introducing technologies to keep up with user voice and data demands. A technology that has gained worldwide adoption for wireless broadband access is WiMAX (Wireless Interoperability for Microwave Access [1]). The technology has been adopted faster in some developing countries to bridge the digital divide and give population internet access. The WiMAX technology has taken advantage of years of advances in research and incorporates the latest technologies and algorithms in its Physical and MAC layers.

The main objective of engineering education is to prepare students to enter the workforce by providing classroom and laboratory knowledge needed to analyze and design complex engineering problems. Recently, a gap is developing in the teaching of wireless communication technology. As the technology advances at an unprecedented rate, most schools are finding hard to keep up with latest technologies in their curriculum and laboratory work. This is due to the rapid pace at which new technologies are introduced, the lack and high-cost of laboratory equipment. Furthermore, some courses are given online and completely lack a hands-on component.

Several educators have taken the responsibility to fill the gap in wireless communications and tie theory to practical real- world problems. In [2], a project-based approach is proposed using Wi-Fi modules. Spread spectrum wireless laboratory is proposed in [3]. The use of optical wireless (infrared) is presented in [4]. A low cost field programmable gate array (FPGA) is discussed in [5]. Wireless sensors design based course is proposed [6]. Other published work focuses on Radio Frequency (RF) and Microwave [7], and antennas [8].

While lab exercises for online courses do exist, these labs are always online, "virtual" labs that simulate certain aspects of practical wireless systems. For example, there are several software systems that model RF propagation, wireless networking, physical layer signal processing, etc. Such software is available through OPNET®, Matlab®, etc. Educators generally accept that even though it is desirable, it is not possible to have a hands-on lab as part of an online course; this is considered as one of the major weaknesses of online education.

The objective of this work is to bridge the gap between the academic education and real world by developing an open-access wireless lab that takes advantage of current deployed wireless and cellular networks and enables students to perform experiments in the field by collecting and observing measurements. The students will be responsible for analyzing the data obtained by their group and interact with other student groups to reinforce theory covered in the course.

This series of experiments is based on a deployed WiMAX network to reinforce theory covered in the classroom including: broadband 4G technology, cellular concepts, cell size, and interference, propagation models, the impact of signal strength, obstacles, and environment on received signal strength, adaptive modulation and dependence on signal strength, data throughput for voice, data and video traffic and dependence on link quality.

The experiments are to be carried out by separate student groups. Note that the results of each group might vary depending on the environment in which the experiments were performed. As a result, the students will be given an opportunity to compare and contrast their results with other groups by uploading their data to a central server. In this way the developed labs foster peer-learning, which we believe is an important pedagogical tool. In the peer-learning process, students learn with and from each other. Indeed, one of the best ways to learn and understand something is to explain it to someone else.

The choice of WiMAX as wireless standard for the lab is motivated by several factors:

- WiMAX is a mature technology with world-wide adoption. This is especially true in some developing and rural areas.
- The ability of the technology to provide access with minimal overhead and set up by users.
- Easy access to CPE link parameters through a PC. This enables students to capture information needed to complete experiments.
- The large coverage area of WiMAX cell enables exercising various channels conditions.
- The abundance of literature and textbooks that describe different aspects of the technology enables students and educators to have a thorough understanding of the concepts.

This paper is organized as follows: In section II, we review course objective and coverage. In section III, we describe experiments procedures and typical results. Finally, the conclusion and observations are presented in section III.

II. Course Coverage and Objectives

Wireless courses taught at senior undergraduate level build on fundamentals learned in senior-level communications and data networking courses. Furthermore, the students would have been already exposed to some of the wireless technical terms and themes such as WiMAX, GSM, HSDPA, 4G, broadband, CDMA, etc. The students familiar with these technical terms expect that a wireless course will help them understand cellular concepts and technology. However, it is the case that often textbooks fall behind on real-world advances in this field. With the introduction of WiMAX and LTE in the last decade and the advent of a new generation of wireless technology every 4-5 years, it is not surprising that instructors and students find information in textbooks obsolete.

This section reviews and introduces fundamental concepts that are covered by senior level wireless courses. The next section will tie these concepts with proposed field experiments.

A. *Cellular Network Concept*

The cellular concept is fundamental in understanding coverage, performance and implementation of a wide area network. The idea behind a cellular network is to divide a coverage area into smaller adjacent, non overlapping areas, called cells. Each cell is served by a transceiver or base station. To increase capacity and reduce power requirements, the overall spectrum allocated for the network is divided in small pieces that can be assigned to each cell. This enables the reuse of the spectrum in adjacent group of cell granted that any carrier frequency is used at farthest cell in the group. The concept of a cellular network is illustrated in Fig. 1.

Figure 1. Cellular network diagram

Cellular networks offer higher capacity, reduced power usage, and reduce interference from adjacent users and cells. Most wireless technologies with wide area coverage rely on the cellular concept to serve their users. This is particularly true for WiMAX.

B. *Wave propagation and propagation models*

Wave propagation and propagation models are fundamental concepts taught in wireless courses. Wave propagation deals with the electromagnetic wave as is it travels between transmitter (base station) and receiver (mobile phone). As the wave travels, it is impacted by the environment. The wave undergoes reflection, diffraction, and scattering. Large scale propagation models attempt to predict average signal power for arbitrary transmitter/receiver separation and environment. These models are critical in planning and deploying a cellular network as they can predict cells performance and coverage for different base station locations. Due to the complexity in predicting received signal strength, propagation models rely on statistical modeling based on field measurements. [9]-[19].

The models are summarized in Table I. The Free Space (FS) path loss is an analytical model that predicts the strength of the signal received when a clear line of sight path exists between transmitter and receiver. The FS model does not account for multipath propagation and cannot be used for point-multipoint radio link. It is included however include it here as a reference. The Stanford University Interim (SUI) is an empirical model recommended by standardizing committee. The constants a, b and c depend on the terrain and are given in Table II. The model contains frequency and height correction factors. The ECC-33 also known as Hata-Okumura extended model is based on Okumura model and is extended original model to frequencies up to 3.5 GHz. The COST-231 model is an extension to the Hata-Okumura model that has a correction factor for environment. The

TABLE I.
SUMMARY OF PROPAGATION MODELS.

<table>
<tr><td>Model</td><td></td></tr>
<tr><td>Free Space</td><td>$PL_{FS} = 32.45 + 20log_{10}(d) + 20log_{10}(f)$</td></tr>
<tr><td>802.16 SUI</td><td>$PL_{SUI} = A + 10\gamma log_{10}\left(\frac{d}{d_0}\right) + X_f + X_h + s$
$A = 20log_{10}(\frac{4\pi d_0}{\lambda})$
$\gamma = a - bh_b + (\frac{c}{h_b})$
$X_f = 6.0log10(\frac{f}{2000})$
$X_h = \begin{cases} -10.8log_{10}\left(\frac{h_r}{2000}\right), & for\ terrian\ A\ and\ B \\ -20.0log_{10}\left(\frac{h_r}{2000}\right), & for\ terrian\ C \end{cases}$
d_0=100m
<table><tr><td>Parameter</td><td>Terrain A (hilly)</td><td>Terrain B (flat)</td><td>Terrain C (rare vegeta-tions)</td></tr><tr><td>a</td><td>4.6</td><td>4.0</td><td>3.6</td></tr><tr><td>b</td><td>0.0075</td><td>0.0065</td><td>0.005</td></tr><tr><td>c</td><td>12.6</td><td>17.1</td><td>20</td></tr></table></td></tr>
<tr><td>ECC-33</td><td>$PL = A_{fs} + A_{bm} - G_b - G_r$
$A_{fs} = 92.4 + 20log_{10}(d) + 20log_{10}(f)$
$A_{bm} = 20.41 + 9.83log_{10}(d) + 7.894log_{10}(f) + 9.56[log_{10}(f)]^2$
$G_b = log_{10}(\frac{h_b}{200})(13.958 + 5.8(log_{10}(d)^2)$
$G_r = [42.57 + 13.7log_{10}(f)][log_{10}(h_r) - 0.585]$
$G_r = 0.759h_r - 1.862$</td></tr>
<tr><td>COST</td><td>$PL = 46.3 + 33.9log_{10}(f) - 13.82log_{10}(h_b) - ah_m + (44.9 - 6.55log_{10}(h_b))log_{10}d + c_m$
$ah_m = 3.20(log_{10}(11.75h_r))^2 - 4.79$
$ah_m = (1.11log_{10}f - 0.7)h_r - (1.5log_{10}f - 0.8$</td></tr>
<tr><td>Ericsson</td><td>$PL = a_0 + a_1log_{10}(d) + a_2log_{10}(h_b) + a_3log_{10}(h_b)log_{10}(d$
$-3.2(log_{10}(11.75h_r)^2) + g(f)$
$g(f) = 44.49log_{10}(f) - 4.78(log_{10}(f))^2$
<table><tr><td>Parameter</td><td>a_0</td><td>a_1</td><td>a_2</td><td>a_3</td></tr><tr><td>Urban</td><td>36.2</td><td>30.2</td><td>12.0</td><td>0.1</td></tr><tr><td>Suburban</td><td>43.20</td><td>68.9 3</td><td>12.0</td><td>0.1</td></tr><tr><td>Rural</td><td>45.95</td><td>100. 6</td><td>12.0</td><td>10.</td></tr></table></td></tr>
<tr><td>COST 231 WALFISH-IKEGAMI</td><td>$PL = L_{FSL} + L_{rts} + L_{msd}$
$L_{FSL} = 32.45 + 20log_{10}(f) + 20\log_{10}(d)$
$L_{rts} = -16.9 - 10\log(w) + 10log_{10}(f) + 20log_{10}(hr) - 10 + 0.353\varphi$
$L_{msd} = -18log_{10}(1 + \Delta_{hbase} + 54 + 18log_{10}(d) + k_f log_{10}(f) - 9log_{10}(B)$
$k_f = \begin{cases} -4 + 0.7(\frac{f}{925} - 1) & \text{for suburban} \\ -4 + 1.5(\frac{f}{925} - 1) & \text{for urban} \end{cases}$</td></tr>
<tr><td colspan="2">d: separation, f frequency, G_b base station antenna gain, G_r receiver gain, h_b base station antenna height, h_r receiver height, c speed of light, φ is the diffraction angle, Δ_{hbase} is the transmitter antenna height above building height, B is the distance between adjacent buildings, w is the building separation</td></tr>
</table>

Hata-Okumura model was developed for 500-1550 MHz. The Cost-231 model extends Okumura-Hata model to frequency range up to 2 GHz. The model calculates path loss for urban, suburban and rural areas. The Ericsson model is based on modified Okumura-Hata model and allows changing parameters according to environments. The environment parameters are summarized in parameters is given in table III. Finally, The COST 231 Walfish-Ikegami model is suitable for flat suburban and urban areas with uniform building height.

C. *Throughput and Adaptive Modulation*

The WiMAX physical layer is based on orthogonal frequency division multiplexing (OFDM). OFDM is an efficient scheme for high data rate transmission in a non-line-of-sight or multipath radio environment. WiMAX supports a variety of modulation and coding schemes and allows for the scheme to change on a burst-by-burst basis per link, depending on channel conditions. The base station uses the channel quality feedback indicator sent by a user to assess downlink quality. Relevant channel-state information fed back by the users can include: Physical CINR, effective CINR and frequency selective sub-channel selection. On the uplink, the base station estimates channel quality based on the received signal quality.

The flexibility of the physical layer of WiMAX enables data rate and performance variation based on the operating parameters. Channel bandwidth, the modulation and coding schemes determine the data rate that can be achieved, in addition to other parameters, such as number of sub channels, OFDM guard time, and oversampling rate. The base station scheduler determines the appropriate data rate (or burst profile) for each burst allocation based on the buffer size, channel conditions at the receiver, and user traffic.

The data rate for various channel bandwidths, modulation and coding schemes is shown in table II [20].

TABLE II.
WIMAX DATA RATE

Parameter		Downlink	Uplink	Downlink	Uplink
System Bandwidth		5 MHz		10 MHz	
FFT Size		512		1024	
Null Sub-Carriers		92	104	184	184
Pilot Sub-Carriers		60	136	120	280
Data Sub-Carriers		360	272	720	560
Sub-Channels		15	17	30	35
Symbol Period, T_S		102.9 microseconds			
Frame Duration		5 milliseconds			
OFDM Symbols/Frame		48			
Data OFDM Symbols		44			
Mod.	Code Rate	5 MHz Channel		10 MHz Channel	
		Downlink Rate, Mbps	Uplink Rate, Mbps	Downlink Rate, Mbps	Uplink Rate, Mbps
QPSK	1/2 CTC, 6x	0.53	0.38	1.06	0.78
	1/2 CTC, 4x	0.79	0.57	1.58	1.18
	1/2 CTC, 2x	1.58	1.14	3.17	2.35
	1/2 CTC, 1x	3.17	2.28	6.34	4.70
	3/4 CTC	4.75	3.43	9.50	7.06
16QAM	1/2 CTC	6.34	4.57	12.67	9.41
	3/4 CTC	9.50	6.85	19.01	14.11
64QAM	1/2 CTC	9.50	*6.85*	19.01	*14.11*
	2/3 CTC	12.67	*9.14*	25.34	*18.82*
	3/4 CTC	14.26	*10.28*	28.51	*21.17*
	5/6 CTC	15.84	*11.42*	31.68	*23.52*

Table 3: Mobile WiMAX PHY Data Rates with PUSC Sub-Channel[4]

http://www.wimaxforum.org/technology/downloads/Mobile_WiMAX_Part1_Overview_and_Performance.pdf

III. Experiment Set up

The design of the experiment is meant to provide flexibility to the students and the instructor. The students can carry the experiments according to their schedule. To obtain the equipments, the department offering the laboratory is encouraged to sign up a memorandum of understanding with service provider. This enables sharing of results and helps answer questions from the network side if needed. Many campuses are often serviced by one or more service providers who will be willing to collaborate on educational projects. The department is expected to sign up for a number of subscriptions for available WiMAX carriers with the understanding that these subscriptions will be used by students to analyze and study the network. During the week of the experiments, the students sign up in groups to borrow the equipments which include: Customer premise equipment (CPE), laptop, and power converter (to power CPE from vehicle outlet).

Depending on the chosen coverage area, the student can choose to collect field data using a vehicle. This ensures large coverage area. Most experiments are expected to be completed in 3-4 hours. The instructor of the lab will determine a safe area either on campus or off campus for students to collect measurements. The locations can be assigned by either map marks on a hard print or landmarks on map software. Depending on the coverage area and network service, the experiments might use a separate CPE or universal serial board (usb) dongle to connect to the network. In the case of a CPE, the equipment can utilize a power converter to supply CPE and laptop devices from the vehicle. The connection of the CPE and laptop is depicted in Fig. 2.

Next, we discuss the procedure and expected results for the objectives laid out in section II:

A. Lab 1 - Cellular Coverage

This experiment studies the cellular coverage. The procedure requires students to collect data about current serving base station using the CPE webpage (available from manufacturer). The webpage displays identification of current serving base station, as well as other parameters related to the connection as shown in Fig. 3. The students capture both the location and identification over an area that covers several base stations. The students should collect large number of measurements to be able to plot the coverage area of base stations identified.

The students can identify cell border by highlighting locations were CPE interchange between two or more stations.

To complement the field measurements, students can use RF planning tools to generate plots of coverage area as shown in Fig. 4.

The students are expected to contrast and compare the results collecting in the field with those obtained through RF planning tool and explain any differences.

B. Lab 2 - Propagation Models

The objective of this experiment is to make students become familiar with RF propagation. The students will be able to experience signal degradation as distance from transmitter increases. They will also be able to see the variation of signal power in any locality caused by the environment. After collecting the measurements, the students are expected to generate a graph that shows sig-

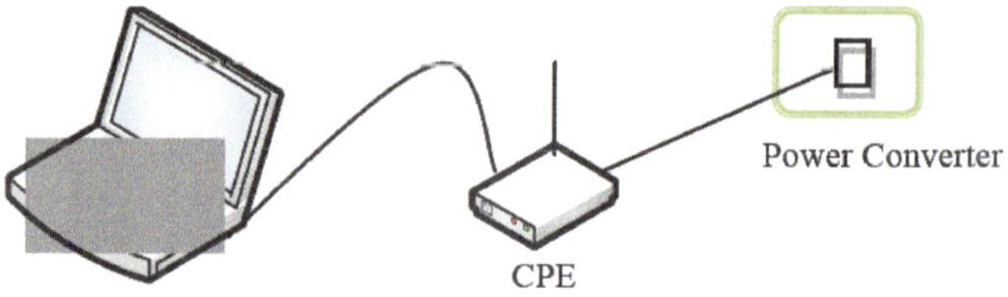

Figure 2. Experiment set up. The CPE uses a power converter plugged into vehicle 12v outlet.

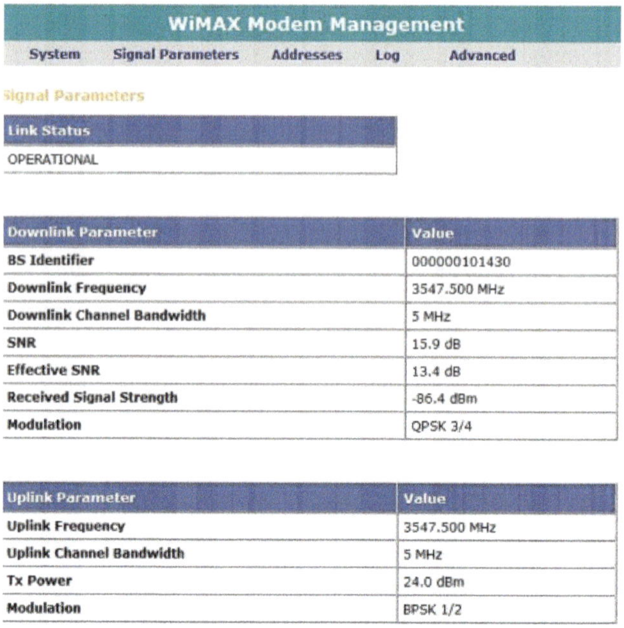

WiMAX Modem Management

System | Signal Parameters | Addresses | Log | Advanced

Signal Parameters

Link Status
OPERATIONAL

Downlink Parameter	Value
BS Identifier	000000101430
Downlink Frequency	3547.500 MHz
Downlink Channel Bandwidth	5 MHz
SNR	15.9 dB
Effective SNR	13.4 dB
Received Signal Strength	-86.4 dBm
Modulation	QPSK 3/4

Uplink Parameter	Value
Uplink Frequency	3547.500 MHz
Uplink Channel Bandwidth	5 MHz
Tx Power	24.0 dBm
Modulation	BPSK 1/2

Figure 3. WiMAX link parameters can be access through web browser by connecting to CPE management server.

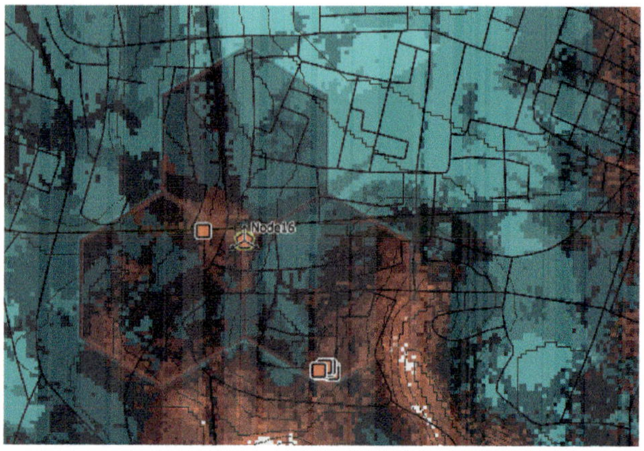

Figure 4. Coverage area prediction using RF planning tool

nal power variation with distance as well as various models used to predict the signals. The students will comment on the applicability of various models and closeness of their fitness to collected data.

The procedure of the experiment is as follows. The team connects the CPE and laptop to power supply. The CPE will attempt to connect to available network. The student will access the configuration page of the link through a browser by entering the CPE webpage. The page displays Received Signal Strength Indicator (RSSI) in dBm and the current serving base station. The students are expected to complete a table with their location, RSSI, and any information about the environment that might interfere with measurement. For each location, the measurement is repeated three times with slight CPE movement. The experiment is completed once all locations data are collected. The experiments can also be conducted with a slight modification. Rather than pre assigning locations for measurements, students are asked to collect measurements at various distance and locations in a coverage area. The students will utilize the base station information in the configuration page to ensure they remain within a coverage area.

After completing the experiment, the students are expected to generate plots of data collected vs. distance and propagation models. A typical graph will look similar to Fig. 5.

The students can also calculate path loss based on base station and receiver gain and loss parameters. The equation below is used to generate Fig. 6.

$$P_R = P_T + G_T + G_R - L_T - L_R + PL \qquad (1)$$

C. Lab 3 - Throughput and Adaptive Modulation

The throughput and adaptive modulation experiment studies the data rate of the link and its dependence on modulation type. As the channel conditions change, the base station changes modulation to minimize bit errors. When the channel condition is favorable, the base station uses efficient modulation such as 64 QAM for transmission. This enable symbols to carry 6 bits of data. As the channel condition deteriorate, the base station switches to more robust modulation such as BPSK which only carries 1 bit per symbol. The base station uses signal power to assess the channel quality. The variation in signal power can be achieved by considering locations obtained in experiment 2.

The set up for the experiment is show in Fig. 7. A free software utility (iPerf®) is installed on the server and client sides. The tool supports TCP (Transmission Control Protocol) and UDP (User Datagram Protocol) traffic. To measure downlink throughput, iPerf® is run as a client on the laptop and as a server on the server machine. The uplink is measured similarly, however, the client is run on the data server and the laptop is run as server. After iPerf® is run, the tool generates the achievable data rate. The experiment is repeated for different RSSI and results are tabulated. The plot in Fig. 8 shows typical results.

In addition to measuring throughput, students are asked to take note of the modulation used as signal strength changes. The plot in Fig. 9 shows the different modulations used by base station as signal to noise ration changes.

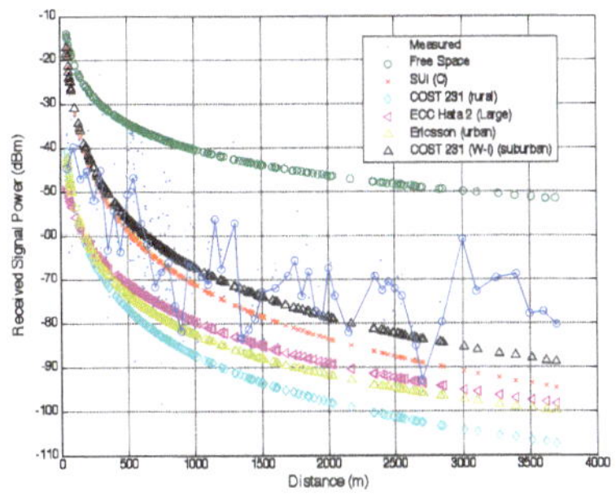

Figure 5. Received signal power vs. distance between transmitted and receiver.

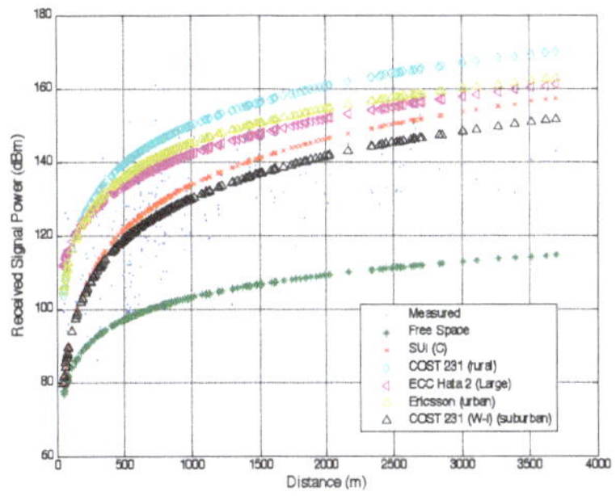

Figure 6. Measured and predicted path loss vs. distance

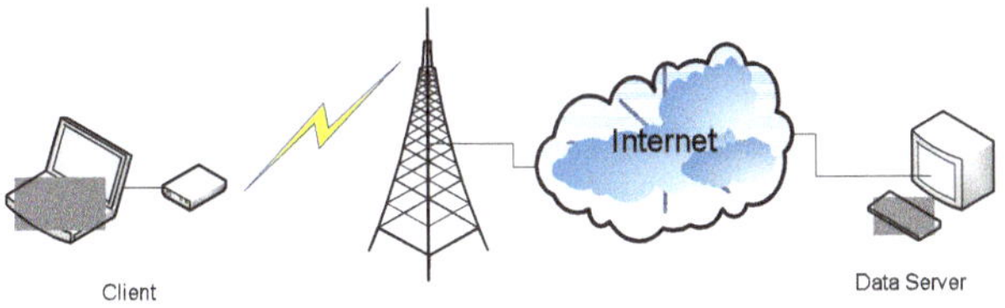

Figure 7. Throughput measurement set up

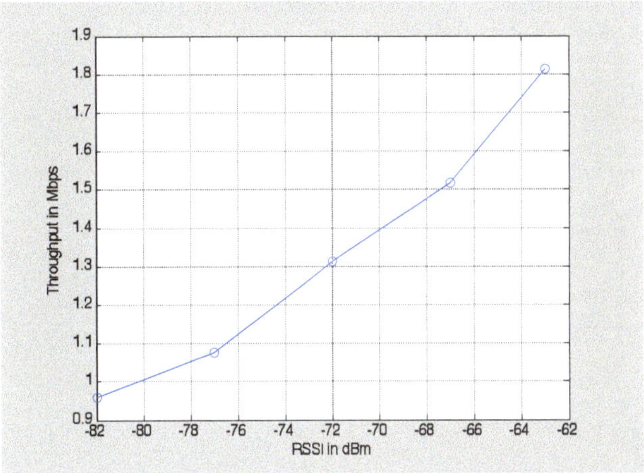

Figure 8. Typical throughput (Mbps) vs. RSSI (dB)

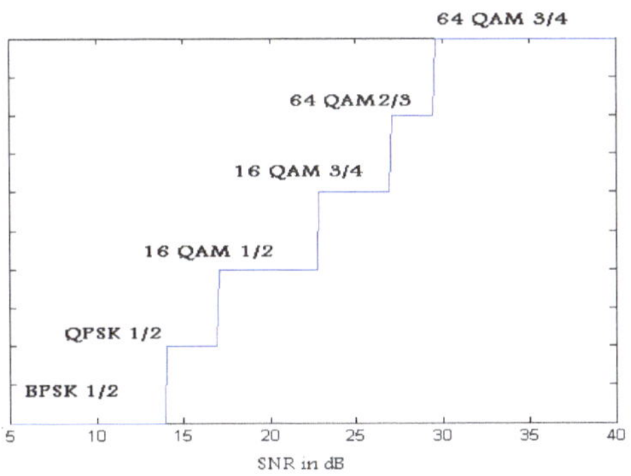

Figure 9. Modulation utilized vs. SNR (dB)

IV. Mapping Course Outcomes

In this section we summarize typical wireless course learning outcomes and map them to experiments and exercises propose in section II. Table III summarizes course outcomes obtained by surveying wireless courses at PSUT and other engineering schools.

By executing the first lab experiments, students are introduced to the cellular concepts. Students can witness firsthand the boundaries of a cell and the carrier frequency used in each cell. By mapping field measurements about cell id, students are made familiar with planning and designing of a cellular system. Also using the equipments to obtain measurement familiarize students with WiMAX or the Wireless standards by observing technical terms used to describe different parameters in the measurements.

The exercises in the second lab focus on frequency reuse and propagation models. Students are expected to compare measured signal strength with propagation models to determine and justify the most accurate model.

The third lab familiarizes students with data rate, quality of service, and coding. Students record data rates for different traffic types and channel conditions. Students also record modulation and coding used for various channel conditions.

TABLE III.
SUMMARY OF COURSE OUTCOMES MAPPED TO EXPERIMENTS

Course Outcomes	Experiment
Understanding of cellular and frequency reuse concepts	Lab 1, Lab 2
Understanding of wireless propagation of Electromagnetic wave and losses	Lab 2
Design and planning of a wireless cellular system	Lab 1, Lab 2
Radio spectrum allocation in multi-user systems	Lab 1
Understanding basic issues related to supporting QoS in voice and data services in cellular networks	Lab 3
Understand how the various signal processing and coding techniques combat channel uncertainties	Lab 3
Wireless systems standards	Lab 1, Lab 2, Lab3

V. CONCLUSION

This paper presented a hands-on low cost wireless communications laboratory based upon deployed broadband wireless network. The experiments are based on several senior projects that studied and analyzed the network performance.

The laboratory allows students to conduct experiments based on real world deployed network. This fosters knowledge of the technology and enables students to contrast and compare theoretical and practical aspects of a wireless network.

Furthermore, the developed lab exercises foster peer-learning as a pedagogical approach. In the future we plan to assess this peer-learning process. In particular, it needs to be investigated is whether peer learning actually occurred, and whether it was effective.

REFERENCES

[1] IEEE Standard 802.16-2004, IEEE Standard for Local and Metropolitan Area Networks, "Part 16: Air Interface for Fixed Broadband Wireless Access Systems." 2004.

[2] Sarkar, N.I.; Craig, T.M.; , "Teaching wireless communication and networking fundamentals using Wi-Fi projects,"*Education, IEEE Transactions on*, vol.49, no.1, pp. 98- 104, Feb.

[3] Cassara, F.A.; , "Wireless communications laboratory," *Education, IEEE Transactions on*, vol.49, no.1, pp. 132-140, Feb. 2006

[4] Padgett, W.T.; Black, B.A.; Ferguson, B.A.; , "Low-frequency wireless communications System-infrared laboratory experiments," *Education, IEEE Transactions on*, vol.49, no.1, pp. 49-57, Feb. 2006

[5] Linn, Y.; , "An Ultra Low Cost Wireless Communications Laboratory for Education and Research," *Education, IEEE Transactions on*, vol.PP, no.99, pp.1, 0,

[6] Frolik, J.; Weller, T.M.; , "Wireless sensor systems: an approach for a multi university design course," *Education, IEEE Transactions on*, vol.45, no.2, pp.135-141, May 2002

[7] C. Furse et al., "Laboratory project in wireless FSK receiver design," IEEE Trans. Educ. vol. 47, no. 1, pp. 18-24, Feb. 2004.

[8] D. Baker, "An innovative project-driven senior-level antennas course," IEEE Trans. Educ., vol. 40, no. 3, pp. 190-194, August. 1997.

[9] J. Milanovic et al, "Comparison of Propagation models accuracy for WiMAX on 3.5 GHz," 14th IEEE International Conference on Electronics, Circuits and Systems, 2007. ICECS 2007.

[10] Pasquale Imperatore, Elio Salvadori, Imrich Chlamtac, "Path loss measurements at 3.5 GHz: A trial test WiMAX based in rural environment," Proc. of the 3rd International Conference on Testbeds and Research Infrastructures for the Development of Networks and Communities, 2007.

[11] E. Moraes et al, "WiMAX Near LOS Measurements and Comparison with Propagation Models",

[12] T. K. Sarkar et al, "A survey of various propagation models for mobile communications," IEEE Antennas and Propagation Magazine, vol 45, no 3, June 2003, pp. 51-79.

[13] V. Erceg et al, "Channel models for fixed wireless application," Tech. rep. IEEE 802.16 Broadband Wireless Access Working Group, January 2001

[14] Y. Okamura, "Field strength and its variability in VHF and UHF land mobile radio services." Rev. Elec. Comm. Lab. No. 9-10, pp825-873, 1968

[15] Electronic Communication Committee (ECC) with the European Conference of Postal and Telecommunication Administration (CEPT), "The analysis of the coexistence of FWA cell in the 3.3-3.8 GHz band," tech. rep., ECC Report 33, May 2003.

[16] M. Hata, "Empirical formula for propagation loss in land mobile radio services," IEEE Trans. Veh. Tech. vol VT-29, August 1980, pp. 317-325.

[17] COST Action 231, "Digital mobile radio towards future generation systems, final report," tech rep., European Communities, EUR 18957, 1999.

[18] Alqudah, Y.A.; Tahat, A.; , "Path loss and propagation models at 3.5 GHz using deployed WiMAX network," *Information Networking (ICOIN), 2011 International Conference on*, vol., no., pp.301-305, 26-28 Jan. 2011

[19] Alqudah, Y.A., "Power Analysis Based on Field Measurements Using 3.5 GHz WiMAX Network", ICOIN 2012, Feb 1-3, 2012, to be published, Bali, Indonesia

[20] Mobile WiMAX – Part I: A Technical Overview and Performance Evaluation, August 2006 [Online].

AUTHORS

Yazan A. Alqudah is with Princess Sumaya University for Technology, Amman, Jordan, (e-mail: Yazan.a.alqudah@ieee.org.)

Todor Cooklev is with Indiana University Purdue University, Fort Wayne, Indiana.

Exploring Accessibility Scenarios for 2020 in Relation with Future ICT Trends on Assistive Technology and Accessibility

C. Kouroupetroglou, D. Tektonidis, A. Koumpis, I. Ingnatiadis
ALTEC Software SA, Thessaloniki, Greece

Abstract—**In this paper we are going to present a set of 5 future scenarios that were developed within the eAccessibility2020 study. The study aims to explore and analyse the referred relationships between the emerging ICT landscape, in the societal and economic context and the development and provision of assistive technologies (AT) and e-Accessibility, within a perspective of 10 years. The scenarios were developed after an initial trend analysis that the study team conducted based on data gathering. The scenarios were developed based on a methodology which defined a set of guides for scenario development and a set of visions for the future of eAccessibility.**

Index Terms—**Accessibility, Assistive technologies, Scenarios, Trends.**

I. Introduction

This paper will present of a set of initial scenarios prepared for the eAccessibility[1] study needs. The scenarios developed were based on a data gathering process and the outcomes from the initial trend analysis that the study team conducted based on the gathered data. The data gathering process of the study is a process and the following trend analysis were based on two main axes: The desk research and interviews with experts.

A. Desk research

Desk research included reading and categorization of a large set of conference and journal research papers, EU funded RTD project deliverables and various trend analysis reports. The collected material was analyzed and categorized into groups according to the technology used, disability/ies addressed, and beneficiaries. Additionally, during the desk research, the study team selected a set of 137 EU funded RTD projects relevant to eAccessibility and eInclusion as a base for other parts of the research. This analysis which is now in its latest stages produced some preliminary outcomes of technologies that seem possible to have an impact on accessibility in the future.

B. Interviews

Out of 137 projects related to eAccessibility and eInclusion the study team has selected a set of 46 projects relevant either to the projects goals or to technologies that were already identified as trends from the desk research. Next, we contacted about 130 persons that participated in those projects by email and asked for their participation in an interview regarding their views for the future of accessibility from their filed of expertise. Almost 30 of them replied and were scheduled for an interview. Until now 23 of them have already been conducted. The interviews objective was twofold. The first one was to get the experts opinions in the future of their area of expertise in relation to eAccessibility. The second was to investigate experts' opinions on the future of eAccessibility in general.

II. Pre-Scenario Development Methodology

In this initial scenario-building attempt we have roughly explored what could be the future in 2020 (and particularly in terms of the available ICTs, societal/ demographic conditions, etc.) Then we have examined the desirable future in terms of eAccessibility (i.e. the technology and market environment that would 'best' fit to the users and society's needs), with an eye on the feasibility aspect of such future (feasibility is posing a set of technology and socio-economic limitations on the 'best' desirable future)

Therefore, there have been three main directions and limitations in the process of scenario building, which are summarised below:

- What is desirable? (user & policy perspective)
- What is feasible? (technology and socio-economic perspective)
- What is the time dimension? 2020 – 2025 when research results reach the market

The scenario development methodology was based on a set of high level theoretical scenarios that were used as guides for the development of more specific scenarios

A. High-level pre-scenarios

Under the scoping phase of the study ideas on initial broad scenarios (high-level scenario guides) have been put forward. These ideas were used (together with the three abovementioned directions/ limitations) as a starting point for the study. They have slightly changed during our work until today (i.e. end of July 2011) and may radically change in the future.

In any case, they will be further broken down to cover also different situations/ aspects of life (e.g. working life, mobility, etc.), the severity of disability, etc. Similarly, the study team may prioritise eAccessibility solutions based on the Maslow hierarchy of (human) needs to ensure they are in line with the potential value they create (e.g. solutions addressing health issues are normally of higher importance than gaming applications).

Also, economic factors such as cost reductions in the private sector (especially among enterprises operating websites or offering software), which may affect the budget allocated to eAccessibility solutions and compliance with relevant standards will be taken into consideration. Finally, issues of high importance such as mainstreaming eAccessibility, etc. will be explored.

1) Technology-driven scenario guide.

Scenarios built based on the first high level scenario guide may focus on a prevalent in 2020 type of ICT applications (such smart mobile applications). The eAccessibility-related user needs will be dictated by these applications (e.g. the way they are used, the interfaces they have, the services they offer, the type of user interaction they require, etc.). Also, another similar scenario may be built upon a technology-breakthrough hypothesis e.g. in brain-machine interaction and thus examine the eAccessibility future in this respect.

2) Disability-driven scenario guide.

Scenarios built based on that high level scenario guide will focus on an expected major disabilities for 2020 (e.g. cognitive disability, etc.). Under this perspective, the scenario will examine how technological and other factors may affect eAccessibility solutions supporting this specific (types of) disabilities.

3) User-group-focused scenario guide.

Scenarios built based on the user-group-focused scenario guide will focus on a major target group for eAccessibility in 2020 such as elderly people. Similarly, the pre-scenario may examine how technological and other factors may evolve to offer a range of eAccessibility solutions on this target group. The potential influence of progress in medicine and health sciences may also be assessed.

B. Visions for the Future of eAccessibility

Apart from having the three high level scenario guides the study team has also produced a set of visions for the future of eAccessibility based on the interaction with user representatives and eAccessibility stakeholders. These visions serve as targets that the scenarios should achieve by following some of the previously presented guides.

1) Vision 1: Accessibility Benefits All

This vision supposes that by 2020 the perception that accessibility is aimed at a minority group will be broken. Therefore all stakeholders will then acknowledge that accessibility is a high value-added activity that benefits all groups. This will result in a variety of eAccessibility solutions which in turn will offer better design choices. Having eAccessibility solutions mainstreamed, career options will be improved and productivity will rise not just for disabled users.

2) Vision 2: Full Spectrum eAccessibility for Elderly Persons

The second vision supposes that by 2020 there will be universal appreciation of the fact that capabilities decline with age for all people. Thus, systems will need to be designed for ease of use by all people regardless of their physical limitations. This will preserve people's quality of life and professional productivity enabling late retirement and high quality of life long after retirement.

3) Vision 3: Publish for All

In this vision we suppose that by 2020 all publications will be available in friendly format for disabled people. This will be achieved by the uptake of accessible, consumer friendly content production models by all major publishing companies. Publishing of accessible content will not include only text but audio and video content as well and all content produced will be compatible with all major standards such as WCAG & DAISY etc.

4) Vision 4: Tools for All

This last vision supposes that by 2020 basic tools such as calendars, agendas, calculators, spreadsheets, databases, etc. Will have far superior functionality based on Universal Design/ Design for All principles. This will be achieved by the wide spread of basic programmatic tools, such components, libraries, APIs etc that will be accessible in first place. This way using core programmatic components that are built with accessibility in mind the production of accessible web apps, mobile apps, desktop apps etc will become far easier. This means that new applications will have even more enhanced ease of use for all users.

C. The five initial scenarios

In the following sections we elaborate on five initial scenarios that can be regarded as indicative of the possible extrapolations that one can make given a specific vision as those described previously and taking into account various parameters related to the use of technology (available or not yet, existing or not mature), the linkage with a particular disability or, finally, a specific user group.

TABLE I.
CLASSIFICATION OF THE FIVE PRE-SCENARIOS WITH RESPECT TO THEIR RELATION TO THE FOUR VISIONS AND THE ORIENTATION REGARDING TECHNOLOGIES, DISABILITIES AND USER GROUP FOCUS.

Scenario:	**Advanced dialog interface**	**Advanced hands-free interaction**	**Automatic real time transcripts and audio description**	**Adaptation of multimodal interfaces**	**Easy development of accessible web apps**
Vision:	Vision 1: Accessibility Benefits All	Vision 1: Accessibility Benefits All	Vision 2: Full Spectrum eAccessibility for Elderly Persons	Vision 3: Publish for All	Vision 4: Tools for All
Technologies used:	Speech recognition, NLP, TTS, Advanced IR	Gaze recognition, face recognition, brain-computer interfaces	Speech recognition, NLP, new metadata enabled video and multimedia formats, Smart home devices such as TV	Mobile device, user profiling, adaptive multimodal interfaces	Reusable components, open source libraries, web apps, adaptive multimodal interfaces
Disability focus:	Visual, mobility, cognitive, elderly	Mobility	Visual, hearing, multiple disabilities.	All disabilities	All disabilities
User group focus:	Elderly, everybody	Motor-impaired, all	Elderly, everybody	All user groups	Application Developers

Figure 1. Scenario: "I am having a conversation with my mobile device"

Given the analysis of trends the high level scenarios and the visions for the future already presented so far, in the following section we attempt to combine them under five scenarios.

These scenarios have been built after a brainstorming exercise which intended to investigate possible ICT trends combinations having in mind the visions for the future and the high level scenarios. Thus, we started combining various weak signals, micro and key ICT trends having a specific disability or user groups in mind, in order to produce possible future solutions for them. Using the visions as targets, the scenarios we developed aim to showcase examples of how these visions could be achieved. These scenarios can be seen as examples of the final scenarios that will be produced in the next stages and evaluated for their validity and plausibility.

The first part of each scenario introduces the technologies exploited and the second part describes the scenario in the form of a story involving imaginative people and situations in the future.

III. Pre-Scenario: Advanced Dialog Interfaces

A. *Technologies exploited*

Dialog interfaces are a reality in many situations nowadays. However, they usually work with specific limited vocabulary and in a specific context. Additionally, mobile devices' capabilities are increasing dramatically by the time leading to more powerful tools in users' hands. Technologies related with dialog interfaces such as speech recognition, TTS and NLP are evolving[2] and we already have signals that more advanced interfaces will appear (see IBM Watson). Combining the increase in mobile devices' capabilities with the advances in dialog interface technologies and with the trend of seamless connectivity through mobile networks such as 3G, 4G etc., we can expect more advanced dialog interfaces without limited vocabularies and able to understand idioms, slung, etc of human language to emerge. The latter combined with more intelligent and possibly personalized information retrieval systems can lead to advanced dialog interfaces that will be close to human to human interaction.

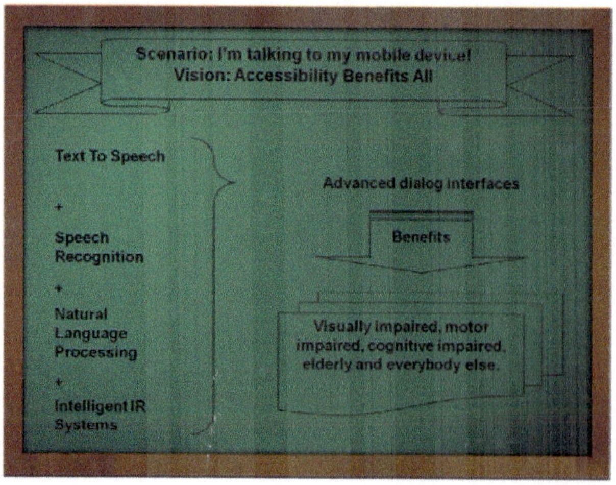

Figure 2. Scenario diagram for advanced dialog interfaces

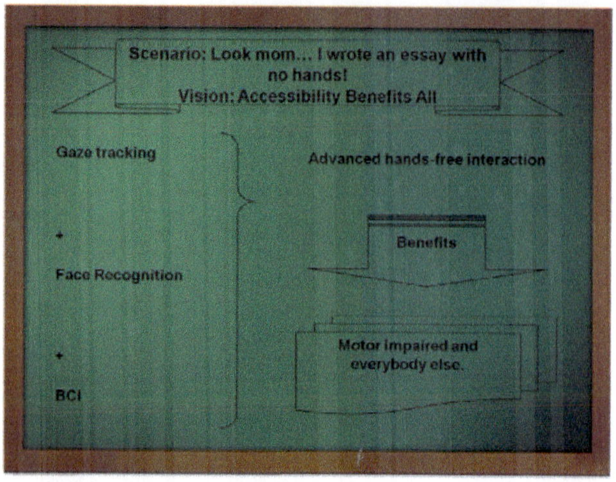

Figure3. Scenario diagram for advanced hands-free interaction

One of the major disabilities benefiting from dialog interfaces are visually impaired people. However, they are not the only disability group that can benefit from such interfaces. Mobility impaired people or even people with cognitive disabilities might also benefit from such interfaces. Additionally, older people will probably be using mobile devices in the future far more than today and being able to operate with them in a more intuitive way will help significantly in their experience and engagement with new technologies. Thus, the elderly is also a user group with significant benefits from this kind of interaction interfaces. It is also easy to understand that such interfaces would also help everybody in their everyday interaction with various devices in cases where they are situationally visually or mobility impaired (e.g. driving a car).

B. *Scenario: "I am having a conversation with my mobile device"*

Mary is a 20 year old blind girl who goes out for shopping. Instead of relying solely on the shopping assistants' opinions' for the new pair of shoes she wants to buy, she tells her phone to ask her friends about the shoes she is trying. The phone automatically uploads the photos to social networks with a related caption such "How do you like it?". After about an hour or two of trying various pairs of shoes and outfits and uploading their photos she asks her phone "What are my friends thinking about the pairs I tried?". The phone and the respective dialog interface realize that she is talking about the latest photos she uploaded and her friends' comments on them, gathers the comments and answers with something like "Most of your friends think the second and fifth pair you tried are cool and awesome and they really didn't like the third pair." The discussion could go on with something like "What is Mark thinking about the fifth pair?" and the phone could answer with something like "He thinks they are great and he likes them the best". So after a discussion with her phone (actually after a discussion with her friends) she could decide on which pair to buy.

IV. Pre-scenario: Advanced hands-free interaction

A. *Technologies exploited*

One of the technologies identified also as a trend from our data gathering process was Brain-Computer Interfaces (BCI)[3]. Using our brain to control various computer functions seems possible in the future. Such devices seem a bit awkward and strange nowadays but mainstreaming of such devices could lead to more user-friendly devices in the form of a baseball cup for example. Users will probably be able to operate them easier than today and this could become a key technology trend in the future. In addition we already have signal of mainstreaming devices and technologies having to do with gesture recognition (i.e. Kinect) which shows the potential of evolvement of such technologies and their enhancement in the core of future device platforms. So an example could be tablets in the future that will have integrated software for gesture and/or face recognition and possibly gaze tracking.

Gestures, face recognition, gaze tracking and brain computer interfaces can all be operated by mobility impaired people and could be really useful for people with serious mobility impairments. Additionally it could also help people that due to their work environment or other circumstances cannot use their hands to interact with an ICT device. The latter could lead to the following use case scenario.

B. *Scenario: "Look mom... I wrote an essay with no hands!"*

Andrew is a 14 years old boy who goes to school in his wheelchair due to his serious mobility impairment. He can use his right hand but with limited mobility allowing him though to make specific gestures. He is wearing a baseball cup that however is not like any other ordinary baseball cup. It uses electrodes to read specific brainwaves and transmits them (through Bluetooth/ WiFi/etc.) to the tablet that is constantly connected in his chair. The tablet's camera is always on and tracks Andrew's eyes movements and possible gesture he makes. He returns from school where his history teacher gave them an assignment on World War II. Additionally, using his baseball cup he opens the browser and starts searching for sources and information on his assignment. Then, using the on-screen keyboard with the eye tracking software he starts writing his assignment while at the same time he uses specific face expressions for styling the text he is writing and controlling other functions of the program. When finished with his assignment he thinks of music and through his BCI baseball cup he logs on to his favourite music service and starts listening to it.

Figure5. Scenario: "Look mom… I wrote an essay with no hands!"

V. Pre-Scenario: Automatic Real Time Transcripts and Audio Description

A. *Technologies exploited*

Another technology trend identified in the analysis is the use of a variety of devices to access web content and services. People are expected to interact seamlessly with any device in their home which in turn uses web content and apps. Such devices could be tablets, electrical appliances, TVs, smart rooms etc.

One very common device used mainly for recreational purposes is the TV. TVs are now starting to integrate operating systems that allow various apps to run on them and be used by their viewers. The TV is becoming another information access point in the house allowing in parallel the use of appropriate applications and services. Videos which are usually displayed on TV are also changing and new formats are coming out such as webVTT[4] which allow the enhancement of various metadata within the video. So, future movies could come in those new enhanced formats. In addition, developments in speech recognition, NLP and speech technologies in general can help in the easier production of such metadata by studios, but they could also facilitate applications for automatic real time transcripts of a video[5] (see closed captions on youtube). Finally these new video formats can also carry video descriptions which through TTS could be utilised and read to people with visual disabilities.

These new multimedia formats enable the simultaneous use of a variety of interactions during their playtime. This means that a movie coming out in these new formats could be easily watched by a deaf person by reading the transcripts (either already embedded or produced in real time) without any problems. In addition, blind persons could also use the same format of a video to watch the same movie but instead of the transcripts they could use the audio description describing the action on the video. This audio description could be enhanced by the producer or produced on the fly using image recognition software. New TV devices with appropriate apps could use all this information in a video to provide the appropriate interaction to each user.

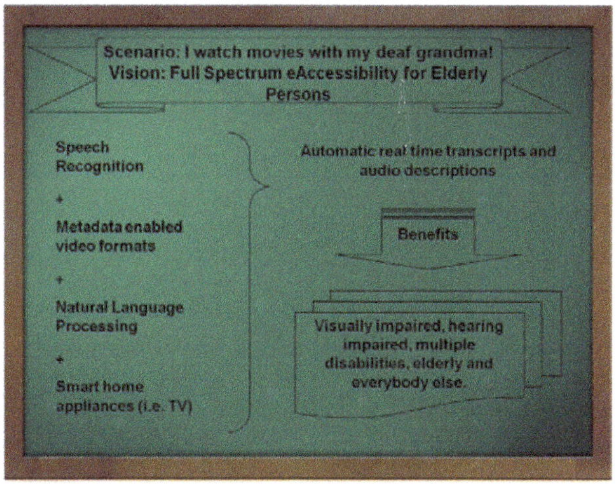

Figure 4. Scenario diagram for automatic real time transcripts and audio descriptions

Figure 6. Scenario: "I watch movies with my deaf grandma"

B. Scenario: "I watch movies with my deaf grandma"

John is a 12 year old boy who loves to watch the movies. Despite his total loss of vision, he uses his new TV with the image recognition software which produces on the fly audio descriptions of his favourite movies. Thus he is able to watch any movie he likes. On the other hand, his 86 year old grandma loves to watch the movies too. However, due to her age she has a hearing impairment and uses an earpiece which she finds very annoying. When she watches the movies on the TV she prefers to turn on the automatic real time transcription feature which places subtitles of what is said in the screen. This weekend John is staying the night at his grandma's place. They decide to watch "Mission Impossible 6". They've heard good critics and they are both interested to watch it. They log on to their favourite video service from grandma's new tv and select the film from the large variety provided. They sit back on the couch and for the next two hours they enjoy the movie. Grandma is watching it on the TV with the transcripts provided by the producer company and John uses his Bluetooth earpiece to hear the audio description while the movie is playing.

VI. Pre-Scenario: Automatic Adaptation of Multimodal Interfaces

A. Technologies exploited

The use of mobile devices is becoming more and more vital day by day. Modern smart phones can contain a lot of information about users, such as documents, music, preferences for various applications and so on. So, having a personal profile on a mobile device seems quite natural for the future of ICT. In addition, mobile phones are enhanced with connectivity sensors which allow them to interact with other machines through WiFi, Bluetooth, NFC etc. Finally, they are also enriched with a variety of sensors, such as GPS, camera, light sensors, movement sensor etc. which allow them to recognise the context where they are used. This means that mobile devices can recognise their context of use and can easily connect to a variety of other ICT devices such as computers, info kiosks, ATMs etc. Additionally, a trend identified in the study is multimodal interfaces[6] which adapt to users' needs. Having a personal profile stored on phone or even in the cloud and communicating it with other devices can make this adaptation much easier and provide users with a seamless interaction among many devices.

Obviously multimodal interfaces can be used by many people with different disabilities and different degrees for disabilities. Having a world where every human-machine interaction is adapted according to users' needs is also a vision for the future. However, the study reveals that apart from having multimodal adaptive interfaces, assistive technologies will always play a significant role since design for all cannot be easily achieved in all cases. This means that a solution of a mobile phone speaking out user's needs to various ICT devices can be used in many ways. It could lead to user interface adaptation according to user needs (enabling screen reader), or it could also lead to user interface adaptation to allow interaction with specific AT used by the user.

Figure 8. Scenario: "My phone speaks out for me"

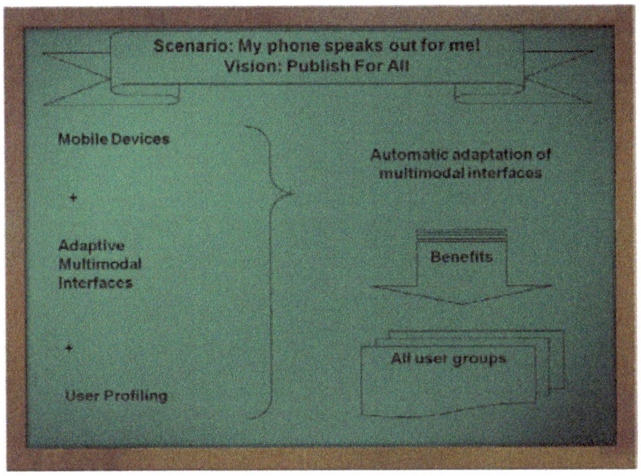

Figure 7. Scenario diagram for automatic adaptation of multimodal interfaces

B. *Scenario: "My phone speaks out for me"*

Mary, the 20year old girl from the first scenario goes to her universities library to use a computer to work on her new assignment. She always carries with her a Bluetooth earpiece and her mobile phone. When sitting in front of the library's computer she places her phone on top of an NFC connector which identifies her to the computer and gives the appropriate information needed for the user's interface adaptation. The computer automatically opens up the screen reader, and since the computer is used in a public library, finds through the phone the Bluetooth earpiece used by Mary and sends the audio input to it. Simultaneously the phone recognises Mary is in a library and goes automatically in silent mode. Mary can now log on to the machine and start working on her assignment.

On the same day, Andrew from the second scenario goes out for shopping and wants to use an ATM to withdraw some money. When approaching with his wheelchair the ATM, his mobile phone automatically recognises through WiFi the ATM and asks Andrew if he wishes to connect with it. Andrew answers yes moving his head, the tablet recognises his gesture, connects with the ATM and displays the ATM's interface on the tablet's screen. Andrew places his card in the ATM and completes the withdrawal using the ATM interface through his tablet, which uses face recognition and eye tracking for the interaction. Now he is free to go shopping wherever he wants.

VII. Pre-Scenario: Easy Development of Accessible Web Apps

A. *Technologies exploited*

This fifth scenario could be tied with Vision 3: Publish for All (which could also be interpreted as "Develop for All) and with Vision 4: Tools for All, because that's what we are aiming too. It could also be tied with legislation about accessibility as a trend.

Our analysis identified ubiquitous connectivity and network services as micro trends together with cloud based services and technologies. Combined with signals from the market such as new web based OS's (see

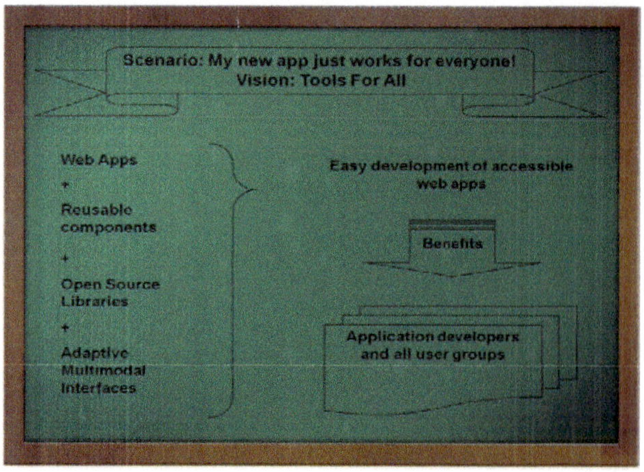

Figure 9. Scenario diagram for easy development of accessible web apps

ChromeOS, JolyCloudetc) it shows that in the coming years there will be an explosion of new web apps that will allow users to use a variety of services through web applications. The development of web apps is becoming easier by the years bringing even more developers to that area. Thus, the need for developing accessible web apps easier would become even more evident[7].

The analysis of our findings shows also that knowing about accessibility and related guidelines doesn't always mean that a developer can make an accessible app. However, a suggestion made by some experts in the interviews was that we should invest on developing core accessible component to be used by developers. For example instead of using a simple calendar widget a developer could have the option to choose between a set of different calendar widgets developed with accessibility in mind. Using such components would allow developers to focus on their app development without having to worry about their accessibility. In addition, it will lead specific developers specializing and really grasping the accessibility issues to deal with them from the core of development structures: reusable components, open source libraries, web apps, adaptive multimodal interfaces.

The benefits from such a dissemination of roles in the development process will be for every user who will need to use a web app using different interaction modalities. It will also make easier the life of developers who will not have to struggle with accessibility aspects in every detail of their application. It could also lead to an easier adoption of the vision of designing for all since the core components in development will be designed for all.

B. *Scenario: "My new app just works for everyone!"*

Peter is a 42 years old web app developer. He develops web applications using a specific set of tools and components that he has learned, tested and grasped really well. In some point of his career he meets George, a blind 35 year old man who tried to use some of his apps but found it very difficult. During their discussion George is telling Peter that he is using another company's applications. The company named "MyD4AApps" is advertising that they use an open source library of basic components that can adapt their interaction modality according to user's needs which are specified during the initial set up of the application. He downloads and starts experimenting with this open source library. After some time reading the documentation and some related tutorials, he builds his first simple web application and sends an email to George to try it out. In some days George returns back with comments not only from himself, but also from a group of other disabled users telling him that the app worked fine with whatever AT and setup they were using. Peter, seeing how easy it is to rely on the library for accessibility issues decides to update the rest of his apps using the new library.

Figure 10. Scenario: "My new app just works for everyone!"

VIII. Further Work

The study's next steps include further production of more scenarios. These scenarios will then be validated by experts on their feasibility and impact in order to be filtered. The best scenarios after the experts' assessment will be used in the final report of the study as suggestions for research policies and specific technologies that the EU should follow and invest on during the coming years to promote and enhance ever further eAccesibility.

References

[1] eAccessibility, http://www.e-accessibility2020.eu/ (accessed on 30/06/2011)

[2] Gilbert, M., & Junlan Feng, "Speech and language processing over the web", *Signal Processing Magazine* IEEE Vol. 25 Issue:3 p18-28, 2008

[3] Brunner, P., Bianchi, L., Guger, C., Cincotti, F. and Schalk, G. , "Current trends in hardware and software for brain - computer interfaces (BCIs)", *J Neural Engineering*, Vol. 8 Number 2, 2011.

[4] WebVTT, http://www.whatwg.org/specs/web-apps/current-work/webvtt.html (accessed on 30/06/2011)

[5] Ogata, J., Masataka, G. and Kouichirou, E.,, "Automatic transcription for a web 2.0 service to search podcasts", In Spoken Language For all, Antwerp, INTERSPEECH, 2007, pp.2617-2620.

[6] Nicu, S., "Multimodal interfaces: Challenges and perspectives", *J. Ambient Intelligence and Smart Environments*, Vol 1, Issue 1, pp23-33, 2009

[7] Mikovec, Z., Vystrcil, J. and Slavik, P., "Web toolkits accessibility study", *SIGACCESS Access. Comput.*, Issue 94, pp. 3-8, June, 2009.

Authors

C. Kouroupetroglou holds a PhD in Applied informatics since 2010 from University of Macedonia in Thessaloniki. His PhD research focused on web accessibility, investigating the impact of a semantic web based mechanism in blind users' browsing experience.

Apart from working in research programs funded by various organisations (Greek Ministry of Education, EU, etc.) he is also teaching in the Alexander Technological Institute (ATEI) of Thessaloniki as a scientific associate.

Dimitris Tektonidis holds a BSc in Software Engineering and a PhD in Computer Science for research in the area of Enterprise Application Integration. His main fields of research are Enterprise Application Integration (EAI), IS interoperability, Service Oriented Architectures (SOA), Web Services, Semantic Web, e-Commerce, e-Government and Enterprise Systems (ERP, CRM, SCM).

Adamantios Koumpis heads the Research Programmes Division of ALTEC S.A., which he founded at 1996 (then as independent division of Unisoft S.A.). Adamantios holds a PhD degree from the University of Kingston, UK and a Bachelor degree from the University of Crete, Greece.

Dr Ioannis Ignatiadis has a PhD in Information Systems from the University of Bath in the UK, as well as undergraduate and MSc degrees in IT and Technology Management.

He has extensive industrial experience as a software consultant in information technology projects for large European multinationals. He also worked as a senior researcher in the implementation of EU Information Society Technologies (IST) projects, notably the LAURA FP5 and the PANDA projects. His research interests evolve around the technical, social and organizational aspects of the development and use of enterprise systems, e-business and service-oriented architecture solutions.

Work reported here has been funded by the European Commission, DG Information Society and Media, ICT addressing societal challenges, ICT for Inclusion as part of the study on e-Accessibility2020 ("Study on Implications from Future ICT Trends on Assistive Technology and Accessibility", SMART 2010/0077).

Using Mobile Technology for Fostering Intellectual Engagement

A.P. Preciado Babb, C. Saar, C. Marcotte, J. Brandon and S. Friesen
University of Calgary, Calgary, Canada

Abstract—**Student engagement, and in particular intellectual engagement, has been identified as a key factor in learning at the high school level. While this type of engagement has an impact on student learning in fields related to science, technology, engineering and mathematics (STEM), it also has a positive impact on the learning of other disciplines such as English, Fine Arts, and Social Sciences. This report describes a set of projects co-designed by teachers and education specialists aimed at engaging students intellectually in each major high school subject area. Student use of mobile technologies, in this case the iPad2, features prominently in each of these projects. Interviews with teachers and students were conducted in order to identify evidence of intellectual engagement, as well as students' interactions with the mobile device. Data from a survey was also used to identify the level of engagement of students involved in these projects.**

Index Terms—**Intellectual engagement, mobile technology, high school, curricular design.**

I. Introduction

In recent years student engagement has acquired increasing attention from researchers and policy makers. Focused originally on high school completion [1], student engagement has been associated with student achievement, positive classroom and school climate, and effective instructional practices [2]. In contrast, disengagement has been identified as a cause of withdrawal from school, and seems to be more prevalent in student populations from low socio-economic areas or where great cultural and linguistic diversity exists [1,3]. The social effects of students disengagement has been researched in several countries, including Canada: "disengagement in and from school is linked to school violence, social exclusion, and a polarization severe enough to pose a threat to social cohesion in Canada" [4]. Student engagement has been considered not only as a strategic process for learning, but also as an accountability outcome unto itself [1]. As we share this perspective, we believe that engagement should be considered beyond final grades or school completion.

Student engagement has been defined generally as "a disposition towards learning, working with others, and functioning in a social institution" [5]. In this paper we focus on three dimensions of engagement: social, institutional, and intellectual [4,5]. *Social engagement* refers to the sense of belonging to school, including the participation in school activities such as sports, arts and community involvement. *Institutional engagement* corresponds to the fulfillment of school requirements. *Intellectual engagement* is described as a serious emotional and cognitive investment in learning characterized by intrinsic motivation and higher order thinking skills—associated with the state of flow [6].

Our focus on engagement entails a shift of educational paradigm. Current conceptions of education have evolved from schools organized to meet the needs of the industrial age [7]. Such conceptions are based on a traditional notion of knowledge being static and the need for schools to sort people into categories based on their final employment destination. Instead of considering knowledge as static content and standard procedures that can be broken down and transmitted to students, new conceptions of knowing and learning place importance on: collaborative knowledge building, scaffolding learning, through ongoing formative assessment, understanding the connections and interactions among different knowledge systems and modes of representation; emphasis on diversity; and foreground process, rather than products [7]. In order to shift the focus from educational paradigms from the industrial age to these new conceptions, educational reforms need to focus on creating the conditions needed to engage all students intellectually [8]. Such reforms also require supporting teachers in the design of learning environments in accordance to these new perspectives.

The Galileo Educational Network Association (GENA) provides professional learning support for teachers to enable them to design tasks and assessment practices that foster student intellectual engagement. For the past two years GENA has worked closely with teachers from a high school in a low-income, culturally diverse district located in a Western Canadian city, to design and enact a variety of discipline-based studies aimed at increasing students' intellectual engagement. This report describes the studies these students and teachers were engaged in, as well as their perceptions of the work and the role of mobile technology in this context. A preliminary version of this research report was presented at the IEEE Global Engineering Education Conference, Berlin, 2013. This refined and extended paper elaborates further on both the methodology and the implications of the research.

II. Student Engagement

Student engagement has been studied internationally for more than 25 years, mostly from the psychological and the social perspectives with a strong focus on measurement using self-reported surveys [9]. Research has consistently shown strong correlations between student engagement and completion of high school as well as improvement of grades. In [2], Yazzie-Mintz and McCormick argued for taking into account students' perceptions of the learning environments and conducted a survey with multiple options and open questions. They concluded that the quantitative data commonly used for measuring student engagement do not provide information on the kind of learning that students have experienced, the depth with which students have engaged in their learning, or what adults might do to help students experience school in a more engaged way: There is a need for qualitative research on student engagement [2].

In the Canadian context, in 2007, an initiative aimed to capture, assess and inspire new ideas for enhancing students' learning experiences was launched: the Tell Them From Me survey (TTFM), designed by Dr. J. Douglas Willms, [4]. In 2008, Penny Milton, Chief Executive Officer of the Canadian Education Association brought two other research and development organizations, GENA and The Learning Bar Inc., to engage in a 3-year research and development project to study, assess and enhance the learning experiences of adolescents in classrooms across Canada. Intellectual engagement was a construct added to The Learning Bar's TTFM student engagement survey instrument. In the province of Alberta, in Western Canada, provincial, district and school leaders have been tracking student engagement using the TTFM survey instrument. Three aspects of student engagement are assessed in the TTFM survey: sense of belonging at school, including participation in school life (social engagement); the extent to which students value schooling outcomes, (institutional engagement); and the extent of their serious emotional and cognitive investment in learning (intellectual engagement). The results of this survey have consistently shown that students: (a) prefer strong relationships with the teachers and their communities; (b) want to explore real problems, issues, questions or ideas that are of real concern to individuals and their communities; and (c) seek learning environments that build interdependent relationships—these results are consistent with [2]. However, a more detailed description of how students experience intellectual engagement is barely reported in the literature. As far as we know, the only two studies that have analyzed qualitative data used open question in surveys ([2] and [4]). The need for more studies that include voices of teachers and students has been a current concern among researchers—e.g. [4].

Focused on the experiences of adolescents in classrooms and schools from across Canada, the TTFM initiative expanded into a framework which put forward a set of teaching principles and strategies shown to improve levels of student engagement: the Teaching Effectiveness Framework (TEF) [10]. This framework is based not only on psychological and social features of engagement, but also on neurological research in education. The most effective learning takes place when learners have reached what Csikszentmihalyi calls a state of 'flow' or intense cognitive and emotional engagement [6]. At this point, the brain begins to make connections and patterns emerge in the information, resulting in powerful illuminations and understanding [10]. This approach contrasts to the mainly psychological and social approaches to student engagement commonly held by other researchers—e.g. [9].

The TEF framework served as a guide for the development and research initiative outlined in this paper. At the risk of an oversimplification, we summarize the core principles that underpin the TEF as they framed our analysis and data presentation. First, *teachers are designers of learning environments that engage students intellectually and academically*: In addition to having disciplinary, curricular, and pedagogical knowledge, teachers thoughtfully and intentionally design learning environments to engage their students. Intellectually engaging environments sponsor curiosity and awaken the desire to know and understand within learners. The result is a deep, personal commitment on the part of students to explore and investigate ideas, issues, problems or questions that have personal relevance, that matter to the discipline being studied or the world beyond the classroom. Second, *work students are asked to undertake is worthy of their time and attention, personally relevant, and deeply connected to their world.* This work engages students in the broader community where they explore real problems and concerns not only relevant to the discipline, but also connected to students' lives inside and outside school. Such explorations foster strong habits of mind where students are invited to formulate plausible theories and defend them through reasoned discourse based on evidence and in light of different viewpoints. The work that students undertake is authentic in that it mirrors the ways of thinking and acting reflective of practitioners in the field—e.g. historians, biologists, journalists, photographers, and architects. Third, *assessment practices improve student learning and guide teaching*. Assessment is a seamless part of the learning process that involves students in examining their work in progress, checking to decide what has been learned and what needs to be learned next. It includes setting and using criteria, having students self-assess, being engaged in peer coaching, receiving feedback about their work, collecting and organizing evidence, and presenting evidence of learning to others. It identifies what has been accomplished, what needs to be done, and informs the next steps for both teacher and students. Four, *teachers foster a variety of interdependent relationships*. Effective learning environments are characterized by a series of interdependent relationships that promote and create a strong culture of learning. The teacher engages students in dialogue by: stimulating discussion, posing questions, provoking thinking, suggesting resources, and helping students determine their next learning steps. Students collaborate with each other as well as with discipline experts both inside and outside of the classroom in order to build their collective understanding about the topic under study. And five, *teachers improve their practice in the company of peers and others.* Teachers improve their practice, and hence their effectiveness, when they have opportunities to learn in the company of their peers. A scholarship of teaching begins to emerge where a regular part of the school discourse centers on the learning of students, and where teachers have access to each other's classrooms, and time to plan collaboratively. Teachers become reflective practitioners who seek out the constructive feedback from peers, mentors and experts in an effort to constantly improve their practice.

An additional aspect of student engagement is the relationship of new generations with technology. The infusion of digital technologies is an essential part in the programs of study, in most Canadian provinces [11]. However, in order to take advantage of this technology in education, roles of, and interactions among, students, teachers, academic subjects, and the technological devices must be reconceptualized [12]. New forms of learning are associated with the daily use of technology. New generations born into high-technologized societies, known as "millennium learners," have different ways of learning and socializing [1,13]. Mobile devices have played an important role in the emergence of these new cultures and now the role of teachers, students and technology at school are shifting in such a fashion that the teacher is no longer the expert, at least not in the use of the mobile technology [12].

III. Research Design

The study presented in this report was based on a three-year participatory design-based research ([14]) in which researchers were actively involved in working with teachers and students. The purpose of this study was to better understand how student learning and engagement are impacted when teachers are provided with intensive professional learning informed by the principles outlined in the TEF.

Design-based research in education is situated in real educational contexts and studies innovations, as they are implemented, involving multiple iterations of design and testing of significant interventions in collaboration with researchers and practitioners. In the last ten years, the adoption of this methodological approach has been increased prominently in educational research ([14]). Due to its focus on the complex phenomena of education, design-based research employs both qualitative and quantitative methods. The term 'participatory' refers to the role of practitioner—e.g. teachers—and researchers in the study. In contrast to other forms of research in which students, and teachers, are studied by researchers, in participatory design-based research both researchers and practitioners play an important role in the design of the innovations. Researchers are co-designer, and teachers are often co-researchers, as they engage also in the development of new insights and knowledge derived from the collaboration.

The first year of this participatory design-based research was mainly devoted to: building trust between teachers and GENA's staff, assessing teaching practices, evaluating levels of student engagement, and developing a strategy for teacher professional development. The preliminary findings reported in this paper correspond to the perceptions of students and teachers of two Grade Ten groups, called the pod, during year two of this research and development initiative[1]. We particularly addressed the following research questions: How did students experience intellectual engagement when involved in the learning environments co-designed with GENA? What were the teachers' experiences, in terms of the TEF, when participating in the second year of this project?

In order to capture the experience of students and teachers in this project, we considered a mainly qualitative approach for this part of the research. We incorporated varied sources of data as a means to better understand the complexities of the school setting: field notes; task designs and unit plans; class observations; semi-open interviews conducted at the end of the year with six students, the teachers involved, and administrators; artifacts of students' work; and the Tell Them From Me (TTFM) survey [4,15]. While the survey and the class observations provided quantitative data regarding the level of engagement, the interviews provided details of students and teachers' perceptions of their engagement in the studies co-developed and enacted with GENA. The TTFM survey served as an external measure of student engagement as a dependent variable. However, the survey accounts for the whole experience at school, not only for those projects co-designed with GENA. In this sense, the results of the survey are considered only as an indirect measure of the results of GENA's initiative. Class observation was adapted from the Class Observation Protocol designed by State Educational Technology Directors Association. Class observation was conducted as a means to identify instructional practices and levels of engagement. Each observation was conducted by at least two researchers, who compared their results thereafter, discussing and negotiating discrepancies. Class observation took place during the first year of the intervention, evidencing a prominent teacher-centered approach to instruction, with a strong emphasis on procedural knowledge and memorization. This observation informed the design of the intervention for the second year.

The TTFM survey was usually administrated twice a year. However, during the academic period 2011 and 2012 it was only administered in November, which was a limitation for this study. During the first part of this academic year, GENA worked intensively only with the Mathematics and Science teachers of the pods.

Students and teachers were interviewed in pairs by the end of the school year. One pair of interviewed teachers were the English Language and Arts and Social Sciences, the other pair consisted of the Mathematics and the Science teachers. Only six students agreed to participate in the interviews and were sorted in three pairs. The Interviews were transcribed and coded in terms of the five core principles of the TEF by at least two researchers. The Nvivo software for qualitative research was used to compare and select excerpts based on both the students' descriptions of their intellectual engagement and the teachers' activities reflecting the TEF core principles. Following the principles of quality for qualitative design, we present excerpts of these interviews as a means of giving participants a voice, to represent multiple perspectives, and to allow the reader to judge the merits of our conclusions ([16,17]). Descriptions of the Projects Designed for Stdents

The teachers involved in this research had expertise in one of the four subjects students are required to study at high school in Alberta, Canada: English Language Arts, Mathematics, Science, and Social Studies. During the second year of this initiative, teachers worked in pairs: Mathematics with Science; and Social Studies with English Language Arts. The later pair merged their two courses into a single-year course: Humanities.

The following projects for students were selected due to their deliberate focus on intellectual engagement. It must also be noted that the appropriate use of digital technologies is infused into each of the five principles outlined in TEF. In this case, the iPad 2 was the mobile technology that featured prominently in the teaching and learning of students involved in this stage of the initiative.

A. Humanities - Mash-ups

As part of a novel study in their Humanities class, students were invited to choose one of the key ideas from the novel *To Kill a Mockingbird*, by Harper Lee, and develop it into a digital collage, or mash-up. They were encouraged to use any combination of media from a variety of sources, including movie trailers, news clips (audio and

[1] Although this initiative was extended to other teachers at this school, we had the opportunity to work with all the teachers in the four main subjects who taught these two groups

video), still images, music, poetry, sound effects, pop-culture representations from television, and radio to represent their interpretation of a key theme or message from the novel. The multimedia features of the iPad played a central role for this project.

The mash-ups were aired for an authentic audience during a public screening when prospective new students and their parents, were visiting the school.

B. Humanities - Romeo and Juliet

Students worked with professional actors to interpret and re-enact scenes from Shakespeare's Romeo and Juliet. Many of the scenes were placed in different contexts and represented fresh and original interpretations, re-created in contemporary contexts proposed by the same students.

The portability and versatility of the iPad tablets enabled students to take on and easily switch between diverse production roles: from scriptwriter, to director, actor, photographer and film editor. It also meant that student production teams could set the stage in almost any setting where today's teens might actually congregate including coffee shops, shopping centers and movie theatres. This approach inspired students to both study the play and also work through their own original interpretations in ways that closely mirrored those employed by contemporary actors and film production companies.

C. Science/Math - Field Study

The mathematics and science teachers collaboratively designed a unit with experts from a biogeoscience station located in a nearby provincial park. Teachers used the Galileo Discipline–Based Inquiry Task Rubric, described in [18], to guide their design of an engaging study. Working in teams, students used the same tools and methods of gathering data that scientists at this field station would use to explore how the environment changes as you move up a slope or in a northerly direction. Once students had collected their data, they were asked to synthesize their findings, compare their results to other studies conducted by scientists in that area, and identify questions for further study. Students presented their findings during a poster session using the same format that scientist from the field station used in presenting their results. The connection with scientists at the field station was essential for the design of the task and to enable students to obtain meaningful feedback about their work as they tested the validity of their findings.

The iPad allowed students to gather, record and analyze data they collected during their visit to the station.

D. Science/Math - Public Service Announcement

After visiting an energy efficient house, students were asked to identify energy issues in the school building and develop a proposal for improving energy efficiency. This project demanded students to provide a sound scientific rationale for the basis of their conclusions relative to the amount of energy saved, as well as determine the real cost of implementing their proposals. Students used the iPad to develop and present their proposal in the form of a short, but compelling, video Public Service Announcement (PSA). Proposals that students presented to the School Board facilities and maintenance directors were both defensible and feasible. As a result a monetary sum was awarded to the school to enable a number of the proposals to be implemented.

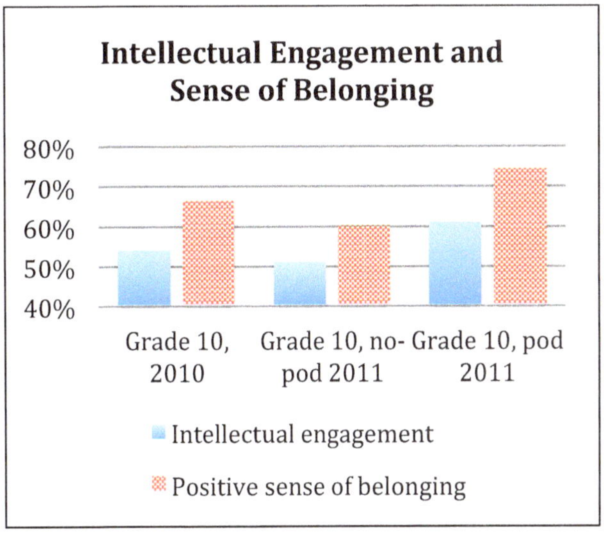

Figure 1. Results from the TTFM survey

E. Mathematics - Expanding Cube

Students used the multimedia capacities of the mobile technology to solve a mathematical problem during several sessions—see [5] for more details on this task. The technology enabled students to conjecture and verify results, and to communicate both their process and final solution in the form of electronic documents that included photographs, tables and both written and oral descriptions of their findings. At the end of the unit students wrote a test on manipulation of polynomials based on the formulae developed through the project.

IV. RESULTS

The TTFM survey indicated an increased level of engagement when compared with other Grade Tens during both the same and the previous year (Figure 1). In particular, the level of intellectual engagement was 11% higher when compared with non-pod students in the same year, which suggests that the intervention impacted positively on student intellectual engagement.

Although it was not a main focus of this research, it is worth noting that social engagement improved as well: The category *Sense of Belonging* was 14% higher in the pod group compared with the other non-pod Grade Ten students in the school, as illustrated in Table 1. The data from student and teacher's interviews provides additional details of how engagement was embodied in the studies students undertook as part of the initiative.

During the interviews conducted at the end of the year, students were asked to identify work they had completed that they were most proud of. The examples cited most frequently were those projects co-designed with GENA: particularly, the Romeo and Juliet, Mash-ups, the Biology field study and PSA projects.

A. Students' perceptions

The following excerpt from a student reflection journal in science and mathematics represents an instance of intellectual engagement: the student was intrinsically motivated and self-directed during the study.

I now understand that an ecological field biologist is a very interesting profession. It's like solving a mystery. You have questions, a purpose, and you try to figure out the answer to the questions. What is causing it? It's all about questioning something and then getting out there and trying to figure it out. It's a hands-on experience. You pick your own tools and YOU decide how you will solve your mystery. It takes patience and constant monitoring, and lots and lots of observations. You also need to be able to interpret information and make connections.

Students were also asked to describe the difference between their courses with their experiences of those same courses in previous years. The following comment contrasts a superficial experience of language arts with a more challenging program that demanded more of students.

We would usually do whole bunch of grammar things and just general learning how to use the language. This year, there was a lot of reading comprehension, like we had to do with To Kill a Mockingbird. We had Romeo and Juliet. We had all of these essentially bits of reading or we would read through the passages of anything that [teachers] gave us and we'd have to comprehend what we read, like what did this character do at this time and how did that affect the story? So, it was a lot less learning how to use the language and more learning to kind of think with it, really, because we know how to use it now, so you get to think.

Students were asked to identify something they did in high school that had meaning and relevance for them. The following two excerpts show how work students undertook was relevant to their personal life in terms of effort invested in the project, personal relevance, authenticity regarding arts, and being proud of their work.

I was sort of thinking of the Mash-up that we did. We spent so much time on it, it's like it was so important it was just...I don't know, it was just all the experience built into it and what we got to do and how it like, related to us. To me, it felt like it was really important

I like the Romeo and Juliet project, that was really fun filming. We just did like a little thing and we have like, we did–I think it was Act 3, Scene 1, the fight scene. We just did like this cat fight it was really fun. It was like a film.

I still like the To Kill a Mocking Bird project. We got– I'm pretty sure I had a fairly good mark on that because we showed (our teacher) that one and he was speechless. Ours was just like, it was completely done when we showed it so it was like, "Oh Yes, let's show it." When he watched it he was like, "Wow, I don't know what to say, it was really good." I was really happy with that one, it took a lot of effort, so much work.

The relevance of the mash-up study went beyond a good mark: it was rewarding, as suggested in the student comments from the previous excerpt. Another aspect of relevance mentioned by students was the involvement of a broader audience, as we can read in the following:

The challenging part is to make my Mash-up appeal to— you want to make it appeal to all age groups, from old to young— explaining how the young shouldn't try to grow up too fast. And reminding the adults that it's OK to have fun, and not be yourself and just be a kid. ... [And we] try to make it seem sentimental and trying to incorporate the "To Kill a Mockingbird" [ideas] into the mash-up. It took my group a while but you notice how Scout, the girl in the movie, she isn't afraid to be a tomboy and just play around and not try to grow up too fast, since she doesn't want to grow up to be a lady. She just wants to be herself. So we tried to incorporate it.

Students were asked also about how assessment was conducted in their courses. In this case, the interviewed student explains how the expectations for the assignments were made explicit.

On the first day [teachers] gave us like an idea and then the next day we would talk it over and then they'd give us more of like what they expected and stuff. They wanted it to be like a lot of details and like appealing and good pictures and stuff.

Moreover, most interviewed students described their preference for performance-based assessments over quizzes and tests. In the following transcripts students explain their preference based on how these studies were more interesting and allowed creativity.

I like the Mash-up better. I do not like tests. It's more interesting and really got me involved and I learned a lot more than on a test.

Same with her, like I prefer the Mash-up because I guess to show my creativity and if I'm having fun I put more effort into the work I've done. So on tests and quizzes, even though I study hard on it, once the test comes there are terms that I couldn't get or I just get lazy and I just guess or I second guess myself.

Only one of the interviewed students preferred assessment in the form of tests and quizzes. However, this student acknowledged that the studies and projects supported a deeper understanding of the course content, as we can read in the following comment:

I think tests and quizzes I do best on, and I think it's just easier for me to express when it's taught that way, rather than say writing your answers for an assignment. ... The assignments and seminars and things helped me with my understanding of the content of the course. But I think tests and quizzes are really, they're the best way for me to determine how much I actually know on the subject.

The previous excerpt can be analyzed in terms of two different types of assessment: summative and formative. The student seems to focus exclusively on summative assessment—to evaluate the state of knowing at the end of a unit or semester regarding content matter. However, an important component of the studies and projects co-designed with GENA was formative, ongoing assessment—to improve student's learning. The student's comment in the excerpt places a value on the assignments in terms of the learning of course content, which is the purpose of formative assessment.

B. Teachers' perception

Teachers' perceptions are presented in relation to each of the five core principles of the TEF. These core principles are interwoven and the excerpts therefore reflect more than one principle. However, one example illustrative of each of the five principles is included in this paper. The interviews were conducted in the corresponding pairs Mathematics-Science, and Humanities.

1) Teachers are designers of learning

Prior to the work conducted in this second year of the intervention, the mathematics teacher used a workbook for the course. In every single class, mathematics teachers had to cover designated pages of the book. However, after the second year of the intervention the teaching practices of the mathematics teacher were more in alignment with those outlined in the TEF as evidenced below:

We don't start at the textbook and work towards the curriculum anymore. We start with the curriculum, start with a big project, and work in the details. It's no longer about starting with the details and getting to the big picture. It starts with the big picture and then filling in the details–and I find that it's actually a much better way to do it because then the students really understand the concept. It's not just the details and they get lost in the concept. It's the other way where they get the concept and then the right details.

One of the Humanities teachers reflected on how the mash-up study provoked greater reflection about the quality of the tasks students were being assigned.

I think the one thing it did make me more conscious of though was thinking deeply about why are we giving students this task? What do we hope for them to accomplish? How does it cover those competencies? Especially in a Humanities class, where English and Social Studies are so closely linked, thinking about can we kill two birds with one stone? If we give them this piece of literature to read and it links to historical globalization but it also requires them to analyze a piece of literature then we've really hit on the competencies for both subject areas.

2) Students engage in worthwhile work

The PSA project engaged students in a real problem that was directly related to their school: energy consumption. The Science teacher explained that students went deeper in their studies by examining the environmental impact of the material of one alternative put forward by students.

I think they certainly went much, much deeper than we expected them to go because they were into it and they want to do it. They were so interested in it that one group started looking at insulation materials and then started looking at, well, what kind of environmental impact it is going to make. So they found this awesome material but it's really expensive and it's not very good for the environment to make. So then they decided, well, they're going to switch to a local company

3) Assessment practices

Assessment for learning was a key feature of each of the projects co-designed during this part of the initiative. In the following script from a Humanities teacher a connection between strong feedback and the personal relevance for students is explained.

When you are constantly giving them that feedback it enables them to feel as though their final product would be something that's worthy of all of their efforts and has made use of that feedback that they had been given. I know for the Mash-up project, a number of the kids, when we saw their first sort of go-around, we were able to say to them, you know you might want to rethink the sound track, or you might want to rethink some of those images, are they really conveying your theme? They were able to take that feedback and then go back and make some changes. That I think resulted in a more meaningful final product.

In the following excerpt a Humanities teacher explains the relevance of peer assessment in the Mash-up project. The script also reflects the expertise in the use of technology of some students in the classroom. The teacher is no longer the expert in the classroom; some students are actually the experts and provide feedback to classmates.

I think that in particular the peer assessment is quite powerful. ... In the case of the Mash-up, I think I said earlier, they are all pretty savvy consumers of technology. ... I think for a lot of them that was probably more meaningful than having their 40-year old teacher say, "I don't really understand what you've tried to accomplish with this set of images."

I think it was scary frankly to do a project where you are asking kids to use technology but you yourself don't really understand how the technology works.

Another component of assessment is the student engagement in the design of the assessment tools, such as performance-based rubrics. In the following excerpt the Mathematics and Science teachers explain the process of developing the assessment rubric for the PSA project.

Then [the project] *was marked on every rubric we had but it's the same rubric that the kids have been working with for all the feedback loops. There must have been at least four or five loops. ... Specifically when it came to PSA's. We looked at exemplary examples and poor examples and really came up with what makes a good PSA, what makes a poor PSA.*

So when we have a project such as PSA we would give them an opportunity to kind of start working [through it and] *assess themselves, and then we'll give them some more chances to improve on the scale, I guess. Then one of us would sit down with them and go through it again. Then they would try to improve a bit and then they might sit down with their friends and try and improve again until they know exactly what they're getting assessed on. They know exactly what our expectations are.*

4) Interdependent relationships

One of the educators from GENA arranged for professional actors to work with the students on their Romeo and Juliet projects. A teacher commented in an interview about the significance of having an expert in the group.

The actors helped the kids write their scene and gave them some notes about how to do their performance. We had time to kind of walk around and give suggestions but again we weren't necessarily saying you have to have this, this and this in your scene. It was very much up to them. They blew us away with their creativity and humor. They were really well-done at the end and they were really proud of them [Romeo and Juliet projects].

The English teacher reported students who were so involved in the creation of poems during a field trip, that they kept doing poetry afterward.

So, having the chance to do outside trips to see what they're learning in the real world. We went to a Spoken Word Festival for English. They were really into that. They were doing slam poetry on the train home.

Both social engagement and institutional engagement seemed to be outcomes of this initiative, as we can read in the following excerpt from the Science teacher.

Another thing that I'll add to that too is our students are much more involved in our school, that there's been multiple times where we would just ask "Do we have any volunteers?" and 20 hands would go up when we only needed five but that's okay. Our students love volunteering. They are always the first ones to step up to the plate and it's part of the relationship that we've built. They know that we need them and we'll give back to them.

5) Teachers practice in the company of their peers

Humanities teachers explained how the process of co-teaching was a process of mutual teaching and learning.

In terms of my development, I think it's really a new teacher's dream to be with an experienced teacher first thing. I went from being absolutely terrified walking in to having someone I trusted and was able and willing to help me. I feel that I learned a ton from [my teaching partner]. *I don't know how she would feel.* [Laughter]

I think it's reciprocal. ... Day one, [Partner teacher] *was getting to know the kids, talking to them, engaging with them, getting to know them as learners and as individuals and bringing creative ideas that I haven't thought of or haven't ever considered. I think it worked out well for both of us. She got to be with somebody a little more experienced. I get to be rejuvenated with her new ideas.*

The Mathematics and Science teachers also commented on their learning when planning as a whole group of teachers from the four major disciplines.

We have common prep times ourselves, the four of us, and we meet at the minimum once a week.

We meet with all four teachers involved with this initiative. It's not that Math and Science are separate from the Humanities; we all get together and talk.

The teachers' excerpts presented in this section demonstrate how the major studies and projects co-designed with GENA were related to the TEF. These excerpts describe ways in which students became intellectually engaged when involved in the projects. Additionally, this evidence suggests that social engagement and institutional engagement could be other outcomes of the initiative.

V. Conclusions

It is very challenging to disrupt conventional teaching approaches. The projects and learning environments described in this report not only served to engage students in authentic tasks related to the major subject areas at high school, but also engaged teachers in designing rich learning tasks and in strengthening their practices.

The preliminary findings reported in this paper offer accounts of intellectual engagement from the perspective of teachers and six interviewed students: representing a contribution to the research on student engagement—as suggested in [4]. The results of the TTFM survey evidence an increase in the levels of both intellectual and social engagement in those students involved in the studies co-enacted with teachers and GENA. The excerpts provided in this paper represent the voices of students and teachers involved in this development and research initiative.

Teachers participating in this professional learning opportunity came to understand the importance of thoughtfully and intentionally designing learning opportunities to intellectually engage students. The mobile technology played a significant role in enabling students to think differently, use digital technology to solve problems, and express their understanding in more richly textured ways than paper and pen may have afforded. Also, the technology in this case required teachers to assume a new role in which they came to realize that they do not hold all the knowledge and expertise. Students, with their expertise and capability using multi-media and iPad Apps, assumed leading roles as experts who supported both peers and their teachers in using these technologies.

The mathematics study, the Expanding Cube, was not mentioned as a memorable project in the interviews. We observed students being engaged during this project and initially expected comments about it in the interviews. Notice that the results of the survey could only be impacted by the work with the Science and Mathematics teachers: no major projects, other than the Expanding Cube, took place before the survey was conducted. A reason for this may be the fact that this math study happened early in the school term and lasted for only one week, while the other studies were conducted more recently and required longer periods of time, sometimes exceeding a month. Another reason might be that students were asked to cite the most memorable and relevant projects. However, there may be an additional explanation. This mathematical task did not involve a broader community, and did not entail connections outside of the classroom: two features described in the TEF. These two factors may be important features of the design of memorable learning experiences.

Although social engagement was not a focus for this study, the evidence suggests that the studies collaboratively designed and enacted in this intervention had a positive impact on this type of engagement as a side effect. This finding is important in that social engagement has been identified as a factor for both student retention at high school and a pre-cursor to the cultivation of socially engaged citizens.

The role of the mobile technology was not limited to facilitate the project co-designed with teachers and GENA specialists. While the use iPads helped students to collect, organize and analyze information, as well as communicate their thinking, a students were able to participate in the classroom as expert, savvy users of mobile technology. Both teachers and students acknowledged the lack, in some cases, of teacher's skills in the use of particular technologies. The increased rate of change in day-to-day technology makes it very difficult, or even impossible, for the teacher to be up to date with the most recent apps and software available, and adaptable, for educational purposes. The teacher is no longer the expert in the classroom in tis regards, which opens the possibility for students to contribute with their knowledge and skills in mobile technology. As argued in [12], the roles of the students and the teacher change allowing a more participatory contribution of students to the class.

Finally, we claim that the work students undertake needs to be relevant, meaningful and authentic–in other words, it needs to be worthy of their time and attention. Too frequently, the work students are asked to do at school does not engage them in personally relevant experiences, nor allow them to experience the life and vitality of rigorous exploration. Both the research field [6,19] and the students themselves [2,5], are clear: The work adolescents want and need to be able to do must be worthy of their time and attention and therefore must be intellectually engaging.

References

[1] L. Taylor, & L. Parsons, "Improving Student Engagement," Current Issues in Education, vol. 14 no. 1, 2011.

[2] E. Yazzie-Mintz, & K. McCormick, "Finding the humanity in the data: Understanding, measuring, and stregthening student engagement," in Handbook of Research on Student Engagement, S. L. Christenson, A. L. Reschly, C. Wylie, Eds. Springer Science+Business Media: New York, 2012.

[3] L. W. Harris, "A phenomenographic investigation of teacher conceptions of student engagement in learning," The Australian Educational Researcher, vol. 35, no. 1, pp. 57-79, April 2008.

[4] J. D. Willms, S. Friesen, & P. Milton, "What did you do in school today? Transforming classrooms through social, academic and intellectual engagement: First national report." Canadian Education Association, Toronto, 2009.

[5] J. D. Willms, "Student engagement at school: A sense of belonging and participation." Organization for Economic Cooperation and Development: Paris, 2003.

[6] M. Csikszentmihalyi, "Flow : The Psychology of Optimal Experience," Harper and Row, New York, 1990.

[7] L. Gilbert, "'Catching the Knowledge wave' Redefining knowledge for the post-industrial age," Education Canada vol. 47 no. 3, pp. 4-8, 2007

[8] J. D. Willms, & S. Friesen, "What did you do in school today? Research series: Report number two: The relationship between instructional challenge and student engagement." Canadian Education Association, Toronto, 2012.

[9] S. L. Christenson, A. L. Reschly, & C. Wylie, Eds. "Handbook of Research on Student Engagement." Springer Science+Business Media: New York, 2012.

[10] S. Friesen, "What did you do in school today? Teaching Effectiveness: A Framework and Rubric." Toronto, Canada: Canadian Education Association, 2009.

[11] Western Northern Canadian Protocol for Collaboration in Education, 2011.

[12] A. P. Preciado Babb, "Incorporating the iPad2 in the Mathematics Classroom: Extending the Mind into the Collective" International Journal of Engineering Pedagogy, vol. 2, no. 2, pp. 23-29, April 2012.

[13] Organisation for Economic Co-operation and Development, "Centre for Educational Research and Innovation (CERI) "New Millennium Learners."

[14] T. Anderson & J. Shattuck, "Design-based research: A decade of progress in education research?," Educational Researcher, vol. 41, no. 1, pp. 16–25, Feb. 2012.

[15] J. D. Willms, & P. Flanagan, "Canadian students 'tell them from me'," Education Canada vol. 47, no. 3 pp. 46-50, 2007.

[16] Lincoln, Y. S. "Emerging criteria for qulity in qualitative and interpretive research," Qualitative Inquiry vol. 1, pp. 275-289, 1995.

[17] Creswll, J. W. "Qualitative Inquiry and research design: Choosing among five approaches," 5th Ed. Thousand Oaks: Sage, 2007.

[18] GENA, "Discipline-based inquiry rubric" GENA, 2011.

[19] D. Jardine, P. Cliford, & S. Friesen, "Back to the basics of teaching and learning: Thinking the world together." 2nd Ed. New York: Routledge Press, 2008.

AUTHORS

A.P. Preciado Babb, C. Saar, C. Marcotte, J. Brandon and **S. Friesen** are with the Galileo Educational Network Association, University of Calgary, Calgary, Canada.

A Short Review on the Trend of Mobile Marketing Studies

Mohammad Ismail[1], Razli Che Razak2
[1] Universiti Teknologi MARA (UiTM), Malaysia
[2] Universiti Utara Malaysia (UUM), Kedah, Malaysia

***Abstract*—The purpose of this paper is to elucidate the major classification of the research in mobile marketing. It provides the trend and common approach in mobile marketing research. The paper aims to highlight the significant area in mobile marketing research and the needs for further understanding of the mobile marketing definition and its phenomenon. This paper has significant implications in terms of understanding the classification and the mobile marketing focusing area. This is a general review based on selected approximately forty-six journal publications which are published between year 2005 until 2010 from selected journal. Based on the selected reviewed literatures, the consumers' adoption research only focuses on a few common areas. There is no agreement among the researchers on definition of mobile marketing and the conceptualization and phenomenon of mobile marketing remain unexplained. The value of this paper is that, it provides the fundamental guidelines for researchers on significant area to focus in the mobile marketing field and it also assists practitioners in indentifying the research trend and major classifications which will assist them in designing the product and services to the right target market.**

***Index Terms*—Mobile marketing, Consumer Adoption, Consumer Acceptance**

I. Introduction

Mobile phone penetration rate and mobile subscribers have tremendously increased. Mobile phone users worldwide have recorded 3 billion mark in late 2007, more than 4 billion in 2008 and are expected to increase more than 5.5 billion by end 2013 [1]. This scenario has increased the amount of people using mobile devices and accelerated the mobile phone to become most influential medium for marketing. Due to recent developments in mobile technology, most marketers are going into this data market in developing their content to reach larger audience and create a new source of revenue. In Asia Pacific alone, mobile data services are projected to grow from USD44 billion in 2008 to USD81 billion in 2012, contributing to 27.7% of total mobile revenues in 2012 [2]. The progress of mobile advertising revenue in the consumers' surrounding have turned the mobile marketing an impressive field to explore the research [3]. Even though there is a plethora of research in this area but the area of mobile marketing is still under development and has attracted considerable number of research among academicians in exploring this filed. Despite of the growing number of research, the academic literature in this area is still at an infant stage [3][4][5]. The purpose of this article is to provide an idea on the major classification of the research in mobile marketing, common focus of the research particularly on consumer's behaviour, common research approaches and reviews the theories that generally used in previous studies. The short review process has been considered on the approximately forty-six selected peer-reviewed journal and proceeding papers published from year 2000 to year 2010 by considering the interest of mobile marketing study which was started in year 2000 and increasingly evolving in the year 2004. The researchers' interest towards empirical study on mobile marketing was started in the year 2005 and the topic is still growing interest recently. Finally, the paper is attained with the discussion which directs towards facilitates future research directions and we end up with the conclusion in the last section.

II. Classification of Mobile Marketing Research

There is no consensus on category or accepted classification of mobile marketing research among researchers[7]. In addition Leppäniemi et al. [7] in their review of mobile marketing research have classified mobile marketing research into three main categories namely consumer, business & management and general category. However, Varnali and Toker [3] have classified mobile marketing research into four main categories namely theory, strategy, consumer behaviour and legal & public policy. Both studies have revealed that the substantial number of publication of mobile marketing studies in consumer behaviour. Furthermore, recent trend of the studies is also focusing on consumer behaviour aspect. This development is in-line with the statement by Bauer et al. [21] which have stated that consumer acceptance is one the core issues in mobile marketing and the consumer acceptance has gained considerable importance as a field of research since the mid 1990s. By adapting the classification by Varnali and Toker [3], this paper has categorized mobile marketing research into four major categories but the study on reviews, definition and conceptualization of mobile marketing has been reclassified as general category.

III. Major Category of Mobile Marketing Research

Basically mobile marketing research has been classified under following categories:

A. General

This paper is classified by the general category which consists of studies related to definition and conceptualization of mobile marketing (e.g., [6], reviews studies (e.g., [7], state of the art studies (e.g., [3], etc. Tähtinen [6] for example has conducted a study in determining the definition and conceptualization of mobile marketing from both perspectives, academic and practitioners. Her study may help the researchers to understand the phenomenon of mobile marketing better than other terms used in describing the mobile marketing phenomenon. The study by Tähtinen [6] has argued that, when comparing the definition by academic and practitioners, mobile marketing looks like to be used to describe only one part of the marketing definition, namely the communication of value. Her study has proposed to use mobile marketing communication, which can differentiate mobile forms from any traditional non-mobile forms of marketing communication. However, the study on this category is still lacking and offers a large opportunity for researchers to explore.

B. Strategy

Basically this study is categorized on the research effort focusing on mobile media or channel and aid practitioners in their expedition to accomplish mobile marketing success (e.g., [8]. Another sub-area in this category is M-marketing tools and applications. Barutçu [9] has revealed that there are seven mobile marketing tools which stand out as the crucial mobile marketing tools namely mobile advertising, mobile sales promotion, mobile entertainment services, location-based mobile services, mobile internet, mobile banking and mobile shopping. Such studies concern on how the tools will be used by consumers (e.g., [10] [11][12][13][14].

C. Consumer Behaviour

This category focuses on individual-level charateristics like demographics, perception, attitude towards mobile marketing and other factors that influence the adoption of mobile marketing. The trend of the studies is more towards consumer acceptance of mobile marketing by focusing on their behavioural intention. This category has grown interests among researchers (e.g., [15][16][17] [18][19][20][30].

D. Legal and Political issues

This category of studies consist of study on issues, trust and privacy in the mobile environment. Risk perception and trust play significant role in studying consumer adoption of mobile marketing. According to Bauer et al. [21] consumers' perceived risk has strongly influenced the consumer willingness to adopt mobile marketing. Consumer perceived risk is likely to become a determinant factor in affecting consumer's participation in mobile services [22]. On the other hand, in terms of consumer policy, the number of academic studies in this field is relatively low and future studies are very much needed [3]. Despite of a number of studies on perceived risk, there are still limited studies concerning on risk perception on mobile marketing. Very little studies are focusing on consumer perceived risk towards mobile marketing services (see [21][22][23].

The following diagram illustrates major category of mobile marketing research. Generally, there are four major categories of mobile marketing research with a few sub- areas. The selected studies concerning on category and the sub-topics from previous studies are figure out in Table I below.

TABLE I.
MAJOR CATEGORY OF RESEARCH IN MOBILE MARKETING

No	Category	Sub Area	Example of Selected Studies
1	General	Definition and conceptualization, Reviews	[3][7]
2	Strategy	M-Marketing strategy, M-marketing tools and applications	[8][10][11][12][13][14]
3	Consumer Behaviour	M-Marketing Acceptance, adoption of M-marketing, attitude towards M-marketing, perception, M-satisfaction and loyalty	[15][16][17][18][19][20][30]
4	Legal and Public Policy	Trust,Privacy, Legal and political issues,	[21][22][23]

Source: Adapted from Varnali and Toker, 2010

IV. THE PRIOR STUDIES APPROACH ON CONSUMER ADOPTION OF MOBILE MARKETING

Generally, most of the previous studies are commonly focusing on a single areas such as mobile advertising, mobile services, mobile data services and mobile coupon, Several studies have examined consumers' attitudes towards mobile marketing (e.g., [5][9] [38]), consumer intention to use mobile data services ([24][25][26] 27]), mobile banking (e.g., [28][29][30]), mobile services (e.g., [31] [17][32][33]), adoption of mobile commerce (e.g., [34][23]) and consumers acceptance of mobile advertising(e.g., [16][12][19]). Another area which gaining more interests among researchers is mobile payment (e.g., [22][33][18]).

However, there are very limited studies focusing on the acceptance of mobile marketing (e.g., [35][36]). Moreover, mostly, the studies on mobile marketing research are concentrating on consumer's intention to use or behavioural intention rather than actual use. Based on the review of selected publication, only a few studies focusing on actual behavior (e.g., [30][37][38]). Prior studies in mobile marketing have provided strong justification for conducting the research on behavioural intention which indicates that behavioural intention is a strong predictor of consumers' behavior (e.g., [31][39][40][41]). Due to that trend, most of the current research is moving towards the same direction by focusing on consumers' behavioural intention (e.g., [15][42][43][32]).

V. RESEARCH APPROACH

Generally, previous studies have applied three approaches in carry out the research in mobile marketing which are personal survey, on-line survey and the combination approaches including interview. Even though each of the approaches has their own benefits and drawback but mostly the researchers applied personal survey approach in performing the research. In addition, personally administered questionnaires allow researcher to collect all the completed responses within a short period of time and it is inexpensive and efficient for collecting the data of large numbers of individuals at the same time [43]. Furthermore, on-line survey is also getting more popular among researchers particularly in mobile service context. Recent studies have shown that online surveys have been utilized by a few researchers in conducting their studies (see [15][16][18][25][33][44].

Overall, previous studies do not limit to one approach only in conducting the studies especially in mobile study context. However, there are very limited studies uses mixed methodologies even though the approach also has the advantage. Driscoll et al. [44] reveal that "mixed method designs have the advantages when exploring complex research questions. The qualitative data provides a deep understanding of survey responses, and statistical analysis can provide detailed assessment of patterns of responses". Table II below demonstrates examples of the research approach employed by a number of previous studies.

TABLE II.
COMMON RESEARCH APPROACH

Source/Author (Example of studies)	**Methodology**
[10][22][29][34][35]36][38][51]	Personal Survey
[15][16][18][21][25][32][38][39]	Online survey
[23][30][42]	Mixed methods

VI. Theories Used by Selected Studies

Several competing models or theories have been broadly used by researchers to predict the intention to adopt mobile services. The models comprise of the theory of reasoned action (TRA) [45], innovation diffusion theory (IDT)[46, theory of planned behaviour (TPB) [47], technology acceptance model (TAM) by [48], and decomposed theory of planned behaviour(DTPB)[49]. Among those theories, TPB, TAM and DTPB are frequently used by researchers in explaining the consumers' intention to adopt the mobile marketing services. This is constant with our prior discussion that numerous researchers concentrate more on consumer behavioural intention rather than the actual usage. Even though, it is very rare to have the studies that employ manifold theories but there are some studies that use multiple theories for study purposes comparison. The innovation diffusion theory [46] describes that five general attributes of innovations diffusion which have revealed to consistently influence consumers' adoption namely (relative advantage, compatibility, complexity,trialability and observability. Most of the previous studies do not fully utilise all the five key attributes of IDT but their studies have utilised some of the key attributes by integrating with other variables. Furthermore, Theory of Planned behaviour is also one of the theories that broadly use in technology acceptance studies. Surprisingly, based on selected reviewed publications, most of the studies that utilize TPB are more towards internet related. On the other hand, Technology Acceptance Model by [50] has been identified consist of two constructs namely perceived usefulness and perceived ease of use as the major determinants of individual attitudes toward the use of computer technology. Most of the studies that utilise TAM model will extend the theory with additional construct. Furthermore, Decomposed Theory of Planned Behaviour combines aspects of innovation diffusion theory [46] with aspects of the theory of planned behaviour [47]. DTPB suggests that attitude, subjective norm and perceived behavioural control will influence the intention to use technology. Previous studies that use DTPB also extend the model towards providing better understanding of behavioural intention to use. The theories also have been utilised in multi-disciplines area such as internet banking, mobile services and information technology related. There are very limited studies have utilised the DTPB in mobile services context (e.g., [10][25]).

VII. Discussion

Mostly, all the research unavoidably deals with study limitations. This study only considered a number of selected journals publication. Since the number of publications in mobile marketing keeps on increased, further research may consider focusing on other journals and more comprehensive analysis of the literatures.

Based on review of selected studies in mobile context, some issues of mobile marketing adoption have been highlighted. First, the consumer behavior studies towards mobile marketing are getting more interest among researchers but the academic literatures on mobile marketing are still at infant stage. More researches on this area are much needed. The explanation of mobile marketing phenomenon is still vague and it could be the main reasons why most of the researchers still do not reach agreement on the definition and classification of mobile marketing. Second, despite of growing number of research on mobile marketing services like mobile payment, mobile advertising, mobile data services, mobile banking, mobile commerce and mobile services,etc, the study on other mobile marketing services is keep on growing such as mobile learning (e.g., [51]), mobile ticketing, mobile booking, etc. The recent trend is most of the studies are using the terms by naming the category based on study context relevancy. The future study may propose the standardization of the mobile marketing services category so that other studies will use similar term and it can be generalized in future study on mobile marketing. Finally, the trend of the research approach in mobile marketing studies is more towards employ on-line survey. Even though personal survey has their advantages but on-line survey is getting trendy among the researchers. This is may be due to this kind of survey can reach globally, fast delivery and it is also convenient. The interesting part is the definition and conceptualization of mobile marketing phenomenon is still remaining unclear. Hence, this will provide a large opportunity for future discussion among researchers.

VIII. Conclusion

Prior studies note that there is no consensus among the researchers on definition of mobile marketing. In fact, the conceptualization and phenomenon of mobile marketing is still not fully explained. Researchers have defined and conceptualized the concept of mobile marketing according to their study context and situation. Hence, the selected review in this article might help the researchers at least to understand the major category of mobile marketing in understanding the definition of mobile marketing. The review of selected studies also may assist the researchers towards applying the research approach in their study by providing common research approach employed in previous studies. In all, despite of substantial amount of research in this field, the academic research in mobile marketing still provides a large research opportunities.

There is still very scant research conducted in the area of trust, risk perception and policy. Varnali and Toker [3] reveal that research in the field of m-acceptance is still inadequate. Since the research on the consumer's acceptance of mobile marketing is still limited, therefore it is a need for researchers to devote more attention and to explore this field especially in different context of study.

REFERENCES

[1] Worldwide Mobile Market Forecasts 2009-2013.

[2] MCMC (2010). *Fact & Figures (Statistics & Record).*

[3] Varnali, A. and Toker, A. (2010), "Mobile marketing research: The state of the art". *International Journal of Information Management*, Vol. 30, pp. 144–151.

[4] Barnes, S. and Scornavacca, E. (2004), "Mobile marketing: the role of permission and acceptance", *International Journal of Mobile Communication*, Vol.2 No.2, pp.128-39.

[5] Carroll, A., Barnes, S., Scornavacca, E. and Fletcher, K. (2007), "Consumer perceptions and attitudes towards SMS advertising: recent evidence from New Zealand". *International Journal of Advertising*, Vol.26 No.1, pp. 79-98.

[6] Tähtinen, J. (2006), "Mobile Advertising or Mobile Marketing. A Need for a New Concept. In: FeBR 2005 -Frontiers of e-Business Research 2005", *Conference Proceedings of eBRF*, 152-164.

[7] Leppäniemi,M., Sinisalo,J. and Karjaluoto,H. (2006), "A Review of Mobile Marketing Research". *International Journal of Mobile Marketing*, Vol. 1 No. 1, pp. 30-40.

[8] Leppäniemi,M. and Karjaluoto,H. (2008), "Mobile marketing: from marketing strategy to mobile marketing campaign implementation". *International Journal of Mobile Marketing*, Vol. 3 No.1, pp. 50-61.

[9] Barutçu,S.(2007), "Attitudes towards mobile marketing tools: A study of Turkish consumers" . *Journal of Targeting, Measurement and Analysis for Marketing*, Vol.16, pp.26-38.

[10] Hsu, T.H.,Wang, Y.S., and Wen,S.C.(2006). "Using the decomposed theory of planned behaviour to analyse consumer behavioural intention toward mobile text message coupons". *Journal of Targeting, Measurement and Analysis for marketing*, Vol. 14 No. 4, pp. 309-324.

[11] Komulainen,H.,Ristola,A., and Still,J.(2006), "Mobile advertising in the eyes of retailers and consumers – empirical evidence from a real-life experiment". *Proceedings of the International Conference on Mobile Business (ICMB'06)*, Copenhagen, pp. 37-45.

[12] Merisavo,M., Kajalo, S., Karjaluoto, H.,Virtanen, V., Salmenkivi,S., Raulas, M., and Leppäniemi, M. (2007), "Emiprical study of the drivers of consumer acceptance of mobile advertising". *Journal Of Interactive Advertising*, Vol. 7 No. 2, pp. 41-50.

[13] Wong,C.C. and Hiew, P.L.(2005), "Diffusion of Mobile Entertainment In Malaysia Drivers and Barriers". *World Academy of Science, Engineering and Technology*, Vol. 11, pp. 135-138.

[14] Vlachos, P.A.and Vrechopoulos,A.P. (2008), "Determinants of behavioral intentions in the mobile internet services market". *Journal of Services Marketing*, Vol. 22 No. 4, pp. 280–291.

[15] Gu, J-C., Lee,S-C. and Suh,Y-H. (2009), "Determinants of behavioral intention to mobile banking". *Expert Systems with Application,* Vol. 36 No. 9, pp. 11605–11616.

[16] Hanley, M., and Becker, M. (2008), "A Cell Phone Usage and Advertising Acceptance Among College Students: A Four-Year Analysis", *International Journal of Mobile Marketing,* Vol. 3 No.1, pp. 67-68.

[17] Lee, R.and Murphy,J. (2006), "The consumption of mobile services by Australian university students". *International Journal of Mobile Marketing*, Vol.1 No.1, pp.13-20.

[18] Schierz, G.P., Schilke,O. and Wirtz, B.W.(2010), "understanding consumer acceptance of mobile payment services: An empirical analysis". *Electronic Commerce Research and Applications*, Vol. 9 No. 3, pp. 209–216.

[19] Yang,C.C. (2007), "Exploring factors affecting consumer intention to use mobile advertising in Taiwan". *Journal of International Consumer Marketing*, Vol 20 No . 1, pp. 33-49.

[20] Zhang,J.and Mao,E.(2008), "Understanding the Acceptance of Mobile SMS Advertising among Young Chinese". *Consumers Psychology & Marketing*, Vol. 25 No. 8, pp. 787–805.

[21] Bauer, H., Barnes, S., Reichardt, T. and Neumann M. M. (2005), " Driving consumer acceptance of mobile marketing: A theoretical framework and empirical study".*Journal of Electronic Commerce Research*, Vol.6 No.3, pp.181-192.

[22] Chen, L.D. (2008), "A model of consumer acceptance of mobile payment". *International Journal of Mobile Communications*, Vol.6 No.1, pp. 32-52.

[23] Wu, J.H., and Wang,S.C.(2005), "What drives mobile commerce? An empirical evaluation of the revised technology acceptance model". *Information&Management*, Vol. 42 No. 5, pp. 719–729.

[24] Hong,S.J.and Tam,K.Y.(2006), "Understanding the adoption of multipurpose information appliances: The case of mobile data services". *Information Systems Research*, Vol. 17 No. 2, pp. 162-179.

[25] Hong,S.J., Thong, Y.L., Moon,J.Y., and Tam,K.Y.(2008), "Understanding the behavior of mobile data services consumers". *Information System Frontier*, Vol.10, pp. 431-445.

[26] Lu,J., Liu,C.,Yu,C-Y. and Wang,K. (2008), "Determinants of accepting wireless mobile data services in China". *Information & Management*, Vol. 45 No.1, pp. 52–64.

[27] Qi,J., Li,L., Li,Y. and Shu,H.(2009), "An Extension of Technology Acceptance Model: Analysis of the Adoption of Mobile Data Services in China". *Systems Research and Behavioral Science*. Vol. 26 No. 3, pp. 391-407.

[28] Laforet,S. and Li,X. (2009), "Consumers' attitudes towards online and mobile banking in China".*International Journal of Bank Marketing*, Vol. 23 No.5, pp. 362-380.

[29] Luo,X., Li,H., Zhang, J. and Shim, J.P. (2010), "Examining multidimensional trust and multi-faceted risk in initial acceptance of emerging technologies: An empirical study of mobile banking services". *Decision Support Systems*. Vol. 49 No. 2, pp. 222-234.

[30] Jaradat,M.I.R. and Twaissi, N.M. (2010), "Assessing the introduction of mobile banking in Jordan using technology acceptance model", *International Journal of Interactive Mobile Technologies*.Vol 4(1),pp.14-21.

[31] Koivumaki,T., Ristola,A., and Kesti,M. (2006), "Predicting consumer acceptance in mobile services: empirical evidence from an experimental end user environment". *International Journal of Mobile Communications*, Vol. 4 No. 4, pp. 418 – 435.

[32] Quan,S., Hao, C. and Jianxin, Y. (2010), "Factors Influencing the Adoption of Mobile Service in China: An Integration of TAM". *Journal of Computers,* Vol. 5 No. 5, pp. 799-806.

[33] Verkasalo,H.(2008), "Dynamics of Mobile Service Adoption". *International Journal of E-Business Research*, Vol. 4 No. 3, pp. 40-63.

[34] Bhatti,T.(2007), "Exploring factors influencing the adoption of mobile commerce". *Journal of Internet Banking and Commerce*, Vol.12 No.3, pp.1-13.

[35] Rohm, A. and Sultan, F. (2006), "An Exploratory Cross-Market Study of Mobile Marketing Accepatance". *International Journal of Mobile Marketing*, Vol. 1 No. 1, pp.4-12.

[36] Sultan, F., Rohm, A.J. and Gao,T.(2009), "Factors Influencing Consumer Acceptance of Mobile Marketing: A Two-country Study of Youth Markets". *Journal of Interactive Marketing*, Vol. 23 No. 4, pp. 308-320.

[37] Pedersen, P. E. (2005), "Adoption of mobile internet services: An exploratory study of mobile commerce early adopters". *Journal of Organizational Computing & Electronic Commerce*, Vol. 15 No. 3, pp. 203–218.

[38] Lu,Y.,Zhou, T. and Wang,B. (2009), "Exploring Chinese users' acceptance of instant messaging using the theory of planned behavior, the technology acceptance model, and the flow theory". *Computers in Human Behavior*, Vol. 25, pp. 29–39.

[39] Karjaluoto,H., and Alatalo, T.(2007), "Consumers' attitudes towards and intention to participate in mobile marketing". *International Journal Service Technology and Management*, Vol. 8 No. 2/3, pp. 155-173.

[40] Kim,D.J., Ferrin,D.L. and Rao,H.R.(2008), "A trustbased consumer decision-making model in electronic commerce: The role of trust, perceived risk, and their antecedents", *Decision Support Systems,* Vol. 44 No. 2, pp. 544–564.

[41] Wang, Y.S., Lin, H.H.and Luarn, P. (2006), "Predicting consumer intention to use mobile services". *Information Systems Journal*, Vol. 16 No. 2, pp. 157-179.

[42] Kim, C., Mirusmonova, M. and Lee. I. (2010), "An empirical examination of factors influencing the intentionto use mobile payment". *Computers in Human Behavior*, Vol.26, pp. 310–322.

[43] Kuo, Y.K. and Yen, S.N. (2009), "Towards an understanding of the behavioral intention to use 3G mobile value-added services". *Computers in Human Behavior*, Vol. 25, pp.103–110.

[44] Driscoll, D.L., Yeboah,A.A.,Salib,A. and Rupert, D.J.(2007), "Merging Qualitative and Quantitative Data in Mixed Methods Research: How To and Why Not". *Ecological and Environmental Anthropology*, Vol. 3 No.1, pp.19-28.

[45] Fishbein, M. and Ajzen, I. (1975), *Belief, Attitude, Intention, Behavior: An Introduction to Theory and Research*, Addison-Wesley, Reading, MA.

[46] Rogers, E. (1995), *Diffusion of Innovations*, 4th ed., Free Press, New York, NY.

[47] Ajzen, I. (1991), "The theory of planned behavior". *Organizational Behavior and Human Decision Processes,* Vol.5, pp.179-211.

[48] Davis, F.D. (1989), "Perceived Usefulness, Perceived Ease of Use and User Acceptance of Information Technology". *MIS Quarterly*, Vol.13 No.3, pp. 319-339.

[49] Taylor,S., and Todd,P.(1995), "Understanding information technology usage: a test of competing models". *Information Systems Research*. Vol. 6 No. 2, pp.144–176.

[50] Davis, F.D., Richard P. Bagozzi, and P.R. Warshaw (1989), "User Acceptance of Computer Technology:A Comparison of Two Theoretical Models". *Management Science*, Vol. 35, pp. 982-1003.

[51] Ismail,I.,Idrus,M.R., Ziden,A. and Rosli,M.(2010), "Adoption of Mobile Learning among Distance Education students in Universiti Sains Malaysia". *International Journal of Interactive Mobile Technologies*, Vol. 4(2), pp.24-28.

AUTHORS

Mohammad Ismail is with the Faculty of Business Management, Universiti Teknologi MARA (UiTM) Malaysia and currently doing his PhD at Universiti Utara Malaysia(e-mail: mohammadismail@perlis.uitm.edu.my).

Assoc. Prof.Dr Razli Che Razak is with the Othman Yeop Business School, Universiti Utara Malaysia, Kedah, Malaysia. Presently, he is attached as a Director at the Centre for International Affairs & Cooperation (CIAC) (e-mail: raz1152@uum.edu.my).

Agent and Mobile Tools for Telehomecare in Developing Countries: An Architecture Approach

Karim Zarour
Lire laboratory, University Constantine 2, Algeria

Abstract—**Nowadays, in the networked health, mobile tools are increasingly important, to provide care, educate patients or to transmit remote information. Telehomecare evolves constantly. It can aid disabled persons, increase patient compliance, enhance caregivers' effectiveness, and connect socially isolated individuals to their care providers. In this perspective, the agent paradigm proposes interesting concepts for the development of telehomecare information systems, such as the control autonomy, decentralization and coordination. This paper describes an architecture of telehomecare based on agents and mobile tools.**

Index Terms—**Telehomecare; Inter-Agent Communication; Electronic Medical Record; Information and Communication Technologies;**

I. Introduction

Information and Communication Technologies (ICT) progress rapidly. The mobile tools have a large impact on the social life of citizens today [1]. Telemedicine in homecare, or telehomecare (THC), uses modern technologies to enable the communication and the transfer of information between the health care provider at the clinical site and the patient at his/her home. In THC, the patients conduct their interactions with their health care providers in their own homes [2]. Telehomecare is an emerging field that will benefit homecare providers and their patients. It does not replace in home care visits [3], but is used primarily to augment current services. Most countries are now facing an urgent requirement to provide appropriate retired home environment solutions for patients and allow them to stay at their own homes. Partial solutions for specific case studies of THC system are described in [4].

According to [5], most developed countries are facing important overall problems regarding healthcare services, such as: (i) increased demand of healthcare due to an increased number of elderly and changed life styles leading to an increase in chronic diseases (ii) demand for increased accessibility of care outside hospitals, moving health services into the patient's own homes (iii) need for increased efficiency, individualization and equity of quality-oriented healthcare with limited financial resources (iv) difficulties of recruiting and retaining personnel in the healthcare services in general and in home and elderly care in particular.

At the same time, there are several problems in homecare service in Algeria, such as: (i) current solutions based on manual note taking are slow (ii) obstacle to real-time data access (iii) medical staff is not up to date on the state of the patient (that curbs the ability of clinical diagnostics and monitoring) (iv) slow and laborious process of homecare (v) time consuming (vi) traditional data storage mode is not conducive to the sharing of resources. Therefore, there cannot be information sharing among all the stakeholders.

Telehomecare uses ICT to deliver patient care to the home [2];[5]. It is an emerging field that will benefit home care providers and their patients. It does not replace in home care visits, but is used primarily to augment current services. Examples of THC are: monitoring a patient's wound care progress, monitoring a newly diagnosed Diabetic patient's blood sugars, monitoring a chronic CHF patient for fluid overload with recording of daily weights. The goal is to inspire individuals to manage their own health at home [6];[7]. According to the literature, several definitions are attributed to the THC.

In the work of [8], THC is the use of technology to deliver patient care in the home or place of residence, providing patient-provider contact without either having to travel. In our context, we take consideration this definition. According to [9], THC is a specialisation of TM where the patient with or without support from family or local care do herself manage in monitoring and treatment under electronic observation from qualified clinical staff. Telehomecare is a subfield within telehealth. It involves the delivery of healthcare services to patients at home through the use of telecommunications technologies, which enable the interaction of voice, video, and health-related data explain [10]. The author of [11] uses the concept of home based eHealth to include both THC and smart homes.

Telehomecare is successful for certain groups of patients [12]:

- Patients requiring more than 2 home visits/week
- Chronic conditions
- Palliative conditions

Generally, users who can benefit from THC are [13]:

- People living alone who are unable to seek help in emergencies (unconsciousness, falls, strokes, myocardial infarction, etc.).
- Elderly or disabled people who suffer from cognitive (e.g. Alzheimer) and/or physical (visual, hearing, mobility, etc.) impairment.
- People who need help in daily life to perform personal care activities (e.g. toileting) and

instrumental activities (cooking healthy meals, dealing with medication, and doing laundry).

- People living in rural and remote communities or in urban communities with inadequate health service provision.
- People who suffer from chronic disease, and who need continuous monitoring (e.g. diabetes)
- People involved in telehealth care undertaking health care at a distance or telemedicine, with physicians practising 'virtual visits'.

Telehomecare in developing countries [14] and particularly in Algeria is still facing many challenges especially in rural areas such as:

- Unavailability of the required ICT infrastructure for THC (e.g. internet connection)
- High cost of THC solutions.
- Lack of patients' awareness and acceptance of receiving healthcare services via THC networks and applications.
- Lack of uniform standards for data-sharing: Unified format and content of data storage are the basis for communication between systems. Information classification and data standards are not strictly uniform.
- Security, safety and privacy of data exchange.

In this paper, we propose a global architecture for THC, adapted to the concrete needs of the homecare in Algeria. This architecture allows the sharing, the fast and supple exchange of information, and opens new perspectives of flexibility.

The remainder of this paper is organised as follows. The architecture of the proposed system is detailed in Section 2 followed by mobile tools in section 3. The prototype implementation is described in section 4 and related work of THC is dedicated in section 5. Finally, a conclusion with a view to the future is given in Section 6.

II. THE PROPOSED ARCHITECTURE

The proposed architecture supports THC (see figure 1). This architecture is based on agents and mobile tools. Each actor (doctor, nurse, etc.) of Radiology Information Systems (RIS), Laboratory Information Systems, etc. is assisted by a software agent named User Agent (UA). The behavior of the agent is more organizational in the sense that it knows how to take into account the other members of homecare organization and to update its organizational knowledge. This architecture aims to facilitate the interoperability, the accessibility to information, and it supplies to homecare actors the same functionalities as if they belong all to a single structure (the same place). Our architecture is independent from any application in the field of patient's homecare (chronic diseases, old persons, etc.). Indeed, the main aims of our architecture to summarize in three points: (1) reducing delays in the management of patient (2) improve the efficiency of the THC team (3) and therefore, improve the service provided to patients and their satisfaction. A Virtual Private network (VPN) is established between all Information systems of the hospital, patient home and the data centre using wireless network links between them. This is achieved by sourcing the services from an Internet Service Provider (ISP).

The measured data are transmitted via wireless connection; the information can be received via smart devices and uploaded in XML format through Internet. After, the system receives, processes and saves the data to the corresponding database location.

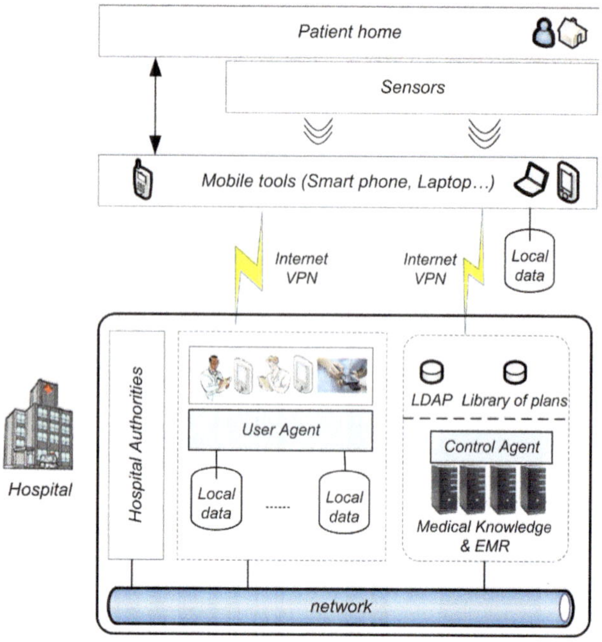

Figure 1. The overall information system architecture

A. *User agent (UA)*

It is a hybrid agent. It assists and helps an actor (doctor, nurse, etc.) that participates in the taking care of patients. Every user agent in the system can communicate with another one without passing by an intermediary. Its components are depicted in figure 2:

User interface. It allows interaction between UA and human agent. This is an interface of assistance for the human agent. This last one has a decision maker's role.

Coordination module. This module takes into consideration all of the overall process of coordination. It takes as input parameters a set of goals (the interpreted results of messages by the user interface and the communication module) and produces a plan which satisfies these purposes. The coordination module consists of a sub module of planning for the orientation and the organization of the local tasks and a sub module of negotiation for task allocation and the resolution of conflicts. For more details see our work in [3].

Individual knowledge module. An agent must have the ability to represent knowledge i.e. to remember and reason above. Thus, the individual knowledge base represents all information and knowledge about the agent himself: its abilities and competencies, the state and the charge of the current task. Accordingly, each UA of system keeps any information relative to its interventions, activities, etc. Example: If X has a body temperature > 40 X has pressure 16-6 and if X has blood sugar level 3.8 ... then hospitalisation.

Homecare-knowledge module. It contains the

information concerning the defined organizational and operational rules in a homecare. What allows having an organizational interoperability between the various partner agents involved in the system. It includes a list of all partner agents. This is knowledge that represents the know-how on the homecare for the good progress of the reasoning process and a coherent behaviour. Consequently, this knowledge base allows achieving the cooperation process and the management of interactions with the other UA. This module also contains information on rights and commitments for the homecare.

Communication module. This module manages the interaction between an UA and the outside world such as other agents (acquaintances). It contains all processes of taking charge of messages from other UA. This module is responsible for all functionalities of sending and receiving messages.

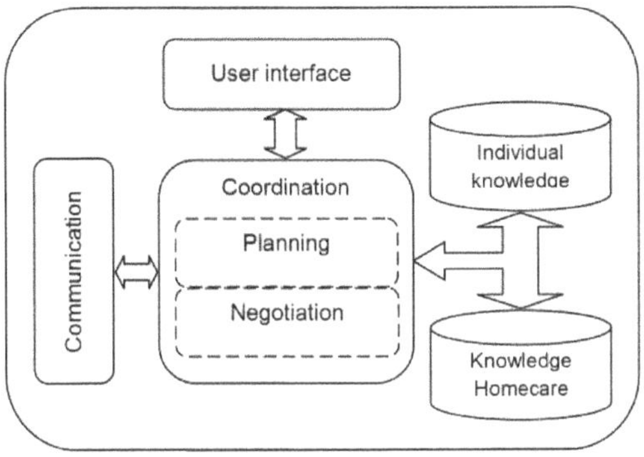

Figure 2. The structure of the user agent

B. *Local data*

Each database is designed to satisfy local needs and not necessarily to interact with other local databases. Indeed, the actors of THC and patients are very attached to their personal notes for their own use. But it is essential to share this information with other stakeholders.

C. *Hospital authorities*

They act on behalf of the hospital. These authorities have a number of powers: recommendation, decision, regulation and sanction. They also ensure some form of control over different medical staff:

LDAP (Lightweight Directory Access Protocol) directory. It is used to store identification data. This choice is based on: (i) quick access, (ii) it is easy to maintain copies and (iii) directory is designed to receive many more requests for reading as for writing.

Library of plans. It contains (i) a set of plans directed by the global purpose (taking care of patients) and sub-purposes (care of the sick, therapy, etc.) useful for the processes of planning and negotiation. These two concepts are the subject of our forthcoming article. (ii) A set of information necessary for actors to realize their tasks. Every plan is characterized by the actions to be executed and the resources to be used. Example: nurse send an activity report to the doctor explaining the actions performed in order to determine the patient's status. The task is to give the patient's daily status (Ex. diabetic elderly).

The plan is (set of actions to be performed):

1. Measure the weight
2. Check blood pressure
3. Check body temperature
4. Check blood sugar level
5. ECG

Furthermore, to better understand the implementation of the plan, and as XML will be used for the description and interpretation of the content of the messages exchanged between the different user agents. XML perfectly illustrates the structure of the plan to execute (see section G).

D. *Control agent*

It is a reactive agent. This agent must manage all associated databases in the data center to make their manipulation uniform. Also, the control agent manages data security and protection of privacy of patients and actors (confidentiality). Thus, it uses authentication and encryption mechanisms. It controls all the formulated queries to databases (radiology data, laboratory data, etc.). The solution of the control agent allows full control of UA, the voluntary or involuntary access to shared medical records which may contain sensitive data. The authentication system allows actors to prove easily their identities and get access to shared databases without threatening the security of system. Control agent has the advantage of supporting the adaptability, reuse and independence.

E. *Shared Databases*

At this level, this is to ensure the entire life cycle of stored medical data of patients. Each database is not only a trace of information, but also a tool for communication, information and coordination between the different structures of the hospital to taking care of patients.

The medical actors will be able to access the THC's databases via Internet with any mobile device (e.g. PDA). The patient will access his medical records using credentials from these devices. Each actor has his own credential key automatically generated and delivered by the hospital authorities to access data. So, databases are composed of:

Medical Knowledge. Relating to disease and medicine.

Electronic Medical Record (EMR). The medical record is an essential tool to share information for cooperation. It consists of an important amount of information. The EMR is stored and capable to be consulted even from distance. The sharing of medical information is vital in medicine concerning at the same time: diagnosis, prognosis, analysis and care continuance. It is a progress factor for better medical care, faster and adapted, and so, systematically transmits information to other participants. To improve information traffic and functioning of our approach, the medical record is primordial, it helps to: (i) Minimize medical errors, (ii) Accelerate knowledge diffusion and (iii) Help decision making.

EMR is brought to evolve and to be updated more frequently. It allows a permanent communication between actors in an interprofessional context in order to cooperate, and to inquire mutually their conclusions' result and actions. Every actor has to find any indication and any information which is useful for understanding situations: to keep information traceability, to orchestrate, to organize interventions, to share and to exchange information to assure the coherence interventions, the continuance and the care quality. EMR guarantees the interoperability among UA in the sense of coordination to collect, enrich and exchange structured medical data.

Furthermore, a virtual collaborative working space reserved for all user actors allows them to work while being geographically distant and by keeping a permanent assistance around the patient.

F. Information exchange

In eHealth is mandatory to use a standardized communication. The development of Telehomecare is also based on standards. They have a significant contribution to interoperability in the THC. Among our contribution's objectives, are to encourage and push the evolution of normalized standards, to support the reusability and the interoperability in the field of distributed medical information. The need to transmit data in a structured form allowing automatic processing has resulted in the development of standard messages. Therefore, it is relatively easy to interconnect systems. Making them interoperable requires an agreement on the structure of exchanged data. But it is not always easy to interconnect systems. For this purpose the international standardization in the field of health plays a vital role to ensure some interoperability in medical information exchange. It should also be noted the predominant place occupied by the XML standard. XML is the standard of data description. This standard promotes the convergence of all standards and norms above [15];[16]. In our proposition, all exchanges of messages and information are based on XML.

G. Inter-Agent Communication

The success of our approach also depends on an effective communication throughout its progress. To analyze data, identify information, maintain knowledge, communicate syntheses, recommendations and decisions-making of agents that are different in their functions and activities, require an unfailing communication. So, it is necessary to assure a permanent communication. Several mechanisms of communication are possible: message exchange, methods invocation and the use of the "blackboard" mechanism. Consequently, standardized inter-agent communication languages should be supplied. In our approach, we use the ACL to formulate messages and the XML to describe the contents of messages. The use of ACL-XML in inter-agent communication permits to achieve a first level of interoperability by surpassing the problem of heterogeneous exchanges among the different actors in the hospital. The success of the information diffusion among actors resides certainly in the communication protocols standardization. The following example illustrates the use of a message in ACL FIPA containing a report structured in XML:

```
(inform
:sender User agent1 (Nurse)
:receiver User agent2 (Doctor)
:language XML
:content (<report>
            <patient>
            <firstNameP>Anonymous</firstNameP>
             <lastNameP>Anonymous</lastNameP>
             <gender>Male</gender>
            ...
            </patient>
            < clinical exam>
                      <weight> 54 </weight>
                      < blood pressure > 14-8 </ blood pressure>
                      < body temperature > 37 </ body temperature >
            </clinical exam>
            <conclusion>
                      <ecg> normal </ecg>
            </conclusion>
          </report>
            )
)
```

III. MOBILE TOOLS

A. Sensors

These devices replace the necessity of manual data gathering. The sensors collect different vital signs. The types of vital signs may include blood pressure, pulse rate, body temperature, etc. There are several practical advantages in using these devices, such as:

- It provides always-on, real-time data collecting;
- It eliminates manual collecting work and possibility of typing errors; and
- It facilitated the deployment process, as wireless networking means no need for cabling or other physical setup.

B. Communication tools

In the last few years, mobile technologies in particular emerged and become very important. Some countries reached an effective penetration of 100% such as Finland. In 2012, there are 35 million users of mobile phones in Algeria that represent 95% of the population [17]. All current phones can access wireless data networks whenever the phone has a signal, through GPRS, 3G, 4G connections or even through Wi-Fi (wireless fidelity), that enables data exchange via the Internet. Telehealth systems in the form of online and mobile tools are already opening up the possibilities for reduced hospitalization and an increased homecare [18]. Indeed, ICT tools can be used to access a wide variety of technological solutions for communication, including text messaging, gathering and monitoring data, diagnosis and treatment at distances, and retrieving electronic health records [19];[5];[20].

Actors of the THC are different in terms of forming, competence or task to be carried out. This variability is translated by a perception mechanism of knowledge and various access modes to information. Then, it is necessary to take into account this variety, to offer to the actors the adapted and efficient interaction modes, a better diffusion, and a data acquisition to interoperate.

Mobile tools such as smart phone constitute a real condition for the THC success. Their uses on behalf of all the participating actors appear to be: the power to share information, knowledge and taking decisions through cooperation.

IV. Prototype Implementation

In order to study the feasibility of the proposed architecture, we implemented a prototype using standards (see figure 3). We used XML technology to represent the information exchanged among agents via the standard communication language ACL-FIPA (the possibility of using standards like HL7 v3 message, CDA, DICOM, etc. encoded in XML [21]). Indeed, the system can interact with other applications in hospital for example through the XML exported files. The platform which gets closer most our criteria is JADE (Java Agent Development Framework) [22]. It is a platform of agents' creation that takes into account FIPA specifications for the Multi-Agent Systems (MAS) interoperability. Figure 4 shows the doctor interface through PDA.

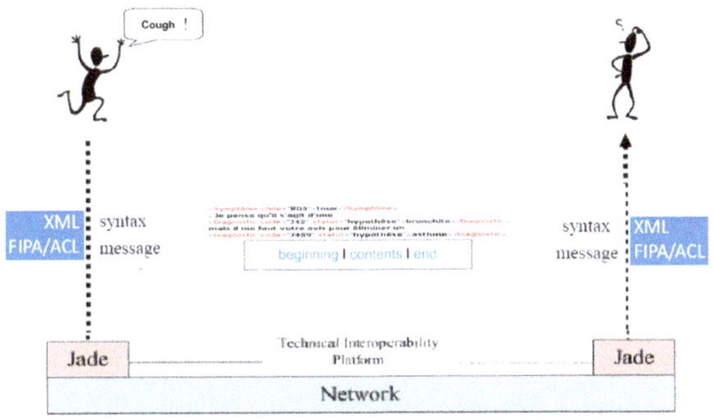

Figure 3. The overall implementation

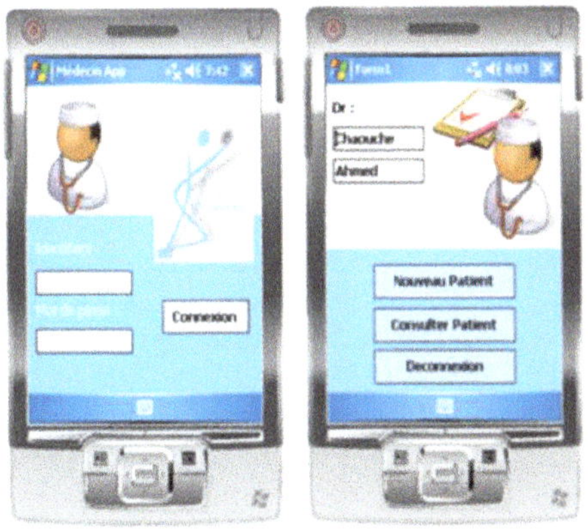

Figure 4. Doctor interface

The following Java code represents the implementation of the `UserAgent` class. These classes are extensions of the basic Agent class defined in JADE.

```
public class UserAgent extends Agent {
 class percevoir extends SimpleBehaviour {
       public percevoir(Agent a)
        {
         super(a);
        }
        public void action() {
           ...
           ACLMessage msg = blockingReceive();
           if (msg!= null){
             switch(msg.getPerformative()) {
                          // Processing…
                                    ...
```

Key benefits of using this architecture:

- Real-time Services (e.g. to support remote medical consultations)
- Health actors and in particular doctors can easily access test results previously made, and information about patients.
- Avoid unnecessary interactions between health actors.
- Efficiency side, health actors spend much less time and effort to search information, which allows to organize activities and to avoid delays in care.
- Patients undergo fewer redundant tests.

Specially reduction in:

- home visits (lack of doctor and nurse)
- hospital stays (limited number of beds and hospitals)

Disadvantages:

- high cost of equipments

V. Related Work

In recent years, many countries have tried to apply information and communication technologies to the healthcare system and especially to THC. . In the study of [23], The status of telehomecare in developing countries in Africa and other regions differs from that in other areas. The problem is accentuated by a lack of infrastructure and health personnel. In the work of [24], an information platform for tele-healthcare services was implemented. The platform deployed a service-oriented architecture, which consisted of HL7 standard messages and web service components. The prototyping system was pretested and evaluated in a homecare department of hospital and a community management center for chronic disease monitoring. In Canada the effectiveness of home tele-healthcare for diabetes is studied and proposed a decision framework to address complex issues of adopting home telehealth technology into practice [25]. In general, it has been concluded that home telehealth systems produce accurate and reliable data and capable of improving patient management [26]. The aim of [27] was to review existing studies describing the use of ICT in home care for communication between patients, family members, and healthcare professionals. The work of Tollinsky [28] described the recent incorporation of "telehomecare" services for older adults living with chronic disease in Northern Ontario communities. Kort and Hoof [29] distinguished three categories of THC technologies: (i) remote telecare, (ii) activity monitoring, and (iii) a category comprising telemedicine and e-health solutions and services. There are numerous barriers to the implementation of Dutch THC technologies projects. The studies in [30] showed that THC allows nurses to treat more patients and to deliver equivalent care.

VI. Conclusion and Future Work

In this paper, we have presented an information system architecture supporting telehomecare. This architecture is based on agents and mobile tools. This architecture allows interoperability and easy data access. Also, it facilitates flexible interaction between doctor and nurse, patient and doctor or patient and nurse through the Internet. Software agents help to ensure the availability of data in healthcare and can be used to support processes, for example, information retrieval. Every agent assists an autonomous actor. These actors are geographically scattered belonging to diverse information's systems of hospital. In the current state we are inclined to study important aspects such as the coordination among different health actors for taking care of patients at home.

In terms of future work, it is desirable to deploy our architecture in a realistic environment to further evaluate and improve our study. Besides, the future work may produce a system with more functions and higher security mechanism in further project.

References

[1] W. Jobe, "Native Apps vs. Mobile Web Apps", *International Journal of Interactive Mobile Technologies*, vol. 7, no. 4, 2013.

[2] G. Demiris, S.M. Speedie, and S. Finkelstein, "Change of Patients Perceptions of TeleHomeCare", *Telemedicine journal and e-Health*, vol.3, pp241-248, 2001.

[3] K. Zarour, N. Zarour, and S. Khalfi, "Inter-agent Interaction in Medical Information System: Case HomeCare", *Journal of Theoretical and Applied Information Technology*, Vol.11, pp.130-142, 2010.

[4] S. Nourizadeh, "Un système de télésanté contextuel avec support de qualité de service pour le maintien à domicile", Published PhD thesis, Institut National Polytechnique Lorraine, France, 2001.

[5] S. Koch, "Home telehealth: Current state and future trends", International Journal of Medical Informatics, vol.75, pp. 565–576, 2006.

[6] D. Pick, Telehomecare.

Ontario telehomecare, "What is Telehomecare?".

[7] M.E. Stachura and E. V. Khasanshina, "Telehomecare and Remote Monitoring:An Outcomes Overview", The Advanced Medical Technology Association, 2007.

[8] T. Tambo, N.H. Petersen, E.B. Pedersen and K. Bejder, "Coherent national IT infrastructure for telehomecare - a case of hypertension measurement, treatment and monitoring", *Engineering and Technology*, vol.47, pp.758-765, 2010.

[9] K. Bowles, A. Baugh, "Applying research evidence to optimize telehomecare, Journal Cardiovasc Nurs", vol.22, pp.5-15, 2007.

[10] G. Demiris, Home based e-health applications, "E-Health: Current Status and Future Trends", *Studies in Health Technology and Informatics*, in Demiris, G. (Ed.), vol. 106, IOS Press, Amsterdam, Berlin, Oxford, Tokyo, Washington DC, 2004, pp. 15-24.

R. Elford, "An Introduction to Telehome Car.

[11] M. Chan, E. Campo, D. Esteve, and J-Y. Fourniols, "Smart homes. — current features and future perspectives", *Maturitas*, vol.64, pp. 90–97, 2009.

[12] R. Hussein and A. Khalifa, "Telemedicine in Egypt: SWOT analysis and future trends", *GMS Medizinische Informatik, Biometrie und Epidemiologie*, vol.8, pp.1-16, 2008.

[13] HL7 Specification.

[14] S. Umer, M. Afzal, M. Hussain, H.F. Ahmad and K. Latif, "Design and implementation of an automation tool for HL7 RIM-to-rational database mapping", *In Proc. of the 10th International HL7 Interoperability Conference (IHIC)*, 2009.

[15] F. Lezzar, A. Zidani and A. Chorfi , "Enabling coordination within medical settings: case of a maternity ward", *Applied Medical Informatics*, vol.32, pp. 39-46, 2013.

[16] A. Venter, R. Burns, M. Hefford and N. Ehrenberg, "Results of a telehealth-enabled chronic care management service to support people with long-term conditions at home", *Journal of Telemedicine and Telecare*, vol.18, pp. 172-175, 2012.

[17] European Commision, What is eHealth? ICT for health. Europe's Information Society.

[18] S. Koch and M. Hägglund, "Health informatics and the delivery of care to older People", *Maturitas*, vol.63, pp. 195-199, 2009.

[19] B. Orgun and J. Vu, "HL7 ontology and mobile agents for interoperability in heterogeneous medical information systems", *Computers in Biology and Medicine*, vol.36, pp. 817–836, 2006.

[20] Jade, Java Agent DEvelopment Framework.

[21] I. Nakajima, Y. Zhao and H. Juzoji, "IP-ECG for Telehomecare."

[22] E-W. Huang, R-S. Rung, S-F. Chiou, F-Y. Liu and D-M. Liou, "Design and Development of a Tele-Healthcare Information System Based on Web Services and HL7 Standards", *Software Tools and Algorithms for Biological Systems, Advances in Experimental Medicine and Biology*, Arabnia H.R. and Tran Q.-N. (eds.), Springer, 2011, pp.599-606.

[23] A. Marilynne, B. Korabek, R. Scott, "Moving Research into Practice: A Decision Framework for Integrating Home Telehealth into Chronic Illness Care", *International Journal of Medical Informatics*, vol.75, pp.786–794, 2006.

[24] P. Guy, J. Mirou, and S. Claude, "Systematic Review of Home Telemonitoring for Chronic Diseases: The Evidence Base", *Journal of the American Medical Informatics Association*, vol.14, pp. 269–277, 2007.

[25] B. Lindberg, C. Nilsson, D. Zotterman, S. Söderberg and L. Skär, "Using Information and Communication Technology in Home Care for Communication between Patients, Family Members, and Healthcare Professionals: A Systematic Review", *International Journal of Telemedicine and Applications*, pp.1-32, 2013.

[26] M. Tollinsky, "Telehomecare programs rolling out across the north", *Northern Ontario Medical Journal*, 2012.

[27] H.S.M. Kort and J. Hoof, "Telehomecare in the Netherlands : barriers to implementation", *International Journal of Ambient Computing and Intelligence*, vol.4, pp. 64-73, 2012.

[28] The Montreal Economic Institute (MEI), "Appendix. Comparison of studies on telehomecare", 2010.

Author

K. Zarour is Associate Professor in the Department of Software Technologies and Information Systems at the University Constantine2, Constantine, Algeria. He received his PhD degree in computer science from the University Mentouri of Constantine, Algeria. His research interests include advanced information systems, particularly medical information systems.

A Mobile Device Based Serious Gaming Approach for Teaching and Learning Java Programming

T. Jordine[1], Y. Liang[2] and E. Ihler[1]
[1] Stuttgart Media University, Stuttgart, Germany
[2] University of the West of Scotland, Paisley, United Kingdom

Abstract—**Most first year computer science students find that learning object-oriented programming is hard. Serious games have ever been used as one approach to handle this problem. But most of them cannot be played with mobile devices. This obviously does not suit the era of mobile computing that intends to allow students to learn programming skills in anytime anywhere. To enhance mobile teaching and learning, a research project started over a year ago and aims to create a mobile device based serious gaming approach along with a serious game for enhancing mobile teaching and learning for Java programming. So far the project has completed a literature review for understanding existing work and identifying problems in this area, conducted a survey for eliciting students' requirements for mobile gaming approach, and then established a mobile-device based serious gaming approach with a developed prototype of the game. This paper introduces the project in details, and in particularly presents and discusses its current results. It is expected that the presented project will be helpful and useful to bring more efficient approaches with new mobile games into teaching object-oriented programming and to enhance students' learning experiences.**

Index Terms—**Java programming; mobile learning; mobile teaching; serious gaming.**

I. INTRODUCTION

In the last few years usage of mobile phones has rapidly increased and become more and more important to our daily life: check e-mails, surf in the Internet, chat with friends, play games, etc. Especially young students are using their mobile phones as modern communication devices and for playing mobile games [1]. Since mobile phones are mobile devices, they can be used by students in anytime anywhere, e.g. during idle times outside their university. In education, this means they also provide new opportunities for teaching and learning programming [2]. The current development of mobile phones (for example, providing more powerful hardware, better and larger touchscreen displays, better surfing ability) has made it even more realistic to use mobile-device based games as a new approach for teaching and learning subjects that are often difficult to teach and learn with traditional approaches in education.

This finding triggered the authors to start a research project for creating a new mobile-device based serious gaming approach for teaching and learning Java programming, as traditional approaches are often insufficient in such teaching and learning and our literature review has found that existing programming learning approaches (see Section II) often use desktop computers and currently there is no approach provided for teaching and learning Java programming using mobile phones, even though Java is one of the most popular object oriented programming language [3], [4] and often taught by universities for programming introduction.

In addition to a literature review, the project also conducted an initial survey (see Section III) recently. The result of the survey has explored the students' willing of having mobile games for learning Java programming with convenience. Both of the literature review and the survey encouraged us to continue with the project and create a new mobile-device based serious gaming approach that supports first year computer science students to learn Java during idle times with a mobile game for teaching and learning Java programming. This includes identification of a suitable game scenario that would keep students motivated when learning object oriented concepts and the Java syntax with the game. The project also faces new challenges (e.g. small screen size, CPU power, etc.) for new mobile game development.

The first prototype (see Section 4) of the mobile game has been completed already so far to support the proposed mobile-device based serious gaming approach.

This paper is structured as follows: Section II will show the existing work that has been done so far in this research area. Section III shows the results of our initial survey that has been conducted at the Stuttgart Media University. The game prototype will be introduced in Section IV. Section V will give an outlook of the planned authoring tool that will allow lecturers to create their own teaching materials. The paper will conclude with a discussion in Section VI and an overview of the future work in Section VII.

II. EXISTING WORK

During the last few decades many programming learning games have been created. This section will look at some of them within the four categories (programming learning games for desktop computers, programming learning games for mobile devices, Java teaching and learning games, authoring tools). The section will also discuss the problems with them for teaching and learning Java programming.

A. Programming learning games for Desktop computers

One of the first interactive programming learning games was Karel the robot [5] in the early 1980's. It provides a simple procedural programming language to control a robot on the screen. Jeroo [6] is similar to Karel the robot, but its object-oriented syntax is more close to C++ or Java, which allows an easier access to those languages. The NetLogo [7] framework was developed to teach undergraduate students object-oriented concepts with a simple syntax. As a feature, it provides a simple editor for creating its own levels. Alice [8] is a programming novice friendly language and does have a puzzle-piece based syntax, which is able to modify 3D objects in a 3D scene. A study that has been conducted at the Monmouth University has shown that the students' success rate was increased from less than 50% to 70% [9]. Scratch [10] has similar features (code puzzle pieces), but it focuses on 2D objects. PlayLogo3D [11] game was developed for early programming education and is based on the LOGO programming language.

B. Programming learning games for mobile devices

Later, the appearance of web techniques like HTML5 and CSS made it possible to play games using mobile device browsers. For example, E-Training DS [12] allows developers to design educational computer science games for mobile gaming devices. Also instructors can choose a predefined set of mini-games and adapt them for their own teaching purposes. As a special feature, a learning management system (e.g. Moodle) can be connected to each generated game to enable instructors or lecturers to design assessments and check students learning progress.

The authors did not find any empirical evaluation of the above mentioned mobile games and frameworks. However, a study [13] that was conducted in a primary school has confirmed that mobile learning games had positive impact on pupils' problem solving skills. By looking at a small study that was conducted in a primary school, it was found out that mobile learning games are having positive impact on pupils' problem solving skills [13].

C. Java teaching and learning games

Some Java learning games have been found for desktop computers. One is RoboCode [14] that plays in the same way like Karel the robot [5], and has a functionality supporting students to compete against each other. Another one is Code Spells [15] that is a role-play game for teaching Java by allowing players to move in a 3D world and requiring them to solve missions with spells. These spells are expressed with Java. Third one, Dream Coders [16] is also a role-play game with which players move in a 2D world, that is incomplete in its game play. Through programming assignments, players can improve playability and fun of the game. Dream Coders provides a framework for adopting an existing assignment and integrate it into courses. Forth one, Greenfoot [17], uses a different approach to provide a framework and an environment for creation of simulation-like, interactive applications on a two-dimensional grid. In terms of Greenfoot, objects displayed on the plane can be browsed through a UML (Unified Modeling Language) class browser. Classes can be instantiated and displayed as game objects on the grid. This Java introduction game frame-work has been used successfully at several universities and schools in order to teach students and pupils the basics of object-oriented programming (e.g. [18]–[20]).

Unfortunately the literature review conducted by the authors has explored that the current serious games for learning Java were only developed for non-mobile devices. This means a lack of the similar games for learning Java with mobile devices.

D. Authoring tools

Authoring tools in general allow course instructors to create their own teaching contents within a serious game for their students. Authoring tools for teaching programming are not found in publications by the authors though some authoring tools found available for other areas stated in the following.

E-Adventure [21] is an authoring tool for educational adventure games that supports desktop games and mobile games. It does not require game creators to have knowledge in game development. It does not particularly focus on learning programming. Any learning game can be produced using it. Nice-Game [22] is comparable to E-Adventure [21] and is a tool for educational content creation.

Second one is Emergo [23] which is a methodology and generic toolkit for developing and delivering scenario-based serious games that aims to teach students in higher education. It makes it easier to create serious games without having specific expertise. Third one is KAR [24] that provides a scenario generator for serious games. It provides an automatic method for generating content about every day activities through combining computer science techniques with the crowd. These scenarios can be used to train every-day activities in medical, military and commerce gaming applications. Forth one is U-Create [25] that attempts to make efficient content creation for interactive setups, mixed reality experiences and location-based services. It uses a graphical tool to create elaborated contents in a fast and easy way. Final one is the Create project [26]. It provides an instructional design-authoring tool that sup-ports entire process from identifying training requirements through the delivery of the trainings. Its developed prototype supports the upfront analysis and design process while enabling any game engine to be used as the development and delivery platform.

III. Initial Survey

In order to find out how the Java programming language and object-oriented concepts are taught at our university an initial survey was set up and completed in November 2013. The survey was split up into two parts:

1. How is Java currently taught at the university and how students perceive the teaching?
2. Is there a need of improvement and would students use a mobile Java learning game as an addition support for their study?

The survey was conducted in the form of online questionnaire at the Stuttgart Media University, Germany. 130 students were contacted and 98 of them actually completed the questionnaire for the survey. The result of the survey is shown and explained in the following.

73% of the respondents were male and the rest were female. More than 75% of them were in age between 21 and 26. 58% of them already had programming experiences before they started study of Java programming.

40% of the respondents do not use their phone for gaming purposes. In the rest 60%, 53% of them use their mobile for gaming up to five hours a week and 7% of them play more than five hours a week.

88% of the respondents used lecture notes and exercises from lab sessions in their study. 48% of them also looked at online tutorials. 4% of them used a learning game. 16% of them used other learning materials like books. When they are asked which of the learning materials or things was most useful to learn for their study, 59% of them regard exercises from lab sessions as most valuable things to do for study; 51% of them regard getting help from their fellow students as most useful thing to do; 42% of them regard accessing online tutorials as additional sources to use. When the respondents are asked if they had problems with learning Java, 43% of them answered "yes".

Furthermore, in order to find out where programming novices would cause the most of problems, the respondents were asked to rank the things that they thought difficult to learn in Java programming. Their answers are as follows (multiple answers were allowed):

TABLE I.
THINGS MOSTLY DIFFICULT TO LEARN

Things mostly difficult to learn	% of respondents
Object concepts, in general	55%
Understanding of the problem (domain) when they had to solve a programming problem	25%
Understanding logical flow in a program	24%
Handling data structures	23%

49% of the respondents had been taught Java with an objects-first approach as a first method [27]; 10% of them had been taught Java in a data structure and algorithm as a first method [28]. 41% of them had been taught with a mixture of both methods.

The survey also investigated if the mark of the mathematics exam is correlated with the mark of Java programming exam. A weak correlation (correlation coefficient: 0.35) was found for this. This means in reverse that students still could get a good mark for their Java programming exam even they did not get a good mark for their mathematics course. The survey also found a similar result for the impact of previous programming experiences on the mark of Java programming exam (correlation coefficient: -0.3).

The survey found that 40% of respondents would like to use a game as a tool for learning Java with their mobile devices, and 41% were undecided. 49% of respondents answered "yes" and 18% answered "no" when they were asked if a mobile Java learning game would help them. When they were asked where they would use a mobile teaching game for Java, most of them answered that they would use it during idle times, e.g. 61% at the bus stop and 51% at home (multiple choice was allowed).

Regarding the features of a mobile Java learning game, the survey found that 79% of the respondents thought of "practical exercises"; 66% considered "a good usability"; and 42% considered "checking their current learning progress" (multiple choice was allowed).

This survey has shown that nearly a half of them have had problems with learning Java in classrooms and students are willing to use their mobile phones for learning Java programming at their own times.

IV. A Mobile Device Based Serious Gaming Approach And A Mobile Game Prototype

The following problems were identified by the conducted literature review and the initial survey:

- *No support for learning Java on mobile devices.* The games mentioned above do not support teaching Java on a mobile phone device. This problem means that a new approach of teaching Java and object-oriented concepts is needed in connection with mobile phones in order to attract young mobile phone users to learn Java.
- *Lack of a suitable help functionality for a mobile programming game.* Some of the existing tools do have an integrated help function (e.g. [6], [8], [14], [17]), but the function is very basic. This problem would decrease play of mobile games because the context sensitive help functionality would give more hints for each specific mission that students need to solve and help students to learn and practice each topic in theory. Especially in programming it is necessary to use the help function to understand the concepts and the specific API (Application Programming Interface).
- *Lack of using game design elements [29] known from traditional video games.* Many of existing approaches and games do not use game design elements (such as high scores, leader boards and bonus quests) as they should do. They often provide a customized programming environment with a code editor where behavior of game objects is coded and a game screen where the coded effects can be observed. This problem would lowed student motivation of using games as, according to existing game patterns [30] it is necessary to add incentives to a learning game in order to keep students motivated. For example, a course-wide high score could keep students motivated to solve a level correctly.
- *Many games for teaching and learning programming do not allow content creation for lecturers.* This problem would leave lecturers away of using games as teaching tools. In order to adapt each level to a specific lecture it is necessary to let games allow lecturers to create their own teaching contents. This can be a key problem for both of students and lecturers to use games if the games do not allow lectures to repeat a specific topic of a lecture not only in a lab session or tutorial but also during idle times in a mobile app.

By addressing the above problems, the second stage of our project tackled the lack of mobile device based games for teaching and learning Java and tried to find a suitable game scenario that would be fun to play and also impart programming knowledge to first year computer science students. This would create a mobile device based serious gaming approach for teaching and learning Java programming and develop a new mobile serious game for implementing the scenario.

This mobile java learning approach was not created to replace existing teaching methods such as lectures and lab sessions; the authors see the game as an extension to them. Based on the findings in existing works and the initial survey a first prototype was implemented. It should give students another opportunity to improve their programming skills, especially during idle times. This means for level design that each level should be playable within five minutes and a suitable game genre that fulfills all requirements that had been found.

After talking to students and experts, the authors have selected a tower defense scenario for the approach and the mobile game as it is flexible enough to add suitable and short missions and also students have known it already. Students will implement the behavior of towers via Java code. Then they will place the towers on the game grid by using programming code.

Figure 1 introduces a prototype of the game and its screens, which are explained in the following sub-sections. Each of the screens in the figure is explained in the following sub-sections.

A. Start screen

The start screen is used to start the game. It provides basic help functionalities and settings. It is planned to add a high score functionality to its next version so that students of a course can compare their games and scores with each other. This can motivate to solve all levels correctly.

B. Mission goals screen

The current goals for implementing the tower are described here. Students can also get hints to solve the level. For example, students have to implement a class where a simple tower is defined by Java code. In order to pass the level, they have to add and set default values to an integer and a string attribute.

C. Coding screen

This screen enables students to write their tower code on mobile phones. They will have the opportunity to check their code during writing, which is indicated by a red or respectively green button. Stu-dents can continue to see the next screen only when the code has passed the check. A mobile optimized Java keyboard is provided on this screen that enables students to write their code word-by-word rather than character-by-character. The words displayed on the keyboard are the words only needed to solve the level.

The code check is divided into two phases as shown in Figure 2. First, an integrated Java parser checks whether the syntax is correct. Then the game checks if the entered code matches the current task, which is done by the game by using regular expressions. It will include a strong help system in the next version of the software that will provide more detailed help and possibly link to external learning resources on internet if an error is found via the game.

D. Object instantiation screen

This screen allows students to instantiate the towers that have been written in the previous screen. A short introduction and help are shown when the screen is opened at the first time. The code would be checked, similarly to the previous screen, and analyzed with parameters. These parameters can be used e.g. to place towers on the game grid. An optimized mobile Java keyboard is also provided here.

E. Game screen

The game screen contains the enemy path, a grid and the instantiated towers. Enemies follow a track to an end point. When enemies cross the end point, the game is

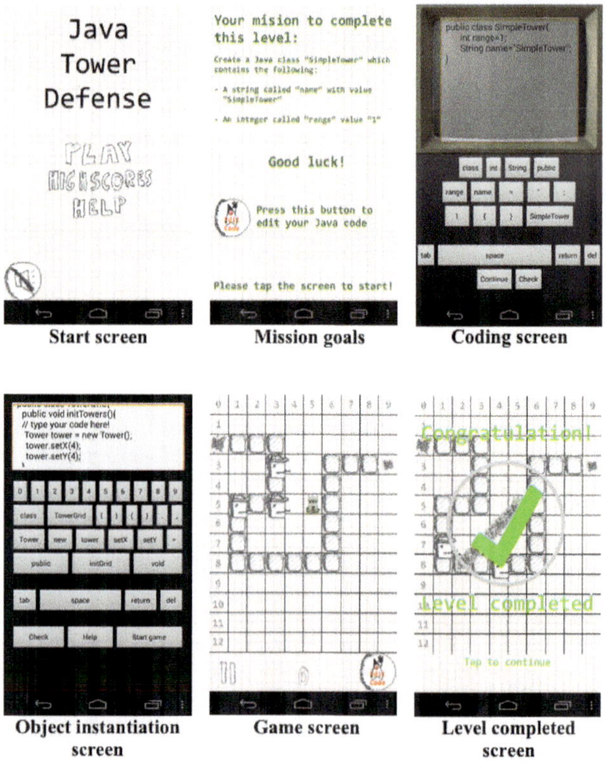

Figure 1. Game screens

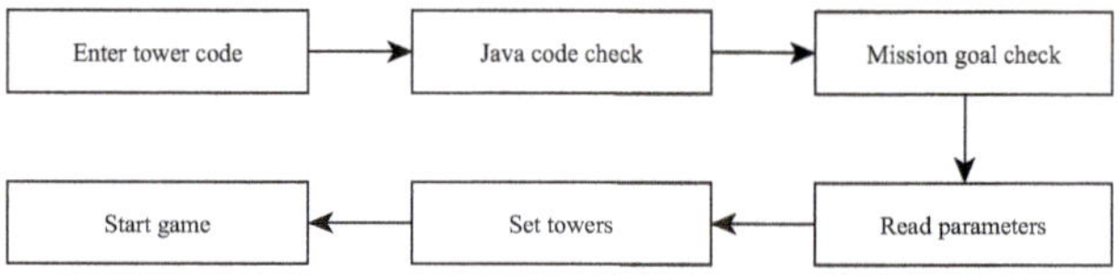

Figure 2. In-game code check workflow

over. In order to prevent this, players have to place and code the towers cleverly. New towers can be added during playing the game. It will provide additional help functionality in this screen.

F. Level completed screen

This screen will be shown when a level has been completed correctly. Only the achieved score is shown at the moment. This screen will be improved with a more advanced scoring system. For example, code quality and other factors will be rated separately.

In addition, the game provides an integrated high score functionality for students that enables them to compete against each other; and an authoring tool that enables Java teachers to create their own contents and teaching materials for their own course.

V. AUTHORING TOOL

The authoring tool enables lecturers to create their own teaching materials for their own courses. In particular, lecturers can use the tool to design their own levels and link the levels to relating learning resources such as hints to solve the level, API documentation or web references for further information. Then they can use the created game in their lectures and lab sessions to help students understand and practice each topic of lectures using their mobile phones inside and outside classes.

This authoring tool enables lecturers to create new levels as long as students require them in the following steps:

1. Java lecturers use the tool to create and design the levels on a desktop computer first.
2. They then to deploy the game with the levels to a cloud server.
3. Then the game and levels on the cloud server are accessed by the registered students' mobile phones.
4. Students now download the levels by running the game on their mobile phones.

The authoring tool will also include a tracking feature that allows lecturers to collect the information that shows the current learning progress of their students. This information can be collected and displayed in terms of a management website that can be accessed via various mobile and desktop browsers.

VI. Discussion

The approach presented in the above is complementary to existing programming teaching methods and will be used as an extension to the methods. Since the project is still in progress more things still need to be considered and done in the project. They are discussed in the following.

Firstly, the developed mobile game prototype will be tested by a wide range of students in universities of Germany and U.K. in order to find out to what extent the approach can support students in learning and understanding Java. A user study has already been set up in two universities of the two countries for this test.

Secondly, the game will be evaluated to find out how a level for the mobile game could be designed for leading a successful learning process for students. Existing work in the game-based learning area [30], [31] have proven the possibility of creating guidelines for content creators/Java instructors. Also other evidence could be found from the ACMs and IEEEs guidelines for teaching computer science [32]. This could be provided and implemented by the proposed authoring tool. Since the proposed game is optimized for mobile platforms, levels must be understandable and playable in a few minutes and not every existing tutorial is suitable for mobile learning. Therefore, it needs to consider how to make the Java code clearly arranged on small mobile phone screens. During the level design, it also needs to use a limited amount of keywords so that a keyword can be mapped to a single button on a screen.

Thirdly, at the moment, the game still has a prototypical user interface, which was useful for the first test mentioned in the above. But, the final version of the game needs to consider that most players are aged between 21 and 26 years (according to the finding of the initial survey). This means, a balance between effective and joyful user interface must be made to the final game.

Fourthly, the current prototype of the game only supports the Android platform as it has a more than 80% market share of mobile phones [33]. But students may use the mobile phones with Windows Phone or Apple iOS. Thus the final game also needs to support these platforms so that teachers can always use the proposed approach for a whole class, no matter which operating system is used by students' mobile phones.

Fifthly, the technical issue that mobile devices are having different screen sizes and resolutions needs to be thought. Initial tests of the game prototype have found that

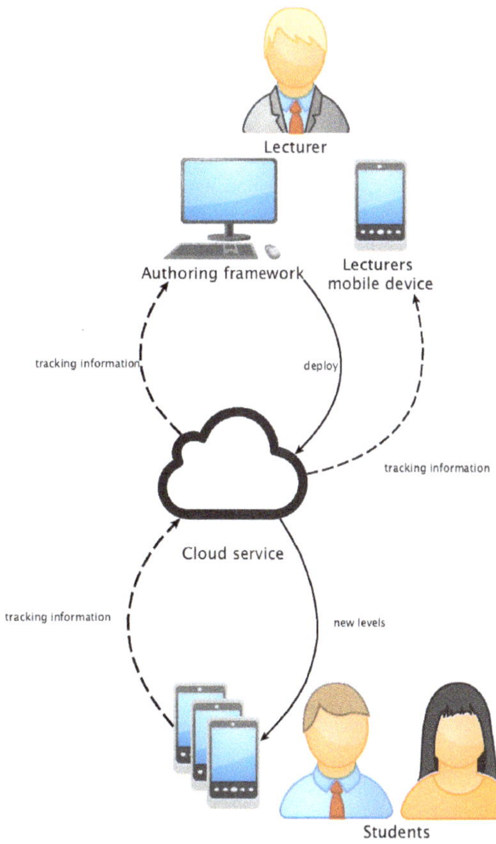

Figure 3. Proposed authoring tool

a display with at least 4.5 inches is required for precisely tapping each key on the coding keyboard.

Finally, it is necessary for the game to provide another functionality that can send students' code to their teacher when they have any doubt about their code at each level.

VII. Conclusion and Future Work

This paper has introduced a new mobile device based serious gaming approach for teaching and learning Java programming. It has shown existing works in Section II and summarized an initial survey in Section III that was conducted at the Stuttgart Media University at an earlier stage of a research project that is currently carried out by the authors. It then presented a prototype of the mobile game in Section IV that was created and implemented by the authors for the project. In Section V, it introduced an authoring tool that would be developed as a part of the game. Then it discussed the remained issues and improvements of the game in in Section VI.

In the next stage of the project a user study will be conducted at two universities of Germany and U.K. with two separate Java programming courses. The results of this study will show to what extent the Java learning game can support students in learning Java programming. This study will be carried out in two phases: the first phase will carry out the test of the current prototype of the game; the second phase will firstly analyze the result of the first test, then revise the code of the prototype, and finally test the revised game towards a final version.

In addition to the user study, a usability test will also be conducted in the next stage in order to find out the pragmatic and hedonistic qualities of the game. This will be done by using the AttraktDiff 2 test [34].

REFERENCES

[1] S. Khalaf, "Apps Solidify Leadership Six Years into the Mobile Revolution," Apr-2014. [Online].

[2] D. Frohberg, C. Göth, and G. Schwabe, "Mobile Learning projects - a critical analysis of the state of the art: Mobile learning projects," *J. Comput. Assist. Learn.*, vol. 25, no. 4, pp. 307–331, Jul. 2009.

[3] "TIOBE Software: Tiobe Index." [Online].

[4] S. O'grady, "The RedMonk Programming Language Rankings: January 2014," *tecosystems*. [Online]. Available.

[5] R. E. Pattis, *Karel the Robot: A Gentle Introduction to the Art of Programming*, 1st ed. New York, NY, USA: John Wiley & Sons, Inc., 1981.

[6] D. Sanders and B. Dorn, "Jeroo: a tool for introducing object-oriented programming," in *Proceedings of the 34th SIGCSE technical symposium on Computer science education*, New York, NY, USA, 2003, pp. 201–204.

[7] M. Dickerson, "Multi-agent simulation and netLogo in the introductory computer science curriculum," *J Comput Sci Coll*, vol. 27, no. 1, pp. 102–104, Oct. 2011.

[8] S. Cooper, W. Dann, and R. Pausch, "Alice: a 3-D tool for introductory programming concepts," in *Journal of Computing Sciences in Colleges*, 2000, vol. 15, pp. 107–116.

[9] K. Johnsgard and J. McDonald, "Using Alice in Overview Courses to Improve Success Rates in Programming I," in *IEEE 21st Conference on Software Engineering Education and Training, 2008. CSEET '08*, 2008, pp. 129–136.

[10] M. Resnick, J. Maloney, A. Monroy-Hernández, N. Rusk, E. Eastmond, K. Brennan, A. Millner, E. Rosenbaum, J. Silver, B. Silverman, and Y. Kafai, "Scratch: programming for all," *Commun ACM*, vol. 52, no. 11, pp. 60–67, Nov. 2009.

[11] I. Paliokas, C. Arapidis, and M. Mpimpitsos, "PlayLOGO 3D: A 3D Interactive Video Game for Early Programming Education: Let LOGO Be a Game," in *2011 Third International Conference on Games and Virtual Worlds for Serious Applications (VS-GAMES)*, 2011, pp. 24–31.

[12] R. Tornero, J. Torrente, P. Moreno-Ger, and B. F. Manjón, "e-Training DS: An Authoring Tool for Integrating Portable Computer Science Games in e-Learning," in *Advances in Web-Based Learning – ICWL 2010*, X. Luo, M. Spaniol, L. Wang, Q. Li, W. Nejdl, and W. Zhang, Eds. Springer Berlin Heidelberg, 2010, pp. 259–268.

[13] J. Sánchez and A. Salinas, "Science Problem Solving Learning Through Mobile Gaming," in *Proceedings of the 12th International Conference on Entertainment and Media in the Ubiquitous Era*, New York, NY, USA, 2008, pp. 49–53.

[14] J. O'Kelly and J. P. Gibson, "RoboCode & problem-based learning: a non-prescriptive approach to teaching programming," *SIGCSE Bull*, vol. 38, no. 3, pp. 217–221, Jun. 2006.

[15] S. Esper, S. R. Foster, and W. G. Griswold, "On the nature of fires and how to spark them when you're not there," in *Proceeding of the 44th ACM technical symposium on Computer science education*, New York, NY, USA, 2013, pp. 305–310.

[16] J. K.-W. Chang, L. H. Dang, J. Pavleas, J. F. McCarthy, K. Sung, and J. Bay, "Experience with Dream Coders: developing a 2D RPG for teaching introductory programming concepts," *J Comput Sci Coll*, vol. 28, no. 1, pp. 227–236, Oct. 2012.

[17] P. Henriksen and M. Kölling, "greenfoot: combining object visualisation with interaction," in *Companion to the 19th annual ACM SIGPLAN conference on Object-oriented programming systems, languages, and applications*, New York, NY, USA, 2004, pp. 73–82.

[18] R. Hijón-Neira, Á. Velázquez-Iturbide, C. Pizarro-Romero, and L. Carriço, "Improving Students Learning Programming Skills with ProGames – Programming through Games System," in *Human-Computer Interaction – INTERACT 2013*, P. Kotzé, G. Marsden, G. Lindgaard, J. Wesson, and M. Winckler, Eds. Springer Berlin Heidelberg, 2013, pp. 579–586.

[19] M. Jonas, "Teaching Introductory Progamming Using Multiplayer Board Game Strategies in Greenfoot," *J Comput Sci Coll*, vol. 28, no. 6, pp. 19–25, Jun. 2013.

[20] T. Vilner, E. Zur, and S. Tavor, "Integrating Greenfoot into CS1: A Case Study," in *Proceedings of the 16th Annual Joint Conference on Innovation and Technology in Computer Science Education*, New York, NY, USA, 2011, pp. 350–350.

[21] P. Lavin-Mera, J. Torrente, P. Moreno-Ger, J. Vallejo-Pinto, and B. Fernández-Manjón, "Mobile Game Development for Multiple Devices in Education," *Int. J. Emerg. Technol. Learn. IJET*, vol. 4, no. 2009, pp. 19–26, 20091024.

[22] E. B. Arias, D. G. Vico, S. A. Herrera, and J. Q. Vives, "A web tool to create educational content with gaming visualization," in *Frontiers in Education Conference (FIE), 2012*, 2012, pp. 1–6.

[23] R. J. Nadolski, H. G. K. Hummel, H. J. van den Brink, R. E. Hoefakker, A. Slootmaker, H. J. Kurvers, and J. Storm, "EMERGO: A methodology and toolkit for developing serious games in higher education," *Simul. Gaming*, vol. 39, no. 3, pp. 338–352, Jan. 2008.

[24] S. Sina, S. Kraus, and A. Rosenfeld, "Using the Crowd to Generate Content for Scenario-Based Serious-Games," *ArXiv14025034 Cs*, Feb. 2014.

[25] S. Sauer, K. Osswald, X. Wielemans, and M. Stifter, "U-Create: Creative Authoring Tools for Edutainment Applications," in *Technologies for Interactive Digital Storytelling and Entertainment*, S. Göbel, R. Malkewitz, and I. Iurgel, Eds. Springer Berlin Heidelberg, 2006, pp. 163–168.

[26] S. E. Kirkle, S. Tomblin, and J. Kirkley, "Instructional design authoring support for the development of serious games and mixed reality training," in *The Interservice/Industry Training, Simulation & Education Conference (I/ITSEC)*, 2005, vol. 2005.

[27] S. Cooper, W. Dann, and R. Pausch, "Teaching objects-first in introductory computer science," in *Proceedings of the 34th SIGCSE technical symposium on Computer science education*, New York, NY, USA, 2003, pp. 191–195.

[28] A. Ehlert and C. Schulte, "Empirical comparison of objects-first and objects-later," in *Proceedings of the fifth international workshop on Computing education research workshop*, New York, NY, USA, 2009, pp. 15–26.

[29] S. Deterding, D. Dixon, R. Khaled, and L. Nacke, "From game design elements to gamefulness: defining 'gamification,'" in *Proceedings of the 15th International Academic MindTrek Conference: Envisioning Future Media Environments*, New York, NY, USA, 2011, pp. 9–15.

[30] S. Kelle, "Game Design Patterns for Learning," Nov. 2012.

[31] "Gameplay Design Patterns Wiki," 2014. [Online].

[32] "IT 2008 Curriculum — Association for Computing Machinery." [Online].

[33] J. Ong, "Android Achieved 85% Smartphone Market Share in Q2," *The Next Web*, 2014. [Online].

[34] M. Hassenzahl, M. Burmester, and F. Koller, "AttrakDiff: Ein Fragebogen zur Messung wahrgenommener hedonischer und pragmatischer Qualität," in *Mensch & Computer 2003*, Springer, 2003, pp. 187–196.

AUTHORS

T. Jordine received the B.Sc. and M.Sc. degree in computer science and media at the Stuttgart Media University in 2009 and 2011. In the beginning of 2013 he started his PhD studies in cooperation with the University of the West of Scotland and the Stuttgart Media University. Besides his research activities he is also an academic assistant at the Stuttgart Media University, Nobelstr. 10, 70569 Stuttgart, Germany (e-mail: jordine@hdm-stuttgart.de)

Y. Liang, is a senior lecturer in the School of Engineering and Computing, the University of the West of Scotland, Paisley PA1 2BE, U.K. (e-mail: ying.liang@uws.ac.uk).

E. Ihler is a professor at the Stuttgart Media University. He is teaching in the area of software engineering and theoretical computer science. Stuttgart Media University, Nobelstr. 10, 70569 Stuttgart, Germany (e-mail: ihler@hdm-stuttgart.de).

Permissions

All chapters in this book were first published in iJIM, by International Association of Online Engineering; hereby published with permission under the Creative Commons Attribution License or equivalent. Every chapter published in this book has been scrutinized by our experts. Their significance has been extensively debated. The topics covered herein carry significant findings which will fuel the growth of the discipline. They may even be implemented as practical applications or may be referred to as a beginning point for another development.

The contributors of this book come from diverse backgrounds, making this book a truly international effort. This book will bring forth new frontiers with its revolutionizing research information and detailed analysis of the nascent developments around the world.

We would like to thank all the contributing authors for lending their expertise to make the book truly unique. They have played a crucial role in the development of this book. Without their invaluable contributions this book wouldn't have been possible. They have made vital efforts to compile up to date information on the varied aspects of this subject to make this book a valuable addition to the collection of many professionals and students.

This book was conceptualized with the vision of imparting up-to-date information and advanced data in this field. To ensure the same, a matchless editorial board was set up. Every individual on the board went through rigorous rounds of assessment to prove their worth. After which they invested a large part of their time researching and compiling the most relevant data for our readers.

The editorial board has been involved in producing this book since its inception. They have spent rigorous hours researching and exploring the diverse topics which have resulted in the successful publishing of this book. They have passed on their knowledge of decades through this book. To expedite this challenging task, the publisher supported the team at every step. A small team of assistant editors was also appointed to further simplify the editing procedure and attain best results for the readers.

Apart from the editorial board, the designing team has also invested a significant amount of their time in understanding the subject and creating the most relevant covers. They scrutinized every image to scout for the most suitable representation of the subject and create an appropriate cover for the book.

The publishing team has been an ardent support to the editorial, designing and production team. Their endless efforts to recruit the best for this project, has resulted in the accomplishment of this book. They are a veteran in the field of academics and their pool of knowledge is as vast as their experience in printing. Their expertise and guidance has proved useful at every step. Their uncompromising quality standards have made this book an exceptional effort. Their encouragement from time to time has been an inspiration for everyone.

The publisher and the editorial board hope that this book will prove to be a valuable piece of knowledge for researchers, students, practitioners and scholars across the globe.

List of Contributors

W. Daher
Al-Qasemi Academic College of Education, Baka, Israel & An-Najah National University, Nablus, Palestine

Dr. J. B. Redd
Iowa State University, Ames, IA, USA

Dr. D. Schmidt
Iowa State University, Ames, IA, USA

A.Setyono
Multimedia University, Cyberjaya, Malaysia
Dian Nuswantoro University, Semarang, Indonesia

M. J. Alam
Multimedia University, Cyberjaya, Malaysia

C. Eswaran
Multimedia University, Cyberjaya, Malaysia

Debopam Acharya
Computer Science, Georgia Southern University, GA U.S.A

Hyo-Joo Han
Information Systems, Georgia Southern University, GA U.S.A

F.Lezzar
LaSTIC laboratory, University of Batna, Algeria

B.Belattar
LaSTIC laboratory, University of Batna, Algeria

A.Zidani
LaSTIC laboratory, University of Batna, Algeria

Hasan Omar Al-Sakran
King Saud University, Saudi Arabia

Qasem Kharma
King Saud University, Saudi Arabia

Irina Serguievskaia
King Saud University, Saudi Arabia

Yoong Cheah Huei
Ngee Ann Polytechnic, Singapore

Thet Swe
Ngee Ann Polytechnic, Singapore

Lee Zheng Xian
Ngee Ann Polytechnic, Singapore

Neo Shou Yee
Ngee Ann Polytechnic, Singapore

Abdul Razaque
University of Bridgeport, Bridgeport, CT, USA

Khaled Elleithy
University of Bridgeport, Bridgeport, CT, USA

F.J. Tapia-Moreno
University of Sonora, Hermosillo, México

H.A.Villa-Martinez
University of Sonora, Hermosillo, México

M. Antic
Metropolitan University, Belgrade, Serbia

S. Jovanovic
Metropolitan University, Belgrade, Serbia

S. Cvetanovic
Metropolitan University, Belgrade, Serbia

Hosam F. El-Sofany
Cairo Higher Institute for Engineering, Computer Science, and Management, Cairo, Egypt

Samir A. El-Seoud
Technische Universität Darmstadt (TU Darmstadt), Darmstadt, Germany

P. A. Correia Pereira
University of Vic, Barcelona, Spain

I.García Medina
Glasgow Caledonian University, Glasgow, UK

M.M. Ramses
University of Colima, Colima, México

E. D. Laura Victoria
University of Colima, Colima, México

A.S. Raúl
University of Colima, Colima, México

D. May
TU Dortmund University, Dortmund, Germany

C. Terkowsky
TU Dortmund University, Dortmund, Germany

T. Haertel
TU Dortmund University, Dortmund, Germany

C. Pleul
TU Dortmund University, Dortmund, Germany

K. Tian
Nagoya University, Nagoya, Japan

M. Endo
Chukyo University, Toyota, Japan

M. Urata
Nagoya University, Nagoya, Japan

K. Mouri
Nagoya City Science Museum, Nagoya, Japan

T. Yasuda
Nagoya University, Nagoya, Japan

D.J. McManus
The University of Alabama, Tuscaloosa, AL U.S.A

H.H. Carr
Auburn University, Auburn, AL U.S.A

B.M. Adams
The University of Alabama, Tuscaloosa, AL U.S.A

M.J. Aitkenhead
The James Hutton Institute, Craigiebuckler, Aberdeen, Scotland, UK

D. Donnelly
The James Hutton Institute, Craigiebuckler, Aberdeen, Scotland, UK

M.C. Coull
The James Hutton Institute, Craigiebuckler, Aberdeen, Scotland, UK

E. Hastings
The James Hutton Institute, Craigiebuckler, Aberdeen, Scotland, UK

O.A. Nykänen
Tampere University of Technology, Tampere, Finland

A.Rivero Rodriguez
Tampere University of Technology, Tampere, Finland

I.Marneanu
Graz University of Technology, Graz, Austria

M. Ebner
Graz University of Technology, Graz, Austria

T. Rößler
Evolaris Next Level GmbH, Graz, Austria

Qardi Hamarsheh
Philadelphia University, Amman, Jordan

Omar Daoud
Philadelphia University, Amman, Jordan

Saleh Saraireh
Philadelphia University, Amman, Jordan

T. Zimmer
University of Bordeaux, Talence, France

M. Billaud
University of Bordeaux, Talence, France

M. Pic
University of Bordeaux, Talence, France

D. Geoffroy
University of Bordeaux, Talence, France

Chad Tossell
US Air Force Research Laboratory, Wright-Patterson AFB, OH, USA

Philip Kortum
Rice University, Houston Texas, USA

Clayton Shepard
Rice University, Houston Texas, USA

Ahmad Rahmati
Apple Corporation, Cupertino, CA, USA

Lin Zhong
Rice University, Houston Texas, USA

Bader M. Al-Fawwaz
Al al-Bayt University, Jordan

Atallah M. Al-Shatnawi
Al al-Bayt University, Jordan

Wafa S. AlSharafat
Al al-Bayt University, Jordan

Tor Gjøsæter
University of Bergen, Norway

A.Ikhlef
Université Constantine 1, Constantine, Algérie

M. Kihel
Université Constantine 1, Constantine, Algérie

B. Boukhezzar
Université Constantine 1, Constantine, Algérie

A.Guerroudj
Université Constantine 1, Constantine, Algérie

N. Mansouri
Université Constantine 1, Constantine, Algérie

Shabnam Shahreen Sifat
Independent University, Bangladesh, Dhaka, Bangladesh

Ali Shihab Sabbir
Independent University, Bangladesh, Dhaka, Bangladesh

Mohammed Amin Almaiah
University Malaysia Terengganu, Terengganu, Malaysia

Masita Masila Abdul Jalil
University Malaysia Terengganu, Terengganu, Malaysia

A. Dirin
Haaga-Helia University of Applied Science, Helsinki, Finland

M. Nieminen
Aalto University, ESPOO, Finland

Athanasios S. Drigas
NCSR DEMOKRITOS, Institute of Informatics and Telecommunications, Net Media Lab, Athens, Greece

Marios A. Pappas
NCSR DEMOKRITOS, Institute of Informatics and Telecommunications, Net Media Lab, Athens, Greece

Ahmed Naddami
Université Hassan 1, Settat, Maroc

Ahmed Fahli
Université Hassan 1, Settat, Maroc

Mourad Gourmaj
Université Hassan 1, Settat, Maroc

Mohammed Moussetad
Université Hassan 2 Mohammedia, Casablanca, Maroc

Enrique Canessa, Marco Zennaro
International Centre for Theoretical Physics (ICTP), Trieste, Italy

Faisal Alkhateeb, Ahmed M. Manasrah, Iyad Abu Doush and Eslam Al Maghayreh
Yarmouk University, Irbid, Jordan

Yazan A. Alqudah
Princess Sumaya University for Technology Amman, Jordan

Todor Cooklev
Indiana University Purdue University, Fort Wayne, Indiana

C. Kouroupetroglou, D. Tektonidis, A. Koumpis, I. Ingnatiadis
ALTEC Software SA, Thessaloniki, Greece

A.P. Preciado Babb, C. Saar, C. Marcotte, J. Brandon and S. Friesen
University of Calgary, Calgary, Canada

Mohammad Ismail
Universiti Teknologi MARA (UiTM), Malaysia

Razli Che Razak
Universiti Utara Malaysia (UUM), Kedah, Malaysia

Karim Zarour
Lire laboratory, University Constantine 2, Algeria

T. Jordine and E. Ihler
Stuttgart Media University, Stuttgart, Germany

Y. Liang
University of the West of Scotland, Paisley, United Kingdom